Microemulsions

Microemulsions

Background, New Concepts, Applications, Perspectives

Edited by

Cosima Stubenrauch
School of Chemical and Bioprocess Engineering,
University College Dublin, Ireland

WILEY
A John Wiley and Sons, Ltd, Publication

This edition first published 2009

Blackwell Publishing was acquired by John Wiley & Sons in February 2007. Blackwell's publishing programme has been merged with Wiley's global Scientific, Technical, and Medical business to form Wiley-Blackwell.

Registered office
John Wiley & Sons Ltd, The Atrium, Southern Gate, Chichester, West Sussex, PO19 8SQ, United Kingdom

Editorial offices
9600 Garsington Road, Oxford, OX4 2DQ, United Kingdom
2121 State Avenue, Ames, Iowa 50014-8300, USA

For details of our global editorial offices, for customer services and for information about how to apply for permission to reuse the copyright material in this book please see our website at www.wiley.com/wiley-blackwell.

Library of Congress Cataloging-in-Publication Data

Microemulsions : background, new concepts, applications, perspectives/edited
by Cosima Stubenrauch. – 1st ed.
p. cm
Includes bibliographical references and index.
ISBN 978-1-4051-6782-6 (hardback : alk. paper)
1. Emulsions. I. Stubenrauch, Cosima.

TP156.E6M5175 2008
660'.294514–dc22

2008013076

A catalogue record for this book is available from the British Library.

Set in 10/12 pt Minion by Aptara Inc., New Delhi, India

1 2009

Microemulsions

Microemulsions

Background, New Concepts, Applications, Perspectives

Edited by

Cosima Stubenrauch
School of Chemical and Bioprocess Engineering,
University College Dublin, Ireland

WILEY
A John Wiley and Sons, Ltd, Publication

This edition first published 2009

Blackwell Publishing was acquired by John Wiley & Sons in February 2007. Blackwell's publishing programme has been merged with Wiley's global Scientific, Technical, and Medical business to form Wiley-Blackwell.

Registered office
John Wiley & Sons Ltd, The Atrium, Southern Gate, Chichester, West Sussex, PO19 8SQ, United Kingdom

Editorial offices
9600 Garsington Road, Oxford, OX4 2DQ, United Kingdom
2121 State Avenue, Ames, Iowa 50014-8300, USA

For details of our global editorial offices, for customer services and for information about how to apply for permission to reuse the copyright material in this book please see our website at www.wiley.com/wiley-blackwell.

Library of Congress Cataloging-in-Publication Data

Microemulsions : background, new concepts, applications, perspectives/edited by Cosima Stubenrauch. – 1st ed.
p. cm
Includes bibliographical references and index.
ISBN 978-1-4051-6782-6 (hardback : alk. paper)
1. Emulsions. I. Stubenrauch, Cosima.

TP156.E6M5175 2008
660'.294514–dc22

2008013076

A catalogue record for this book is available from the British Library.

Set in 10/12 pt Minion by Aptara Inc., New Delhi, India

1 2009

Contents

Contributors

Jürgen Allgaier Forschungszentrum Jülich GmbH, Institut für Festkörperforschung, 52425 Jülich, Germany

Raquel Antón Universidad de Los Andes, Facultad de Ingeniería, Lab FIRP, Av. Don Tulio Febres Coordero, Tercer piso. Mérida, Edo. Mérida 5101, Venezuela

Ignác Capek Polymer Institute, Slovak Academy of Sciences, Dúbravská cesta 9, 84236 Bratislava, and Trencin University, Faculty of Industrial Technologies, 020 32 Puchov, Slovakia

Carlos C. Co Chemical and Materials Engineering, University of Cincinnati, Cincinnati, OH 45221-0012, USA

Abhijit A. Date Department of Pharmaceutics, Bombay College of Pharmacy, Kalina, Santacruz (E.), Mumbai 400098, India

Sandra Engelskirchen Institut für Physikalische Chemie, Universität zu Köln, Luxemburger Str. 116, 50939 Köln, Germany

Ana Forgiarini Universidad de Los Andes, Facultad de Ingeniería, Lab FIRP, Av. Don Tulio Febres Coordero, Tercer piso. Mérida, Edo. Mérida 5101, Venezuela

Henrich Frielinghaus Forschungszentrum Jülich GmbH, Jülich Centre for Neutron Science, Lichtenbergstrasse 1, 85747 Garching, Germany

Feng Gao Chemical and Materials Engineering, University of Cincinnati, Cincinnati, OH 45221-0012, USA

Franz-Hubert Haegel Forschungszentrum Jülich GmbH, Institut für Chemie und Dynamik der Geosphäre, ICG-4 Agrosphäre, 52425 Jülich, Germany

Thomas Hellweg Universität Bayreuth, Lehrstuhl Physikalische Chemie I, Room 1.1 02 03, Universitätsstraße 30, D-95440 Bayreuth, Germany

Matthias Hloucha	Henkel KGaA, VTR Physical Chemistry, Henkelstrasse 67, 40191 Düsseldorf, Germany
Krister Holmberg	Chalmers University of Technology, Department of Chemical and Biological Engineering, SE-41296, Göteborg, Sweden
Ingegärd Johansson	Akzo Nobel Surfactants Europe, SE-44485 Stenungsund, Sweden
Björn Lindman	Physical Chemistry 1, University of Lund, P.O. Box 124, S-221 00 Lund, Sweden
Juan Carlos Lopez	Universidad de Los Andes, Facultad de Ingeniería, Lab FIRP, Av. Don Tulio Febres Coordero, Tercer piso. Mérida, Edo. Mérida 5101, Venezuela
Laura Márquez	Universidad de Los Andes, Facultad de Ingeniería, Lab FIRP, Av. Don Tulio Febres Coordero, Tercer piso. Mérida, Edo. Mérida 5101, Venezuela
Satya P. Moulik	Centre for Surface Science, Department of Chemistry, Jadavpur University, Kolkata 700032, India
Vandana B. Patravale	Department of Pharmaceutical Sciences and Technology, Institute of Chemical Technology, Nathalal Parikh Marg, Matunga, Mumbai 4000019, India
Animesh K. Rakshit	Department of Natural Sciences, West Bengal University of Technology, BF 142, Sector 1, Salt Lake, Kolkata 700 064, India
Wolfgang von Rybinski	Henkel KGaA, VTR Physical Chemistry, Henkelstrasse 67, 40191 Düsseldorf, Germany
Jean-Louis Salager	Universidad de Los Andes, Facultad de Ingeniería, Lab FIRP, Av. Don Tulio Febres Coordero, Tercer piso. Mérida, Edo. Mérida 5101, Venezuela
Reinhard Schomäcker	Technical University of Berlin, Institute of Chemistry, Section of Technical Chemistry, Secretary TC 8, Straβe des 17. Juni 124-128, 10623 Berlin, Germany
Thomas Sottmann	Institut für Physikalische Chemie, Universität zu Köln, Luxemburger Str. 116, 50939 Köln, Germany
Reinhard Strey	Institut für Physikalische Chemie, Universität zu Köln, Luxemburger Str. 116, 50939 Köln, Germany
Cosima Stubenrauch	School of Chemical and Bioprocess Engineering, Centre for Synthesis and Chemical Biology (CSCB), SFI-Strategic Research Cluster in Solar Energy Conversion, University College Dublin, Belfield, Dublin 4, Ireland

Preface

Although microemulsions were first described by Winsor in 1954, the 'Chemistry and Technology of Microemulsions' can be regarded as a relatively novel research area. The fact that microemulsions were not used in large-scale applications was due primarily to the lack of knowledge regarding their phase behaviour and microstructure and to the large overall surfactant concentration that is generally needed to formulate a microemulsion. Three achievements, however, fundamentally changed this situation. In the 1980s, it was systematic studies (Chapter 1) and new sophisticated techniques (Chapter 2) that allowed us to understand and thus to tune the properties of microemulsions, including the optimisation of their efficiency. Second, with the help of this new fundamental knowledge it was subsequently found that it is with surfactant mixtures, oil mixtures and additives such as alcohols or electrolytes that microemulsions with special properties can be formulated (Chapter 3). Last but not least, the addition of polymers to microemulsions turned out to have significant effects depending on the amount and/or polymer structure of the polymer. For example, adding amphiphilic block copolymers one can formulate highly efficient microemulsions with total surfactant concentrations of less than 1 wt.% (Chapter 4).

On the basis of the knowledge described in the first four chapters we are now able to use microemulsions for specific applications. The fact that an organic and an aqueous phase coexist in a thermodynamically stable mixture allows us to use one of the phases as reaction medium while the second phase serves as reservoir for the reactants or vice versa (Chapter 5). Moreover, the discrete water droplets of a water-in-oil microemulsion can be used as templates for the synthesis of metallic nanoparticles (Chapter 6). The wide variety of applications for which microemulsions are potential candidates is mirrored in the fact that studies with non-aqueous microemulsions are becoming increasingly important (Chapter 7). These research activities show very convincingly that the general concept of formulating a microemulsion is not restricted to traditional water–oil systems. Last but not least, because of the knowledge we have gained so far we are now able to use microemulsions for highly sensitive applications such as cosmetic (Chapter 8) and pharmaceutical products (Chapter 9) as well as for large-scale applications (Chapter 10).

Having read the first ten chapters, one might gain the impression that most of the 'microemulsion mysteries' have been solved during the course of time and that applying microemulsions in fields other than those mentioned in the book is just a question of 'creative thinking'. Unfortunately, or indeed fortunately, that is not the case! Examples will be given that highlight the challenges and perspectives we are currently faced with

(Chapter 11). I hope that these challenges will be dealt with and solved in the future so that microemulsions will be considered a versatile tool for all kinds of applications including sensitive cosmetic and pharmaceutical products, large-scale processes and the design of new composite materials.

I would like to thank all contributors for their time, their effort and their patience regarding my wish to make the book as consistent as possible in terms of structure and design. I would like to dedicate this book to my scientific mentors, namely Prof. Gerhard Findenegg and Prof. Reinhard Strey, who taught me how to work scientifically and to ask the right questions at the right time. I also thank Sarahjayne Sierra from Blackwell Publishing for her continuous support. I do hope that this book will become a reference book not only for experts in this research area but also for the next generation of scientists.

Cosima Stubenrauch
Dublin, Ireland

Some Thoughts about Microemulsions

Björn Lindman

Microemulsions emerged as an area of scientific research in a circumventional way. Strong research efforts were directed to this type of systems long before the term *microemulsion* was coined. The term microemulsion was selected because of a fundamental misunderstanding of the nature of these systems; they were considered like emulsions to be a type of dispersed system. During a long period of time there was no agreed definition on what should constitute a microemulsion, but the term was used broadly to include several types of surfactant systems. However, these initial confusions and disagreements contributed to the creation of a strong and vital research field, now occupying a large and increasing number of researchers both in academia and industry.

A thorough scientific account of microemulsions is certainly very timely both since our fundamental understanding has matured into a considerable consensus and since interesting applications emerge on a broad scale. How this understanding has been achieved makes us better understand the systems, in particular in relation to alternative pictures, which have been put forward on the quite long 'microemulsion journey'. The development of our understanding has by no means been linear but has involved steps both forward and backwards. Having followed the developments not from the start but for a considerable time, I wish here to give some personal reflections.

The 1980s were certainly a period of reaching a general consensus about one important aspect of microemulsions, namely that of thermodynamic stability. It was also a period when we obtained increasing evidence for its microstructure. It is striking that authors then normally found it important to stress what they meant by the term 'microemulsion'. Thus, the first sentence of many papers reads like 'Microemulsions are thermodynamically stable fluid mixtures of water, oil, and amphiphiles/surfactants'. Normally, we do not need to emphasise what we mean with a concept so this practice points to a previous confusion and a need to take a stand in a controversial issue.

For all systems we characterise as physico-chemists, the fundamental issue we deal with is that of whether we have a thermodynamically stable system or not. However, in the case of microemulsions, looking back we can see that it were the spectacular properties of microemulsions that called attention, while issues of whether the system was kinetically or thermodynamically stable were not in focus. Therefore, in the early work, a phase diagram approach, already established for surfactant systems in general, was not applied.

The second question we address as physico-chemists would be that of the arrangement of atoms and molecules, i.e. that of structure. While earlier workers naturally focused on ways to obtain microemulsions and study their stabilities and macroscopic properties, even quite late, microstructure was not much considered or even taken for granted; here, the term microemulsion is much to blame as for many it directly implied a structure analogous to that of emulsions, i.e. a structure of droplets of one liquid dispersed in another.

In general, it is fruitful to classify phases with regard to the degree of order. For surfactant systems, we can distinguish between long- and short-range order and disorder, respectively. Short-range disorder implies that the molecules are in a liquid state, while short-range order implies a crystalline solid-like state. Long-range order describes the relative distribution of the surfactant aggregates. In a micellar solution, for example, the distance between micelles is not fixed and we have a long-range disorder. When the micelles crystallise into a cubic or hexagonal lattice we have a fixed distance between aggregates, i.e. a long-range order. The same holds true for lamellar phases, where the spacing between the lamellaes is fixed. The corresponding long-range order is manifested in the diffraction behaviour.

The introduction of microemulsions in the scientific literature is normally ascribed to Schulman – although such systems had appeared in the patent literature before – and he and his co-workers produced a considerable fraction of the early work regarding their preparation and properties [1–8]. Other major contributors in the early period of microemulsions were Winsor [9, 10], Friberg [11–14] and Shinoda [15–22]; it can also be mentioned that Ekwall [23, 24], although not using the term microemulsion, made pioneering work on similar types of systems.

In the earlier days the way to obtain a microemulsion was by titrating a milky emulsion with a medium-chain alcohol such as pentanol or hexanol, later termed co-surfactant. While, as pointed out by Friberg [25], Schulman first called these systems micellar solutions, he later advocated the idea that they were disperse systems, i.e. only kinetically stable. A break-through in our understanding of microemulsions was due to the determination of phase diagrams, which was done extensively by Friberg, Shinoda and their co-workers. These authors prepared microemulsions with non-ionic surfactants, which was essential since for these surfactants only three components were needed and thus the description of the phase behaviour became manageable. Later extensive further work on phase diagrams contributed much to clarify the existence range of microemulsions for a wide range of surfactants, and to relate phase behaviour to molecular interactions; most important work here came from the groups in Göttingen (Kahlweit, Strey) [26, 27] and Yokohama (Shinoda, Kunieda) [28, 29].

As already mentioned, for a long period of time, the microstructure of microemulsions was considered to be that of droplets of one liquid dispersed in another, i.e. either water-in-oil (w/o-) or oil-in-water (o/w-) microemulsions. While this picture was easy to understand for water-rich or oil-rich systems, it became problematic for microemulsions with similar volume fractions of the two solvents. Even more intriguing from a microstructural point of view was the discovery by Friberg and Shinoda of systems with a continuous transition from water-rich to oil-rich systems. Suggestions of a coexistence of oil and water droplets were made by others. However, contradicting our general understanding of surfactant self-assembly structures, they were immediately rejected.

Friberg was certainly the one who made the most important contributions to establish the thermodynamic stability of microemulsions, providing key phase diagrams and being very active in refuting arguments of kinetic stability in the scientific literature and at conferences. He also at an early stage realised the problem of microstructure. This was particularly striking for the so-called middle-phase microemulsions, i.e. microemulsions in equilibrium with both oil and water. Friberg argued that a structure containing different curvatures of the surfactant aggregates could not be ruled out [14]. Shinoda, who made equally ground-breaking contributions to explaining microemulsion stability on the basis of phase diagrams, also provided important discussions on the microstructure of what he termed the 'surfactant phase' and argued for closely planar surfactant films, i.e. zero curvature [22]. The suggested structure basically was one of a thermally disrupted lamellar phase.

It is interesting to note that Ekwall [24], although not directly addressing the problem of microemulsion structure, much earlier addressed the same problem in his studies of ternary surfactant systems. He noted that in many cases a lamellar liquid crystalline phase forms at intermediate mixing ratios while in others there could be a continuous region from water to an organic solvent (immiscible with water). As an example he wrote (translating from Swedish): 'A third type of transition is indicated between solutions of reversed and normal micelles. Whether the mentioned micellar transitions in a homogeneous phase go directly from reversed to normal micelles and vice versa, or if they perhaps pass through an intermediate state with layered structure is still an open question. On the whole, this part of the research area offers many unsolved problems, which deserve a systematic study'.

The solution to the problem came in the late 1970s with the pioneering work of Scriven [30], introducing the bicontinuous structures based on minimal surfaces. Scriven's work, which included considerations of other surfactant phases (e.g. bicontinuous cubic phases), considerably stimulated the field and his ideas, based on theoretical arguments, were soon confirmed by experimental work, using mainly self-diffusion, electron microscopy and neutron scattering measurements.

The ideas of the relevance of phase diagrams and thermodynamic stability as well as the bicontinuous structure were certainly not accepted immediately and many publications until well into the 1990s caused confusion as some authors still took droplet structures for granted. A title for a paper [31] in *Nature* as late as 1986 entitled 'Occurrence of liquid-crystalline mesophases in microemulsion dispersions' illustrates both the slow acceptance and the ignorance of previous work on phase diagrams.

Our own involvement in microemulsion research was very much influenced by the contacts with the Swedish masters in the field of phase behaviour, Ekwall and Friberg, and at a later stage Shinoda, as well as by our previous experience of studying molecular interactions and association phenomena for other types of surfactant systems. Regarding the stability issue, we found it useful to suggest a definition [32] of a microemulsion as 'a system of water, oil and amphiphile which is a single isotropic and thermodynamically stable liquid solution'. While this definition certainly provided nothing new, we felt it contributed to eliminate some confusion.

As seen above, the entry into the microemulsion field via studies of surfactant systems in general, in many different ways facilitated the work. For myself, I came into contact with Ekwall's phase diagram work at an early stage. My interest into microstructure started with cubic liquid crystalline phases [33]. Reading the literature, I found out that there was an important contradiction between two of the leaders in the surfactant field, Luzzati [34–36] and Winsor [10, 37], regarding the structure of cubic phases, in particular regarding the build-up by discrete aggregates or connected surfactant aggregates. According to Winsor, all cubic phases must be built up of discrete spherical aggregates; a main piece of evidence was the narrow NMR signals (long spin–spin relaxation times), which would exclude any extended structures (rod micelles give broad signals). On the other hand, Luzzati deduced from X-ray studies structures with infinitely connected surfactant aggregates, thus bicontinuous structures or a 'mixture' model with both discrete and infinite aggregates. Both Winsor's and Luzzati's ideas were in direct conflict with a monotonic change in aggregate structure with surfactant concentration, which we nowadays call changes in the 'critical packing parameter' or spontaneous curvature of the surfactant film. Having learnt the new spin-echo NMR technique for self-diffusion with Hertz in Karlsruhe [38] and the radiotracer self-diffusion approach with Brun and Kamenka in Montpellier [39, 40], I could clearly see how powerful self-diffusion would be for surfactant systems. A phase diagram of dodecyl trimethylammonium chloride by Balmbra and Clunie [41] with two cubic phases appeared to be ideal for testing the novel approach to microstructure. A brief study with Bull [42] giving differences in surfactant diffusion by orders of magnitude between the two cubic phases, could directly prove that one was built up of discrete micelles while the other was bicontinuous. The cubic phase, which is more dilute in surfactant, was thus found to be characterised by very slow surfactant diffusion and thus must consist of (more or less stationary) discrete aggregates. In the more concentrated cubic phase, surfactant diffusion was found to be more than one order of magnitude faster. This, from other starting points surprising, finding could only be understood if the surfactant molecules could diffuse freely over macroscopic distances. Thus, surfactant aggregates had to be connected over macroscopic distances.

The distinction between discrete 'droplet' and bicontinuous structures, starting for the cubic phases before Scriven's suggestion about bicontinuous microemulsion structures, became central also in the subsequent studies on microemulsions. It was very clear from work by Ekwall, Friberg, Shinoda and others that surfactant self-assembly systems (including liquid crystalline phases and isotropic solutions) can be divided into those which have discrete self-assembly aggregates and those where the aggregates are connected in one, two or three dimensions. Regarding lamellar phases, the two-dimensional connectivity was appreciated already at a very early stage. The general acceptance of connectivity for these anisotropic phases contrasted sharply with gaining a consensus in the scientific community about the bicontinuity of solution phases. This is related partly to the fact that contrary to these anisotropic phases, it has been much more difficult to structurally characterise the different isotropic phases found in simple and complex surfactant systems. Indeed, in particular for microemulsions, various interpretations can be found in literature of investigations carried out with different techniques. The fact that the same results have sometimes been interpreted in completely opposite ways illustrates the difficulties of interpreting experimental findings. In fact, very few experimental observations allow a distinction between discrete and connected structures. The first real verification was thus due to observations of molecular self-diffusion over macroscopic distances. Later cryogenic transmission electron microscopy [43, 44] has developed into a very important tool for imaging different surfactant phases, as have also scattering techniques [45].

Thus, by measuring oil and water self-diffusion coefficients, it was quite easy to establish whether oil or water or none of them are confined to discrete domains, i.e. to 'droplets'. In the first work on microemulsion structure by self-diffusion [46], using both tracer techniques and NMR spin-echo measurements, it was clearly shown that microemulsions can indeed be bicontinuous over wide ranges of composition, which is manifested by both oil and water self-diffusion being rapid, i.e. not much slower than the self-diffusion of the neat liquids.

Microemulsions are multi-component systems with typically at least 3–5 components. In the first study, using both radiotracer and classical NMR methodology, each component had to be studied in a separate experiment on a separate sample with suitable component labelling. Both the labelling and the huge experimental efforts considerably slowed down progress. However, by using a Fourier transformation in the NMR spin-echo experiment, Stilbs and his student Moseley showed it to be possible in a single fast experiment to measure the self-diffusion coefficients of all components even for a complex multi-component solution [47, 48]. This was immediately seen as the remedy to answer questions related to the microstructure of microemulsions [49, 50].

The self-diffusion approach relies on the fact that molecular displacements over macroscopic distances are very sensitive to confinement and thus to microstructure. For example, we found that at the same composition (water, oil, surfactant), the ratio between water and oil self-diffusion coefficients could differ by a factor of 100 000. This also illustrates that the microstructure is primarily determined by the spontaneous curvature of the surfactant film and not by the oil-to-water ratio. Contributions to a better understanding of microemulsion structures with FT spin-echo NMR self-diffusion starting with Stilbs, included also Nilsson, Olsson, Söderman, Khan, Guéring, Monduzzi, Ceglie, Das and many others in Lund. In this work [49–63], the access to suitable systems was very important. Here, the contacts with Friberg, Shinoda, Strey and Langevin played a central role.

International meetings have been instrumental in providing a forum for scientific discussions about microemulsions and thus to the progress of the field. Many important and memorable events can be mentioned but in the author's opinion the first meeting in the now well-established biannual series of conferences denoted 'Surfactants in Solution' under the general chairmanship of Kash Mittal was a significant step forward. This meeting in Albany, NY, in 1976 was attended by Friberg, Shinoda, Scriven as well as by Schulman pupils like Prince and Shah. At this conference, Scriven [64] presented his bicontinuous structure and Friberg [65, 66] presented novel phase diagrams establishing the thermodynamic stability of microemulsions. Microemulsions have continued to be an important part of this series of meetings and probably the discussion was particularly intense during the meetings in Lund in 1982 and in Bordeaux 1984. Regarding our own work, the possibility of summarising and discussing our findings [67] at the large conference of the International Association of Colloid and Interface Scientists (IACIS) in Hakone, Japan, in 1988 marked a break-through in general acceptance. Starting from the 14th Surfactants in Solution Symposium in Barcelona in 2002, The Kash Mittal Award for 'outstanding achievements in colloid science' is awarded. The present author received this first prize for his research on microstructure in surfactant systems. The two other Kash Mittal Awards went to Barry Ninham (2004) and Eric Kaler (2006); both have made pioneering contributions to microemulsions.

Thus the microemulsion field continues to be a very active field both scientifically and in applications, as is amply shown by the different contributions in this timely book. Here, several important novel aspects are discussed in depth, like effects of polymers on microemulsions and the use of microemulsions as reaction media for organic synthesis and for the preparation of nanomaterials. That microemulsions constitute just one type of self-assembled surfactant systems continues to be an important consideration. As illustrated above, the important early developments were always promoted by an understanding of other phases.

References

1. Hoar, T.P. and Schulman, J.H. (1943) Transparent water-in-oil dispersions; the oleopathic hydromicelle. *Nature,* **152,** 102–103.
2. Schulman, J.H. and Riley, D.P. (1948) X-ray investigation of the structure of transparent oil–water disperse systems. 1. *J. Colloid Sci.,* **3,** 383–405.
3. Bowcott, J.E. and Schulman, J.H. (1955) Emulsions- control of droplet size and phase continuity in transparent oil–water dispersions stabilized with soap and alcohol. *Z. Electrochem.,* **59,** 283–290.
4. Schulman, J.H. and Friend, J.A. (1949) Light scattering investigation of the structure of transparent oil–water disperse systems. 2. *J. Colloid Sci.,* **4,** 497–509.
5. Schulman, J.H., Stoeckenius, W. and Prince, L.M. (1959) Mechanism of formation and structure of microemulsions by electron microscopy. *J. Phys. Chem.,* **63,** 1677–1680.
6. Zlochower, I.A. and Schulman, J.H. (1967) A study of molecular interactions and mobility at liquid/liquid interfaces by NMR spectroscopy. *J. Colloid Interface Sci.,* **24,** 115.
7. Prince, L.M. (1967) A theory of aqueous emulsions. 1. Negative interfacial tension at oil/water interface. *J. Colloid Interface Sci.,* **23,** 165.
8. Prince, L.M. (1969) A theory of aqueous emulsions. 2. Mechanism of film curvature at oil/water interface. *J. Colloid Interface Sci.,* **29,** 216.
9. Winsor, P.A. (1954) *Solvent Properties of Amphiphilic Compounds.* Butterworths, London.
10. Winsor, P.A. (1968) Binary and multicomponent solutions of amphiphilic compounds. Solubilization and the formation, structure and theoretical significance of liquid crystalline solutions. *Chem. Rev.,* **68,** 1.
11. Gillberg, G., Lehtinen, H. and Friberg, S.E. (1970) NMR and IR investigation of conditions determining stability of microemulsions. *J. Colloid Interface Sci.,* **33,** 40.
12. Sjöblom, E. and Friberg, S.E. (1978) Light-scattering and electron microscopy determinations of association structures in W-O microemulsions. *J. Colloid Interface Sci.,* 67, 16–30.
13. Rance, D.G. and Friberg, S.E. (1977) Micellar solutions versus microemulsions. *J. Colloid Interface Sci.,* **60,** 207–209.
14. Friberg, S.E., Lapczynska, I. and Gillberg, G. (1976) Microemulsions containing non-ionic surfactants – importance of PIT value. *J. Colloid Interface Sci.,* **56,** 19–32.
15. Saito, H. and Shinoda, K. (1967) Solubilization of hydrocarbons in aqueous solutions of non-ionic surfactants. *J. Colloid Interface Sci.,* **24,** 10.
16. Saito, H. and Shinoda, K. (1970) Stability of W/O type emulsions as a function of temperature and of hydrophilic chain length of emulsifier. *J. Colloid Interface Sci.,* **32,** 647.
17. Shinoda, K. (1967) Correlation between dissolution state of non-ionic surfactant and type of dispersion stabilized with surfactant. *J. Colloid Interface Sci.,* **24,** 4.
18. Shinoda, K. (1970) Thermodynamic aspects of non-ionic surfactant–water systems. *J. Colloid Interface Sci.,* **34,** 278.
19. Shinoda, K. and Ogawa, T. (1967) Solubilization of water in nonaqueous solutions of non-ionic surfactants. *J. Colloid Interface Sci.* **24,** 56.
20. Shinoda, K. and Friberg, S.E. (1975) Microemulsions- colloidal aspects. *Adv. Colloid Interface Sci.,* **4,** 281–300.
21. Shinoda, K. and Arai, H. (1964) Correlation between phase inversion temperature in emulsion and cloud point in solution of non-ionic emulsifier. *J. Phys. Chem.,* **68,** 3485.

22. Shinoda, K. (1983) Solution behaviour of surfactants. The importance of surfactant phase and the continuous change in HLB of surfactant. *Prog. Colloid Polymer Sci.*, **68**, 1–7.
23. Ekwall, P., Danielsson, I. and Mandell, L. (1960) Assoziations und Phasengleichgewichte bei der Einwirkung von Paraffin-Alkoholen auf wässrige Lösungen von Assoziationskolloiden. *Angew. Chemie-Int. Ed.*, **72**, 119–120.
24. Ekwall, P. (1967) Association and ordered states in systems of amphiphilic substances. *Svensk Kemisk Tidskrift*, **79**, 605.
25. Friberg, S. and Lindman, B. (1999) Microemulsions – a historical overview. In P. Kumar and K.L. Mittal (eds), *Handbook of Microemulsion Science and Technology*. Marcel Dekker, New York, pp. 1–12.
26. Kahlweit, M., Lessner, E. and Strey, R. (1983) Influence of the properties of the oil and the surfactant on the phase-behaviour of systems of the type H_2O–oil–nonionic surfactant. *J. Phys. Chem.*, **87**, 5032–5040.
27. Kahlweit, M. (1982) The phase behaviour of the type H_2O–oil–nonionic surfactant-electrolyte. *J. Colloid Interface Sci.*, **90**, 197–202.
28. Kunieda, H. and Shinoda, K. (1980) Solution behaviour and hydrophile–lipophile balance temperature in the Aerosol OT-isooctane-brine system-correlation between microemulsions and ultralow interfacial tensions. *J. Colloid Interface Sci.*, **75**, 601–606.
29. Kunieda, H. and Shinoda, K. (1982) Phase behavior in systems of non-ionic surfactant–water–oil around the hydrophile–lipophile balance temperature (HLB-temperature). *J. Dispersion Sci. Technol.*, **3**, 233–244.
30. Scriven, L.E. (1976) Equilibrium bicontinuous structure. *Nature*, **263**, 123–125.
31. Tabony, J. (1986) Occurrence of liquid-crystalline mesophases in microemulsion dispersions. *Nature*, **320**, 339–341.
32. Danielsson, I. and Lindman, B. (1981) The definition of microemulsion. *Colloids Surf.*, **3**, 391–392.
33. Fontell, K. (1974) X-ray diffraction by liquid crystals- amphiphilic systems. In G.W. Gray and P.A. Winsor (eds), *Liquid Crystals and Plastic Crystals*. Ellis Horwood Publishers, Chichester, pp. 80–109.
34. Luzzati, V. and Spegt, P.A. (1967) Polymorphism of lipids. *Nature*, **215**, 701.
35. Tardieu, A. and Luzzati, V. (1970) Polymorphism of lipids. A novel cubic phase-A cage-like network of rods with enclosed spherical micelles. *Biochim Biophys Acta*, **219**, 11.
36. Luzzati, V., Tardieu, A., Gulik-Krzywicki, T., Rivas, E. and Reiss-Husson, F. (1968) Structure of cubic phase of lipid–water systems. *Nature*, **220**, 485.
37. Gray, G.W. and Winsor, P.A. (1976) Generic relationships between non-amphiphilic and amphiphilic mesophases of fused type. Relationship of cubic mesophases (plastic crystals) formed by non-amphiphilic globular molecules to cubic phases of amphiphilic series. *Adv. Chem. Ser.*, **152**, 1–12.
38. Hertz, H.G., Lindman, B. and Siepe, V. (1969) Translational motion and hydration of the symmetrical tetraalkylammonium ions in aqueous solution. *Ber. Bunsenges. Phys. Chem.*, **73**, 542–549.
39. Lindman, B. and Brun, B. (1973) Translational motion in aqueous sodium *n*-octanoate solutions. *J. Colloid Interface Sci.*, **42**, 388–399.
40. Kamenka, N., Lindman, B. and Brun, B. (1974) Translational motion and association in aqueous dodecyl sulphate solutions. *Colloid Polymer Sci.*, **252**, 144–152.
41. Balmbra, R. and Clunie, J. (1969) Cubic mesomorphic phases. *Nature (London)*, **222**, 1159.
42. Bull, T. and Lindman, B. (1975) Amphiphile diffusion in cubic lyotropic mesophases. *Mol. Cryst. Liquid Cryst.*, **28**, 155–160.
43. Jahn, W. and Strey, R. (1988) Microstructure of microemulsions by freeze fracture electron microscopy. *J. Phys. Chem.* **92**, 2294–2301.

44. Talmon, Y. (1996) Transmission electron microscopy of complex fluids: The state of the art. *Berichte Bunsen-Ges. Phys. Chem.* **100**, 364–372.
45. Lichterfeld, F., Schmeling, T. and Strey, R. (1986) Microstructure of microemulsions of the system H_2O-*n*-tetradecane-$C_{12}E_5$. *J. Phys. Chem.*, **90**, 5762–5766.
46. Lindman, B., Kamenka, N. Kathopoulis, T.-M., Brun, B. and Nilsson, P.-G. (1980) Translational diffusion and solution structure of microemulsions. *J. Phys. Chem.*, **84**, 2485–2490.
47. Stilbs, P. and Moseley, M.E. (1979) Nuclear spin-echo experiments on standard Fourier-transform NMR spectrometers – Application to multi-component self-diffusion studies. *Chem. Scripta.*, **13**, 26–28.
48. Stilbs, P. (1987) Fourier transform pulsed-gradient spin-echo studies of molecular diffusion. *Progress NMR Spectroscopy*, **19**, 1–45.
49. Stilbs, P., Moseley, M.E. and Lindman, B. (1980) Fourier transform NMR self-diffusion measurements on microemulsions. *J. Magn. Resonance*, **40**, 401–404.
50. Lindman, B., Stilbs, P. and Moseley, M.E. (1981) Fourier transform NMR self-diffusion and microemulsion structure. *J. Colloid Interface Sci.*, **83**, 569–582.
51. Chatenay, D., Guéring, P., Urbach, W., Cazabat, A.M., Langevin, D., Meunier, J., Léger, L. and Lindman, B. (1987) Diffusion coefficients in microemulsions. In K.L. Mittal and P. Bothorel (eds), *Surfactants in Solution*, Vol. 6. Plenum, New York, pp. 1373–1381.
52. Nilsson, P.G. and Lindman, B. (1982) Solution structure of nonionic surfactant microemulsions from NMR self-diffusion studies. *J. Phys. Chem.*, **86**, 271–279.
53. Guéring, P. and Lindman, B. (1985) Droplet and bicontinuous structures in cosurfactant microemulsions from multi-component self-diffusion measurements. *Langmuir*, **1**, 464–468.
54. Olsson, U., Shinoda, K. and Lindman, B. (1986) Change of the structure of microemulsions with the HLB of nonionic surfactant as revealed by NMR self-diffusion studies. *J. Phys. Chem.*, **90**, 4083–4088.
55. Ceglie, A., Das, K.P. and Lindman, B. (1987) Effect of oil on the microscopic structure in four-component cosurfactant microemulsions. *J. Colloid Interface Sci.*, **115**, 115–120.
56. Lindman, B., Shinoda, K., Jonströmer, M. and Shinohara, A. (1988) Change of organized solution (Microemulsion) structure with small change in surfactant composition as revealed by NMR self-diffusion studies. *J. Phys. Chem.*, **92**, 4702–4706.
57. Shinoda, K., Araki, M., Sadaghiani, A., Khan, A. and Lindman, B. (1991) Lecithin-based microemulsions: Phase behavior and microstructure. *J. Phys. Chem.*, **95**, 989–993.
58. Das, K.P. Ceglie, A., Lindman, B. and Friberg, S. (1987) Fourier-transform NMR self-diffusion studies of a nonaqueous microemulsion system. *J. Colloid Interface Sci.* **116**, 390–400.
59. Ceglie, A., Das, K.P. and Lindman, B. (1987) Microemulsion structure in four-component systems for different surfactants. *Colloids Surf.*, **28**, 29–40.
60. Khan, A., Lindström, B., Shinoda, K. and Lindman, B. (1986) Change of the microemulsion structure with the hydrophile–lipophile balance of the surfactant and the volume fractions of water and oil. *J. Phys. Chem.*, **90**, 5799–5801.
61. Kamenka, N., Haouche, G., Brun, B. and Lindman, B. (1987) Microemulsions in zwitterionic surfactant systems: Dodecylbetaine. *Colloids Surf.*, **25**, 287–296.
62. Lindman, B. and Olsson, U. (1996) Structure of microemulsions studied by NMR. *Ber. Bunsenges. Phys. Chem.*, **100**, 344–363.
63. Shinoda, K. and Lindman, B. (1987) Organized surfactant systems: Microemulsions. *Langmuir*, **3**, 135–149.
64. Scriven, L.E. (1977) Equilibrium bicontinuous structures. In K.L. Mittal (ed), *Micellization, Solubilization, and Microemulsions*, Vol. 2. Plenum, New York, pp. 877–893.
65. Friberg, S., Buraczewska, I. and Ravey, J.C. (1977) Solubilization by non-ionic surfactants in the HLB-temperature range. In K.L. Mittal (ed), *Micellization, Solubilization, and Microemulsions*, Vol. 2. Plenum, New York, pp. 901–911.

66. Friberg, S. and Buraczewska, I. (1977) Microemulsions containing ionic surfactants. In K.L. Mittal (ed), *Micellization, Solubilization, and Microemulsions*, Vol. 2. Plenum, New York, pp. 791–799.
67. Lindman, B., Shinoda, K., Olsson, U., Anderson, D., Karlström, G. and Wennerström, H. (1989) On the demonstration of bicontinuous structures in microemulsions. *Colloids Surf.*, **38**, 205–224.

Chapter 1

Phase Behaviour, Interfacial Tension and Microstructure of Microemulsions

Thomas Sottmann and Cosima Stubenrauch

1.1 Introduction

Microemulsions are macroscopically isotropic mixtures of at least a hydrophilic, a hydrophobic and an amphiphilic component. Their thermodynamic stability and their nanostructure are two important characteristics that distinguish them from ordinary emulsions which are thermodynamically unstable. Microemulsions were first observed by Schulman [1] and Winsor [2] in the 1950s. While the former observed an optically transparent and thermodynamically stable mixture by adding alcohol, the latter induced a transition from a stable oil-rich to a stable water-rich mixture by varying the salinity. In 1959, Schulman *et al.* [3] introduced the term 'micro-emulsions' for these mixtures which were later found to be nano-structured.

The extensive research on microemulsions was prompted by two oil crises in 1973 and 1979, respectively. To optimise oil recovery, the oil reservoirs were flooded with a water–surfactant mixture. Oil entrapped in the rock pores can thus be removed easily as a microemulsion with an ultra-low interfacial tension is formed in the pores (see Section 10.2 in Chapter 10). Obviously, this method of tertiary oil recovery requires some understanding of the phase behaviour and interfacial tensions of mixtures of water/salt, crude oil and surfactant [4]. These in-depth studies were carried out in the 1970s and 1980s, yielding very precise insights into the phase behaviour of microemulsions stabilised by non-ionic [5, 6] and ionic surfactants [7–9] and mixtures thereof [10]. The influence of additives, like hydro- and lyotropic salts [11], short- and medium-chain alcohols (co-surfactant) [12] on both non-ionic [13] and ionic microemulsions [14] was also studied in detail. The most striking and relevant property of microemulsions in technical applications is the low or even ultra-low interfacial tension between the water excess phase and the oil excess phase in the presence of a microemulsion phase. The dependence of the interfacial tension on salt [15], the alcohol concentration [16] and temperature [17] as well as its interrelation with the phase behaviour [18, 19] can be regarded as well understood.

From the late 1980s onwards, the research on microemulsions turned to the understanding of the fascinating microstructure of these mixtures. Microemulsions are created by a surfactant film forming at the microscopic water/oil interface. Different methods such as NMR self-diffusion [20, 21], transmission electron microscopy (TEM) [20, 22] and scattering techniques (small angle X-ray scattering (SAXS) [23] and small angle neutron scattering (SANS) [16, 24]) provided some of the larger pieces in the puzzle of the manifold structure of microemulsions [25]. A recent overview of the state of the art of microemulsions, which contains the basic features of microemulsions as well as their theoretical description, is given in Ref. [26].

The research on microemulsions currently concentrates on even more complex mixtures. By adding amphiphilic macromolecules the properties of microemulsions can be influenced quite significantly (see Chapter 4). If only small amounts of amphiphilic block copolymers are added to a bicontinuous microemulsion a dramatic enhancement of the solubilisation efficiency is found [27, 28]. On the other hand, the addition of hydrophobically modified (HM) polymers to droplet microemulsions leads to a bridging of swollen micelles and an increase of the low shear viscosity by several orders of magnitude [29].

Within the last 30 years, microemulsions have also become increasingly significant in industry. Besides their application in the enhanced oil recovery (see Section 10.2 in Chapter 10), they are used in cosmetics and pharmaceuticals (see Chapter 8), washing processes (see Section 10.3 in Chapter 10), chemical reactions (nano-particle synthesis (see Chapter 6)), polymerisations (see Chapter 7) and catalytic reactions (see Chapter 5). In practical applications, microemulsions are usually multicomponent mixtures for which formulation rules had to be found (see Chapter 3). Salt solutions and other polar solvents or monomers can be used as hydrophilic component. The hydrophobic component, usually referred to as oil, may be an alkane, a triglyceride, a supercritical fluid, a monomer or a mixture thereof. Industrially used amphiphiles include soaps as well as medium-chained alcohols and amphiphilic polymers, respectively, which serve as co-surfactant.

The fact that microemulsions have gained increasing importance both in basic research and in industry is reflected in the large number of publications on microemulsions. A survey of paper titles reveals that the number of papers on the subject of microemulsions increased within the last 30 years from 474 in 1976–1985 to over 2508 in 1986–1995 and to 6691 in 1996–2005.[1] The fact that microemulsions also provide the potential for numerous practical applications is mirrored in the number of patents filed on this topic. A survey of patents on microemulsions[2] shows an increase from 159 in 1976–1985 to over 805 in 1986–1995 and to 2107 in 1996–2005. In the following the basic properties of microemulsions will be presented concentrating on the close connection between the phase behaviour and the interfacial tensions as well as on the fascinating microstructure.

1.2 Phase behaviour

The primary aim of microemulsion research is to find the conditions under which the surfactant solubilises the maximum amounts of water and oil, i.e. the phase behaviour has to be studied. As the effect of pressure on the phase behaviour is (in general) rather weak [30], it is sufficient to consider the effect of the temperature. Furthermore, it has been shown that simple ternary systems consisting of water, oil and non-ionic *n*-alkyl polyglycol ethers (C_iE_j) exhibit all properties of complex and technically relevant systems [6]. Therefore, we will first describe the phase behaviour of ternary non-ionic microemulsions.

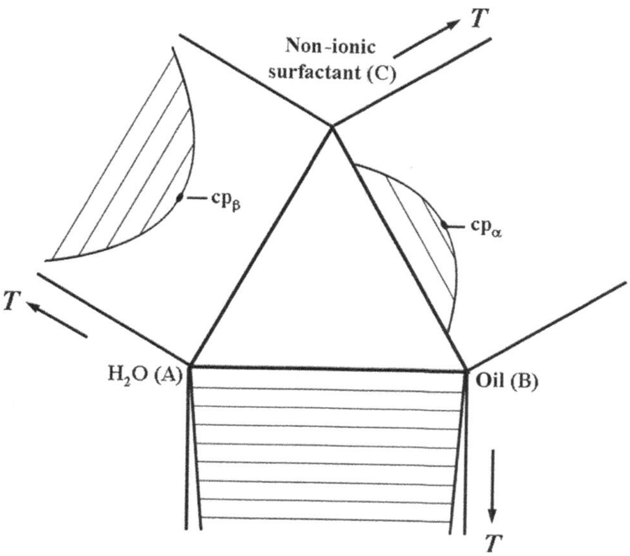

Figure 1.1 Schematic view of the phase behaviour of the three binary systems water (A)–oil (B), oil (B)–non-ionic surfactant (C), water (A)–non-ionic surfactant (C) presented as an 'unfolded' phase prism [6]. The most important features are the upper critical point cp_α of the B–C miscibility gap and the lower critical point cp_β of the binary A–C diagram. Thus, at low temperatures water is a good solvent for the non-ionic surfactant, whereas at high temperatures the surfactant becomes increasingly soluble in the oil. The thick lines represent the phase boundaries, while the thin lines represent the tie lines.

1.2.1 Microemulsions with alkyl polyglycol ethers

One successful approach to understanding the complex phase behaviour of microemulsions is to consider first the phase diagrams of the corresponding binary base systems [6]. In the case of ternary non-ionic microemulsions these are the three binary systems: water (A)–oil (B), oil (B)–non-ionic surfactant (C) and water (A)–non-ionic surfactant (C). For thermodynamic reasons, each of these systems shows a lower miscibility gap with an upper critical point. Figure 1.1 shows the unfolded phase prism with schematic diagrams of the three binary systems. The phase diagram of the binary water (A)–oil (B) system is the simplest of the three. The upper critical point of its lower miscibility gap lies well above the boiling point of the mixture, i.e. water and oil are almost immiscible between the melting and boiling point. The phase diagram of the binary oil (B)–non-ionic surfactant (C) system is almost as simple. Its upper critical point cp_α usually lies not far from the melting point of the mixture and depends on the nature of both oil and surfactant. In general, the lower the more hydrophilic the oil is and the more hydrophobic the surfactant is. The phase diagram of the binary water (A)–non-ionic surfactant (C) system is the most complex of the three. The lower miscibility gap (not shown in Fig. 1.1) lies far below the melting point of the mixture and plays no role in the following considerations. At ambient temperatures and above the critical micelle concentration (cmc) the surfactant molecules self-assemble. Additionally, concentrated and diluted liquid crystalline phases can be found [31] (not shown in Fig. 1.1). At higher temperatures most of the systems show an

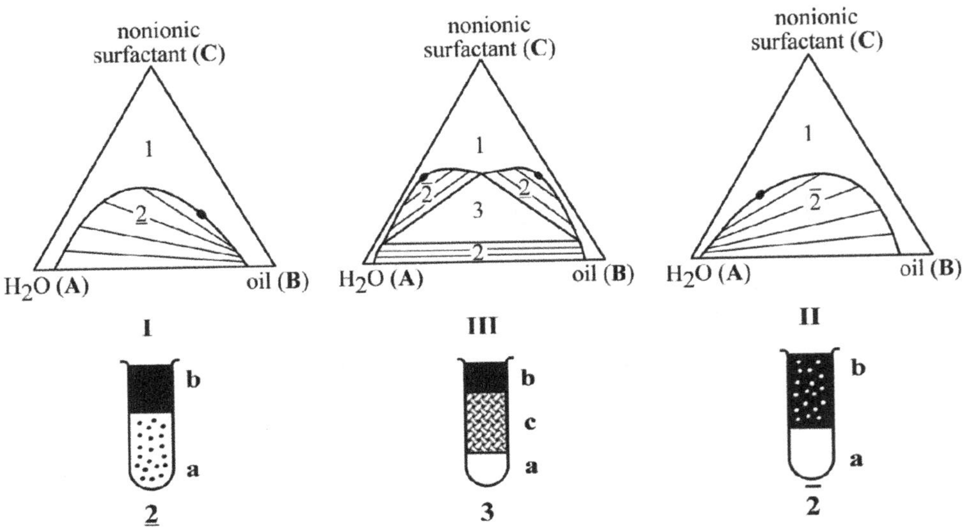

Figure 1.2 Isothermal Gibbs triangles of the system water (A)–oil (B)–non-ionic surfactant (C) at different temperatures. Increasing the temperature leads to the phase sequence $\underline{2}$–3–$\overline{2}$. A large miscibility gap can be found both at low and high temperatures. While at low temperatures a surfactant-rich water phase (a) coexists with an oil-excess phase (b), a coexistence of a surfactant-rich oil phase (b) with a water-excess phase (a) is found at high temperatures. At intermediate temperatures the phase behaviour is dominated by an extended three-phase triangle with its adjacent three two-phase regions. The test tubes illustrate the relative change in phase volumes.

additional upper (closed) miscibility gap with a lower critical point cp_β. The shape of this loop depends on the nature of the surfactant and plays an important role in the phase behaviour of the ternary system.

1.2.1.1 Phase inversion

From Fig. 1.1, it can be anticipated that the temperature-dependent phase behaviour of the ternary system is a result of the interplay between the lower miscibility gap of the B–C mixture and the upper miscibility gap of the A–C mixture. At low temperatures the non-ionic surfactant is mainly soluble in water, while it is mainly soluble in oil at high temperatures. Thus, an increase in temperature turns a non-ionic surfactant from hydrophilic into hydrophobic. Figure 1.2 shows this behaviour in the form of the related Gibbs phase triangles. At low temperatures the phase behaviour is dominated by a large miscibility gap. The negative slope of the tie lines indicates that a non-ionic surfactant-rich water phase (a) coexists with an oil-excess phase (b). This situation is denoted as $\underline{2}$ or Winsor I (see Fig. 1.2 (left)). Increasing the temperature one observes (Fig. 1.2, centre) an extended three-phase triangle with its adjacent three two-phase regions. Within the three-phase triangle (denoted as 3 or Winsor III) a surfactant-rich microemulsion (c) coexists with an excess water (a) and oil phase (b). The symmetric form of the triangle implies the solubilisation of equal amounts of water and oil. A further increase of the temperature

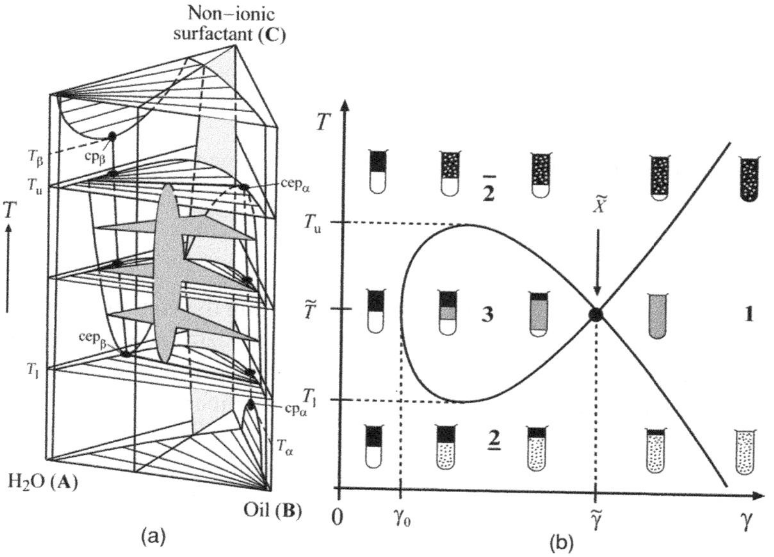

Figure 1.3 (a) Schematic phase prism of the system water–oil–non-ionic surfactant showing the temperature-dependent phase behaviour. A convenient way to study these systems is to measure the phase behaviour at constant oil/(water + oil) ratios as function of temperature T and surfactant mass fraction γ (3 phase region = dark grey, 1 phase region = light grey). (b) Schematic $T(\gamma)$-section at a constant oil/(water + oil) volume fraction of $\phi = 0.5$. Assigned are the minimal mass fraction $\tilde{\gamma}$ of surfactant needed to solubilise water and oil, the mass fraction γ_0 of surfactant which is solubilised monomerically in water and oil, the lower (T_l), upper (T_u) and mean ($\tilde{T}$) temperature of the three-phase body. Again the test tubes illustrate the relative volume of the phases.

again leads to the formation of an extended miscibility gap (see Fig. 1.2 (right)). Here, the positive slope of the tie lines indicates that a non-ionic surfactant-rich oil phase (b) coexists with a water-excess phase (a). This situation is denoted as $\bar{2}$ or Winsor II. The test tube shown below each Gibbs phase triangle illustrates the relative change in phase volumes for mixtures containing equal volumes of water and oil.

Stacking the isothermal Gibbs triangles on top of each other results in a phase prism (see Fig. 1.3(a)), which represents the temperature-dependent phase behaviour of ternary water–oil–non-ionic surfactant systems. As discussed above, non-ionic surfactants mainly dissolve in the aqueous phase at low temperatures ($\underline{2}$). Increasing the temperature one observes that this surfactant-rich water phase splits into two phases (a) and (c) at the temperature T_l of the lower critical endpoint cep$_\beta$, i.e. the three-phase body appears. Subsequently, the lower water-rich phase (a) moves towards the water corner, while the surfactant-rich middle phase (c) moves towards the oil corner of the phase prism. At the temperature T_u of the upper critical endpoint cep$_\alpha$ a surfactant-rich oil phase is formed by the combination of the two phases (c) and (b) and the three-phase body disappears. Each point in such a phase prism is unambiguously defined by the temperature T and two composition variables. It has proved useful [6] to choose the mass fraction of the oil in the mixture of water and oil

$$\alpha = \frac{m_B}{m_A + m_B} \tag{1.1}$$

and that of the surfactant in the mixture of all three components

$$\gamma = \frac{m_C}{m_A + m_B + m_C}. \tag{1.2}$$

Knowing the densities of the components for calculating the volumes one can also use the volume fractions ϕ and ϕ_C, respectively. A simple and extremely useful procedure to obtain an overview of the phases occurring in such a phase prism is to measure the phase diagram at a constant oil/water ratio as a function of the temperature T and the surfactant mass fraction γ ($T(\gamma)$-section). Such a section through the phase prism is highlighted in Fig. 1.3(a) (3 phase region = dark grey, 1 phase region = light grey) and shown schematically in Fig. 1.3(b). It permits easily to determine the *phase inversion temperature* (PIT), at which the hydrophilic–lipophilic balance (HLB) is achieved.

Figure 1.3(b) shows such a $T(\gamma)$-section at a constant oil/(water + oil) volume fraction of $\phi = 0.5$. As can be seen, the phase boundaries resemble the shape of a fish. Starting with the binary water–oil system, two phases, namely a pure water phase and a pure oil phase, coexist over the entire experimentally accessible temperature range. Small amounts of added surfactant molecules dissolve monomerically in the two phases. Being amphiphilic, the surfactant molecules preferentially adsorb at the macroscopic interface. At a mass fraction γ_0 both excess phases and the macroscopic interface are saturated with the surfactant molecules and the amphiphilic molecules are forced into the microscopic water/oil interface leading to topologically ordered interfacial films in solutions, i.e. the 'real' microemulsions. Looking at these mixtures microscopically, we find at low temperatures an amphiphilic film that forms oil-swollen micelles in a continuous water phase (a). This oil-in-water (o/w) microemulsion coexists with an oil-excess phase (b) ($\underline{2}$). At high temperatures the inverted situation ($\bar{2}$) is found. Here, a water excess phase (a) coexists with a water-in-oil (w/o) microemulsion in which the amphiphilic film forms water-swollen micelles in a continuous oil phase (b). At intermediate temperatures the surfactant is almost equally soluble in both solvents and a locally planar amphiphilic film is formed. Here, three phases (3), i.e. a surfactant-rich bicontinuously structured (for details see below) phase (c), an excess oil and water phase coexist. Microscopically, the observed trend of the phase behaviour from $\underline{2}$ over 3 to $\bar{2}$ with increasing temperature can be attributed to a gradual change of the mean curvature H of the amphiphilic film [25, 32]. While at low temperatures the film is curved around the oil ($H > 0$) it curves around water at high temperatures ($H < 0$) (see Section 1.4, Fig. 1.18).

Considering now the variation of the phase behaviour with increasing mass fraction γ of surfactant one can see that the volume of the respective microemulsion phase increases (see test tubes in Fig. 1.3(b)) until the excess phases vanish and a one-phase microemulsion is found. The optimal state of the system is the so-called $\tilde{X}$-point where the three-phase body meets the one-phase region. It defines both the minimum mass fraction $\tilde{\gamma}$ of surfactant needed to solubilise water and oil, i.e. the efficiency of the surfactant, as well as the corresponding temperature $\tilde{T}$, which is a measure of the PIT.

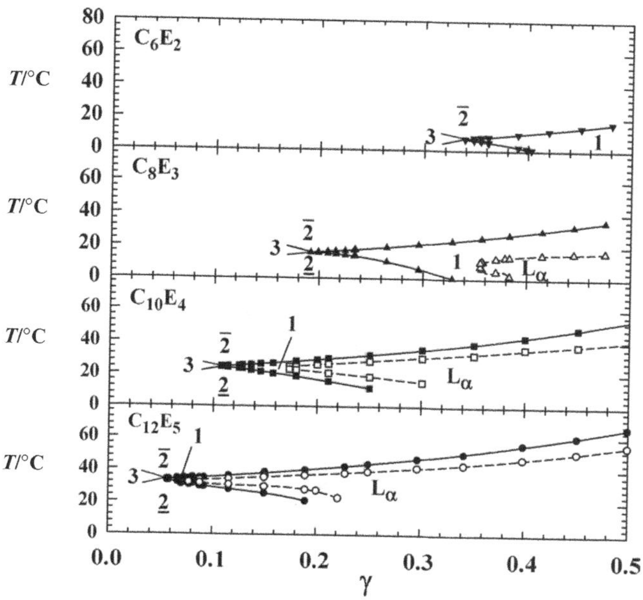

Figure 1.4 $T(\gamma)$-sections through the phase prism of the systems H_2O–*n*-octane–C_6E_2, C_8E_3, $C_{10}E_4$ and $C_{12}E_5$ at an oil/(water + oil) volume fraction of $\phi = 0.5$. In order to determine the respective $\tilde{X}$-point the phase boundaries are measured only for surfactant mass fractions $\gamma > \tilde{\gamma}$. An increase of both the hydrophobic chain length i and the size of the hydrophilic head group j shifts the $\tilde{X}$-point to lower values of $\tilde{\gamma}$, i.e. the efficiency increases. Simultaneously the stability range of the bicontinuous one phase microemulsion shrinks dramatically due to the increased extension of the lamellar mesophase (L_α). (From Ref. [26], reprinted with permission of Elsevier.)

1.2.1.2 Efficiency

One of the central questions of microemulsion formulation has been, and still is, the quest for high efficiency, i.e. finding microemulsions in which a minimum amount of surfactant is necessary for solubilising oil in water or vice versa. A rapid method for quantifying the efficiency of a system is to determine the $\tilde{X}$-point by recording a $T(\gamma)$-section at an oil/(water + oil) volume fraction $\phi = 0.5$. In this fashion the optimal state ($\tilde{X}$-point) can be determined extrapolating the phase boundaries from $\underline{2}$ to 1 (turbid to clear) and 1 to $\overline{2}$ (clear to turbid), which makes the exact determination of the three-phase region dispensable. In Fig. 1.4, it is demonstrated in which way the $\tilde{X}$-point and, consequently, the one-phase microemulsion region ($\gamma > \tilde{\gamma}$) are influenced by the chain length of the surfactant [26]. The figure shows the $T(\gamma)$-section of four H_2O–*n*-octane–*n*-alkyl polyglycol ether (C_iE_j) systems at an oil/(water + oil) volume fraction of $\phi = 0.5$. Starting with the H_2O–*n*-octane–C_6E_2 system (Fig. 1.4, top) it can be seen that a surfactant mass fraction of $\tilde{\gamma} = 0.334$ is needed for the solubilisation of equal volumes of water and *n*-octane. Using the surfactant C_8E_3 instead of C_6E_2 only 19 wt.% of surfactant is needed to solubilise water and *n*-octane. A further increase of the chain length of the surfactant to $C_{10}E_4$ and $C_{12}E_5$ shifts the $\tilde{X}$-point to $\tilde{\gamma} = 0.099$ and $\tilde{\gamma} = 0.048$, respectively. Thus, enlarging both the alkyl chain i and the head group size j (number of ethylene oxide groups) of the surfactant from

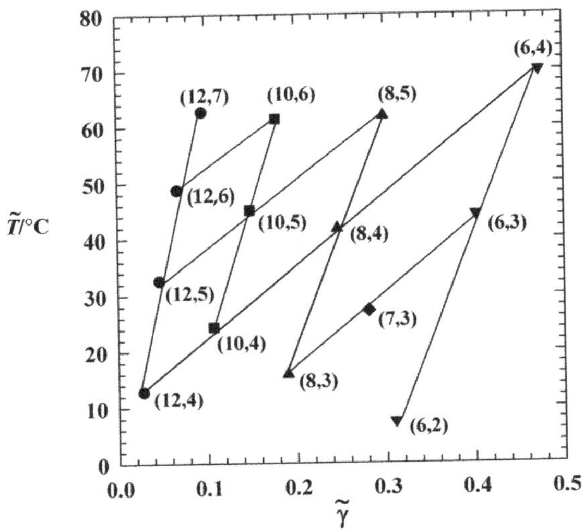

Figure 1.5 $\tilde{X}$-points of the systems H_2O–*n*-octane–C_iE_j at an oil/(water + oil) volume fraction of $\phi = 0.5$ [34]. The individual systems are characterised by the (*i, j*) pairs. While an increase of the hydrophobic chain length *i* leads mainly to a decrease of $\tilde{\gamma}$, an increase of the number of oxyethylene groups *j* increases mainly $\tilde{T}$. (From Ref. [34], reprinted with permission of the Royal Society of Chemistry.)

C_6E_2 to $C_{12}E_5$ leads to an enormous increase in efficiency. This increase in efficiency is a result of the increasing amphiphilicity of the surfactant molecules forcing them into the microscopic water/oil interface.

All four systems show the phase sequence characteristic of non-ionic microemulsions, namely $\underline{2} \rightarrow 3 \rightarrow \overline{2}$ at intermediate and $\underline{2} \rightarrow 1 \rightarrow \overline{2}$ at larger surfactant mass fractions. However, the lamellar liquid crystalline phase L_α (surrounded by a two-phase coexistence region, not shown), which is not present in the C_6E_2 system, occurs in the C_8E_3 system where it is embedded in the one-phase region of the microemulsion. Increasing the amphiphilicity of the surfactant even further leads to an extension of the L_α phase that nearly covers the entire one-phase region and thus limits the existence of the one-phase bicontinuous microemulsion to a very small region. As these liquid crystalline phases are often highly viscous and thus tend to complicate the handling of water–oil-surfactant systems their formation is undesirable in technical applications. An alternative and new road to the formulation of highly efficient microemulsion is the addition of small amounts of amphiphilic block copolymers to medium-efficient microemulsions [27, 33] (see Chapter 4).

In general, the $\tilde{X}$-point gives the efficiency of the surfactant and the PIT provides an excellent criterion for choosing the appropriate surfactant for the formulation in question. In Fig. 1.5, a synopsis of the $\tilde{X}$-points of 14 different H_2O–*n*-octane–C_iE_j systems at an oil/(water + oil) volume fraction of $\phi = 0.5$ is shown in a $\tilde{T}(\tilde{\gamma})$ plot [34]. The hydrophobic chain length *i* is varied between 6 and 12, the number of ethylene oxide groups *j* between 2 and 7. An increase of the hydrophobic chain length *i* renders the surfactant more hydrophobic. Thus, the $\tilde{X}$-point shifts to lower temperatures. Concomitantly, $\tilde{\gamma}$ decreases strongly, i.e. the surfactant becomes more efficient. An increasing number of ethylene oxide groups j shifts the $\tilde{X}$-point to higher temperatures due to an increasing hydrophilicity of the surfactant and $\tilde{\gamma}$ increases slightly. Furthermore, the whole grid of the $\tilde{X}$-points varies systematically with the chain length k of the n-alkane (not shown in Fig. 1.5 for the sake of clarity). Kahlweit *et al.* found [11] that with increasing k the $\tilde{X}$-point shifts to higher temperatures and $\tilde{\gamma}$ increases, i.e. the surfactant becomes less efficient. Recently, analogous trends of the $\tilde{X}$-point with k have been observed for both polymerisable n-alkyl methacrylates [35] and triglycerides [36].

1.2.1.3 Monomeric solubility

In one-phase microemulsions the surfactant molecules partition between the microscopic water/oil interface and the microemulsion sub-phases (e.g. in swollen micelles or bicontinuous oil- and water-rich domains) in which they are dissolved monomerically. They also dissolve monomerically in coexisting excess phases and adsorb at the macroscopic interfaces between the phases. The significance of this fact is that these parts of the surfactant are not available for the micro-emulsification of water and oil. Thus, for technical applications surfactants with high amphiphilicity but small monomeric solubilities in both solvents are desirable.

The monomeric solubility of the surfactant in the water $\gamma_{\mathrm{Cmon,a}}$ can be easily determined from surface tension measurements [37]. An interesting method to obtain $\gamma_{\mathrm{Cmon,b}}$ is provided by the macroscopic phase behaviour through the determination of the mass fraction of surfactant γ_0 (see Fig. 1.3), i.e. the monomerically dissolved surfactant in both excess phases. Therefore, the volume fraction of the middle phase V_c/V has to be measured as a function of the mass fraction of surfactant γ at a constant $\phi = 0.5$ and the mean temperature $\tilde{T}$ of the three-phase body [34, 38, 39]. Plotting V_c/V versus γ yields γ_0 at $V_c/V = 0$ and $\tilde{\gamma}$ at $V_c/V = 1$. Then the monomeric solubility in the oil is calculated from

$$\gamma_{\mathrm{Cmon,b}} = \frac{\gamma_0 + \gamma_{\mathrm{Cmon,a}}(\alpha(1-\gamma_0) - 1)}{\gamma_0 + \alpha(1-\gamma_0) - \gamma_{\mathrm{Cmon,a}}}. \tag{1.3}$$

Figure 1.6 shows the monomeric solubility $\gamma_{\mathrm{Cmon,b}}$ in n-octane calculated according to Eq. (1.3) at the respective mean temperature $\tilde{T}$ of the three-phase body [34]. For the calculations the monomeric solubility $\gamma_{\mathrm{Cmon,a}}$ in water was set equal to 0.03, 0.02, 0.01, 0.006 and 0.002 for C_6E_2, C_6E_3, C_6E_4, C_7E_3 and C_8E_j, respectively. For longer chain surfactants $\gamma_{\mathrm{Cmon,a}} < 0.001$ was neglected [40, 41]. For the sake of visual clarity, a grid of lines was drawn through the data points at constant i and j to even out the experimental error.

As can be seen, the $\tilde{T}(\gamma_{\mathrm{Cmon,b}})$ plot shows the same pattern as the $\tilde{T}(\tilde{\gamma})$ plot, i.e. the monomeric solubility $\gamma_{\mathrm{Cmon,b}}$ in n-octane decreases with increasing hydrophobic chain length i and increases slightly with increasing number of ethylene oxide groups j. These findings suggest that both monomeric solubilities are correlated with the efficiency of the surfactant to solubilise water and oil. Having the monomeric solubility of the surfactants in both water and oil at hand the mass fraction γ_i of the surfactant molecules which reside

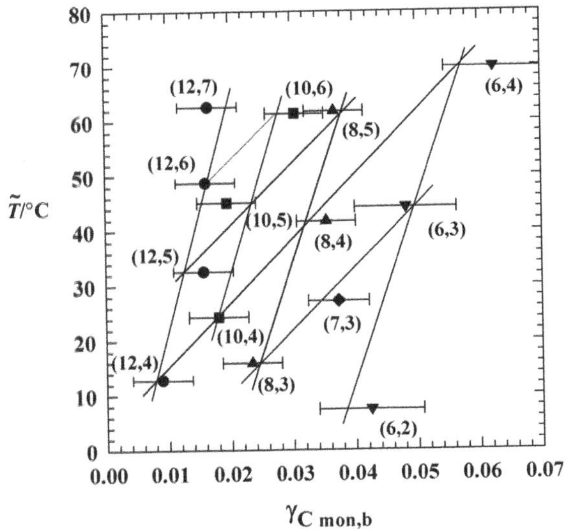

Figure 1.6 Monomeric solubility $\gamma_{Cmon,b}$ of 14 different surfactants in *n*-octane at the mean temperature $\tilde{T}$ obtained from the determination of γ_0 (see Fig. 1.3) [34]. Note the similar patterns of the $\tilde{T}(\gamma_{Cmon,b})$ plot and the $\tilde{T}(\tilde{\gamma})$ plot (see Fig. 1.5), respectively. (From Ref. [34], reprinted with permission of the Royal Society of Chemistry.)

at the microscopic water/oil interface can be calculated according to

$$\gamma_i = \gamma - \frac{w_A \gamma_{Cmon,a}}{1 - \gamma_{Cmon,a}} - \frac{w_B \gamma_{Cmon,b}}{1 - \gamma_{Cmon,b}}, \tag{1.4}$$

where w_A and w_B are the weight fractions of water and oil, respectively. The parameter γ_i is a measure for the size of the specific area of the interface S/V ($S/V \sim \gamma_i$), for the characteristic length ξ ($\xi \sim \gamma_i^{-1}$) of the structures [42–44], and for the interfacial tension σ_{ab}($\sigma_{ab} \sim \gamma_i^2$) between water- and oil-rich phases [45, 46] (for details see Sections 1.3 and 1.4).

The facts presented so far show that the general phase behaviour, the location of the three-phase body (i.e. $\tilde{\gamma}$, $\tilde{T}$(PIT)) and the monomeric solubilities ($\gamma_{Cmon,a}$, $\gamma_{Cmon,b}$) depend sensitively but systematically on the chemical nature of the components. Furthermore, the striking similarities that many systems share suggest that, as in the corresponding state description for real gases, suitable parameters exist which scale the phase behaviour of all microemulsions. Systematic measurements of the extension of the three-phase body identified $\tilde{\gamma}(\phi = 0.50)$, T_l and T_u as the relevant parameters for a corresponding state description of microemulsions [47]. These parameters also determine the phase behaviour far on the water- and oil-rich side of the phase prism which is particularly interesting for technical applications.

1.2.1.4 Water- and oil-rich microemulsions

The phase behaviour of water-rich and oil-rich microemulsions can be studied most conveniently by considering vertical sections through the phase prism at a constant surfactant/(water + surfactant) mass fraction

$$\gamma_a = \frac{m_C}{m_A + m_C} \tag{1.5}$$

and a constant surfactant/(oil + surfactant) mass fraction

$$\gamma_b = \frac{m_C}{m_B + m_C}, \tag{1.6}$$

respectively. Starting from the binary systems A–C or B–C, the temperature-dependent phase behaviour is measured as a function of the mass fraction of oil w_B or water w_A, respectively. A schematic drawing of $T(w_B)$- and $T(w_A)$-sections performed at constant mass fractions γ_a and γ_b, respectively, is seen in Fig. 1.7(a). The variation of the phase behaviour in these sections is discussed by means of the system H_2O–n-octane–$C_{10}E_5$. Figure 1.7(b) shows the section on the water-rich side ($T(w_B)$ at $\gamma_a = 0.10$), while the corresponding section ($T(w_A)$ at $\gamma_b = 0.10$) on the oil-rich side of the phase prism is shown in Fig. 1.7(c).

Looking first of all at the phase boundaries of the $T(w_B)$-section one observes that the $1 \rightarrow \overline{2}$ phase boundary starts at $w_b = 0$ near the critical point of the miscibility gap of the binary water–$C_{10}E_5$ system. Upon the addition of n-octane this near-critical boundary descends steeply and runs through a minimum as the weight fraction of oil w_b is increased further. Simultaneously, the $\underline{2} \rightarrow 1$ phase boundary ascends monotonically with increasing w_B. This phase boundary indicates, for a given temperature, the maximum amount of oil that can be solubilised in a one-phase oil-in-water (o/w) microemulsion and is thus called the *emulsification failure boundary* (*efb*). With increasing temperature the capability of the surfactant to solubilise oil is strongly increased. Close to the lower critical endpoint temperature T_l the one-phase region closes like a funnel. It terminates at the intersection of the lower oil emulsification failure and the upper near-critical phase boundary.

At the oil-rich side, the phase behaviour is inverted temperature-wise as can be seen in the $T(w_A)$-section provided in Fig. 1.7(c). Thus, the near-critical phase boundary $\underline{2} \rightarrow 1$ starts at low temperatures from the lower n-octane–$C_{10}E_5$ miscibility gap (below $<0°C$) and ascends steeply upon the addition of water. With increasing w_A, this boundary runs through a maximum and then decreases down to the upper critical endpoint temperature T_u. The emulsification failure boundary $1 \rightarrow \overline{2}$ starts at high temperatures and low values of w_A, which means that only small amounts of water can be solubilised in a water-in-oil (w/o) microemulsion at temperatures far above the phase inversion. Increasing amounts of water can be solubilised by decreasing the temperature, i.e. by approaching the phase inversion. At T_u the *efb* intersects the near-critical phase boundary and the funnel-shaped one-phase region closes.

From the above considerations, it can be concluded that $T(w_B)$- and $T(w_A)$-sections provide an easy method to determine the location of emulsification failure boundaries which are of particular interest if the optimal formulation for an industrial application has to be found. Furthermore, these sections yield the lower and upper temperature of

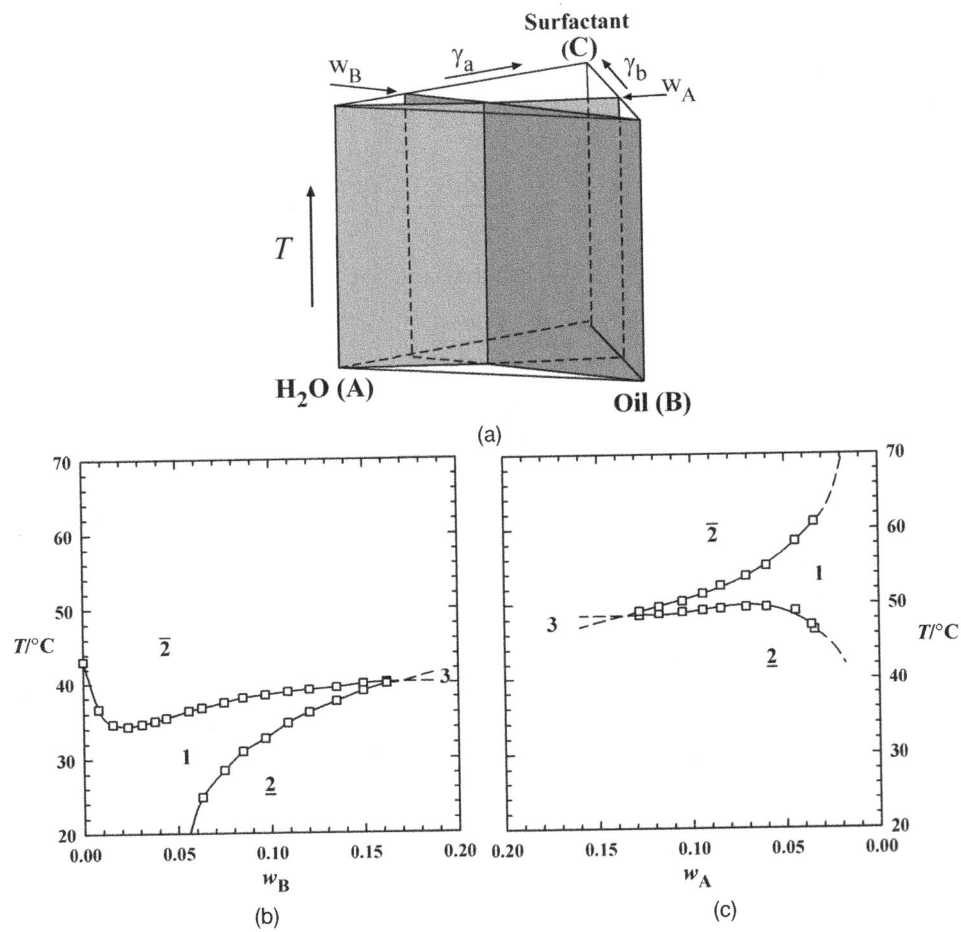

Figure 1.7 Vertical sections $T(w_B)$ and $T(w_A)$ through the phase prism which start at the binary water–surfactant ($w_B = 0$) and the binary oil–surfactant ($w_A = 0$) corner, respectively. These sections have been proven useful to study the phase behaviour of water- and oil-rich microemulsions. (a) Schematic view of the sections $T(w_B)$ and $T(w_A)$ performed at a constant surfactant/(water + surfactant) mass fraction γ_a and at a constant surfactant/(oil + surfactant) mass fraction γ_b, respectively. (b) $T(w_B)$ section through the phase prism of the system H_2O–n-octane–$C_{10}E_5$ at $\gamma_a = 0.10$. Starting from the binary system with increasing mass fraction of oil w_B, the oil emulsification boundary ($\underline{2} \rightarrow 1$) ascends, while the near-critical phase boundary ($1 \rightarrow \overline{2}$) descends. (c) $T(w_A)$ section through the phase prism of the system H_2O–n-octane–$C_{10}E_5$ at $\gamma_b = 0.10$. The inverse temperature behaviour is found on the oil-rich side: With increasing fraction of water w_A the water emulsification boundary ($1 \rightarrow \overline{2}$) descends, whereas the near-critical phase boundary ($\underline{2} \rightarrow 1$) ascends.

the three-phase body (T_l and T_u) and allow distinguishing between weak and strong surfactants if one considers the shapes of the near-critical phase boundaries [41]. While for weak surfactant systems the boundary decreases down to T_l (water-rich side) and increases up to T_u (oil-rich side), in strong surfactant systems the near-critical phase boundary has a minimum (water-rich side) and a maximum (oil-rich side), respectively. These extrema originate from additional two-phase regions in the form of closed loops appearing at temperatures below T_l and above T_u in the Gibbs triangle [37, 41, 48, 49]. The origin of the loops (the separation of a micellar phase into two phases which become homogeneous again upon swelling with a solute) was not understood for a long time. Recently, Safran *et al.* attributed the origin of the loops to the demixing of a connected network of swollen cylindrical micelles into a dense connected network in equilibrium with a dilute phase [50–52]. This description also explains why the loops appear only in strongly structured and not in weakly structured microemulsions.

Having discussed the general phase behaviour of microemulsions by studying simple ternary non-ionic systems of the type the water–*n*-alkane–*n*-alkyl polyglycol ether (C_iE_j) we will now apply this knowledge to more complex systems relevant in technical applications. It will be shown that the insight gained by studying the temperature dependence of ternary non-ionic microemulsions can easily be adapted to systems containing technical-grade non-ionic surfactants, *n*-alkylpolyglucosides, ionic surfactants as well as mixtures of non-ionic and ionic surfactants.

1.2.2 Microemulsions with technical-grade non-ionic surfactants

In industrial applications, technical-grade surfactants which are usually mixtures of homologues and (or) isomers are used instead of pure surfactants. Common non-ionic technical-grade surfactants are ethoxylated alcohols or ethoxylated alkyl phenols. In contrast to the pure C_iE_j surfactants, which were discussed above, the technical-grade surfactants have a broad distribution of the ethoxylation degree and a residual amount of non-reacted alcohol. However, the chain length of the alcohol is rather narrowly distributed. Several studies on microemulsions formulated with technical-grade surfactants have shown that surfactant blends can be treated as a single (pseudo) component [39, 53–55]. Thus, the phase behaviour of such a pseudo-ternary system can also be characterised by $T(\gamma)$-sections through the phase prism as was described above.

In order to show the effect of technical-grade surfactants on the phase behaviour of microemulsions, $T(\gamma)$-sections of the systems H_2O–*n*-octane–$C_{12}E_6$ and the technical grade analogue DA-6 (dodecyl-alcohol-6) are shown for comparison in Fig. 1.8 [56]. As can be seen, the $C_{12}E_6$ system shows the well-known phase behaviour of ternary non-ionic microemulsions with a horizontal three-phase region that touches the horizontal one-phase region at the $\tilde{X}$-point. On the other hand, the phase boundaries of the system containing the technical-grade surfactant are strongly distorted, especially at low γ. Despite this distortion the two systems behave in a similar way. Both systems are rather efficient and show an extended L_α phase within the one-phase region. However, the technical-grade DA-6 system solubilises water and *n*-octane slightly more efficiently than the pure $C_{12}E_6$ surfactant.

The distortion of the phase boundaries in the system with the technical-grade surfactant can be explained with the broad distribution of the ethoxylation degree of DA-6 and the resulting different monomeric solubilities of every specific homologue in water $\gamma_{Cmon,a}$ and oil $\gamma_{Cmon,b}$. Taking into account only $\gamma_{Cmon,b}$ (because $\gamma_{Cmon,a} << \gamma_{Cmon,b}$ for non-ionic surfactants (see Fig. 1.6)), the lower ethoxylated, more hydrophobic homologues of the surfactant DA-6 tend to dissolve in the oil-excess or sub-phase (e.g. in oil-swollen

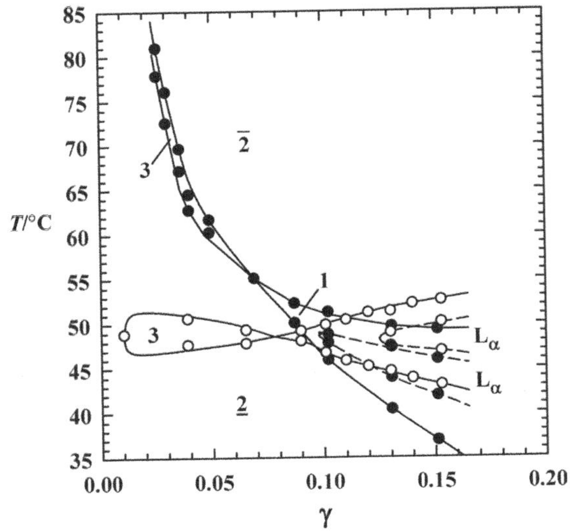

Figure 1.8 $T(\gamma)$-section through the phase prism of the systems H_2O–*n*-octane–$C_{12}E_6$ and the technical grade analogue DA-6 at an oil/(water + oil) volume fraction of $\phi = 0.5$. In contrast to the horizontal fish-type phase diagram of the $C_{12}E_6$ system the phase boundaries of the technical-grade surfactant system are distorted towards low γ. This effect is due to the broad distribution of the ethoxylation degrees of DA-6 and the higher monomeric oil solubility of the hydrophobic homologues in *n*-octane. (Figure redrawn with data from Ref. [56].)

micelles or bicontinuous oil-rich domains). Thus, the remaining surfactant mixture in the amphiphilic film is effectively more hydrophilic than the base-surfactant DA-6. Decreasing the surfactant mass fraction γ by adding water and oil one extracts increasing amounts of the more hydrophobic fractions of the surfactant DA-6 from the amphiphilic film, which accordingly becomes increasingly hydrophilic. Following the properties of ternary non-ionic microemulsions (see Fig. 1.5), the phase behaviour shifts to higher temperatures with decreasing γ, explaining the large distortion of the phase boundaries, i.e. an increasing HLB with decreasing γ. The distortion of the phase boundaries can also be discussed in terms of the mean curvature H of the amphiphilic film (see Section 1.4, Fig. 1.18). Upon decreasing γ, the fraction of surfactant molecules with large head groups increases within the film and leads to an amphiphilic film which is increasingly curved around the oil. Accordingly, within the technical-grade surfactant systems the mean curvature H of the amphiphilic film as well as the phase behaviour both depend not only on the temperature, but also on the composition of the film. In the following the dependence of the phase behaviour on the composition of the mixed amphiphilic film will be discussed in more detail.

1.2.3 Microemulsions with alkylpolyglucosides

The formulation of non-toxic, biodegradable microemulsions is of enormous importance in the cosmetic and pharmaceutical industries. One class of biodegradable surfactants which can be used to formulate such non-toxic microemulsions are the non-ionic alkylpolyglucosides (C_mG_n, where m is the number of carbons within the hydrophobic chain and n is the number of glucose units) [57]. However, having six hydroxyl groups in one glucose unit these sugar surfactants are usually rather hydrophilic. Thus, hydrophobic amphiphiles, like alcohols [58–61] or hydrophobic C_iE_j-surfactant [62] have to be added to these rather hydrophilic C_mG_n-surfactants to formulate microemulsions. Furthermore, the weak temperature dependence of the strong hydration of the hydroxyl groups causes the rather weak temperature sensitivity of the C_mG_n-microemulsions. Thus, temperature is not the appropriate parameter to tune the mean curvature of the amphiphilic film and with it the phase behaviour of the system. Instead, the mixing of two surfactants of different hydrophilicity is the appropriate method to drive the quaternary system through the phase inversion.

In general, the phase behaviour of such a quaternary system containing H_2O, oil, a hydrophilic C_mG_n- and a hydrophobic co-surfactant is rather complex. At constant temperature and pressure it has to be represented within a phase tetrahedron (see Fig. 1.9(a)). As for the ternary temperature-sensitive microemulsions (see Fig. 1.1) an insight into the phase behaviour of a quaternary system can be gained by considering the phase diagrams of the corresponding ternary base systems. In the following the phase behaviour of the quaternary system H_2O–n-octane–n-octyl-β-D-glucopyranoside (β-C_8G_1)–1-octanol (C_8E_0) system will be discussed as an example. Systematic studies have shown that all ternary base systems (= faces of the phase tetrahedron) show extensive miscibility gaps at $T = 25°C$ [61]. Here, the phase behaviour of the two-side systems H_2O–n-octane–β-C_8G_1 and H_2O–n-octane–C_8E_0 are of major interest. Within the former system the β-C_8G_1 molecules prefer the water phase, i.e. a $\underline{2}$ miscibility gap is formed. In contrast, the latter system shows a $\overline{2}$ behaviour, i.e. the C_8E_0 molecules reside mainly in the oil phase. Since on top of this there is the demixing tendency of the third ternary-side system H_2O–C_8E_0–β-C_8G_1 the formation of a three-phase region is induced inside the phase tetrahedron. Figure 1.9(a) illustrates this situation schematically by means of a $w_D(w_C)$-section through the phase tetrahedron at a constant oil/(water + oil) fraction. As can be seen, a typical fish-type phase diagram is found if the ratio of co-surfactant (D) in the surfactant (C) plus co-surfactant (D) mixture

$$\delta = \frac{m_D}{m_C + m_D} \tag{1.7}$$

is varied. The monomeric solubilities of surfactant and co-surfactant in water and oil, i.e. $\gamma_{Cmon,a}$, $\gamma_{Dmon,a}$, $\gamma_{Cmon,b}$ and $\gamma_{Dmon,b}$ as well as the HLB-plane (HLB = hydrophilic–lipophilic–balance) are also shown in Fig. 1.9(a). As has already been mentioned above, the HLB-plane indicates the compositions at which the mean curvature H of the amphiphilic film is zero, i.e. the system is driven through the phase inversion. Thus, the three-phase triangle (shown in dark grey) has to lie in the HLB-plane.

In order to determine $w_D(w_C)$-sections through the phase tetrahedron experimentally a sample containing the desired amounts of water, oil and surfactant has to be titrated with the co-surfactant. Figure 1.9(b) shows such a $w_D(w_C)$-section for the H_2O–n-octane–n-octyl-β-D-glucopyranoside (β-C_8G_1)–1-octanol (C_8E_0) system at $\phi = 0.50$ and $T = 25°C$ [61]. As can be seen, the phase sequence $\underline{2}$–3–$\overline{2}$ is found with increasing 1-octanol content at low mass fractions of β-C_8G_1, while at higher mass fractions of β-C_8G_1, the $\underline{2}$–1–$\overline{2}$

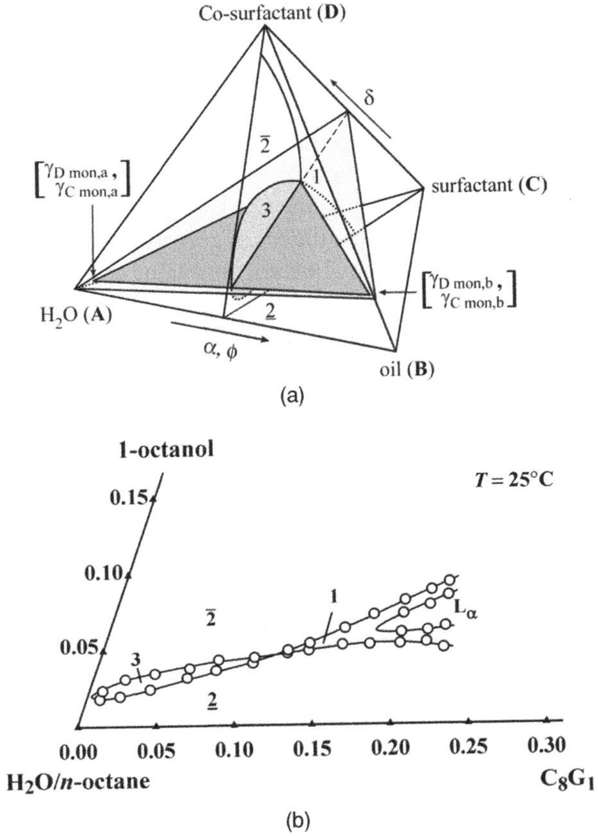

Figure 1.9 (a) Schematic phase tetrahedron of a quaternary water (A)–oil (B)–surfactant (C)–co-surfactant (D) system at constant temperature and pressure [39]. Shown is a section at a constant oil/(water + oil) ratio and the HLB-plane (HLB = hydrophilic–lipophilic–balance). The latter indicates the compositions at which the curvature of the amphiphilic film is zero. Note that the three-phase triangle (shown in dark grey) lies in the HLB-plane. (b) Section through the phase tetrahedron for the quaternary system H_2O–*n*-octane–*n*-octyl-β-D-glucopyranoside (C_8G_1)–1-octanol (C_8E_0) at a constant oil/(water + oil) volume fraction of $\phi = 0.5$ and $T = 25°C$ [61]. The system is driven through the phase inversion by adding C_8E_0. (Figure redrawn with data from Ref. [61].)

sequence is observed. For even higher mass fractions a lamellar phase appears. In these quaternary systems the location of the $\tilde{X}$-point is typically given by the ratio δ and the overall mass fraction surfactant (i.e. surfactant plus co-surfactant)

$$\gamma = \frac{m_C + m_D}{m_A + m_B + m_C + m_D}. \tag{1.8}$$

The $\tilde{X}$-point for the system under consideration lies at $\tilde{\delta} = 0.276$ and $\tilde{\gamma} = 0.161$, which shows that the surfactant/co-surfactant mixture β-C_8G_1/C_8E_0 solubilises water and *n*-octane with a medium efficiency. Comparing the $w_D(w_C)$-section performed through the phase tetrahedron with the $T(\gamma)$-section through the phase prism (see, e.g. Fig. 1.3) one sees that a fish-type phase diagram is found in both quaternary temperature-insensitive and ternary temperature-sensitive microemulsions. Thus, the temperature T may simply be replaced by the fraction δ of co-surfactant in the mixture of surfactant and co-surfactant. However, it turned out that a quantitative description of the quaternary temperature-insensitive systems can only be obtained if the composition of the amphiphilic film

$$\delta_i = \frac{m_{D,i}}{m_{D,i} + m_{C,i}} \tag{1.9}$$

is used instead of δ, where $m_{C,i}$ and $m_{D,i}$ correspond to the mass of surfactant and co-surfactant molecules residing in the mixed amphiphilic film. To determine δ_i the monomeric solubilities (see Fig. 1.6) of 1-octanol in n-octane and β-C_8G_1 in H_2O have to be known, while the solubilites of 1-octanol in H_2O and β-C_8G_1 in n-octane can be neglected. The former can be determined individually from the phase behaviour applying the analysis of Kunieda and co-workers [63, 64].

The phase behaviour of the quaternary system can thus be tuned by varying the composition of the amphiphilic film δ_i. Starting from the ternary system H_2O–n-octane–β-C_8G_1 at $\phi = 0.50$ and at a mass fraction $\gamma = 0.10$ of β-C_8G_1 an oil-in-water (o/w) microemulsion forms that coexists with an excess oil phase ($\underline{2}$). As one adds the 1-octanol it partitions between the oil-excess or sub-phase (e.g. in oil-swollen micelles) and the amphiphilic film. Thus, on the one hand, the alcohol acts as a co-solvent making the oil more hydrophilic. On the other hand, the alcohol mixes into the amphiphilic film making it increasingly hydrophobic. Although the mean curvature H of the amphiphilic film is lowered by both effects, the latter is predominant since the OH-group of the alcohol is small compared to the large head groups of the sugar surfactant. Increasing the concentration of 1-octanol further, the film is enriched in 1-octanol and decreases its curvature until it inverts to form a water-in-oil (w/o) microemulsion (see Fig. 1.18). Accordingly, the tuning parameter δ_i in quaternary temperature-insensitive n-alkylglycoside systems plays the same role as the temperature in the ternary water–oil–C_iE_j systems. That this is indeed the case can be shown by scaling the phase behaviour. The corresponding scaling parameters for the quaternary temperature-insensitive microemulsions are $\tilde{\gamma}_i(\phi = 0.50)$, the lower limit $\delta_{i,l}$ and the upper limit $\delta_{i,u}$ of the three-phase body [61].

1.2.4 Microemulsions with ionic surfactants

In the preceding sections, the phase behaviour of rather 'simple' ternary and quaternary non-ionic microemulsions have been discussed. However, the first microemulsion found by Schulman more than 50 years ago was made of water, benzene, hexanol and the ionic-surfactant potassium oleate [1, 3]. Winsor also used the ionic-surfactant sodium decylsulphate and the co-surfactant octanol to micro-emulsify water/sodium sulphate and petrol ether [2]. In the last 30 years, in-depth studies on ionic microemulsions have been carried out [7, 8, 65, 66]. It turned out that nearly all ionic surfactants which contain one single hydrocarbon chain are too hydrophilic to build up microemulsions. Such systems can only be driven through the phase inversion if an electrolyte and a co-surfactant is added to the mixture (see below and Fig. 1.11).

However, using double-chain ionic surfactants, e.g. sodium-bis-ethylhexylsulfosuccinate (AOT) [9, 67] and didodecyl dimethyl ammonium bromide (DDAB) [68], no co-surfactant is necessary to tune the mean curvature of the amphiphilic film from positive to negative. In the following the quaternary (pseudo-ternary) system H_2O/NaCl (A)–*n*-decane (B)–AOT (D) will be discussed to show the main features of ionic microemulsions. Subsequently, the knowledge gained for alkylpolyglucoside microemulsions (see Section 1.2.3) will be applied to understand the complex behaviour of five component ionic mixtures.

1.2.4.1 Quaternary AOT microemulsions

AOT has become the most widely studied amphiphile to formulate ionic microemulsions because only traces of an inert electrolyte shifts the phase inversion of the H_2O–oil–AOT system to ambient temperatures. In general, an ionic amphiphile gets more hydrophilic with increasing temperature which is in contrast to the non-ionic alkyl polyglycol ethers mentioned above (see Section 1.2.1). This increase in hydrophilicity can be attributed to the increasing effective degree of dissociation of the counterions at higher temperatures. The opposite effect is achieved by the addition of electrolytes to ionic microemulsions. The salt ions compete with the counterions and the head groups for water of hydration and decrease the hydrophilicity of the surfactant. This combination of adding salt and increasing the temperature can be used to tune the mean curvature of the amphiphilic film and with it the phase behaviour of the quaternary AOT microemulsion [67].

Considering the phase behaviour of the system H_2O/NaCl (A)–*n*-decane (B)–AOT (D) as an example, the temperature-dependent phase behaviour of the system can be represented as a first approximation in an upright Gibbs phase prism, if the mixture of H_2O and NaCl (often referred to as brine) is treated as a pseudo-component. It holds for the mass fraction of NaCl in the H_2O/NaCl mixture

$$\varepsilon = \frac{m_{\text{salt}}}{m_{\text{salt}} + m_{\text{water}}}. \tag{1.10}$$

As for the systems discussed above, the binary base systems are considered first to understand the phase behaviour of the pseudo-ternary ionic microemulsion. Brine (H_2O/NaCl) (A) and *n*-decane (B) are practically immiscible over the experimentally accessible temperature range. The binary system *n*-decane (B)–AOT (D) shows a lower miscibility gap that lies below the melting point of *n*-decane. Thus, a complete miscibility of *n*-decane and AOT exists between the melting and boiling points of *n*-decane. However, adding traces of water the situation changes and an upper miscibility gap appears [67]. As for the ternary non-ionic microemulsions the pseudo-binary system (H_2O/NaCl) (A)–AOT (D) plays the decisive role in the phase behaviour of the AOT microemulsion. As regards the binary system H_2O–AOT the lower miscibility gap with an upper critical point is covered by a lamellar phase that extends to low amphiphile concentrations. Adding NaCl one observes that the lamellar phase is pushed back, while simultaneously the miscibility gap is enlarged towards higher temperatures [69].

The phase behaviour of the pseudo-ternary ionic mixture can again be explained with the interplay of the three binary base systems. At low temperatures the ionic surfactant is preferentially soluble in oil, at high temperatures in brine. Thus, having added the adequate amount of salt, an increase in temperature turns the ionic surfactant from hydrophobic at low into hydrophilic at high temperatures. On the microscopic level the temperature-dependent behaviour of ionic microemulions can be explained with changes of the mean curvature H of the amphiphilic film. Starting with a film which is curved around the water the degree of dissociation of the counterions increases, if the temperature is increased. This increasing degree of dissociation leads to stronger repulsive interactions between the ionic head groups on the water-side of the amphiphilic film. Accordingly, the mean curvature of the film inverts from being curved around the water ($H < 0$) to being curved around the oil ($H > 0$) as the temperature is increased.

In Fig. 1.10(a), the temperature dependence of the phase behaviour is shown by means of a $T(\gamma)$-section through the phase prism of the pseudo-ternary system H_2O/NaCl–n-decane–AOT at $\varepsilon = 0.006$ and an oil/(water + oil) mass fraction $\alpha = 0.50$. As can be seen, the phase boundaries of the ionic system also resemble the shape of a fish with a three-phase region at ambient temperatures. The three-phase region touches the one-phase region at $\tilde{T} = 29.0°C$ and $\tilde{\gamma} = 0.061$, which shows a rather good efficiency of the double-chain ionic surfactant AOT in solubilising brine and n-decane. At lower temperatures a water-in-oil (w/o) microemulsion coexists with a water excess phase ($\overline{2}$), whereas at higher temperatures an o/w microemulsions coexists with an oil excess phase ($\underline{2}$). Thus, a phase sequence of $\overline{2}$, 3, $\underline{2}$ is observed in the ionic system which is inverse to the $\underline{2}$, 3, $\overline{2}$ sequence found in non-ionic microemulsions. Despite this inverse temperature dependence the overall behaviour of both types of systems is rather similar. This analogy between the properties of pseudo-ternary ionic and ternary non-ionic microemulsions is another indication that the behaviour of microemulsions obviously follows a general pattern irrespective of the components they are made of.

The variation of the phase behaviour as a function of the salinity is shown in Fig. 1.10(b) in the form of an $\varepsilon(\gamma)$-section through the phase tetrahedron of the quaternary H_2O/NaCl–n-decane–AOT system at $\alpha = 0.50$ and a constant temperature of $T = 40°C$. In order to compare the variation of the phase behaviour with temperature and salinity a rectangular representation is used also for the $\varepsilon(\gamma)$-section through the phase tetrahedron. As can be seen, the phase boundaries also resemble the shape of a fish in this isothermal $\varepsilon(\gamma)$-section. However, with increasing mass ε fraction of salt the phase sequence $\underline{2}$, 3, $\overline{2}$ is found which is inverse to the $\overline{2}$, 3, $\underline{2}$ sequence observed with increasing temperature.

The dependence of the phase behaviour on the salinity can also be explained by the interplay of the three binary base systems. At a constant temperature (e.g. $T = 40°C$) and an amount of salt $\varepsilon = 0.006$ the ionic surfactant is preferentially soluble in brine ($\underline{2}$, see also Fig. 1.10(a)). Increasing ε makes AOT less hydrophilic, i.e. will enlarge the brine (A)–AOT (D) miscibility gap towards higher temperatures so that the AOT molecules become preferentially soluble in oil ($\overline{2}$). The mean curvature of the amphiphilic film is curved around the oil at low ε. As the addition of salt leads to an increasing screening of the electrostatic interactions between the ionic head groups on the water-side of the amphiphilic film, the mean curvature of the film inverts from being curved around the oil ($H > 0$) to being curved around the water ($H < 0$) with increasing ε. Thus, the tuning parameter salinity (ε) in quaternary ionic microemulsions plays the same role as the temperature in ternary systems water–oil–C_iE_j.

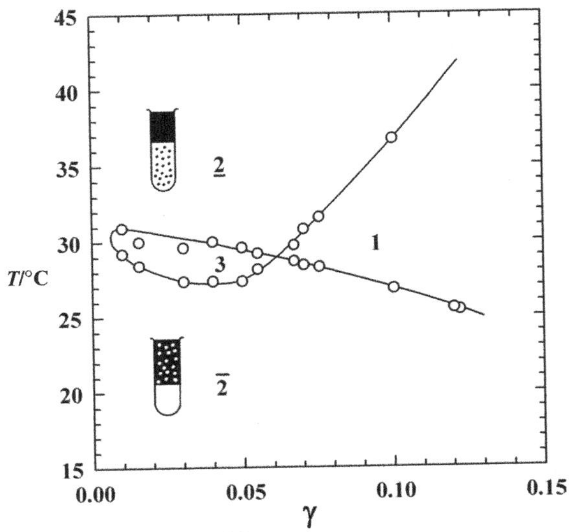

(a)

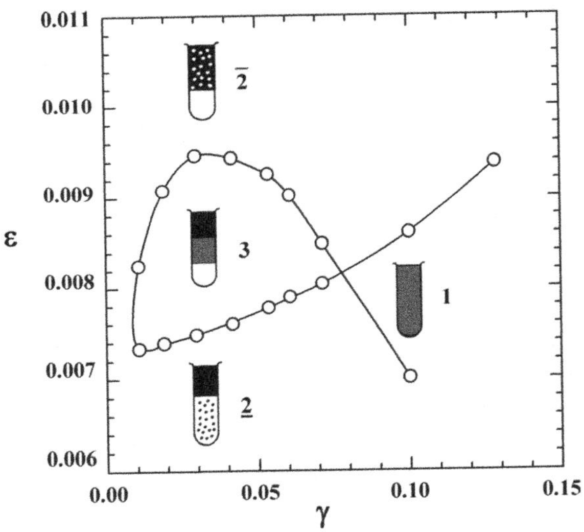

(b)

Figure 1.10 Phase behaviour of the ionic system H_2O/NaCl–*n*-decane–sodium-bis-ethylhexylsulfosuccinate (AOT) at a constant oil/(water + oil) mass fraction of $\alpha = 0.50$ [67]. (a) $T(\gamma)$-section performed at a constant salt mass fraction in water of $\varepsilon = 0.006$. The phase boundaries resemble the shape of a fish (general pattern of microemulsions!). Note that regarding the temperature dependence the phase sequence is inverted compared to that of non-ionic microemulsions. (b) $\varepsilon(\gamma)$-section through the phase tetrahedron of the quaternary system H_2O–NaCl–*n*-decane–sodium-bis-ethylhexylsulfosuccinate (AOT) at a temperature of $T = 40°C$. In this isothermal section the phase boundaries again resemble the shape of a fish. However, with increasing mass fraction of salt ε the phase sequence $\underline{2}$, 3, $\overline{2}$ is found due to the increasing screening of the electrostatic interactions.

1.2.4.2 Quinary SDS microemulsions

Ionic surfactants with only one alkyl chain are generally extremely hydrophilic so that strongly curved and thus almost empty micelles are formed in ternary water–oil–ionic surfactant mixtures. The addition of an electrolyte to these mixtures results in a decrease of the mean curvature of the amphiphilic film. However, this electrolyte addition does not suffice to drive the system through the phase inversion. Thus, a rather hydrophobic co-surfactant has to be added to invert the structure from oil-in-water to water-in-oil [7, 66]. In order to study these complex quinary mixtures of water/electrolyte (brine)–oil–ionic surfactant–non-ionic co-surfactant, brine is considered as one component. As was the case for the quaternary sugar surfactant microemulsions (see Fig. 1.9(a)) the phase behaviour of the pseudo-quaternary ionic system can now be represented in a phase tetrahedron if one keeps temperature and pressure constant.

As an example, the phase behaviour of the system H_2O/NaCl–*n*-decane–sodium dodecyl sulphate (SDS)–1-butanol (C_4E_0) will be discussed at a rather large salinity of $\varepsilon = 0.10$ and $T = 20°C$. Again, the systems representing the faces of the phase tetrahedron are considered first in order to understand the complex behaviour of the pseudo-quaternary system. Of major interest are the two-side systems H_2O/NaCl–*n*-decane–SDS and H_2O/NaCl–*n*-decane–C_4E_0. Both systems show miscibility gaps. Within the former system the SDS molecules prefer the water phase, i.e. a $\underline{2}$ miscibility gap is formed. In contrast, the latter system shows a $\overline{2}$ behaviour, i.e. the C_4E_0 molecules reside mainly in the oil phase. Since there is an additional demixing tendency in the third ternary-side system H_2O/NaCl–C_4E_0–SDS the formation of a three-phase region is induced inside the phase tetrahedron.

Equivalently to the quaternary sugar surfactant microemulsions the $w_D(w_C)$-sections through the tetrahedron are obtained experimentally by titrating a sample containing the desired amounts of brine, *n*-decane and SDS with the co-surfactant C_4E_0. Figure 1.11 shows such a section at $\phi = 0.58$, $\varepsilon = 0.10$ and $T = 20°C$. As can be seen, the phase boundaries obtained resemble the shape of the fish. At low mass fractions of SDS, the phase sequence $\underline{2}$, 3, $\overline{2}$ is found with increasing 1-butanol content. At higher mass fractions of SDS, the $\underline{2}$, 1, $\overline{2}$ sequence is observed. For even higher mass fractions a lamellar phase appears.

From Fig. 1.11 it is obvious that the phase behaviour of pseudo-quaternary ionic microemulsions follows the general patterns of microemulsions, which is mainly determined by the variation of the mean curvature H of the amphiphilic film. Starting from the pseudo-ternary system without 1-butanol, a small amount of the oil is already solubilised in the SDS-micelles due to the screening of the repulsive interaction between the ionic head group obtained by the addition of NaCl. Thus, an oil-in-water (o/w) microemulsion forms that coexists with an excess-oil phase ($\underline{2}$). Adding 1-butanol it partitions between the bulk oil phase and the bulk water phase as well as the amphiphilic film. Enriching the film with 1-butanol lowers H until the curvature inverts, i.e. a water-in-oil (w/o) microemulsion forms that coexists with an excess-water phase ($\overline{2}$). However, having driven the system through phase inversion (from $\underline{2}$ to $\overline{2}$) by adding 1-butanol the system can be tuned back to $\underline{2}$ by keeping the fraction δ_i of 1-butanol in the amphiphilic film constant and increasing the temperature.

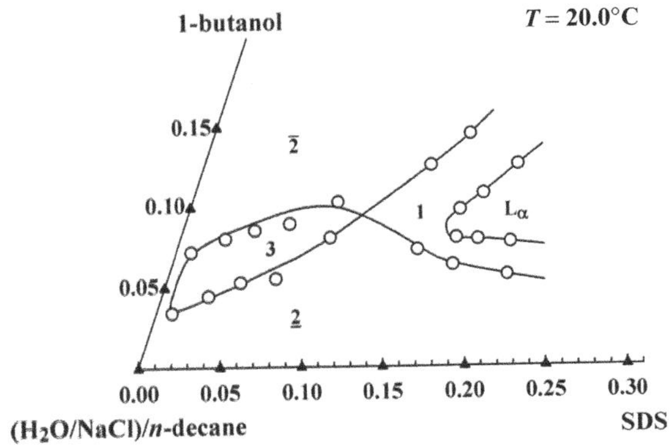

Figure 1.11 Section through the phase tetrahedron of the pseudo-quaternary system H_2O/NaCl–*n*-decane–sodium dodecyl sulphate (SDS)–1-butanol (C_4E_0) at $\phi = 0.58$, $\varepsilon = 0.10$ and $T = 20°C$ [26]. Note that the pseudo-quaternary ionic system can be driven through phase inversion by adding C_4E_0 as was the case for the quaternary alkylpolyglucoside microemulsions. (From Ref. [26], reprinted with permission of Elsevier.)

1.2.5 *Microemulsions with non-ionic and ionic surfactants*

In the previous section a quinary ionic microemulsion was tuned through the phase inversion by adding a short-chain alcohol as a non-ionic co-surfactant to a single-tailed ionic surfactant. In the following the short-chain alcohol is replaced by an ordinary long-chain non-ionic surfactant. It was discussed above that the temperature dependence of the phase behaviour of ionic (see Section 1.2.4) and non-ionic microemulsions (see Section 1.2.1) is inverse. Thus, one can expect that at a certain ratio δ of non-ionic and ionic surfactants the inverse temperature trends compensate so that a temperature-insensitive microemulsion forms. It goes without saying that this property is extremely relevant in technical applications, where often mixtures of non-ionic and ionic surfactants are used.

In order to locate the composition where most of the properties of the complex quinary (pseudo-quaternary) mixture are expected to be temperature-insensitive, time-consuming studies of the phase behaviour have to be performed. Such studies were carried out with the quinary system H_2O–NaCl–*n*-decane–$C_{12}E_4$–AOT [10]. The result is shown in Fig. 1.12 in the form of a $T(\gamma)$-section through the phase prism at $\phi = 0.60$ and $\varepsilon = 0.006$ considering H_2O/NaCl and $C_{12}E_4$/AOT, respectively, as a pseudo one-component system. In this study, a mass fraction of $\delta = 0.60$ of AOT in the $C_{12}E_4$/AOT mixture was chosen to obtain an almost temperature-insensitive phase behaviour. Note that only the phase boundaries of the one-phase region are determined experimentally, whereas the extension of the three-phase region is shown schematically. As can be seen, the phase boundaries around the $\tilde{X}$-point are very steep, which indicates the temperature insensitivity. Thus, preparing a mixture of H_2O/NaCl–*n*-decane–$C_{12}E_4$–AOT at $\phi = 0.60$, $\delta = 0.60$, $\varepsilon = 0.006$ and an overall surfactant mass fraction of $\gamma = 0.08$ a one-phase microemulsion is obtained between 0 and 75°C.

In conclusion, the extensive study of the phase behaviour of different microemulsions provides detailed knowledge about the phase behaviour of microemulsions. As discussed above, it is emphasised that microemulsions show striking similarities in the phase

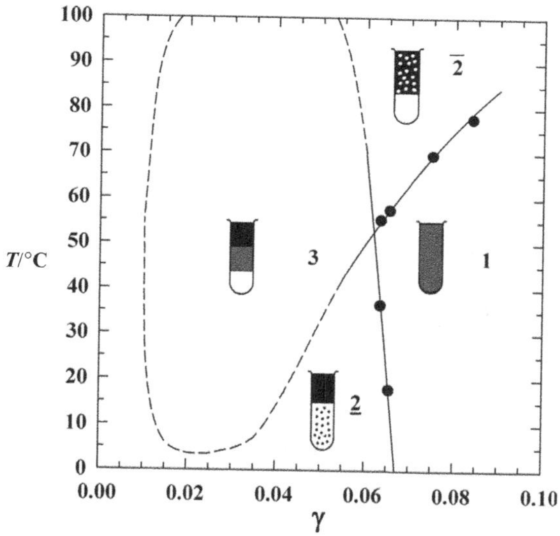

Figure 1.12 $T(\gamma)$-section of the quinary system H_2O/NaCl–*n*-decane–$C_{12}E_4$/AOT at constant $\phi = 0.60$, a mass fraction of $\delta = 0.60$ of AOT in the $C_{12}E_4$/AOT mixture and a salt mass fraction in water of $\varepsilon = 0.006$. Note that the steepness of the phase boundaries indicates their temperature-insensitivity. Furthermore, the $C_{12}E_4$/AOT mixture provides an efficient solubilisation ($\tilde{\gamma} = 0.06$) of the oil *n*-decane in water and vice versa. (From Ref. [26], reprinted with permission of Elsevier.)

behaviour irrespective of whether they are stabilised by pure non-ionic surfactants, technical-grade non-ionic surfactants, *n*-alkylpolyglucosides, ionic surfactants, or by mixtures of non-ionic and ionic surfactants. It turned out that several tuning parameters can be chosen to drive the system through the phase inversion, i.e. to obtain a zero-mean curvature of the amphiphilic film. These parameters can be the temperature T, the salinity ε, or the ratio δ_i of two different surfactants in the amphiphilic film.

1.3 Interfacial tension

Perhaps the most striking property of a microemulsion in equilibrium with an excess phase is the very low interfacial tension between the macroscopic phases. In the case where the microemulsion coexists simultaneously with a water-rich and an oil-rich excess phase, the interfacial tension between the latter two phases becomes ultra-low [70, 71]. This striking phenomenon is related to the formation and properties of the amphiphilic film within the microemulsion. Within this internal amphiphilic film the surfactant molecules optimise the area occupied until lateral interaction and screening of the direct water–oil contact is minimised [2, 42, 72]. Needless to say that low interfacial tensions play a major role in the use of microemulsions in technical applications [73] as, e.g. in enhanced oil recovery (see Section 10.2 in Chapter 10) and washing processes (see Section 10.3 in Chapter 10). Suitable methods to measure interfacial tensions as low as 10^{-3} mN m^{-1} are the sessile or pendent drop technique [74]. Ultra-low interfacial tensions (as low as 10^{-5} mN m^{-1}) can be determined with the surface light scattering [75] and the spinning drop technique [76].

As the latter is comparatively simple to use it can be regarded as the most suitable method to measure low and ultra-low interfacial tensions. In the following the general features of interfacial tensions in microemulsion systems are presented. The dramatic decrease of the water/oil interfacial tension upon the addition of surfactant, the correlation of interfacial tension and phase behaviour, the variation of the water/oil interfacial tension with the respective tuning parameter and the scaling of the interfacial tension will be discussed in detail. All data presented have been determined using the spinning drop technique [17].

1.3.1 Adsorption of the surfactant

The starting point of this discussion is the pure water–oil system. In the absence of surfactant the interfacial tension σ_{ab} of the water/oil interface is generally 30–50 mN m^{-1}. Adding small amounts of surfactant the molecules will either adsorb at the water/oil interface or be monomerically distributed between the water- (a) and oil-rich (b) phase. For ionic surfactants such as AOT the monomer is mainly soluble in the water phase ($\gamma_{\mathrm{Cmon,a}} >> \gamma_{\mathrm{Cmon,b}}$), while for non-ionics such as $C_{12}E_5$ the monomer is mainly soluble in the oil phase ($\gamma_{\mathrm{Cmon,a}} << \gamma_{\mathrm{Cmon,b}}$, see Section 1.2.1). The adsorption of the surfactants at the macroscopic water/oil interface and the increase of the monomer concentrations causes the interfacial tension σ_{ab} to drop to a value which may vary from a few mN m^{-1} to ultra-low values of 10^{-3}–10^{-4} mN m^{-1} [70]. A further addition of surfactant leads to a complete saturation of the water-rich and oil-rich phases as well as the water/oil interface with surfactant molecules so that the unfavourable water/oil contact is nearly perfectly screened. Above this concentration the excess surfactant molecules form aggregates in either the water, oil or a third phase (between the lower and upper limit of the three-phase body). Thus, the monomeric concentration of the surfactant in the water- and oil-rich phases as well as the water/oil interfacial tension stays practically constant.

Figure 1.13 shows the variation of the water/oil interfacial tension σ_{ab} as a function of the surfactant mass fraction γ (on a logarithmic scale) for a microemulsion at the mean temperature $\tilde{T}$ (i.e. PIT) of the three-phase body and equal volumes of water (A) and oil (B), i.e. $\phi = 0.5$, schematically [26]. As can be seen, σ_{ab} decreases from ~50 mN m^{-1} to almost zero if γ is increased (drawn line). At surfactant mass fractions above γ_0 a lens of a third phase, which is the microemulsion (c), is formed. The three test tubes illustrate the situation without surfactant (left), with only partially screened water/oil contact (centre) and at $\gamma > \gamma_0$ (right). Thus, as in aqueous surfactant solutions, the distinct discontinuity in the slope of the $\sigma_{ab}(\log\gamma)$-curve is an indication of the onset of aggregation. Below γ_0 the slope $(\delta\sigma_{ab}/\delta\log\gamma)$ is proportional to the interfacial concentration Γ_C of the surfactant which is given by the appropriate Gibbs equation [77]

$$\Gamma_C = -\frac{1}{2.303RT}\frac{\partial\sigma_{ab}}{\partial\log\gamma}\bigg|_{T,p} \tag{1.11}$$

with R = gas constant. As is indicated by the dashed line in Fig. 1.13, the slope of the curve becomes practically constant already at concentrations well below γ_0 for most surfactant systems, whereas σ_{ab} continues to decrease rather steeply. This behaviour could

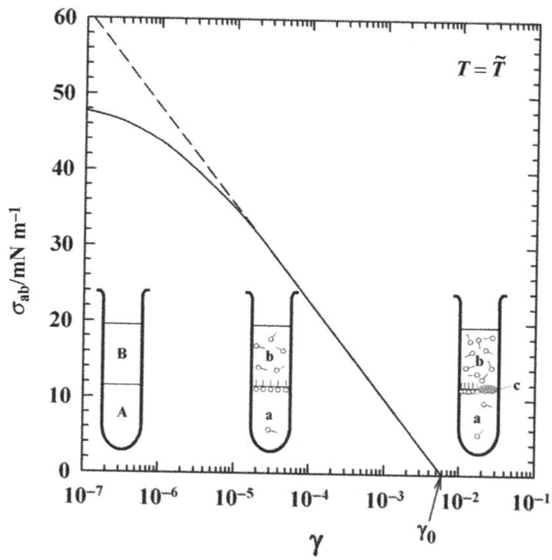

Figure 1.13 Schematic representation of the water/oil interfacial tension σ_{ab} (drawn line) as function of the non-ionic surfactant mass fraction γ at the mean temperature $\tilde{T}$ of the three-phase body. Starting from equal volumes of water (A) and oil (B), i.e. $\phi = 0.50$, the interfacial tension σ_{ab} decreases from 50 mN m^{-1} to values as low as 10^{-4} mN m^{-1}. After having crossed the monomeric solubility γ_0 of the surfactant in the water- and oil-rich phase, σ_{ab} remains constant. The test tubes illustrate the situation without surfactant (left), with only partially screened water/oil contact (centre) and at $\gamma > \gamma_0$, where the microemulsion phase (c) exist in form of a lens (right). (From Ref. [26], reprinted with permission of Elsevier.)

be interpreted as a consequence of the strong adsorption of the surfactants at the interface which saturates the water/oil interface well below γ_0. However, this would still raise the question why hardly measurable changes of Γ_C lead to a strong decrease of σ_{ab}. Knowing the interfacial concentration Γ_C in a saturated water/oil monolayer, the area per molecule $a_C = (N_A\Gamma_C)^{-1}$ can be determined [78, 79]. Another method to obtain reliable values of a_C in the water/oil interface is the analysis of experimentally more demanding SANS measurements [80] (see Chapter 2).

1.3.2 Interfacial tension and phase behaviour

From the above, it is clear that a pre-requisite of low water/oil interfacial tensions is the complete saturation of the water-rich and oil-rich phases as well as the water/oil interface by surfactant molecules. Of course, this pre-requisite is fulfilled if one of the phases considered is a microemulsion. Furthermore, since the pioneering work of Lang and Widom [81] it is known that if a system is driven through phase inversion the interfacial tensions may become ultra-low. However, about 20 years ago, a number of experimental investigations were devoted to clarifying the origin of the ultra-low interfacial tensions [15, 17, 39, 71, 81–85]. In order to understand this correlation between phase behaviour and interfacial

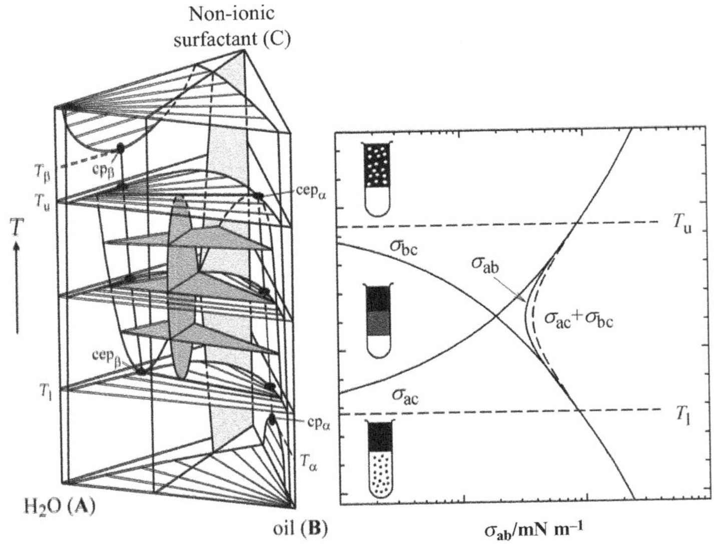

(a) (b)

Figure 1.14 Schematic phase prism (a) and interfacial tensions (b) as function of temperature for the system water–oil–non-ionic surfactant. The minimum of the water/oil interfacial tension $\tilde{\sigma}_{ab}$ at $\tilde{T}$ is a consequence of the phase behaviour. Increasing the temperature the aqueous phases separates into the phases (a) and (c) at the critical endpoints cep_β whereas the phases (b) and (c) merge into a single oil-rich phase at cep_α. Thus, the interfacial tensions σ_{ac} and σ_{bc} show an opposite temperature dependence, becoming zero at T_l and T_u, respectively. Note that the interfacial tensions are plotted on a log-scale.

tensions, let us consider as an example the temperature dependence of both properties in ternary non-ionic microemulsion systems.

Figure 1.14(a) shows the phase prism of the system water–oil–non-ionic surfactant (already shown in Fig. 1.3) together with the temperature dependence of the interfacial tensions (Fig. 1.14(b)). As discussed in Section 1.2.1, at low temperatures, non-ionic surfactants mainly dissolve in the aqueous phase and form an oil-in-water (o/w) microemulsion (a) that coexists with an oil-excess phase (b). Thus, for temperatures below the temperature T_l the interfacial tension σ_{ab} refers to the interface between an o/w-microemulsion and an oil-rich excess phase. As the temperature is increased, the o/w-microemulsion separates into two phases (a) and (c) at the temperature T_l which, in turn, leads to the appearance of the three-phase body. Thus, three different interfacial tensions occur within the three-phase body, namely the interfacial tension between the water-rich and the surfactant-rich phase σ_{ac}, between the oil-rich and the surfactant-rich phase σ_{bc}, and between the water-rich and the oil-rich phase σ_{ab}. However, the latter can only be measured if most of the surfactant-rich middle phase (c) is removed, which then floats as a lens at the water/oil interface. Increasing the temperature one observes that the three-phase body vanishes at the temperature T_u, where a water-in-oil (w/o) microemulsion is formed by the combination of the two phases (c) and (b). Therefore, at temperatures above T_u the interfacial tension σ_{ab} refers to the interface between a w/o-microemulsion and a water-rich excess phase.

From the temperature dependence of the phase behaviour the qualitative shape of the three interfacial tension curves can be deduced. As the two phases (a) and (c) are identical at the critical tie line at T_l the interfacial tension σ_{ac} has to start from zero and increases monotonically with increasing temperature. Whereas the interfacial tension σ_{bc} decreases (monotonically) with increasing temperature and vanishes at T_u, because the two phases (c) and (b) become identical at the critical tie line at T_u. This opposite temperature dependence of σ_{ac} and σ_{bc} results in a minimum if one considers the sum of the two, $\sigma_{ac} + \sigma_{bc}$. In order to assure the stability of the water/oil interface

$$\sigma_{ab} \leq \sigma_{ac} + \sigma_{bc} \tag{1.12}$$

must hold [83]. Otherwise a thin layer of the middle phase would form between the water- and oil-rich excess phase, which is the case for well-structured microemulsions (see Section 1.2.1, Fig. 1.7 and Section 1.4) only near the critical endpoints [41, 86]. Consequently, also σ_{ab} has to pass through a minimum at the mean temperature of the three-phase body $T_m = (T_u + T_l)/2$, i.e. at the PIT. Thus, the minimum of the water/oil interfacial tension $\overline{\sigma}_{ab}$ can be found at the same temperature where the solubilisation of water and oil is obtained with the minimum amount of surfactant $\tilde{\gamma}$.

Furthermore, knowing σ_{ab}, T_l and T_u, the relative location of the individual σ_{ac}- and σ_{bc}-curves is fixed. Near the critical endpoint temperatures T_l and T_u even a quantitative description of the interfacial tensions σ_{ac} and σ_{bc} can be obtained applying the scaling laws

$$\sigma_{ac} = \sigma_{ac,0}\varepsilon^{\mu} \quad \text{and} \quad \sigma_{bc} = \sigma_{bc,0}\varepsilon^{\mu}, \tag{1.13}$$

where $\mu = 1.26$ is the critical exponent [82, 83, 85], $\sigma_{ac,0}$, $\sigma_{bc,0}$ are the critical amplitudes and $\varepsilon = |T_i - T|/T_i$ is the distance from T_l and T_u, respectively.

1.3.3 Tuning parameters for the interfacial tension σ_{ab}

As was mentioned earlier, it is above all the water/oil interfacial σ_{ab} that plays an important role in technical applications. Thus, much work has been carried out to obtain the variation of σ_{ab} as a function of the respective tuning parameter, i.e. temperature T [17, 84, 87, 88], salinity ε[15, 89] and co-surfactant to surfactant ratio δ[16, 90]. In the following the variation of the water/oil interfacial as a function of temperature and composition of the amphiphilic film (see Section 1.2.3) is discussed by way of example.

Figure 1.15(a) shows the variation of the interfacial tension σ_{ab} with the temperature for the system water–n-octane–$C_{10}E_4$ [17] and σ_{ab} as a function of the composition of the amphiphilic film $\delta_{V,i}$ ($\delta_{V,i}$ is the volume fraction and can be calculated by replacing m in Eq. (1.9) with V) in the quaternary system H_2O–n-octane–β-C_8G_1–C_8E_0 at $T = 25°C$ (Fig. 1.15(b)) [90]. In both cases a log-scale is used for the interfacial tension because of the strong variation over several orders of magnitudes.

As can be seen independently of the parameter used to drive the system through the phase inversion the shape of the interfacial tension curves is similar. Because of the fundamental link of the interfacial tension and phase behaviour discussed above, both systems show

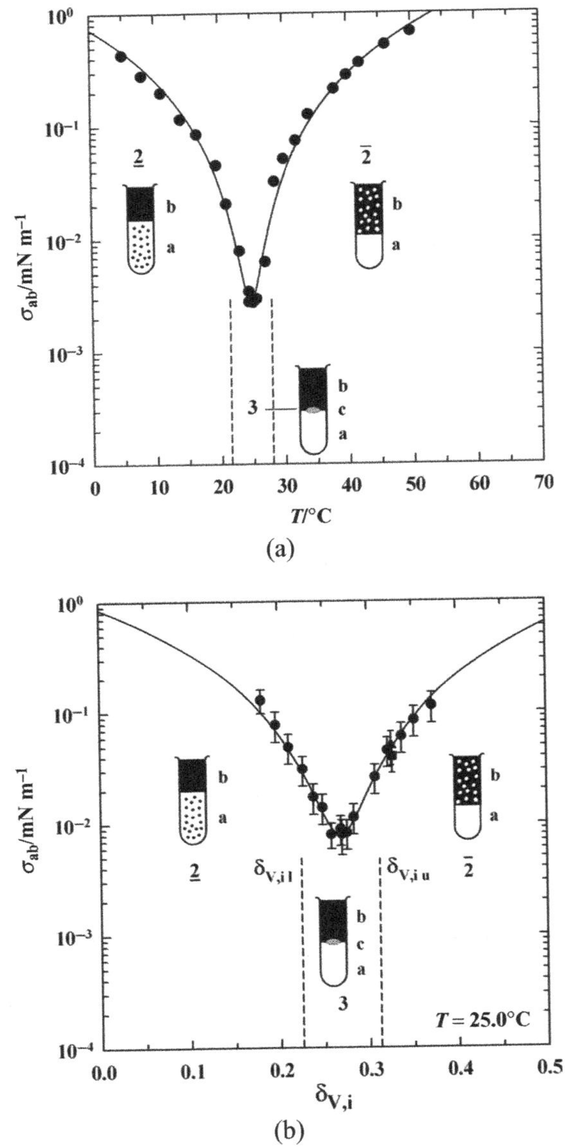

Figure 1.15 Water/oil interfacial tension σ_{ab} (plotted on log-scale) as function of the relevant tuning parameter. (a) Variation of σ_{ab} with temperature T, exemplarily shown for the water–n-octane–$C_{10}E_4$ system [17]. (b) Variation of σ_{ab} with the composition of the amphiphilic film $\delta_{V,i}$ in the quaternary system H_2O–n-octane–β-C_8G_1–C_8E_0 at $T = 25°C$ [90]. Both systems show that the water/oil interfacial tension runs through a distinct minimum in the middle of the three-phase region. The full line is calculated considering the bending energy difference between a curved amphiphilic film in the microemulsion and the flat film of the macroscopic interface [96].

an extreme minimum of the interfacial tension $\overline{\sigma}_{ab}$ in the three-phase region around $T_m = (T_u + T_l)/2$ and $\delta_{V,i,m} = (\delta_{V,i,l} + \delta_{V,i,u})/2$, respectively. The minimum of the water/oil interfacial tension is found to be ultra-low, i.e. $\overline{\sigma}_{ab} = 0.003$ mN m^{-1} in the $C_{10}E_4$ system and $\overline{\sigma}_{ab} = 0.008$ mN m^{-1} in the β-C_8G_1 system. Increasing the distance from the three-phase body the interfacial tension between the microemulsion and an excess-phase ($\underline{2}$- or $\overline{2}$-state) increases up to $\sigma_{ab} \approx 1$ mN m^{-1}.

A quantitative description of the variation of the water/oil interfacial tension with the respective tuning parameter can be obtained via the bending energy difference between a curved amphiphilic film in the microemulsion and the flat film of the macroscopic interface [25, 91]. The bending energy approach is based on Helfrich's mechanical model which describes vesicles by an ensemble of fluctuating amphiphilic films [92]. Later this membrane model was used to describe the properties of microemulsions [93–95]. The parameters which characterise the properties of the amphiphilic film are the bending rigidity κ, the saddle splay modulus $\overline{\kappa}$ and the spontaneous curvature of the film H_0 (see also Section 1.4 and Chapter 2). Interestingly, the drawn lines in Fig. 1.15, calculated from the analysis of the interfacial tension measurement in terms of bending energy, describe the data points quantitatively within the experimental error. Thus, the analysis of the macroscopic interfacial tension measurements is one of the few methods to determine the microscopic parameters κ and $\overline{\kappa}$ [17, 25, 91]. For more details the reader is referred to the quoted literature.

In Fig. 1.16, the variation of the water/oil interfacial tension with temperature is shown for four representative systems, namely water–*n*-octane–C_6E_2, C_8E_3, $C_{10}E_4$ and $C_{12}E_5$. In

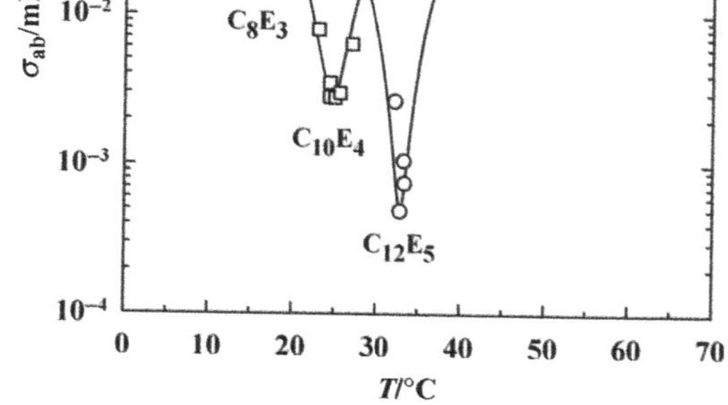

Figure 1.16 Temperature dependence of the water/oil interfacial tension σ_{ab} (plotted on log-scale) for some representative water–*n*-octane–C_iE_j systems. Note that the minimum of the interfacial tension curves $\overline{\sigma}_{ab}$ decreases substantially by increasing both the hydrophobic chain length *i* and the size of the hydrophilic head group *j* of the surfactants. The shift on the temperature scale stems from the shift of the phase behaviour. The full line is again calculated from an analysis of interfacial tensions in terms of the bending energy model [96]. (Figure redrawn with data from Ref. [17].)

this series the surfactant size is increased from C_6E_2 to $C_{12}E_5$ by increasing both i and j. As can be seen, the interfacial tension curves shift to higher temperatures, as do the three-phase bodies (see Fig. 1.4). Even more striking is the strong decrease of the minimum of the interfacial tensions $\overline{\sigma}_{ab}$ with increasing chain length of the surfactant shifting from system to system by one order of magnitude to lower values [17]. But, although the curves sharpen as the surfactant chain becomes longer, the shape remains similar. The full line which is calculated from the analysis of the interfacial tension experiment in terms of bending energy describes the data points again quantitatively within the experimental error. The value of the bending rigidity κ obtained from this analysis increases (as expected) with the surfactant chain length from values of about 0.6 kT to 1.1 kT [17].

1.3.4 Scaling of the interfacial tension σ_{ab}

The similar shape of the interfacial tension curves – which is obviously independent of the tuning parameter – suggests a scaling of the σ_{ab}-curves. The steepness of the interfacial tension curves around the centre of the three-phase body (i.e. around T_m or $\delta_{V,i,m}$) seems to correlate directly with the height of the three-phase body ($\Delta T = T_u - T_l$ or $\Delta\delta_{V,I} = \delta_{V,i,u} - \delta_{V,i,l}$). Thus, after centring the $\sigma_{ab}(T)$-curves by subtracting T_m or $\delta_{V,i,m}$ the axis of the respective tuning parameter can be normalised by $\Delta T/2$ or $\Delta\delta_{V,i}/2$, respectively. A reasonable normalisation of the interfacial tension axis is obtained using the minimum of the interfacial tension $\overline{\sigma}_{ab}$ [96]. Apart from this, however, one can follow Volmer's method [45, 46] to correlate $\overline{\sigma}_{ab}$ with the volume fraction of surfactant in the amphiphilic film of the optimum microemulsion, i.e. at the $\tilde{X}$-point at $\phi = 0.5$. Vollmer argued that colloidal dispersions should become thermodynamically stable if the interfacial free energy times the area of the colloidal object is provided by the thermal energy kT, i.e.

$$\sigma_{ab}\xi^2 \approx kT, \tag{1.14}$$

where ξ is the characteristic length scale of the colloidal object. Applied to microemulsions this relation holds for various types of structures [25, 87, 97] including the optimum microemulsion, which was found to have a bicontinuous structure [20, 21, 98, 99] (see also Fig. 1.20, Section 1.4). Since for bicontinuous microemulsions the characteristic length scale ξ is inversely proportional to the volume fraction $\phi_{C,i} + \phi_{D,i}$ of surfactant and co-surfactant (if present) in the amphiphilic film [42–44] (see also Section 1.4) one obtains for the minimum interfacial tension

$$\overline{\sigma}_{ab} \propto \left(\phi_{C,i} + \phi_{D,i}\right)^2, \tag{1.15}$$

which has been confirmed experimentally [37, 71]. This finding suggests to normalise the interfacial tension by the squared volume fraction $\left(\phi_{C,i} + \phi_{D,i}\right)^2$ of surfactant and co-surfactant (if present) in the amphiphilic film at the $\tilde{X}$-point at $\phi = 0.5$.

Figure 1.17 shows the scaling of interfacial tension curves for four ternary water–n-octane–C_iE_j systems (see Fig. 1.16) and the quaternary system water–n-octane–β-C_8G_1–C_8E_0 (see Fig. 1.15). As can be seen, the scaled $\sigma_{ab}(T)$-curves collapse onto one single curve, irrespective of the tuning parameter. However, some rather small, but systematic deviations

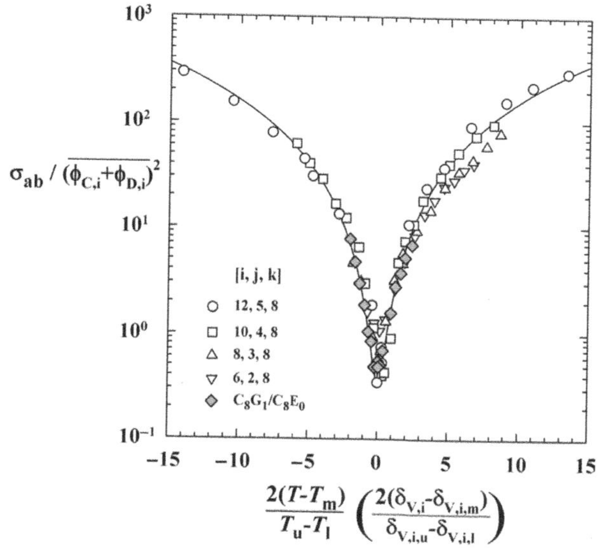

Figure 1.17 Scaling of the variation of the water/oil interfacial tension with the respective tuning parameter for four ternary water–*n*-octane–C_iE_j systems (see Fig. 1.16) and the quaternary system water–*n*-octane–β-C_8G_1–C_8E_0 (see Fig. 1.15). The tuning parameter T ($\delta_{V,i}$) is reduced by subtracting the mean temperature T_m (composition $\delta_{V,i,m}$) of the three phase body and normalising by $(T_u - T_l)/2$ $((\delta_{V,i,u} - \delta_{V,i,l})/2)$. Dividing σ_{ab} by the squared volume fraction $\overline{(\phi_{C,i} + \phi_{D,i})}^2$ of surfactant and co-surfactant (if present) in the amphiphilic film of the optimum microemulsions ($\tilde{X}$-point at $\phi = 0.5$) the data of all systems lie on top of each other which emphasises the general patterns of microemulsions. The full line is again calculated from an analysis of interfacial tensions in terms of the bending energy model [96].

remain, above all in the three-phase region. These deviations were eliminated eventually by a more detailed analysis which has been used to calculate the full line in Fig. 1.17 [96]. To conclude this section it should be emphasised that the minimum in the water/oil interfacial tension at the centre of the three-phase body enables the optimal solubilisation of water and oil, i.e. with the minimum amount of surfactant $\tilde{\gamma}$. This correlation between phase behaviour and interfacial tension also holds for technical applications. For example, the removal of hexadecane from synthetic tissue reaches a maximum within the three-phase region (see Fig. 8.12 in Chapter 8) [100, 101]. Furthermore, the interfacial tension curves can be scaled with the same tuning parameters as the phase behaviour.

1.4 Microstructure

Most of the recent applications of microemulsions depend on the fact that microemulsions, though macroscopically homogeneous, are heterogeneous on the sub-microscopic scale. Topologically ordered interfacial films are formed by the surfactant molecules which are forced into the microscopic water/oil interface because of their amphiphilicity. The nature and properties of these microscopic interfacial films are essential for microemulsions as a whole and, in particular, for the most interesting feature of microemulsions, i.e. their microstructure. In the 1950s, Winsor [2] and Schulman [102] suggested that microemulsions are always spherical, and that a layered, lamellar structure exists as an exceptional phenomenon in the middle phase. In 1976, Scriven [98] put forward the crucial idea of the bicontinuous structure of the surfactant-rich middle phase, which 10 years later was proven with the help of NMR self-diffusion measurements [20, 21] and the direct visualisation by freeze-fracture electron microscopy (FFEM) [20, 22]. Further studies of the microstructure by NMR self-diffusion, TEM and scattering techniques (SAXS and SANS) revealed droplet-like and wormlike microemulsions, sample spanning networks and bicontinuous structures. Furthermore, liquid crystalline phases such as the cubic (V), hexagonal (H) and lamellar phases (L_α) exist and compete with these complex fluids. It has been realised that the main parameter determining the microstructure is the mean curvature of the amphiphilic interfacial film. Thus, controlling the curvature is the ultimate goal in order to be able to choose any desired structure.

1.4.1 Mean curvature of the amphiphilic film

The mean curvature of the amphiphilic film is given by

$$H = \frac{1}{2}(c_1 + c_2), \tag{1.16}$$

where $c_1 = 1/R_1$ and $c_2 = 1/R_2$ are the principal curvatures at a certain point on the film. By definition curvatures are positive if the amphiphilic film tends to enclose oil (o/w-microemulsions) and negative if it tends to enclose water (w/o-microemulsions). Parameters on which the curvature of the amphiphilic film depends are the temperature, the composition of the amphiphilic film, the salinity, etc. The mean curvature H, which can be determined experimentally by scattering techniques [25] (see Chapter 2), is closely related to the spontaneous curvature H_0, which is the curvature the interfacial film will adopt if no external forces, thermal fluctuations or conservation constraints exist. Both H_0 and the Gaussian curvature $K = c_1 c_2$ are important parameters in Helfrich's bending energy [92].

Figure 1.18 schematically shows the variation of the mean curvature H of the amphiphilic film for the temperature-sensitive ternary water–oil–C_iE_j systems (Fig. 1.18(a)) and the temperature-insensitive quaternary water–oil–C_nG_m–alcohol systems (Fig. 1.18(b)) by means of a wedge-shaped representation. As discussed above, for temperature-sensitive ternary systems one finds oil-in-water (o/w) microemulsions at low and water-in-oil (w/o) microemulsions at high temperatures due to a change of the mean curvature H of the amphiphilic film. At low temperatures the size of the surfactant head group is larger than that of the hydrophobic chain which curves the amphiphilic film around the oil. With increasing temperature, the size of the surfactant head group shrinks due to a dehydration, whereas the size of the hydrophobic chain increases due to an increasing number of chain conformations and the increasing penetration of oil molecules. Thus, H changes gradually from $H > 0$ to $H < 0$, i.e. from oil-in-water (o/w) to water-in-oil structures (w/o) structures via a locally planar amphiphilic film, i.e. $H = 0$ (Fig. 1.18(a)).

In Sections 1.2.3 and 1.3.3, it was shown that in temperature-insensitive quaternary C_nG_m systems the composition of the amphiphilic film instead of the temperature has to

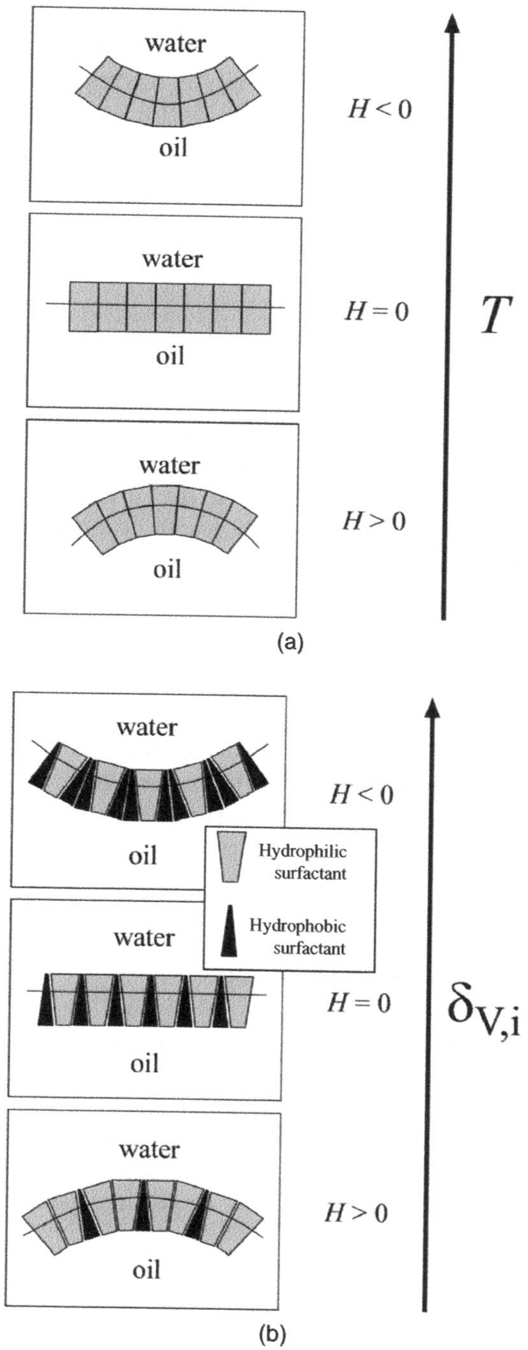

Figure 1.18 Mean curvature H of a non-ionic surfactant film at the water/oil interface as a function of temperature T (a) [26] and composition of the internal interface $\delta_{V,i}$ (b) [90]. The decrease in H with increasing T is mainly due to the shrinking size of the head group, while the decrease in H with increasing $\delta_{V,i}$ is due to the smaller head group area of the alcohol compared to the sugar surfactant. In order to illustrate this behaviour, a wedge-shaped representation has been chosen.

be varied to tune the phase behaviour and the interfacial tensions of the system. Recalling the results for these systems, we may conclude that the variation of the composition of the amphiphilic film changes its curvature. Figure 1.18(b) schematically shows this change of the curvature with increasing fraction of alcohol in the amphiphilic film. Knowing that the head group area of alcohols [80] is smaller than that of the sugar surfactants [103], one observes that an increasing fraction of alcohol in the mixed interfacial film causes a decrease of the mean curvature from $H > 0$ for o/w-microemulsions to $H < 0$ for w/o-microemulsions. In other words, the composition of the amphiphilic film is the tuning parameter of the mean curvature in quaternary temperature-insensitive systems.

Having understood the variation of the curvature of the amphiphilic film qualitatively the next step is to determine the variety and length scale of the microstructure and with it the underlying curvature of the amphiphilic film quantitatively. However, to gain a comprehensive insight into the structure pattern of microemulsions several different experimental methods have to be employed. Direct and local information about the occurring types of nano-structures can be provided by transmission electron microscopy [20, 99, 104, 105] (see Section 1.4.2). Statistical information about frequently occurring distances can be obtained from scattering techniques (see Chapter 2). Detailed information about the length scale of the microstructure can be gained from SAXS [106] and SANS [24]. Because of the relatively large wavelength, static light scattering (SLS) usually provides [107] only very unspecific information. What is somewhat more useful is dynamic light scattering (DLS) [108], which yields the diffusion coefficient of the structural domains. Furthermore, indirect methods like NMR self-diffusion [109] and electric conductivity [110] measurements provide valuable information on the connectivity of the microstructure and the transition from one type of structure to another. Each of the techniques provides a piece in the puzzle of the structure of microemulsions. In the following only results obtained by transmission electron microscopy will be discussed, while results obtained by scattering techniques are described in Chapter 2.

1.4.2 Transmission electron microscopy

In order to use electron microscopy to visualise the microemulsion structure, the problem of the fixation of the liquid mixtures has to be solved. The method of choice is to solidify the microemulsion structure via cryofixation. However, given that the phase behaviour as well as the curvature of the amphiphilic film (see Fig. 1.18) and with it the microstructure of most microemulsions show a strong temperature-dependence it has to be ensured that the cooling rate should be as high ($>10^4$ K/s) and the reorganisation kinetics of the microstructure as slow as possible.

Three different techniques, namely FFEM [20, 22], Cryo-Direct Imaging (Cryo-DI) [104] and freeze-fracture direct imaging (FFDI) [105], can be used to visualise the structure of microemulsions. In FFEM the samples are prepared in a protected fashion in a sandwich. They are then rapidly frozen, fractured, shadowed with metal, and replicated with a thin carbon film. The replica of the fractured surface, the morphology of which is controlled by the sample's microstructure, is then studied by a TEM. In contrast to FFEM, in Cryo-DI thin films of the sample are rapidly frozen but immediately, without replication, trans-

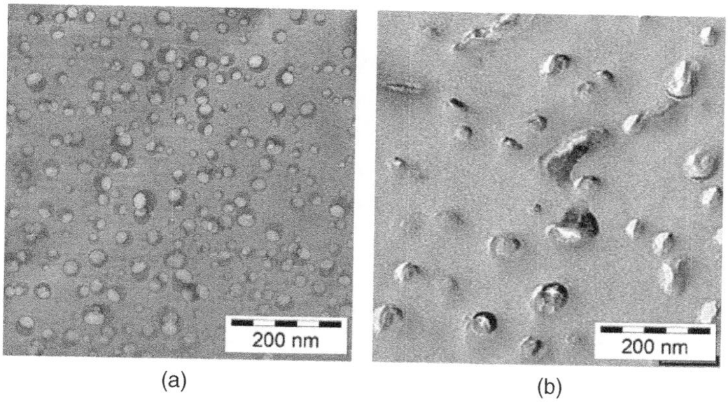

Figure 1.19 Micrographs of microemulsion droplets of the o/w-type in the system H_2O–*n*-octane–$C_{12}E_5$ prepared near the emulsification failure boundary at $\gamma_a = 0.022$, $w_B = 0.040$ and $T = 26.1°C$. (a) Freeze-fracture direct imaging (FFDI) picture showing dark spherical oil droplets of a mean diameter $<d> = 24 \pm 9$ nm in front of a grey aqueous background. Note that each oil droplet contains a bright domain of elliptic shape which is interpreted as voids. (b) The freeze-fracture electron microscopy (FFEM) picture supports the FFDI result. Each fracture across droplets which contain bubbles shows a rough fractured surface. (From Ref. [26], reprinted with permission of Elsevier.)

ferred to a low temperature stage within the microscope and imaged directly. In order to obtain these thin films (which have to be thin enough to allow for the electrons to transverse the sample) the sample has to be blotted prior to the vitrification. However, using this blotting technique a change of the concentration of the sample and a shearing of the internal microstructure are unavoidable. The recently developed FFDI method which is a hybrid of FFEM and Cryo-DI has solved the problems of the direct imaging technique. Like in FFEM, the sandwich method is used. However, after the sample is vitrified and fractured it is not shadowed and replicated but directly imaged. Thus, the FFDI technique avoids some experimental artifacts produced by the blotting of the sample using Cryo-DI. A disadvantage of the FFEM and FFDI techniques is the noticeably smaller cooling rate compared to the Cryo-DI-method, which can be attributed to the preparation and slower vitrification of the sample sandwich. Despite all sources of error which could be encountered during the sample preparation, reliable images of microemulsions can be obtained. In the following images of H_2O–*n*-octane–$C_{12}E_5$ microemulsions will be shown for which an o/w-, a w/o- and a bicontinuous microemulsion could have been visualised using both the FFEM and FFDI techniques.

Figure 1.19 shows micrographs [26, 111] of this ternary microemulsion prepared at low temperatures ($T < T_l$), where the amphiphilic film should be curved around the oil (see Fig. 1.18(a)), i.e. $H > 0$. To be more accurate the sample was prepared within the one-phase region near the emulsification failure boundary (see Fig. 1.7(b)) at the water-rich side ($\gamma_a = 0.022$, $w_B = 0.040$ and $T = 26.1°C$). Both the FFDI (Fig. 1.19(a)) and the FFEM picture (Fig. 1.19(b)) prove the existence of *n*-octane-swollen micelles in a water matrix. The FFDI picture shows dark spherical oil droplets of a mean diameter $<d> = 24 \pm 9$ nm in front of a bright aqueous background. Surprisingly, each oil droplet contains an elliptically shaped bright domain. These domains could be an artifact of the sample

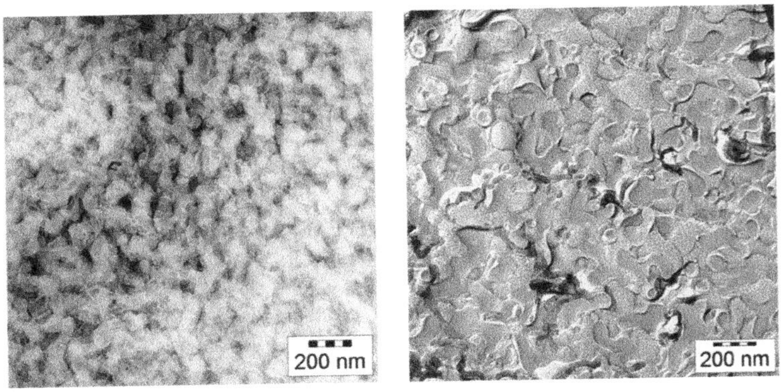

Figure 1.20 Micrographs of a bicontinuous microemulsion of the system H_2O–n-octane–$C_{12}E_5$ prepared near the $\tilde{X}$-point at $\phi = 0.50$, $\gamma = 0.06$ and $T = 32.4°C$. (a) Freeze-fracture direct imaging (FFDI) picture showing particularly in the middle of the image a sponge-like bicontinuous structure consisting of 'white' and 'black' domains. Note that the colours are inverted. (b) The freeze-fracture electron microscopy (FFEM) picture supports the FFDI result. (From Ref. [105], reprinted with permission of the American Chemical Society.)

preparation. Cooling the oil-in-water microemulsion rapidly from $T = 26.1°C$ to liquid nitrogen temperature $T = -196°C$, first the water is vitrified. Because of the different dependencies of the density of water and n-octane on temperature the still liquid n-octane droplets are now entrapped in a rigid matrix and contract. Thus, a differential stress is created that leads to the rupture of the fluid oil droplets and the resulting voids could be the elliptically shaped bright domains which can be seen in each n-octane droplet in the FFDI picture. The FFEM picture of the same sample shown in Fig. 1.19(b) supports the FFDI result. In fractures across the droplets one can see that the droplets in part exhibit planar fractures but in other places show a rough surface. Although the mean diameter of the oil-in-water droplets is difficult to determine from the FFEM picture it is obvious that their size is comparable to the FFDI result.

Micrographs of a bicontinuous microemulsion prepared near the $\tilde{X}$-point ($\phi = 0.50$, $\gamma = 0.06$, $T_m \approx \tilde{T} = 32.62°C$) are shown in Fig. 1.20. Again the FFDI (Fig. 1.20(a)) and the FFEM micrograph (Fig. 1.20(b)) are taken from the same sample [105]. Looking at the picture taken with the conventional FFEM technique (see also Ref. [112]), one can easily distinguish oil-rich and water-rich domains because of the texture of the oil domains, which stems from the shadowing of the fractured surface with tantalum (Ta) and tungsten (W) [22]. It is caused by the differing nucleation probabilities and surface mobilities on the various substrates and should not be mistaken for the real microstructure. As can be seen, the fracture through the water domains is in most cases planar, whereas for the oil domains the fracture follows the amphiphilic film. This difference leads to a three-dimensional impression of the oil-domains. Furthermore, one clearly sees water-rich and oil-rich domains which are mutually intertwined in a sponge-like fashion showing many saddle-shaped structures. Typically, the two principal curvatures appear to be almost equal but of opposite signs, i.e. $c_1 = -c_2$. As a consequence, the mean curvature H of the amphiphilic film can be around 0, while the Gaussian curvature K is negative. In other

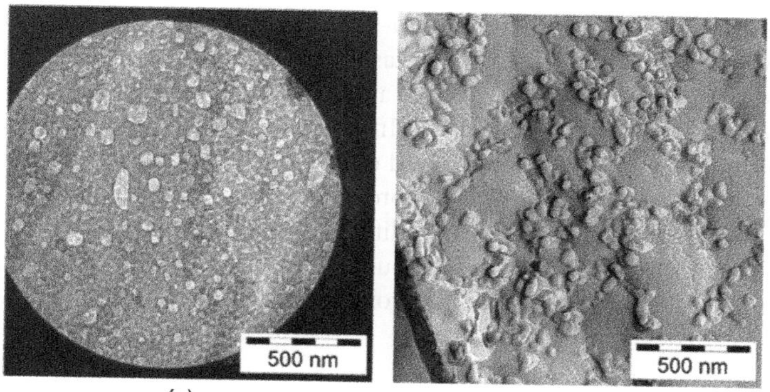

Figure 1.21 Micrographs of microemulsion droplets of the w/o-type in the system H_2O/NaCl–n-octane–$C_{12}E_5$ prepared near the emulsification failure boundary at a $\gamma_b = 0.050$, $w_B = 0.100$, $\varepsilon = 0.006$ and $T = 36.3°C$. (a) Freeze-fracture direct imaging (FFDI) picture showing bright water droplets of a mean diameter $<d> = 44 \pm 13$ nm against a dark oily background. (b) The freeze-fracture electron microscopy (FFEM) picture supports the FFDI result. The mean diameter of the water droplets is $<d> = 47 \pm 8$ nm. (From Ref. [105], reprinted with permission of the American Chemical Society.)

words, the structure is water-continuous and oil-continuous at the same time, a situation which is called *bicontinuous*. A similar situation is seen in the FFDI micrograph, i.e. in Fig. 1.20(a). Note that the colours are inverted to get images in which the details of the structure can be recognised more easily. In the middle of the image a sponge-like structure can be seen consisting of 'white' and 'black' domains. As this is a direct image through the sample, different numbers of layers of the structure are seen at each position. Comparing the length scale of the bicontinuous structures (i.e. the diameter ($d \approx 50$ nm) of the water and n-octane domains) visible in the FFEM and FFDI micrographs, one finds a good agreement between the two methods and the SANS [80] performed on a sample of similar composition.

Increasing the temperature and turning to the oil-rich side of the phase prism, one obtains micrographs of a water-in-oil microemulsion (see Fig. 1.21) [105]. The sample was prepared within the one-phase region near the water emulsification failure boundary (see Fig. 1.7(c)) of the ternary mixture H_2O/NaCl–n-octane–$C_{12}E_5$ at $\gamma_b = 0.050$, $w_A = 0.100$, $\varepsilon = 0.006$ and $T = 36.3°C$. The image obtained with the help of the conventional FFEM technique (Fig. 1.21(b)) shows water droplets in a continuous oil phase. As already mentioned, it is the decoration of the oil that allows distinction between the water- and oil-rich domains. Evaluating the FFEM image one obtains a mean diameter of the water droplets of $<d> = 47 \pm 8$ nm. Looking at the FFDI image (Fig. 1.21(a)), one clearly sees bright water droplets in the dark, textured n-octane matrix. For the size of the water droplets a mean diameter of $<d> = 44 \pm 13$ nm is found, which is in perfect agreement with FFEM and the SANS results [25]. Thus, comparing the FFEM and the FFDI image one clearly sees that the results are not only qualitatively but also quantitatively the same.

In conclusion, the transmission electron microscopy images show that for mixtures of water, oil and long-chain non-ionic surfactants the structure gradually changes with increasing temperature from discrete oil-in-water micelles via a bicontinuous network to discrete water-in-oil micelles. Thus, as was already deduced from the variation of the phase behaviour, the mean curvature of the amphiphilic film changes from $H > 0$ at low temperatures to $H < 0$ at high temperatures. In between, at the PIT, $H = 0$ which is reflected in the bicontinuous structure. While a good deal of qualitative insight into the manifold structural properties is nicely gained, a more quantitative determination of parameters such as the length scale of the structure is difficult to infer from such images. As already mentioned above, detailed information about the length scale of the microstructure can be gained from scattering techniques, i.e. from SAXS and SANS, which are described in Chapter 2.

1.4.3 Estimation of length scales and overview of microstructure

Knowing the shape (topology) of the microstructure one can obtain an estimate of the length scales from the composition of the microemulsion. Thus, the diameter of the domains ($d = \xi$) in a bicontinuously structured microemulsion can be calculated from

$$\xi = a\frac{v_C}{a_C}\frac{\phi(1-\phi)}{\phi_{C,i}}, \tag{1.17}$$

where v_C and a_C are the volume and the area of the surfactant molecule, $\phi_{C,i}$ is the volume fraction of surfactant in the amphiphilic film, and a is a pre-factor, which depends on the model used to describe the bicontinuous structure. The model of Debye *et al.* [43] predicts $a = 4$, the Voronoi tessellation of Talmon and Prager [44] leads to $a = 5.84$ and the model of cubes by De Gennes and Taupin [42] yields $a = 6$. Experimentally, a somewhat larger factor of $a \approx 7$ is found from the analysis of SANS measurements [26, 80]. A rough estimation of the length scales of almost symmetric ($\phi = 0.5$) bicontinuous microemulsions can be obtained by

$$\xi \approx 1.5\ \text{nm}/\gamma. \tag{1.18}$$

To obtain Eq. (1.18) four things need to be assumed: (i) the monomeric solubility of the surfactant in water and oil can be neglected, (ii) $\delta = v_C/a_C \approx 1$ nm, (iii) $a = 6$ and (iv) the density of all components is the same. Furthermore, the radius of the spherical droplets can be calculated by

$$r_0 = 3\frac{v_C}{a_C}\frac{\phi_i + \phi_{C,i}}{\phi_{C,i}}\frac{1+p^2}{1+3p^2}, \tag{1.19}$$

where ϕ_i denotes the volume fraction of the component solubilised in the respective micelles and p the polydispersity of an assumed Gaussian distribution of radii. Assuming again that $\delta = v_C/a_C \approx 1$ nm and that the density of all components is the same and neglecting the polydispersity and the monomeric solubility of the surfactant, one finds

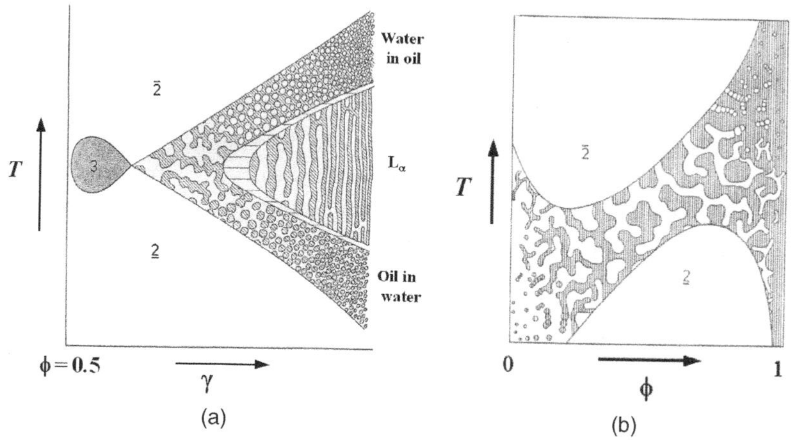

Figure 1.22 Schematic overview of the microstructure of non-ionic microemulsions as deduced from TEM, SANS, NMR-diffusiometry and electric conductivity together with the underlying phase behaviour. The shaded regions represent the oil and the white regions represent water. (a) $T(\gamma)$-section at $\phi = 0.5$ [25]. The variation of the curvature of the amphiphilic film with temperature becomes apparent by the change of microstructure from o/w- to w/o-droplet structures. Around $\tilde{T}$ bicontinuous structures can be found at low γ, whereas the lamellar phase L_α exists at higher γ. (b) $T(\phi)$-section through the phase prism at a constant $\gamma > \tilde{\gamma}$ (Shinoda cut) [114]. Within the homogeneous channel the microstructure changes from discrete o/w-structures on the water-rich side and low temperatures to discrete w/o-structures on the oil-rich side and high temperatures. (Figure redrawn with data from Ref. [17] and Ref. [18].)

that Eq. (1.19) simplifies to

$$r_0 = 3\frac{w_i + \gamma}{\gamma}\text{nm}, \tag{1.20}$$

with w_i being the weight fraction of the solubilised water or oil, respectively.

Studying the microstructure of microemulsions extensively by several different methods like TEM, SAXS, SANS, NMR diffusometry and electric conductivity, one can gain profound insights into their structure pattern. With respect to the ternary non-ionic microemulsion, the temperature dependence of the microstructure is presented in Fig. 1.22 in terms of two different sections through the prism. In Fig. 1.22(a), the microstructures existing within the extended one-phase region behind the $\tilde{X}$-point are drawn into the $T(\gamma)$-section (see Fig. 1.3) [25]. Starting at the $\tilde{X}$-point, i.e. close to the three-phase region, the structure of the microemulsion is bicontinuous with a zero-mean curvature of the amphiphilic film ($H = 0$), but a negative Gaussian curvature ($K < 0$). An increase of the surfactant mass fraction γ leads to a shrinking of the structure because the total area of the internal interface increases. At high surfactant concentrations the lamellar phase is observed, with a zero curvature structure, i.e. $H = 0$ and $K = 0$ ($c_1 = c_2 = 0$). Moving both γ- and temperature-wise away from the $\tilde{X}$-point one observes a transition to oil-in-water and water-in-oil droplets at low and high temperatures, respectively. Furthermore, the droplet size decreases as one moves further away from the $\tilde{X}$-point.

In order to study the variation of the microstructure as a function of the oil/(water + oil) volume fraction ϕ it has proved useful to perform a so-called Shinoda cut [113] ($T(\phi)$ section) through the phase prism at a constant mass fraction of surfactant $\gamma > \tilde{\gamma}(\phi = 0.50)$. Figure 1.22(b) shows a schematic drawing of this cut [114]. Within the one-phase channel on the water-rich side, the mixtures consist of stable dispersions of oil droplets in water that transform into a branched tubular oil network with rising temperature. Increasing ϕ, one finds sponge-like bicontinuous structures around the mean temperature of the three-phase body $T_m \approx \tilde{T}(\phi = 0.50)$, i.e. the PIT, if ϕ is varied between 0.2 and 0.8. On the oil-rich side, at high temperatures water droplets are found to be dispersed in a continuous oil-phase. These droplets transform into a branched tubular water network with decreasing temperature. Accordingly, the mean and Gaussian curvature of the amphiphilic film varies with temperature and composition. As mentioned before, other parameters to tune the curvature of the amphiphilic film are the salinity and the composition of the amphiphilic film, respectively.

1.5 Conclusion

The wide range of applications as well as the steadily increasing number of papers and patents on microemulsions already show their significance for many branches of chemistry and suggest that microemulsions will become even more significant in the future. The first chapter of this book dealt with the phase behaviour as well as the associated interfacial tensions and microstructures of microemulsions. The fact that these features are similar irrespective of what solvents and amphiphiles are used is a strong indication of the existence of a general microemulsion pattern. This general pattern is also mirrored in the fact that various tuning parameters lead to the same general observations.

We will conclude this chapter by summarising the most relevant properties of microemulsions. The system of choice is H_2O–*n*-octane–$C_{12}E_5$ as it has been studied very extensively [25]. Recalling the transmission electron microscopy images (see Fig. 1.20), one can observe a truly bicontinuous structure at the mean temperature of the three-phase body $T = T_m \approx \tilde{T}(\phi = 0.50)$. This result is supported by similar self-diffusion coefficients D of water and *n*-octane obtained from NMR self-diffusion measurements [115]. Figure 1.23 (top) shows the self-diffusion coefficients D plotted versus the temperature. As can be seen, the values of D for water and oil are nearly equal at the mean temperature (PIT) of the three-phase body, which is further evidence of the bicontinuity. Note that plotting the reduced self-diffusion coefficients D/D_0 (D_0 = self-diffusion coefficient of the pure solvents at the respective temperature) versus T indeed leads to equal D/D_0 values for both solvents at the PIT. At the PIT the length scale ξ of the structure shows a distinct maximum because of the optimal solubilisation of water and oil. Calculating the mean curvatures from the length scales by setting $H = 1/\langle r \rangle$ and $H = -1/\langle r \rangle$ for o/w- and w/o-microemulsions as well as $H = 0$ for the bicontinuous structure an almost linear decrease of the curvature is found with increasing temperature. As is seen in Fig. 1.23, H changes sign at $T = T_m$ [25, 26]. One further consequence of the length scale ξ reaching a maximum and the mean curvature changing its sign is that the interfacial tension σ_{ab} between the water and oil phases passes through a minimum at the PIT. Thus, knowing the variation of the curvature with the appropriate tuning parameter one cannot only adjust

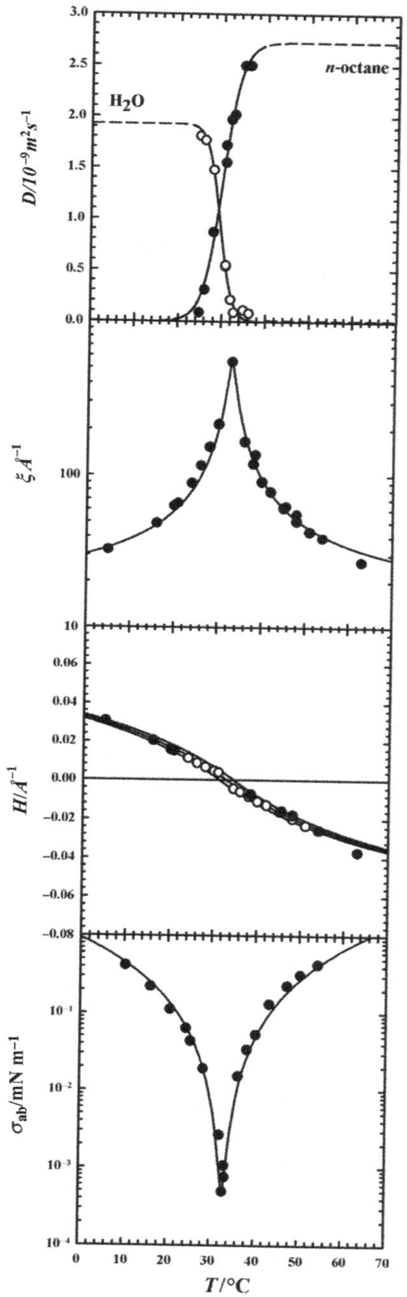

Figure 1.23 Variation of the water (shown as hollow symbols) and n-octane (shown as filled symbols) diffusion coefficients D_A and D_B [115], the length scale ξ [25], the mean curvature H and the water/oil interfacial tension σ_{ab} as function of the temperature for the system H_2O–n-octane–$C_{12}E_5$. Note that at the mean temperature of the three-phase body $\tilde{T}$ the diffusion of water and oil molecules is equal (points to bicontinuity), the length scale runs through a maximum, the curvature change sign and the water/oil interfacial shows an extreme minimum.

the desired shape and length scale of the structure but also the optimal solubilisation of water and oil via the ultra-low water–oil interfacial tension.

Acknowledgement

Thomas Sottmann wishes to thank Prof. Strey for his continuous financial and professional support. His thoughts and suggestions have been of great value. During the last 20 years of joint research and deep discussion a close friendship has evolved.

Notes

1. ISI Web of Knowledge, Science Citation Index Expanded.
2. European Patent Office, esp@cenet.

References

1. Schulman, J.H. and Hoar, T.P. (1943) Transparent water-in-oil dispersions: The oleopathic hydromicelle. *Nature*, **152**, 102.
2. Winsor, P.A. (1954) *Solvent Properties of Amphiphilic Compounds.* Butherworth & Co., London.
3. Schulman, J.H., Stoeckenius, W. and Prince, L.M. (1959) Mechanism of formation and structure of microemulsions by electron microscopy. *J. Phys. Chem.*, **63**, 1677–1680.
4. Sharma, M.K. and Shah, D.O. (1985) Macro- and microemulsions in enhanced oil recovery. In D.O. Shah (ed), *Macro- and Microemulsions Theory and Applications*, Vol. 272. American Chemical Society, Washington, DC, pp. 149.
5. Shinoda, K. (ed) (1967) *Solvent Properties of Surfactant Solutions*, Vol. 2. Marcel Dekker, New York, pp. 27–63.
6. Kahlweit, M. and Strey, R. (1985) Phase behavior of ternary systems of the type H_2O–oil–non-ionic amphiphile (microemulsions). *Angew. Chem. Int. Ed.*, **24**, 654–668.
7. Bellocq, A.M., Biais, J., Clin, B., Gelot, A., Lalanne, P. and Lemanceau, B. (1980) 3-Dimensional phase diagram of the brine–toluene–butanol–sodium dodecyl-sulfat. *J. Colloid Interface Sci.*, **74**, 311–321.
8. Langevin, D. (1986) Microemulsions and liquid crystals. *Mol. Cryst. Liq. Cryst.*, **138**, 259–305.
9. Eicke, H.F. (1982) The microemulsion concept in nonpolar surfactant solutions. In I.D. Robb (ed), *Mikroemulsions.* Plenum, New York, pp. 17–32.
10. Kahlweit, M. and Strey, R. (1988) Phase behavior of quinary mixtures of the type H_2O–oil–non-ionic amphiphile–ionic amphiphile–salt. *J. Phys. Chem.*, **92**, 1557–1563.
11. Kahlweit, M., Strey, R., Firman, P., Haase, D., Jen, J. and Schomäcker, R. (1988) General patterns of the phase behavior of mixtures of H_2O, nonpolar solvents, amphiphiles, and electrolytes. 1. *Langmuir*, **4**, 499.
12. Penders, M.H.G.M. and Strey, R. (1995) Phase behavior of the quaternary system H_2O/*n*-octane/C8E5/*n*-octanol: Role of the alcohol in microemulsions. *J. Phys. Chem.*, **99**, 10313–10318.
13. Schubert, K.-V. and Kaler, E.W. (1996) Non-ionic microemulsions. *Ber. Bunsenges. Phys. Chem.*, **100**, 190–205.
14. Lekkerkerker, H.N.W., Kegel, W.K. and Overbeek, J.Th.G. (1996) Phase behavior of ionic microemulsions. *Ber. Bunsenges. Phys. Chem.*, **100**, 206–217.

15. Aveyard, R., Binks, B.P., Clark, S. and Mead, J. (1986) Interfacial tension minima in oil–water–surfactant systems. *J. Chem. Soc. Faraday Trans.*, **82**, 125–142.
16. Langevin, D. (1992) Low interfacial tensions in microemulsions. In S.H. Chen, J.S. Huang and P. Tartaglia (eds), *Structure and Dynamics of Strongly Interacting Colloids and Supramolecular Aggregates in Solutions.* Kluwer Academic Publishers, Dordrecht, pp. 325–349.
17. Sottmann, T. and Strey, R. (1997) Ultralow interfacial tensions in water–*n*-alkane–surfactant systems. *J. Chem. Phys.*, **106**, 8606–8615.
18. Shinoda, K. and Friberg, S. (1975) Microemulsions – Colloidal aspects. *Adv. Colloid Interface Sci.*, **4**, 281–300.
19. Schechter, R.S., Wade, W.H., Weerasooriya, U., Weerasooriya, V. and Yiv, S. (1985) Synthesis and performance of isomer-free secondary alkane sulfonate surfactants. *J. Dispersion Sci. Techn.*, **6**, 223.
20. Bodet, J.F., Bellare, J.R., Davis, H.T., Scriven, L.E. and Miller, W.G. (1988) Fluid microstructure transition from globular to bicontinuous in midrange microemulsion. *J. Phys. Chem.*, **92**, 1898–1902.
21. Lindman, B., Shinoda, K., Olsson, U., Anderson, D., Karlström, G. and Wennerström, H. (1989) On the demonstration of biconinuous structures in microemulsions. *Colloids Surfaces*, **38**, 205–224.
22. Jahn, W. and Strey, R. (1988) Microstructure of microemulsions by freeze fracture electron microscopy. *J. Phys. Chem.*, **92**, 2294–2301.
23. Lichterfeld, F., Schmeling, T. and Strey, R. (1986) On the microstructure of microemulsions of the system H_2O–*n*-Tetradecane–$C_{12}E_5$. *J. Phys. Chem.*, **90**, 5762–5766.
24. Chen, S.H. (1986) Small angle neutron scattering studies of the structure and interaction in micellar and microemulsions. *Ann. Rev. Phys. Chem.*, **37**, 351–399.
25. Strey, R. (1994) Microemulsion microstructure and interfacial curvature. *Colloid Polym. Sci.*, **272**, 1005.
26. Sottmann, T. and Strey, R. (2005) Microemulsions. In J. Lyklema (ed), *Fundamentals of Interface and Colloid Science*, Vol. V. Academic Press, New York.
27. Jakobs, B., Sottmann, T., Strey, R., Allgaier, J., Willner, L. and Richter, D. (1999) Amphiphilic block copolymers as efficiency boosters for microemulsions. *Langmuir*, **15**, 6707.
28. Jakobs, B., Sottmann, T., Strey, R., Allgaier, J., Willner, L. and Richter, D. (2004) Method for increasing the efficiency of surfactants with simultaneous suppression of lamellar mesophases and surfactants with an additive added thereto. Patent No. US 6,677,293.
29. Bagger-Jörgensen, H., Coppola, L., Thuresson, K., Olsson, U. and Mortensen, K. (1997) Phase behavior, microstructure, and dynamics in a non-ionic microemulsion on addition of hydrophobically end-capped poly(ethylene oxide). *Langmuir*, **13**, 4204–4218.
30. Sassen, C.L., Cassielles, A.G., De Loos, T.W. and De Swaan-Arons, J. (1992) The influence of pressure and temperature on the phase behavior of the system $H_2O + C12 + C7E5$ and relevant binary subsystems. *Fluid Phase Equilibria*, **72**, 173–187.
31. Laughlin, R.G. (1994) *Aqueous Phase Behavior of Surfactants.* Academic Press, New York.
32. Safran, S.A. (1992) Microemulsions: An ensemble of fluctuating interfaces. In S.-H. Chen, J.S. Huang, P. Tartaglia (eds), *Structure and Dynamics of Strongly Interacting Colloids and Supramolecular Aggregates in Solutions.* Kluwer Academic Publishers, Dordrecht, pp. 237–263.
33. Sottmann, T. (2002) Solubilization efficiency boosting by amphiphilic block copolymers in microemulsions. *Curr. Opin. Coll. Interface Sci.*, **7**, 57–65.
34. Burauer, S., Sachert, T., Sottmann, T. and Strey, R. (1999) On microemulsion phase behavior and the monomeric solubility of surfactant. *Phys. Chem. Chem. Phys.*, **1**, 4299–4306.
35. Lade, O., Beizai, K., Sottmann, T. and Strey, R. (2000) Polymerizable non-ionic microemulsions: Phase behavior of H_2O–*n*–Alkyl Methacrylate–*n*–Alkyl Poly Ethylenglycol Ether (C_iE_j). *Langmuir*, **16**, 4122–4130.

36. Engelskirchen, S., Elsner, N., Sottmann, T. and Strey, R. (2007) Triacylglycerol microemulsions stabilized by alkyl ethoxylate surfactants – A basic study: Phase behaviour, interfacial tensions and microstructure. *J. Colloid Interface Sci.*, **312**, 114–121.
37. Kahlweit, M., Strey, R. and Busse, G. (1990) Microemulsions – A qualitative thermodynamic approach. *J. Phys. Chem.*, **94**, 3881.
38. Burauer, S., Sottmann, T. and Strey, R. (2000) Non-ionic microemulsions with cyclic oils – Oil penetration, efficiency and monomeric solubility. *Tenside Surfactants Detergents*, **37**, 8–16.
39. Kunieda, H. and Shinoda, K. (1985) Evaluation of the hydrophile–lipophile balance (HLB) of non-ionic surfactants. 1. Multisurfactant mixtures. *J. Colloid Interface Sci.*, **107**, 107–121.
40. Kahlweit, M., Strey, R. and Firman, P. (1986) Search for tricritical points in ternary systems: Water–oil–non-ionic amphiphile. *J. Phys. Chem.*, **90**, 671.
41. Kahlweit, M., Strey, R. and Busse, G. (1993) Weakly to strongly structured mixtures. *Phys. Rev. E*, **47**, 4197–4209.
42. De Gennes, P.G. and Taupin, C. (1982) Microemulsions and the flexibility of oil/water interfaces. *J. Phys. Chem.*, **86**, 2294–2304.
43. Debye, P., Anderson, H.R. and Brumberger, H. (1957) Scattering by an inhomogeneous solid. 2. The correlation function and its application. *J. Appl. Phys.*, **28**, 679–683.
44. Talmon, Y. and Prager, S. (1978) Statistical thermodynamics of phase-equilibria in microemulsions. *J. Chem. Phys.*, **69**, 2984–2991.
45. Volmer, M. (1927) Zur Theorie der lyophilen Kolloide. *Z. Phys. Chem.*, **125**, 151–157.
46. Volmer, M. (1957) Die kolloidale Natur von Flüssigkeitsgemischen in der Umgebung des kritischen Lösungspunktes I. *Z. Phys. Chem.*, **206**, 181–193.
47. Sottmann, T. and Strey, R. (1996) Evidence of corresponding states in ternary microemulsions of water–alkane–CiEj. *J. Phys. Condens. Matter*, **8**, A39.
48. Kilpatrick, P.K., Gorman, C.A., Davis, H.T., Scriven, L.E. and Miller, W.G. (1986) Patterns of phase-behavior in ternary ethoxylated alcohol–normal alkane–water mixtures. *J. Phys. Chem.*, **90**, 5292–5299.
49. Strey, R. (1996) Water–non-ionic surfactant systems, and the effect of additives. *Ber. Bunsenges. Phys. Chem.*, **100**, 182.
50. Tlusty, T., Safran, S.A., Menes, R. and Strey, R. (1997) Scaling laws for microemulsions governed by spontaneous curvature. *Phys. Rev. Lett.*, **78**, 2616–2619.
51. Tlusty, T., Safran, S.A. and Strey, R. (2000) Topology, phase instabilities, and wetting of microemulsion networks. *Phys. Rev. Lett.*, **84**, 1244–1247.
52. Zilman, A.G. and Safran, S.A. (2002) Thermodynamics and structure of self-assembled networks. *Phys. Rev. E*, **66**, 051107.
53. Graciaa, A., Lachaise, J., Sayous, J.G., Grenier, P., Yiv, S., Schechter, R.S. and Wade, W.H. (1983) The partitioning of complex surfactant mixtures between oil–water microemulsion phases at high surfactant concentrations. *J. Colloid Interface Sci.*, **93**, 474–486.
54. Kunieda, H. and Ishikawa, N. (1985) Evaluation of the hydrophile lipophile balance (HLB) of non-ionic surfactants. 2. Commercial surfactant systems. *J. Colloid Interface Sci.*, **1985**, 122–128.
55. Sottmann, T., Lade, M., Stolz, M. and Schomäcker, R. (2002) Phase behavior of non-ionic microemulsions prepared from technical-grade surfactants. *Tenside Surf. Det.*, **39**, 20–28.
56. Jakobs, B., Sottmann, T. and Strey, R. (2000) Efficiency boosting with amphiphilic block copolymers – a new approach to microemulsion formulation. *Tenside Surf. Det.*, **37**, 357–364.
57. Stubenrauch, C. (2001) Sugar surfactants – aggregation, interfacial and adsorption phenomena. *Curr. Opin. Coll. Interface Sci.*, **6**, 160–170.
58. Fukuda, K., Söderman, O., Lindman, B. and Shinoda, K. (1993) Microemulsions formed by alkyl polyglycosides an alkyl glycerol ether. *Langmuir*, **9**, 2921–2925.

59. Kahlweit, M., Busse, G. and Faulhaber, B. (1995) Preparing microemulsions with alkyl monoglycosides and the role of *n*-alkanols. *Langmuir*, **11**, 3382–3387.
60. Stubenrauch, C., Paeplow, B. and Findenegg, G.H. (1997) Microemulsions formed by octyl monglycoside and geraniol. Part 1: The role of the alcohol in the interfacial layer. *Langmuir*, **13**, 3652–3658.
61. Sottmann, T., Kluge, K., Strey, R., Reimer, J. and Söderman, O. (2002) General patterns of the phase behavior of mixtures of H_2O, alkanes, alkyl glucosides, and cosurfactants. *Langmuir*, **18**, 3058–3067.
62. Ryan, L.D., Schubert, K.-V. and Kaler, E.W. (1997) Phase behavior of microemulsions made with *n*-alkyl monoglucosides and *n*-alkyl polyglycol ethers. *Langmuir*, **13**, 1510–1518.
63. Kunieda, H. and Yamagata, M. (1993) Mixing of non-ionic surfactants at water oil interfaces in microemulions. *Langmuir*, **9**, 3345–3351.
64. Yamaguchi, S. and Kunieda, H. (1997) Determination of a three-phase tie triangle (the hydrophile-lipophile balance plane) in a composition tetrahedron: Evaluation of the composition of adsorbed mixed-surfactant and the monomeric solubilities of short-chain surfactant. *Langmuir*, **13**, 6995–7002.
65. Billman, J.F. and Kaler, E.W. (1990) Structure and phase-behavior in 5-component microemulsions. *Langmuir*, **6**, 611–620.
66. Lekkerkerker, H.N.W., Kegel, W.K. and Overbeek, J.T.G. (1996) Phase behavior of ionic microemulsions. *Ber. Bunsenges. Phys. Chem.*, **100**, 206–217.
67. Kahlweit, M., Strey, R., Schomäcker, R. and Haase, D. (1989) General patterns of the phase behavior of mixtures of H_2O, nonpolar solvents, amphiphiles, and electrolytes. 2. *Langmuir*, **5**, 305–315.
68. Chen, S.J., Evans, D.F. and Ninham, B.W. (1984) Properties and structure of 3-component ionic microemulsions. *J. Phys. Chem.*, **88**, 1631–1634.
69. Chen, S.H., Chang, S.-L. and Strey, R. (1990) Structural evolution within the one-phase region of a three-component microemulsion: Water-*n*-decane-sodium-bis-ethylhexylsulfosuccinate (AOT). *J. Chem. Phys.*, **93**, 1907–1918.
70. Wade, W.H., Morgan, J.C., Schechter, R.S., Jacobson, J.K. and Salager, J.L. (1978) Interfacial tension and phase behavior of surfactant systems. *Soc. Petroleum Eng.*, **18**, 242–252.
71. Kahlweit, M., Strey, R., Haase, D. and Firman, P. (1988) Properties of the three-phase bodies in H_2O-oil-non-ionic amphiphile mixtures. *Langmuir*, **4**, 785–790.
72. Prince, L.M. (1977) *Microemulsions: Theory and Practice.* Academic Press, New York.
73. Schwuger, M.-J., Stickdorn, K. and Schomäcker, R. (1995) Microemulsions in technical processes. *Chem. Rev.*, **95**, 849–864.
74. Ambwani, D.S. and Fort, T. (1979) Pendant drop technique for measuring liquid boundary tensions. In R.J Good and R.R Stromberg (eds), *Surface and Colloid Science*, Vol. II. Plenum, New York.
75. Langevin, D. and Meunier, J. (1977) Light scattering by liquid interfaces. In H.Z. Cummins and P. Pike (eds), *Photon Correlation Spectroscopy and Velocimetry*, Plenum, New York.
76. Vonnegut, B. (1942) Rotating bubble method for the determination of surface and interfacial tensions. *Rev. Sci. Instr.*, **13**, 6–9.
77. Gibbs, J.W. (1928) *The Collected Works of J.W. Gibbs*, Vol. 1. Longmans Green, London.
78. Aveyard, R., Binks, B.P. and Fletcher, P.D.I. (1990) Surfactant molecular geometry within planar and curved monolayers in relation to the microemulsion phase behavior. In D.M. Bloor and E. Wyn-Jones (eds), *The Structure, Dynamics and Equilibrium Properties of Colloidal Systems.* Kluwer Academic Publishers, Dordrecht/Boston/London, pp. 557–581.
79. Rosen, M.J. and Murphy, D.S. (1991) Effect of the nonaqueos phase on interfacial properties of surfactants. 2. Individual and mixed non-ionic surfactants in hydrocarbon/water systems. *Langmuir*, **7**, 2630–2635.

80. Sottmann, T., Strey, R. and Chen, S.H. (1997) A SANS study of non-ionic surfactant molecules at the water–oil interface: Area per molecule, microemulsion domain size and rigidity. *J. Chem. Phys.*, **106**, 6483.
81. Lang, J.C. and Widom, B. (1975) Equilibrium of 3 liquid-phases and approach to tricritical point in benzene–ethanol–water ammonium sulfate mixtures. *Physica A*, **81**, 190–213.
82. Cazabat, A.M., Langevin, D., Meunier, J. and Pouchelon, A. (1982) Critical-behavior in microemulsions. *J. Phys. Lett.*, **43**, L89–L95.
83. Widom, B. (1987) Phase-transitions in surfactant solutions and in their interfaces. *Langmuir*, **3**, 12–17.
84. Fletcher, P.D.I. and Horsup, D.I. (1992) Droplet dynamics in water-in-oil microemulsions and macroemulsions stabilized by non-ionic surfactants – correlation of measured rates with monolayer bending elasticity. *J. Chem. Soc. Faraday Trans.*, **88**, 855–864.
85. Bonkhoff, K., Hirtz, A. and Findenegg, G.H. (1991) Interfacial-tensions in the three-phase region of non-ionic surfactant + water + alkane systems – critical point effects and aggregation behavior. *Physica A*, **172**, 174–199.
86. Schubert, K.-V., Strey, R., Kline, S. and Kaler, E.W. (1994) Small-angle neutron scattering near the Lifshitz lines: Transition from weakly structured mixtures to microemulsions. *J. Chem. Phys.*, **101**, 5343–5355.
87. Kahlweit, M., Jen, J. and Busse, G. (1992) On the stability of microemulsions. 2. *J. Chem. Phys.*, **97**, 6917–6924.
88. Lee, L.T., Langevin, D., Meunier, J., Wong, K. and Cabane, B. (1990) Film bending elasticity in microemulsions made with non-ionic surfactants. *Progr. Colloid Polym. Sci.*, **81**, 209–214.
89. Binks, B.P., Meunier, J., Abillon, O. and Langevin, D. (1989) Measurement of film rigidity and interfacial tensions in several ionic surfactant-oil-water microemulsions. *Langmuir*, **5**, 415–421.
90. Kluge, K., Stubenrauch, C., Sottmann, T. and Strey, R. (2001) Temperature-insensitive microemulsions formulated from octyl monoglucoside and alcohols: Potential candidates for applications. *Tenside Surf. Det.*, **38**, 30–40.
91. Leitao, H., Somoza, A.M., Telo da Gama, M.M., Sottmann, T. and Strey, R. (1996) Scaling of the interfacial tension of microemulsions: A phenomenological description. *J. Chem. Phys.*, **105**, 2875–2883.
92. Helfrich, W. (1973) Elastic properties of lipid bilayers – theory and possible experiments. *Z. Naturforsch.*, **28c**, 693.
93. Safran, S.A. and Tlusty, T. (1996) Curvature elasticity models of microemulsions. *Ber. Bunsenges. Phys. Chem.*, **100**, 252–263.
94. Morse, D.C. (1997) Entropy and fluctuation of monolayers, membranes, and microemulsions. *Curr. Opin. Coll. Interface Sci.*, **2**, 365–372.
95. Gompper, G. and Schick, M. (1994) *Self-assembling Amphiphilic Systems*. Academic Press, New York.
96. Leitao, H., Telo da Gama, M.M. and Strey, R. (1998) Scaling of the interfacial tension of microemulsions: A Landau theory approach. *J. Chem. Phys.*, **108**, 4189.
97. Kahlweit, M. and Reiss, H. (1991) On the stability of Microemulsions. *Langmuir*, **7**, 2928–2933.
98. Scriven, L.E. (1976) Equilibrium bicontinuous structure. *Nature*, **263**, 123–125.
99. Jahn, W and Strey, R. (1987) Images of bicontinuous microemulsions by freeze fracture electron microscopy. In J. Meunier, D. Langevin and N. Boccara (eds), *Physics of Amphiphilic Layers*. Springer, Berlin, Heidelberg, p. 353.
100. Benson, H.L., Cox, K.R. and Zweig, J.E. (1985) *Happi*, 50.
101. Kahlweit, M., Strey, R. (1987) Phase behaviour of H_2O–oil–non-ionic amphiphile ternary systems. In M. Clausse (ed), *Microemulsions*, Vol. 24. Marcel Dekker, New York, p. 1.
102. Bowcott, J.E. and Schulman, J.H. (1955) Emulsions – control of droplet size and phase continuity in transparent oil–water dispersions stabilized with soap and alcohol. *Z. Elektrochem.*, **59**, 283.

103. Reimer, J., Söderman, O., Sottmann, T., Kluge, K. and Strey, R. (2003) Microstructure of alkyl glucoside microemulsion: Control of curvature by interfacial composition. *Langmuir*, **19**, 10692–10702.
104. Talmon, Y. (1999) Cryogenic temperature transmission electron microscopy in the study of surfactant systems. In B.P. Binks (ed), *Modern Characterization Methods of Surfactant Systems*. Marcel Dekker, New York, pp. 147–178.
105. Belkoura, L., Stubenrauch, C. and Strey, R. (2004) Freeze fracture direct imaging: A hybrid method in preparing specimen for Cryo-TEM. *Langmuir*, **20**, 4391–4399.
106. Glatter, O. (1982) Data treatment. In O. Glatter and O. Kratky (eds), *Small Angle X-ray Scattering*. Academic Press, London, pp. 119–163.
107. Chu, B. (1974) *Laser Light Scattering*. Academic Press, New York.
108. Berne, B.J. and Pecora, R. (1996) *Dynamic Light Scattering*. Wiley, New York.
109. Lindman, B. and Olsson, U. (1996) Structure of microemulsions studied by NMR. *Ber. Bunsenges. Phys. Chem.*, **100**, 344–363.
110. Eicke, H.F., Shepherd, J.C.W. and Steinmann, A. (1976) Exchange of solubilized water and aqueous-electrolyte solutions between micelles in apolar media. *J. Colloid Interface Sci.*, **56**, 168–176.
111. Burauer, S. (2002) *Elektronenmikroskopie Komplexer Fluide*. TENEA, Berlin.
112. Burauer, S., Belkoura, L., Stubenrauch, C. and Strey, R. (2003) Bicontinuous microemulsions revisited: A new approach to freeze fracture electron microscopy (FFEM). *Colloids Surf. A*, 159–170.
113. Shinoda, K. and Saito, H. (1968) The effect of temperature on phase equilibria and the types of dispersions of the ternary system composed of water, cyclohexane, and non-ionic amphiphile. *J. Colloid Interface Sci.*, **26**, 70–74.
114. Kahlweit, M., Strey, R. and Schomäcker, R. (1989) Microemulsions as liquid media for chemical reactions. In W. Knoche and R. Schomäcker (eds), *Reactions in Compartmentalized Liquids*. Springer, Berlin, Heidelberg, pp. 1–10.
115. Kahlweit, M., Strey, R., Haase, D., Kunieda, H., Schmeling, T., Faulhaber, B., Borkovec, M., Eicke, H.F., Busse, G., Eggers, F., Funck, T., Richmann, H., Magid, L., Söderman, O., Stilbs, P., Winkler, J., Dittrich, A. and Jahn, W. (1987) How to study microemulsions. *J. Colloid Interface Sci.*, **118**, 436–453.

Chapter 2

Scattering Techniques to Study the Microstructure of Microemulsions

Thomas Hellweg

2.1 Introduction

During the last 30 years, a lot of knowledge about structure formation and self-assembly processes in systems containing amphiphiles was gained using scattering experiments. Amphiphiles form a large variety of microstructures: micelles, either spherical [1], cylindrical [2, 3], worm-like [4, 5] or disc-like, mono or bilayers, vesicles, etc. The self-assembly process takes place in water, in organic solvents (referred to as 'oils' in the following), or when both water and oil are present. In the latter case, the surfactant can help to solubilise the two otherwise immiscible liquids. Some of the phases obtained in such a ternary system are called microemulsions. Two typical microstructures are droplets (oil in water, i.e. o/w-droplet microemulsions, or water in oil, i.e. w/o-droplet microemulsions) or so-called bicontinuous phases. These phases consist of extended surfactant layers, which separate oil and water domains such that they are continuous [6] (see Chapter 1), and hence have a fascinating 'sponge'-like microstructure similar to structures obtained, e.g. during spinodal decomposition in binary melts.

The type of structure observed is closely related to the spontaneous curvature C_0 of the surfactant assemblies [7]. By using an analogy with liquid crystals, which can also adopt layered structures, Helfrich [8] introduced the concept of the elastic-free energy associated with thermally excited deviations from the spontaneous curvature of the microstructures. This elastic-free energy per unit area is given by

$$F = \frac{1}{2}\kappa(C_1 + C_2 - 2C_0)^2 + \overline{\kappa}C_1C_2. \tag{2.1}$$

C_1 and C_2 are the principal curvatures of the surfactant layer. The spontaneous curvature C_0 of the layer is by convention positive if it is curved against water. For $C_0 = 0$, flat aggregates are the most favourable shape for the interfacial film. The elastic constants κ and $\overline{\kappa}$ are related, respectively, to deviations from the mean curvature and from the Gaussian curvature. This Helfrich bending-free energy allows explaining the behaviour of the systems for which $C_0 \approx 0$ [9]. Whereas the meaning of κ is rather obvious, the influence and meaning of $\overline{\kappa}$ is more difficult to assess. If $\overline{\kappa}$ is negative, the second term of

Eq. (2.1) is smaller if C_1 and C_2 have the same sign, i.e. favouring spheres or vesicles. If $\bar{\kappa}$ is positive, structures with saddles (regions where C_1 and C_2 have opposite signs) are favoured: these are the 'sponge-like' bicontinuous microemulsion phases. However, the actual structure adopted by the microemulsion forming system is determined not only by the two terms of Eq. (2.1), but also by the many other contributions to the total free energy of the system: entropy of mixing, repulsive or attractive interaction between aggregates (e.g. in the case of charged surfactants) etc. It must also be mentioned that the Helfrich description is only strictly valid for surfactant layer thicknesses much smaller than the inverse curvatures.

The strength of scattering methods with respect to investigations of microemulsions is that a lot of theoretical works directly relate to the results of elastic and inelastic scattering experiments to the parameters used in the Helfrich approach. However, a large variety of other experimental methods can also be used to gain information on κ, $\bar{\kappa}$ and the curvatures of the interfacial film. Numerous measurements of κ can be found in the literature. Commonly used experimental approaches are the analysis of the peak shape of elastic small-angle neutron and X-ray scattering spectra of lamellar phases [10–12] and the investigation of shape fluctuations of giant vesicles by means of video-microscopy [13, 14]. Quasi-elastic light [15, 16] and neutron scattering (spin-echo) were also used for studying κ values of lamellar phases [17, 18]. Moreover, the analysis of thermal fluctuations at macroscopic oil–water interfaces of low interfacial tension by ellipsometry leads to κ [19, 20]. If only the structure is of interest, freeze-fracture electron microscopy (FFEM) [21], cryogenic transmission electron microscopy (cryo-TEM) [22–25] and the newly developed cryogenic freeze-fracture direct imaging technique (cryo-FFDI) [26] are very valuable and of growing importance in the study of microemulsions. In the nineties a major problem of all works treating the bending elasticity of the surfactant layer was that different methods did not always lead to the same values of the bending elastic constants [27, 28]. This problem is meanwhile solved at least for droplet microemulsions [29]. It should be pointed out here that droplet structures only occur close to the *emulsification failure boundary* (*efb*), i.e. most of the structures in the L_1 or L_2 phase are non-spherical (e.g. rod-like or worm-like micelles). However, these structures will not be discussed in the present text.

This chapter is organised as follows. Section 2.2 will describe how different scattering techniques can be combined to learn something about the structure and dynamics in droplet microemulsions (o/w- and w/o-droplet microemulsions). This section is followed by Section 2.3 which deals with scattering experiments on bicontinuous microemulsions. For those readers who are interested in the basic principles and some general ideas that are important for all scattering experiments, this chapter contains an extended appendix comprising rather fundamental knowledge about the different scattering techniques used to study microemulsions. Since lamellar phases by definition do not belong to microemulsions they are not discussed here. However, the principles which govern the scattering behaviour of lamellar phases (L_α) are of course the same as for microemulsions and especially the bicontinuous microemulsions are strongly related to the L_α phases. Readers interested in this subject are referred to [16, 18, 30–33].

2.2 Scattering from droplet microemulsions

2.2.1 General outline

Size and shape of droplet microemulsions can be investigated using a combination of different scattering techniques. Small angle neutron scattering (SANS) is an excellent tool to study the shape, size (in terms of the radius of gyration R_g) and polydispersity of microemulsion droplets. This is due to the size of the droplets in o/w- and w/o-droplet microemulsions, respectively, which usually is in the nm range. The translational diffusion of the droplets can be studied using dynamic light scattering (DLS) (photon correlation spectroscopy, PCS) [3, 34–41]. Finally, neutron spin-echo spectroscopy (NSE) gives direct access to the shape fluctuations of the droplets.

These droplet shape fluctuations are governed by both κ and $\bar{\kappa}$. The problem was theoretically treated by Milner and Safran [42–44] and values of the bending elastic constants κ and $\bar{\kappa}$ can be obtained by using experimental methods which are able to resolve fluctuations of the surfactant layer in a microemulsion.

The first direct studies of droplet shape fluctuations were done with neutron spin-echo techniques [45]. These experiments led to unexpectedly high values of κ, which were about one order of magnitude larger than those computed on the basis of static determination methods and to $\bar{\kappa}$ values close to -2κ [18, 46, 47]. This was somewhat surprising because the condition $2\kappa + \bar{\kappa} = 0$ corresponds to the stability limit for spheres. It was shown later that one should rather find $\kappa + 2\bar{\kappa} = 0$ [48]. It was also shown that the approximation of equal oil and water viscosities, which was assumed in Ref. [43], leads to systematic errors and that the accuracy of the analysis of the quasi-elastic neutron scattering data can be improved by using double-exponential fits where some of the parameters are fixed to values obtained by means of PCS [29, 49] or NMR self-diffusion measurements [50]. The results obtained for κ and $\bar{\kappa}$ were of the order of kT and $-0.5\ kT$, respectively, and in satisfying agreement with a large number of direct independent determinations: $2\kappa + \bar{\kappa}$ from interfacial tension measurements, and κ from ellipsometry experiments, or the analysis of droplet polydispersity (elastic scattering data) [51, 52]. They were also in agreement with other more indirect determinations, e.g. based on the determination of the persistence length of the surfactant layer, which is the characteristic size in a bicontinuous microemulsion [20], and models relating interfacial tension to microemulsion sizes [53, 54].

2.2.2 Quasi-elastic scattering from droplets: theory

In this subsection, the theoretical background for SANS and neutron spin-echo measurements carried out with o/w- and w/o-droplet microemulsions will be presented. According to Milner, Safran and others, shape fluctuations in droplet microemulsions can be described in terms of spherical harmonics [42–44]. This offers the possibility to calculate a dynamic structure factor $S(q,\omega)$ or its Fourier transform, i.e. the intermediate scattering function $I(q,t)$ for the problem, which can be used to analyse dynamical measurements by neutron spin-echo spectroscopy [45]. For the scattering from thin shells $I(q,t)$ was calculated [43] and can be written as

$$I(q,t) = \langle \rho_q(t)\rho_{-q}(0)\rangle \exp(-D_0 q^2 t)\left[4\pi[j_0(qR)]^2 + \sum f_l(qR)\,\langle u_l(0)u_l(t)\rangle\right]. \tag{2.2}$$

This leads to [18]

$$I(q,t) = \left\langle \exp(-\Gamma_1 t)V_s^2(\Delta\rho)^2\left[f_0(qR) + \sum_{l\geq 2}\frac{2l+1}{4\pi}f_l(qR)\times\left\langle |u_l|^2\right\rangle \exp(-\Gamma_l t)\right]\right\rangle_R, \tag{2.3}$$

which is a sum of at least two exponential decays if terms corresponding to $l > 2$ are omitted. The first contribution represents the translational motion of the particles and the second represents the shape fluctuations of the particles. Γ_1 is the relaxation rate of the corresponding mode. For the relaxation time τ_2 of the modes corresponding to the second-order spherical harmonic ($l = 2$), the expression

$$\tau_2^{-1} = \frac{1}{\eta R^3}\left[4\kappa - \bar{\kappa} - \frac{kT(\ln\phi - 1)}{4\pi}\right]\frac{1}{Z(2)} \tag{2.4}$$

was derived [44] for spherical droplets, which only exist at the *efb*. At this boundary, the microemulsion droplets are swollen to their maximum size. $Z(l)$ is given by

$$Z(l) = \frac{(2l+1)(2l^2+2l-1)}{l(l+1)(l+2)(l-1)}, \tag{2.5}$$

which leads for $l = 2$ to 55/24. In these expressions, the viscosities of oil and water were assumed to be equal. However, in real systems they are usually different and a more rigorous approach for the calculation of $Z(l)$ has to be used [55], namely

$$Z(l) = \frac{[(2l^2+4l+3)E + 2l(l+2)][2(l^2-1)E + 2l^2 + 1]}{(l-1)l(l+1)(l+2)(2l+1)(E+1)}, \tag{2.6}$$

where E is the ratio of the viscosities of the interior of the droplets and the continuous phase. In the limit of equal viscosities and for $l = 2$, Eq. (2.6) leads to a ratio of 52.5/24. There is a small difference compared to Eq. (2.5), which is probably due to different approximations in the hydrodynamic calculations. For a wavelength of 6 Å, a Fourier time window of about 20 ns is accessible (this depends of course on the used NSE machine and on the setup). This is well suited for the study of the droplet motions expected in microemulsions. As a result one obtains directly the normalised intermediate scattering function, which should correspond to the form given in Eq. (2.3).

Unfortunately, in a typical NSE experiment the number of points obtained as a representation of this function is in most cases too small and the error of the individual points is too high to allow for a fit with more than two or three adjustable parameters or for an analysis using Laplace transformation – and maximum entropy methods [56]. Without information from additional experiments it is only possible to compute an effective diffusion coefficient from the data by using a first- or second-order cumulant analysis [57].

In the first studies of microemulsions using NSE spectroscopy [18, 45], the q variations of the effective diffusion coefficients were afterwards fitted by expressions of the form given below

$$\Gamma_{\mathrm{eff}} = A(q)Dq^2 + \sum_{l\geq 2} B_l(q)\langle|u_l|^2\rangle \Gamma_l, \tag{2.7}$$

which are obtained by a cumulant expansion of the theoretical intermediate scattering function for the dynamics in droplet microemulsions (Eq. (2.3)). In these works, the translational diffusion coefficient D in this expression was calculated using the droplet radius R_g derived from SANS experiments [18, 45, 46]. This radius is smaller than the hydrodynamic radius and the diffusional contribution to the decay is therefore slightly overestimated by this approach. Hence, a direct measurement of D with PCS or NMR is preferable.

All modes corresponding to values of $l > 2$ do not contribute significantly to the second term in Eq. (2.7) when q is not too high ($qR \leq 10$) and can be omitted. For the first cumulant Γ_{eff}, which is averaging the total dynamics in the system, an apparent proportionality to q^3 is predicted [43]. This is in line with the observations for Zimm dynamics of polymers [58]. On the basis of this reasoning another more straightforward approach to the analysis of NSE data will be presented in the following. Assuming that only the mode corresponding to $l = 2$ contributes significantly to the decay, the intermediate scattering functions from the NSE experiment will be fitted directly using a double exponential function of the following form

$$\frac{I(q,t)}{S(q)} = a\exp(-D_0 q^2 t) + (1-a)\exp(-\Gamma t), \tag{2.8}$$

with

$$\tau_2^{-1} = \Gamma - D_0 q^2, \tag{2.9}$$

for the relaxation time τ_2 of the mode $l = 2$. The diffusion coefficient D_0 used here can be obtained from DLS experiments. Therefore, $I(q,t)/S(q)$ contains only two adjustable parameters. According to Eq. (2.3), the amplitude a (or $(1-a)$, respectively) should be proportional to

$$a \sim \frac{f_0(qR)}{f_0(qR) + \frac{5}{4\pi} f_2(qR)\langle|u_2|^2\rangle}. \tag{2.10}$$

The obtained τ_2 is expected to be independent from the scattering angle. This is also similar to Zimm polymer dynamics where constant frequencies are obtained for the different bending modes of the chain, when the different contributions to the intermediate scattering functions can be separated. In reality, there might be a slight q-dependence observable because of the rising contribution of modes corresponding to $l > 2$ with increasing q. These cannot be properly resolved from the experimental NSE data at present.

2.2.3 Small angle neutron scattering from droplets

The theoretical description in terms of spherical harmonics also yields a relation between the size polydispersity index p of the microemulsion droplets and the bending elastic constants [43]. The quantity p is accessible by SANS [51, 52, 59–61]. For polydisperse shells as obtained by using deuterated oil and heavy water for the preparation of the microemulsion (contrast variation), one can account for the droplet polydispersity by applying an appropriate form factor, e.g. containing a Gaussian function to model the size distribution [52, 59, 62]. A possible often-used choice is the following form factor

$$F(q) = 16\pi^2(\Delta\rho)^2(\delta^2/q^2)\exp(-q^2t^2)\,[f_1(q) + f_2(q) + f_3(q) + f_4(q)] \qquad (2.11)$$

with

$$f_1(q) = \tfrac{1}{2}q^2t^4(1 + \cos(2qR_0)\exp(-2\sigma^2q^2))$$
$$f_2(q) = qt^2(R_0\sin(2qR_0) + 2q\sigma^2\cos(2qR_0))\exp(-2\sigma^2q^2)$$
$$f_3(q) = \tfrac{1}{2}R_0^2(1 - \cos(2qR_0)\exp(-2\sigma^2q^2))$$
$$f_4(q) = \tfrac{1}{2}\sigma^2(1 + 4qR_0\sin(2qR_0)\exp(-2\sigma^2q^2) + \cos(2qR_0)(4\sigma^2q^2 - 1)\exp(-2\sigma^2q^2)).$$

Here, t is a parameter describing the thickness of the surfactant layer and σ (or p) contains the information about the size polydispersity of the microemulsion drops. R_0 is the mean value of the shell inner and outer radii. Besides this approach other distribution functions were also already applied to model the droplet polydispersity [52]. The absolute scattering intensity $I(q)$, which is the experimentally observed quantity is given by

$$I(q) \propto NF(q)S(q). \qquad (2.12)$$

In this relation, N is the number density of the scattering microemulsion droplets and $S(q)$ is the static structure factor. Equation (2.12) is only strictly valid for the case of monodisperse spheres. However, for the case of low polydispersities the occurring error is small [63, 64]. $S(q)$ describes the interactions between and the spatial correlations of the droplets. These are in general well approximated by hard sphere interactions in microemulsion systems [65]. The influence of inter-particle interactions as described by $S(q)$ can be estimated at least for $S(0)$ using the Carnahan–Starling expression [52, 64, 66]

$$S(0) = \frac{(1 - \phi_{hs})^4}{(1 - 2\phi_{hs})^2 - \phi_{hs}^3(1 - 4\phi_{hs})}. \qquad (2.13)$$

In this equation, ϕ_{hs} is the hard sphere volume fraction which is about 14% larger in o/w-droplet microemulsions of non-ionic surfactant than the dispersed volume fraction. This is caused by the water penetration in the surfactant layer [64]. $S(q)$ approaches unity for q values smaller than the minimum of $I(q)$. This behaviour occurs even for fairly high volume fractions in non-ionic surfactant systems (see for example Fig. 8 in Ref. [64]). Seeing that the value of the radius is fixed by the position of the minimum of $I(q)$, the approximation of $S(q) \approx 1$ in Eq. (2.12) does not lead to a significant error in the determination of R_0 if the low q part of the experimental curve is not taken into

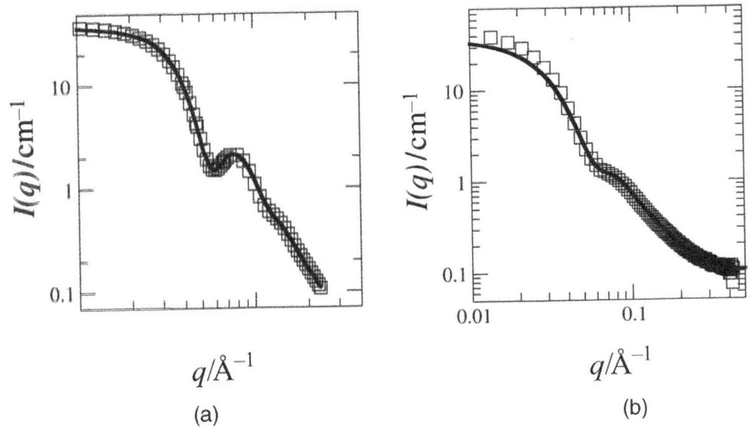

Figure 2.1 Comparison of SANS curves obtained for the system D_2O/*n*-octane-d_{18}/$C_{10}E_4$ on the (a) water-continuous (o/w-droplet microemulsion) and (b) the oil-continuous (w/o-droplet microemulsion) side, respectively. The solid lines in both plots are from factor curves according to Eq. (2.11). Usually, the polydispersity is slightly higher for w/o-droplet microemulsions. (Figures redrawn with data from Ref. [67].)

account for the fitting procedure. Figure 2.1 shows SANS data for a water- and an oil-continuous droplet microemulsion. Both microemulsion phases were found in the system D_2O/*n*-octane-d_{18}/$C_{10}E_4$. The two samples have compositions close to the EMF (see Fig. 1.7 in Chapter 1; the lower boundary between the single phase and the two phase region in the diagram on the water-rich side; on the oil-rich side the EMF is the upper phase boundary) and hence, contain spherical droplets. Microemulsion droplets are only found to be spherical close to this phase boundary. In other areas of the single phase region the structure can even be worm-like [4, 5]. Because of the use of deuterated solvents for the preparation of the samples the curves in Fig. 2.1 were obtained for shell contrast. The solid lines are fits according to Eq. (2.11) and for sample (Fig. 2.1(b)) a polydispersity index of $p = 0.335$ is obtained. The polydispersity index is related to the two bending elastic constants by

$$p^2 = \sigma^2 = \frac{\langle |u_0|^2 \rangle}{4\pi} = \frac{kT}{8\pi(2\kappa + \bar{\kappa}) + 2kT(\ln\phi - 1)}. \tag{2.14}$$

For the example of the oil-continuous microemulsion this leads to $2\kappa + \bar{\kappa} = 0.66kT$ [67] (see Fig. 2.1(b)). Figure 2.2 shows another example for typical SANS curves as obtained in shell contrast. The investigated systems were all of the type D_2O/*n*-decane-d_{22}/C_iE_j and the solid lines are again fits according to Eq. (2.11). The obtained polydispersity index is then used to calculate the sum $2\kappa + \bar{\kappa}$ for the three different surfactants. The found values are $1.2kT$, $2.28kT$ and $3.42kT$ for the surfactants C_8E_3, $C_{10}E_4$ and $C_{12}E_5$, respectively [52]. Note that besides the approach presented above, the 'model free' indirect Fourier transformation (IFT) method can also be applied to obtain information about the radius and shape of the droplets in a microemulsion [41, 68].

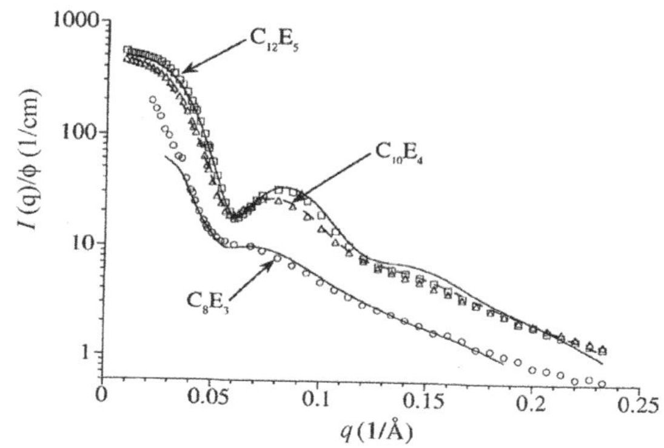

Figure 2.2 SANS curves obtained for three systems of the type D_2O/n-decane-d_{22}/C_iE_j on the water-continuous (o/w-droplet microemulsion) side of the phase diagram. The solid lines are again fits with Eq. (2.11). (From Ref. [52], reprinted with permission of the American Physical Society.)

2.2.4 *Examples*

2.2.4.1 O/W-droplet microemulsions

As already pointed out the first work directly measuring the deformation dynamics in an o/w-droplet microemulsion using NSE was published by Huang *et al.* [45]. In this work, a microemulsion based on the surfactant AOT was studied and it was shown that the intermediate scattering functions contain information about the centre of mass diffusion and in addition also contributions from the deformation dynamics. The intermediate scattering functions obtained in this work are shown in Fig. 2.3.

The experimental data for this system were fitted using single exponential functions and the obtained relaxation rate was used to calculate the effective diffusion coefficient D_{eff}. In Fig. 2.4, the effective diffusion coefficients as obtained from the intermediate scattering curves in Fig. 2.3 are plotted versus q^2. At low q values, D_{eff} is equal to the centre of mass

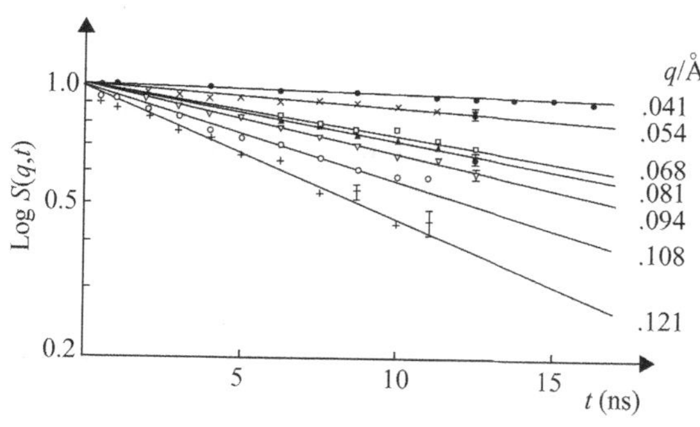

Figure 2.3 Intermediate scattering functions obtained for an AOT-based o/w-droplet microemulsion using NSE. The solid lines are fits with a single exponential function yielding an effective diffusion coefficient. Note that this was the first NSE study of a microemulsion showing the calculation of κ on the basis of the intermediate scattering functions. (From Ref. [45], reprinted with permission of the American Physical Society.)

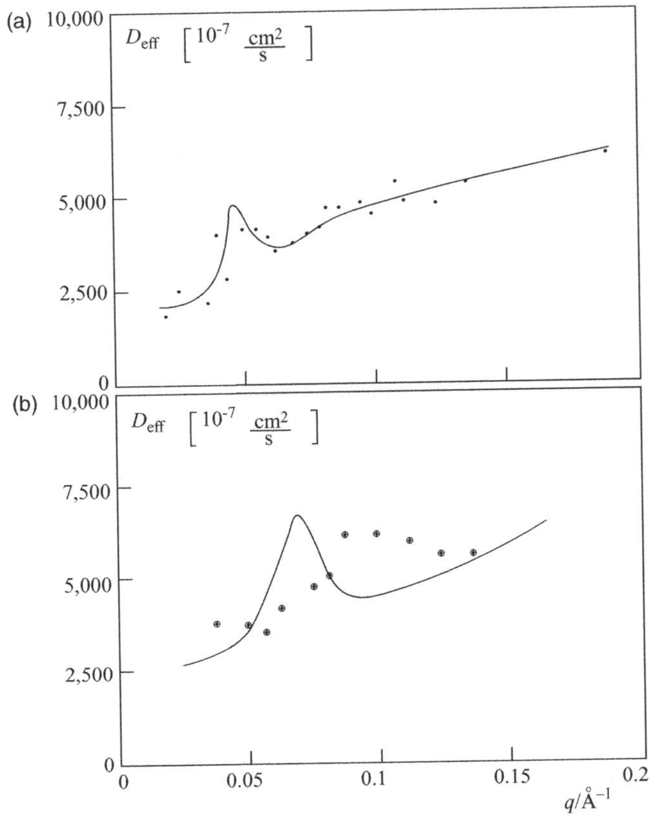

Figure 2.4 Effective diffusion coefficient D_{eff} as a function of q for an AOT-based o/w-droplet microemulsion. (From Ref. [45], reprinted with permission of the American Physical Society.)

diffusion coefficient of the droplets. However, for higher q values, Huang *et al.* found an increase of D_{eff} reaching a maximum at the minimum of the droplet form factor. This effect arises from the q dependence of the amplitude of the droplet deformations, which reaches a maximum at the form factor minimum. The solid line in Fig. 2.4 represents a fit using the expression in Eq. (2.7). However, in this first work $\bar{\kappa}$ was omitted and moreover, also the difference of the viscosities of the employed oil and water were partly neglected.

After these first experiments it took 11 years until this problem was studied again exploiting the unique possibilities of NSE with respect to contrast variation and energy resolution [29]. The studied microemulsion was an o/w-droplet microemulsion in the system H_2O/n-octane/$C_{10}E_5$. It turned out that the NSE data can be analysed using a double exponential fit according to Eq. (2.8), when the translational diffusion coefficient is already measured in advance using PCS. The same approach was also successfully applied to study another water-continuous microemulsion in the system H_2O/n-dodecane/$C_{10}E_5$ [49]. Since the approach works as well for oil-continuous systems an extended example for the approach will be discussed in the next subsection.

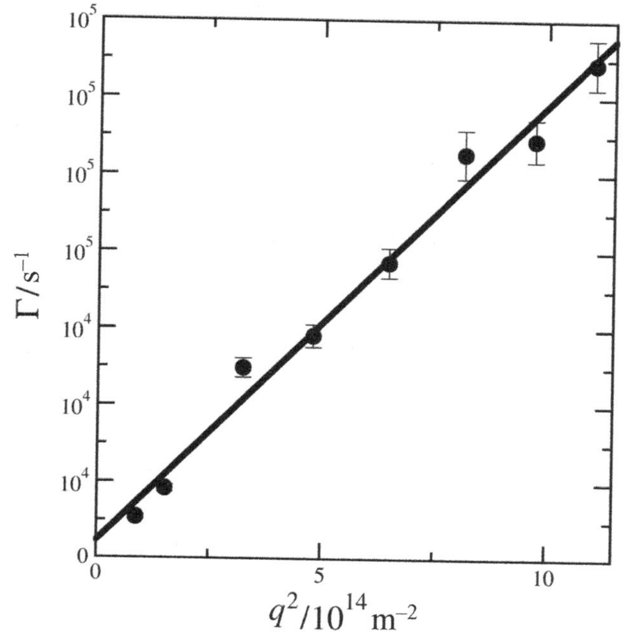

Figure 2.5 Relaxation rates Γ of the intensity correlation functions as a function of q^2 obtained via a photon correlation spectroscopy experiment. The sample was a w/o-droplet microemulsion made of D_2O/*n*-octane-d_{18}/$C_{10}E_4$. On the oil-continuous side of the phase diagram the scattered light intensity is usually low leading to rather large errors of the individual data points. Nevertheless, from the slope of the linear fit the translational diffusion coefficient is obtained. (Figure redrawn with data from Ref. [67].)

2.2.4.2 W/O-droplet microemulsions

In this subsection, the results for droplet microemulsions arising from the combination of NSE, PCS and SANS will be presented in some more detail. The system D_2O/*n*-octane-d_{18}/$C_{10}E_4$ on the oil-continuous side of the phase diagram will serve as an example [67]. In Fig. 2.5, the relaxation rates Γ as obtained from PCS are plotted versus q^2. All given Γ values are averages from at least three independently measured intensity correlation functions (square of the intermediate scattering function). However, the error of the individual points is still rather high for a DLS experiment. This is typical for PCS measurements of oil-continuous microemulsions. Usually, for these systems the scattering contrast is rather low for light leading to low scattered intensities and long measurement times. Also, the interface of the colloidal droplets can be rather diffuse on the oil-rich side of the phase diagram, which is due to the high oil solubility of the C_iE_j surfactants. Nevertheless, the slope of the linear fit in Fig. 2.5 still yields the translational diffusion coefficient for the droplets with a rather small error of about ±6% and via the Stokes–Einstein equation also the hydrodynamic radius.

The translational diffusion coefficient from the PCS measurements is then used in Eq. (2.8) to reduce the number of adjustable parameters. In Fig. 2.6, the experimentally obtained intermediate scattering functions for four scattering angles are shown. The solid lines in this figure are fits using Eq. (2.8). The relaxation τ_2 for the deformation mode with

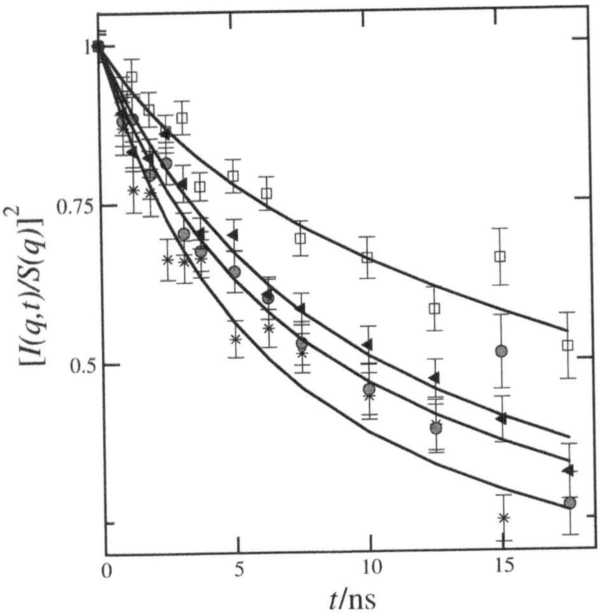

Figure 2.6 Measured intermediate scattering functions of a w/o-droplet microemulsion for the system D_2O/n-octane-d_{18}/$C_{10}E_4$. The four curves were obtained at four different q values close to the minimum of the droplet form factor. The solid lines are double exponential fits with only two adjustable parameters. The translational diffusion coefficient was determined using PCS (see Fig. 2.5) and used as input for the analysis of the NSE data. (Figure redrawn with data taken from Ref. [67].)

$l = 2$ can be calculated from the obtained relaxation rates according to

$$\tau_2^{-1} = \Gamma - D_0 q^2. \tag{2.15}$$

In Fig. 2.7, the results for the oil-continuous system D_2O/n-octane-d_{18}/$C_{10}E_4$ and for two water-continuous microemulsions are shown. At least for two of the examples (the oil-continuous $C_{10}E_4$ system and the water-continuous C_8E_3 system), the obtained relaxation rate is clearly constant in q. This is the expected behaviour. Only the amplitude should change with q. Equation (2.4) directly relates τ_2 to $4\kappa - \bar{\kappa}$. Hence, knowing also the polydispersity index p it is now possible to calculate κ and $\bar{\kappa}$. For the case of the oil-continuous droplet system D_2O/n-octane-d_{18}/$C_{10}E_4$, $\kappa = 0.66kT$ and $\bar{\kappa} = -1.11kT$ was found using this approach [67].

2.3 Scattering from bicontinuous microemulsions

This type of microemulsion is usually obtained when the system contains similar amounts of oil and water. In the phase diagram (see Fig. 1.3 in Chapter 1) this is the middle phase found in the three-phase region (hatched area in the left scheme of Fig. 1.3 in Chapter 1, 'fish' body). On the right-hand side of the X-point the system forms a single phase. In this region of the phase diagram the bicontinuous phase extends over the whole

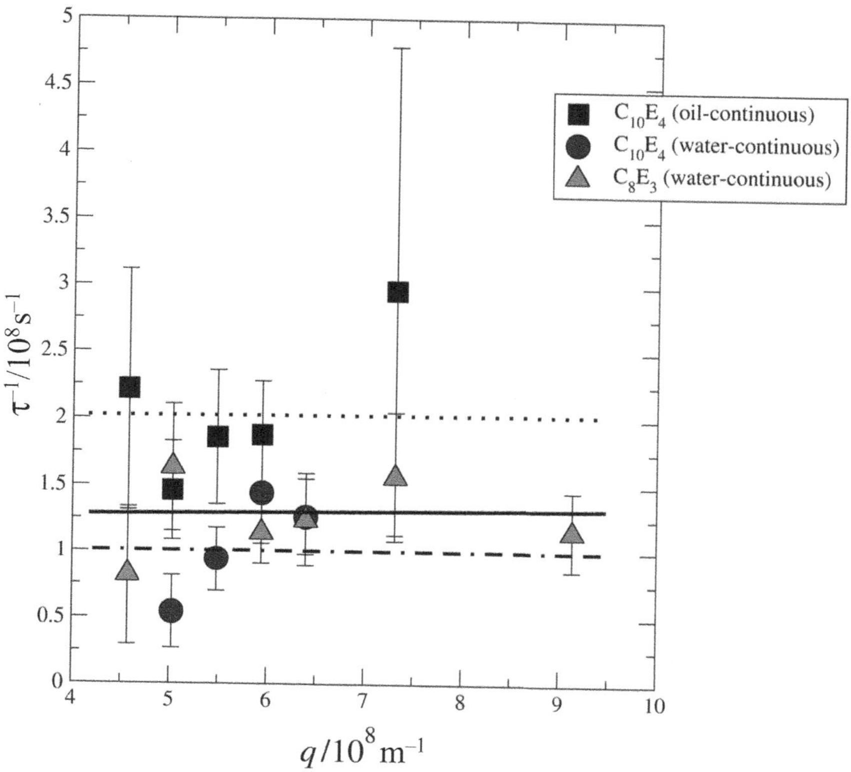

Figure 2.7 Relaxation times τ for the deformation modes as obtained from the analysis of the NSE intermediate scattering functions using DLS data for the determination of the translational diffusion coefficient. The lines indicate the average relaxation time for the different samples.

sample volume. The characteristic structural feature of bicontinuous microemulsions is the sponge-like structure formed by the surfactant layer separating the two immiscible liquids (see Fig. 2.8). This interesting structure exhibits a very unique static and dynamic scattering behaviour, which will be discussed in the following four subsections.

2.3.1 Small angle scattering from bicontinuous microemulsions

Since their structure is not well ordered, bicontinuous microemulsions exhibit a rather broad structure factor peak in SANS or SAXS experiments. This scattering behaviour was first quantitatively analysed by Teubner and Strey [70] using the expansion of a phenomenological Landau free energy equation with respect to an intermediate or rather long-range order parameter. This leads to the following expression for the scattered intensity

$$I(q) \propto \frac{8\pi\phi_{\text{oil}}\phi_{\text{water}}/\xi}{\left(q_{\text{max}}^2 + \xi^{-2}\right)^2 - 2\left(q_{\text{max}}^2 - \xi^{-2}\right)^2 q^2 + q^4} \tag{2.16}$$

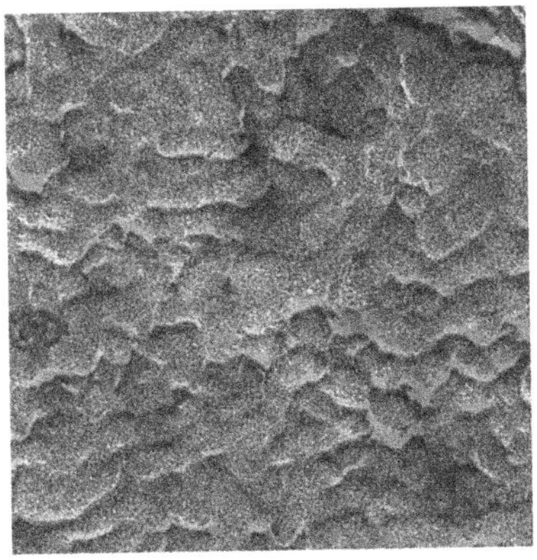

Figure 2.8 Freeze-fracture scanning electron microscopy image from a bicontinuous microemulsion. The system was prepared with the technical surfactant Marlowet IHF and perchloroethylen as oil. (From Ref. [69], unpublished work.)

where $q_{max} = 2\pi/d$, ϕ_{oil} and ϕ_{water} are the volume fractions of oil and water in the bicontinuous system, respectively, d is the domain size and ξ is the correlation length of intermediate range collective fluctuations in the system or in other words a kind of persistence length of the interface between water and oil (size of approximately stiff patches). The persistence length of the film in bicontinuous microemulsions is given by

$$\frac{\xi_{\kappa}}{\delta} = \exp\left(\frac{4\pi\kappa_0}{3kT}\right) \quad (2.17)$$

with $\delta = v_c/a_c$, v_c = volume of one surfactant molecule and a_c = area per surfactant molecule [71]. Hence, this approach is different compared to a standard Ornstein–Zernicke description for fluctuations in a system, where only one length scale is of importance. Compared to the Ornstein–Zernicke behaviour, the additional length scale d arises from the mean 'pore' size of the sponge-like structure of the microemulsion. Often ξ is approximately half of d.

One problem of this approach is the omission of short-range fluctuations of the interfacial film. Only long-range fluctuations are taken into account [72]. However, the short length scale undulations in a real system lead to an apparent increase of the interfacial area and therefore to a shift of the Porod region of the experimental scattering curves towards higher q values compared to the Teubner–Strey equation. Hence, Eq. (2.16) often yields a nice description of the structure factor peak of the experimental curves, but fails for the higher q part of the data. For a satisfying fit of the entire scattering curve, Eq. (2.16) has to be corrected for the diffuse scattering from the undulations and also for the very local roughness of the interfacial film [72, 73]. Such a correction can be obtained using an empirical approach published by Beaucage [74]. This approach was developed to account for the scattering from fractal structures [74]. Employing this correction leads to the following equation for $I(q)$

$$I(q) \propto \left[\left(\frac{8\pi\phi_{\text{oil}}\phi_{\text{water}}/\xi}{(k^2+\xi^{-2})^2 - 2(k^2-\xi^{-2})^2 q^2 + q^4}\right) + \frac{G\, erf^{12}(1.06qGR_g/\sqrt{6})}{1.5q^4R_g^4}\right] \times \exp(-\sigma^2 q^2) + b_g, \tag{2.18}$$

where R_g is an additional length scale describing the average size of a single domain and G is the amplitude of the fractal term from the Beaucage approach. In addition, this equation contains a Gaussian factor to account for the very local roughness of the interfacial film. Finally, b_g is the constant incoherent background. Since this approach is only justified by the fact that it sometimes works and by hand waving arguments about local roughness and diffuse interfaces, it is usually preferable to only rely on the Teubner–Strey approach for the analysis of small angle scattering curves from bicontinuous microemulsions.

2.3.2 Neutron spin-echo studies of bicontinuous microemulsions

Only a small number of studies have addressed the problem of direct measurements of the dynamics of the surfactant layer in a bicontinuous microemulsion [16, 73, 75, 76]. The Zilman–Granek (ZG) model, which assumes membrane Zimm dynamics on an ensemble of 'free' membrane patches [75] is expected to be applicable to the problem. In the framework of this model the intermediate scattering functions should then be describable by a stretched exponential function of the type

$$I(q,t)/S(q) = A\exp(-(\Gamma_q t)^{2/3}). \tag{2.19}$$

The exponent is fixed to 2/3 similar to the Zimm polymer dynamics. Within the approximations used in [75] to arrive at a closed analytical expression, the slope and the elastic constant κ are connected by

$$\Gamma_q = 0.025\gamma_\kappa \left(\frac{k_BT}{\kappa}\right)^{1/2} \frac{k_BT}{\eta_0} \tag{2.20}$$

with $\gamma_\kappa = 1$ for large κ and η_0 = effective viscosity of the fluid separating the interfacial surfactant layers. However, due to the approximations made in this approach the κ values obtained for bicontinuous microemulsions are usually too high. Especially, for samples where κ is expected to be of the order of kT. In Ref. [77], the problem was solved numerically leading to

$$I(q,t) = \frac{2\pi\xi^2}{a^4}\int_0^1 d\mu \int_0^{r_{\max}} dr\, r\, J_0\left(qr\sqrt{1-\mu^2}\right) \times \exp\left(\frac{-kT}{2\pi\kappa}q^2\mu^2 \int_{k_{\min}}^{k_{\max}} dk[1 - J_0(kr)\exp(-\omega(k)t)]/k^3\right). \tag{2.21}$$

Table 2.1 Results computed by Teubner and Strey for the data from Ref. [78] with ϕ_s = volume fraction of the surfactant, ξ = correlation length and d = domain size

Sample number	ϕ_s	ξ/nm	d/nm
1	0.181	7.2	17.5
2	0.237	6.7	12.7
3	0.323	5.5	8.4

The respective fits are shown in Fig. 2.9.

Here, μ is the cosine of the angle between $\vec{q}$ and the surface normal, $k_{\min}$ is a cut-off describing the most extended bending mode, which still fits into the persistent surface area with size ξ, and J_0 is the Bessel function of the order 0. This approach was already used to describe experimental intermediate scattering functions [76] and an example will be given in the next subsection. The same method also applies to lamellar phases since only persistent areas are assumed and no further assumptions about the geometry of the surfactant layer are made.

2.3.3 Examples

2.3.3.1 Structure

In their original work, Teubner and Strey have analysed data of different authors using fits to Eq. (2.16). Figure 2.9 is reproduced from the original Teubner–Strey article and shows SANS data measured by Kotlarchyk *et al.* [78]. The curves correspond to increasing surfactant volume fractions ϕ_s and the solid lines are fits using Eq. (2.16). The data are nicely described by the model and reliable values for the adjustable parameters were obtained (see Table 2.1). An increase of the surfactant concentration means growing interfacial area and hence, it makes sense that the domain size d and the correlation length ξ decrease with growing ϕ_s. A relation between κ and d can be found in [27], namely

$$d = \frac{2\delta}{\psi}\left(1 + \frac{kT}{4\pi\kappa}\ln\left(c\sqrt{\frac{\kappa}{kT}}\frac{d}{2a}\right)\right). \tag{2.22}$$

Here c is a cut-off constant, a is related to the head group area of the surfactant [71], δ is the thickness of the membrane and ψ the total membrane volume fraction. For more details see [79]. Figure 2.10 shows the problem already discussed in subsection 2.3.1. Often for rather soft interfacial films the surface area is underestimated by the Teubner–Strey formula. This leads to deviations of the data shown in Fig. 2.10 and the solid line labelled with TS. The Beaucage correction obviously leads to a better fit of the high q part (black solid line in Fig. 2.10).

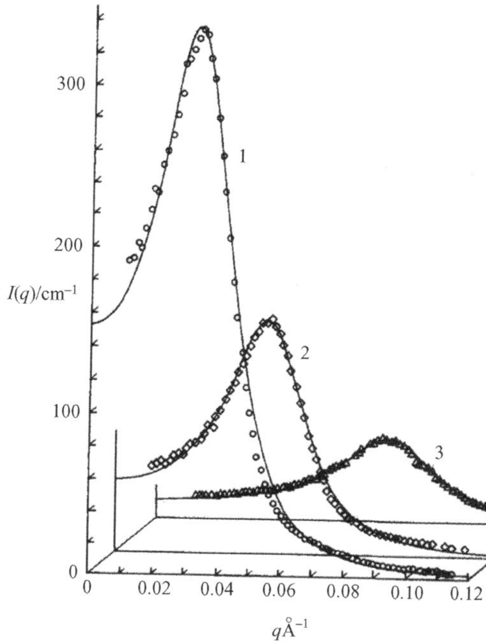

Figure 2.9 Figure from the original article of Teubner and Strey [70]. In this figure SANS curves measured for bicontinuous D_2O/*n*-decane/AOT microemulsions at three different AOT concentrations were analysed using Eq. (2.16). The data used by Teubner and Strey were measured by Kotlarchyk *et al.* [78]. The resulting domain sizes *d* and correlation lengths ξ are listed in Table 2.1. (From Ref. [70], reprinted with permission of the American Institute of Physics.)

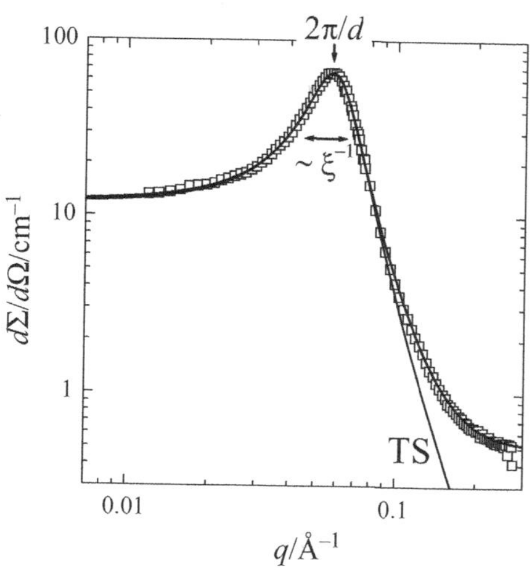

Figure 2.10 Experimental scattering data for a typical bicontinuous microemulsion. The data were described with the Teubner–Strey model. In addition, the Beaucage correction was used for high *q* values. (From Ref. [72], reprinted with permission of the American Chemical Society.)

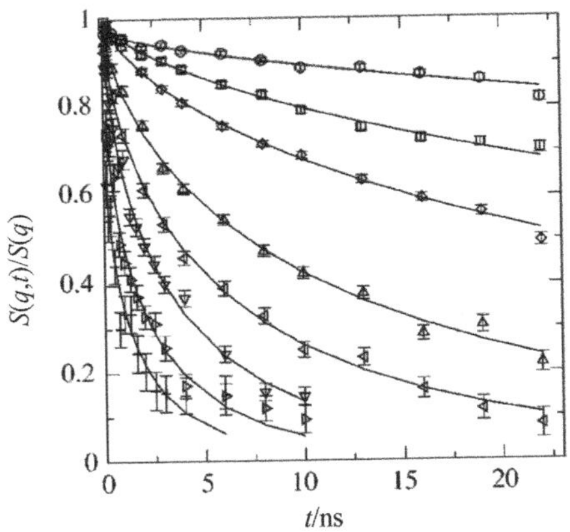

Figure 2.11 Experimental data for a bicontinuous microemulsion containing a homopolymer. The solid lines are fits using Eq. (2.19). Each symbol indicates a different q value. (From Ref. [76], reprinted with permission of the American Institute of Physics.)

2.3.3.2 Dynamics

In Figs. 2.11 and 2.12, NSE data for a homopolymer containing bicontinuous microemulsion are given [76]. The system is based on D_2O/n-decane-d_{22}/$C_{10}E_4$. Equation (2.19) derived by Zilman and Granek is able to fit all the experimental intermediate scattering functions with an acceptable error. This is obvious looking at Fig. 2.11. However, using then the relation in Eq. (2.20) leads to rather high values for κ. In Fig. 2.12, the full solution (Eq. (2.21)) is used for a simultaneous fit of the complete set of $I(q,t)$ functions. For high q values, the description is satisfying and also the calculated values for κ are in the correct range and seem to be reliable. However, in this case the description of the low q data is rather bad (see Fig. 2.12). Obviously, there is an additional contribution from a

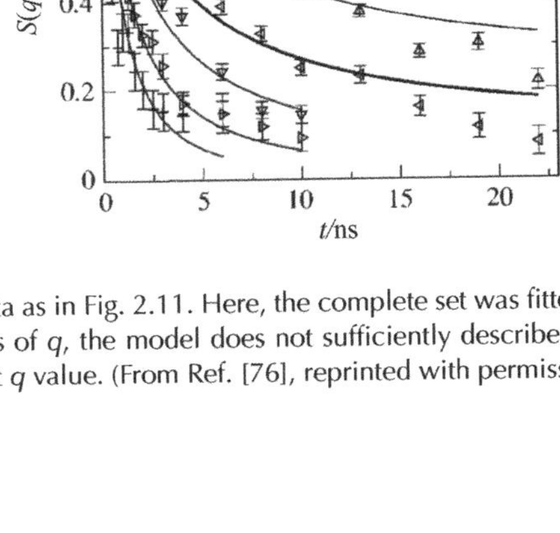

Figure 2.12 The same data as in Fig. 2.11. Here, the complete set was fitted using the expression given in Eq. (2.21). At low values of q, the model does not sufficiently describe the experimental data. Each symbol indicates a different q value. (From Ref. [76], reprinted with permission of the American Institute of Physics.)

slow mode present in the data, which is not yet properly taken into account. For additional information on these systems the reader is also referred to Chapter 4.

2.4 Summary

The sections above clearly show that scattering methods were successfully employed to investigate different important aspects of structure and dynamics in microemulsion phases. Especially, the models for droplet microemulsions are meanwhile very sophisticated and scattering experiments can routinely be used for the investigation of these phases. In case of bicontinuous systems the situation is slightly different. The small angle scattering curves from these systems can be described using the Teubner–Strey approach. However, the high q part of these SANS spectra still remains to be described in a way that correlates physical parameters of the system with its scattering curves. Concerning the dynamics of the bicontinuous systems a complete description covering all occurring time scales is still missing.

2.5 Appendix

2.5.1 *General remarks*

When radiation interacts with a sample scattering or diffraction occurs due to spatial and temporal correlations in the sample. In this section, the basic quantities and correlation functions will be introduced. In elastic and quasi-elastic scattering experiments the most important quantity is the magnitude of the so-called scattering vector given by

$$q = \frac{4\pi}{\lambda}\sin\frac{\theta}{2} \tag{2.23}$$

with λ being the wavelength of the used radiation (e.g. neutrons or light) in the scattering medium and θ the scattering angle. q has the dimension of a reciprocal length and is a measure for the spatial resolution of a scattering experiment, which is schematically shown in Fig. 2.13. In this figure, the squares indicate the spatial resolution in the respective q-range. At low values of q, the overall size and shape of the colloidal particle is seen. Moreover, for higher concentrations also inter-particle correlations will be visible (the so-called structure factor). At high values of q, the internal structure of the particles can be resolved. The choice of the appropriate scattering method, which has to be used to probe a specific property of a colloidal particle depends on the relationship length-scale to q-range. Static light scattering (SLS) and especially DLS are well suited to study the overall size and shape of colloidal particles. In SANS experiments the accessible q-range is usually 0.007–0.4 per $Å^{-1}$. This is two to three orders of magnitude smaller compared to light scattering experiments. Therefore, SANS and also NSE are well suited to study local (internal) structures and movements inside colloidal systems. SLS and SANS are elastic scattering methods (also called static) because they monitor the time averaged scattering

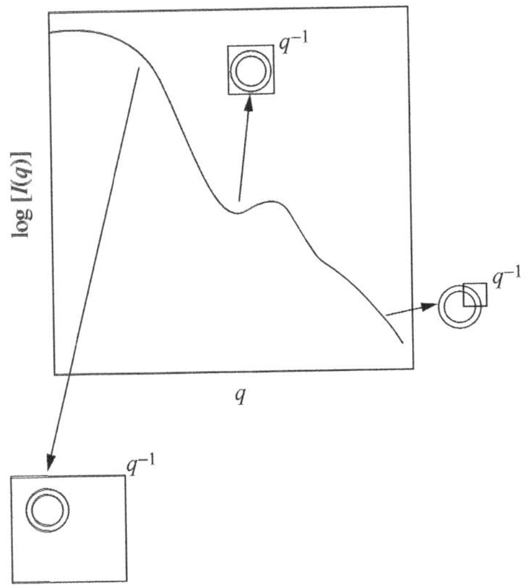

Figure 2.13 Schematic elastic scattering curve of a spherical colloid (e.g. a microemulsion droplet) in solution. As a rule of thumb q^{-1} is an approximate measure for the spatial resolution of the used scattering experiment. At low values of q (i.e. in the Guinier region of the scattering curve) the overall size and shape of the particles as well as correlations between different particles can be monitored (typically by static and dynamic light scattering). At high values q, the internal structure of the particles, i.e. the local structure of the interfacial film is resolved (e.g. by neutron or X-ray small angle scattering and neutron spin-echo spectroscopy (NSE)).

intensity (see Fig. 2.13). Photon correlation spectroscopy (DLS) and NSE are inelastic (or better quasi-elastic) scattering techniques. These methods scrutinise the energy transfer between the sample and the used radiation.

As can be seen in Fig. 2.14, the initial wavelength distribution (e.g. from a laser or a neutron source) is broadened by energy transfer with the sample (Doppler broadening) or in other words the dynamic structure factor $S(q,\omega)$ can be monitored this way. DLS and NSE both directly measure the Fourier transform of $S(q,\omega)$ the intermediate scattering function $S(q,t)$ (see Fig. 2.14). A lot of books and reviews concerned with scattering techniques have been published [80–89] and for a detailed description of the different methods the reader is referred to these publications. In the following, an introduction to the formalism of correlation functions will be given and afterwards the scattering methods used in the study of microemulsions will be briefly discussed.

2.5.2 *Space and time correlation functions*

Elastic and inelastic scattering experiments measure spatial and temporal correlations in the scattering medium and are therefore strongly related to the respective correlation functions describing the sample. This section will introduce some of the important correlation functions [90–92].

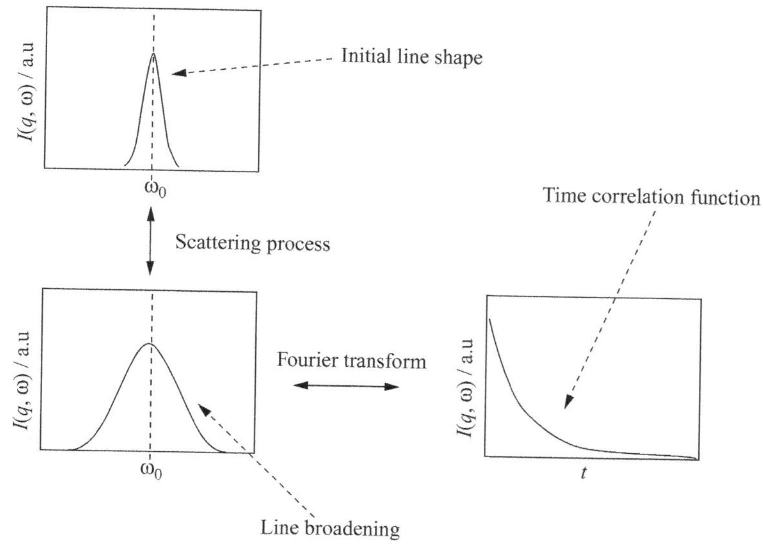

Figure 2.14 Scheme of the important processes and relationships in quasi-elastic scattering experiments. The used radiation exhibits an initial line shape, which is changed (broadened) due to the energy exchange with the thermally excited modes in the sample, e.g. centre of mass diffusion.

2.5.2.1 *Spatial correlations*

The easiestexample for establishing the relationship between the structure of a sample and its scattering behaviour is the case of monoatomic liquid [93]. In this section, the static neutron scattering function (static structure factor) will be derived. In a simple homogeneous monoatomic liquid the probability to find a certain atom in volume element d^3r at the position $\vec{r}$ is given by $P(\vec{r})d^3r$. Since the liquid is homogeneous $P(\vec{r} = \frac{1}{V}$ and

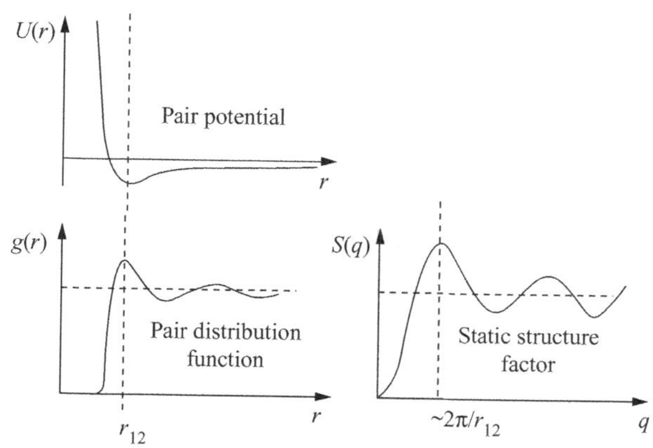

Figure 2.15 Scheme of the pair potential *U*(*r*), the pair correlation function *g*(*r*) and the static structure factor *S*(*q*).

the number density at the position $\vec{r}$ is given by

$$\rho(\vec{r}) = NP(\vec{r}) = \frac{N}{V} = \rho_0. \tag{2.24}$$

Here, N is the number of atoms. Based on this the probability to find a specific atom at r_1 and a second one at r_2 is $P(r_1,r_2)$. For non-interacting particles $P(r_1,r_2)$ can be factorised in

$$P(r_1, r_2) = P(r_1)P(r_2). \tag{2.25}$$

If a distance-dependant interaction between the atoms of the liquid occurs this can be described in terms of the pair correlation function $g(r_{12})$

$$g(r_{12}) = \frac{P(r_1,r_2)}{P(r_1)P(r_2)} \tag{2.26}$$

Using the pair distribution function $n(r_1,r_2) = N(N-1)(P(r_1,r_2)$ and the lines for large N $(N(N-1) \approx N^2$ lead to

$$g(r_{12}) = \frac{n(r_1,r_2)}{\rho_0^2}. \tag{2.27}$$

In the limit $r \rightarrow 0$, $g(r)$ vanishes due to the excluded volume of the atoms. For large values of r the pair correlation function approaches 1. The differential neutron cross-section, $(\frac{d\sigma}{d\Omega})_{\text{coherent}}$ can now be written as

$$\left(\frac{d\sigma}{d\Omega}\right)_{\text{coherent}} = |\overline{b}|^2 \left\langle \sum_{i,j=1}^{N} \exp(i\vec{q}(r_i - r_j)) \right\rangle. \tag{2.28}$$

The above expression is strictly valid only for a mono-isotopic liquid. When this condition is not fulfilled, an additional constant incoherent term would appear [92]. The $\langle\ \rangle$ brackets denote the average which can be evaluated using the pair correlation function leading to the following derivation

$$\left(\frac{d\sigma}{d\Omega}\right)_{\text{coherent}} = |\overline{b}|^2 \left(1 + \left\langle \sum_{i \neq j}^{N} \exp(i\vec{q}(r_i - r_j)) \right\rangle\right) \tag{2.29}$$

$$= |\overline{b}|^2 \left(N + \int_V d^3r_1 \int_V d^3r_2 n(r_1, r_2) \exp(i\vec{q}(r_i - r_j))\right) \tag{2.30}$$

$$= |\overline{b}|^2 \left(N + \rho_0^2 V \int_V d^3r_{12} g(r_{12}) \exp(i\vec{q}(r_i - r_j))\right) \tag{2.31}$$

$$= |\overline{b}|^2 N \left(1 + \rho_0 \int_V d^3r_{12} g(r_{12}) \exp(i\vec{q}(r_i - r_j))\right) \tag{2.32}$$

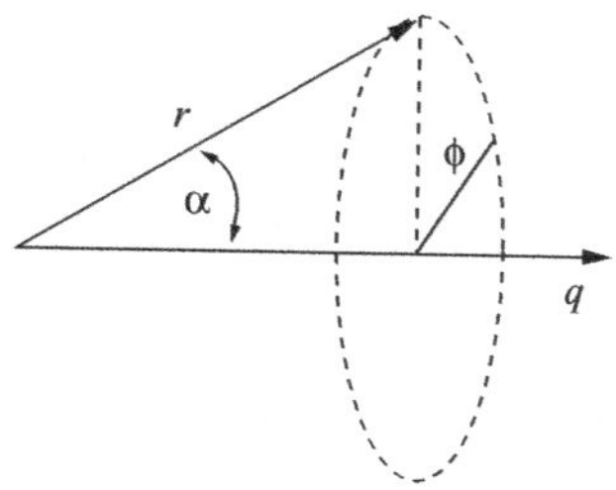

Figure 2.16 Placing the scattering centre 1 in the origin of the coordinate system the polar coordinate description of $\vec{r}$ is obtained. The radial average is first calculated over ϕ and then over α [90].

Placing now the atom 1 in the origin and expressing $d\vec{r}$ by

$$d\vec{r} = r^2 \sin\alpha \, dr \, d\alpha \, d\phi \tag{2.33}$$

one obtains (see Fig. 2.16)

$$\left(\frac{d\sigma}{d\Omega}\right)_{\text{coherent}} = \left|\bar{b}\right|^2 N\left(1 + \rho_0 \int_0^\infty dr \int_0^\pi d\alpha \int_0^{2\pi} d\phi \times \exp(iqr\cos\alpha) g(r) r^2 \sin\alpha\right). \tag{2.34}$$

Using now the substitution $x = \cos\alpha$ leads to

$$\left(\frac{d\sigma}{d\Omega}\right)_{\text{coherent}} = \left|\bar{b}\right|^2 N\left(1 + \rho_0 \int_0^\infty dr \int_{-1}^{+1} dx \int_0^{2\pi} d\phi \times \exp(iqr\cos\alpha) g(r) r^2\right). \tag{2.35}$$

since $dx = -d\alpha \sin\alpha$ and the sine factor drops out. Averaging now about ϕ and making use of the Euler equation one arrives at

$$\left(\frac{d\sigma}{d\Omega}\right)_{\text{coherent}} = \left|\bar{b}\right|^2 N\left(1 + \rho_0 \int_0^\infty dr \int_{-1}^{+1} dx[2\pi r^2 g(r)(\cos qrx + i\sin qrx)]\right). \tag{2.36}$$

Separating the two terms in the argument of the integral leads to

$$\left(\frac{d\sigma}{d\Omega}\right)_{\text{coherent}} = \left|\bar{b}\right|^2 N\left(1 + \rho_0 \int_0^\infty dr \int_{-1}^{+1} dx[2\pi r^2 g(r)(\cos qrx)] + i\int_{-1}^{+1} dx \sin qrx\right). \tag{2.37}$$

A symmetric integral of a sine function is zero and one obtains

$$\left(\frac{d\sigma}{d\Omega}\right)_{\text{coherent}} = \left|\bar{b}\right|^2 N\left(1 + \rho_0 \int_0^\infty dr \left[4\pi r^2 g(r)\frac{\sin qr}{qr}\right]\right) \tag{2.38}$$

after substitution of x with $\cos\alpha$. The term in the big parentheses is the so-called static structure factor, $S(q)$, of the scattering sample and hence, is the Fourier transform of the pair correlation function. For large q, the oscillations of the integral part decay to zero and the high q limit of $S(q)$ is 1. At low values of q, the static structure factor provides information about density fluctuations in the system. The relation between the pair potential, the pair distribution function and $S(q)$ is graphically shown in Fig. 2.15.

The above derivation can also be applied to colloidal or polymer-based liquids and is then used to calculate the so-called form factors of soft matter samples. The major difference between a monoatomic liquid and a polymer chain in the melt or in solution is that the total structure factor consists of two parts. The first is the inter-particle structure factor and the second the intra-particle structure factor. This second part is also often called the particle form factor $P(q)$. Using Eq. (2.38) it is straightforward to calculate $P(q)$ for a given soft matter sample. A good example is the form factor of a single polymer coil in a melt [88, 92]. The pair correlation function of such a coil is given by

$$g(r_{ij}) = \left(\frac{3}{2\pi\langle r_{ij}{}^2\rangle}\right)\exp\frac{-3r_{ij}{}^2}{2\langle r_{ij}{}^2\rangle}. \tag{2.39}$$

Using this expression for $g(r_{ij})$ in Eq. (2.38) leads to

$$P(q) = \frac{1}{N}\sum_{i,j=1}^{N} 4\pi\int r_{ij}d^3r_{ij}\frac{\sin qr_{ij}}{qr_{ij}}g(r_{ij}) \tag{2.40}$$

This equation can be simplified to

$$P(q) = \sum_{i,j=1}^{N}\exp\frac{q^2}{6}|i-j|l^2. \tag{2.41}$$

Here, $|i-j|l^2 = \langle r_{ij}^2\rangle$ is the mean squared monomer distance between the monomers i and j. N is the number of monomers (degree of polymerisation) of the coil. This expression is only valid for $\frac{2\pi}{q} >$ the size of a monomer, since the monomers are treated as a kind of 'big' atom here. In terms of the difference $k = |i-j|$ the above equation can now be converted into

$$P(p) = \frac{1}{N}\left[1 + 2\sum_{k=1}^{N}\left(1-\frac{k}{N}\right)\exp-\frac{q^2}{6}kl\right] \tag{2.42}$$

Converting this sum into an integral leads to the well-known Debye function $D(x)$

$$P(q) = \frac{2}{x^2}(e^{-x} - 1 + x) = D(x). \tag{2.43}$$

In this expression x is given by

$$x = \frac{q^2Nl^2}{6}. \tag{2.44}$$

The Debye function is the scattering form factor of a single polymer chain in a melt [88, 92]. Experimentally, this function can be observed using the method of contrast variation in SANS. A calculation similar to the one presented above can be done for soft matter objects of arbitrary shape and a lot of form factors were already derived. A good review for different form factors was given by Pedersen [94].

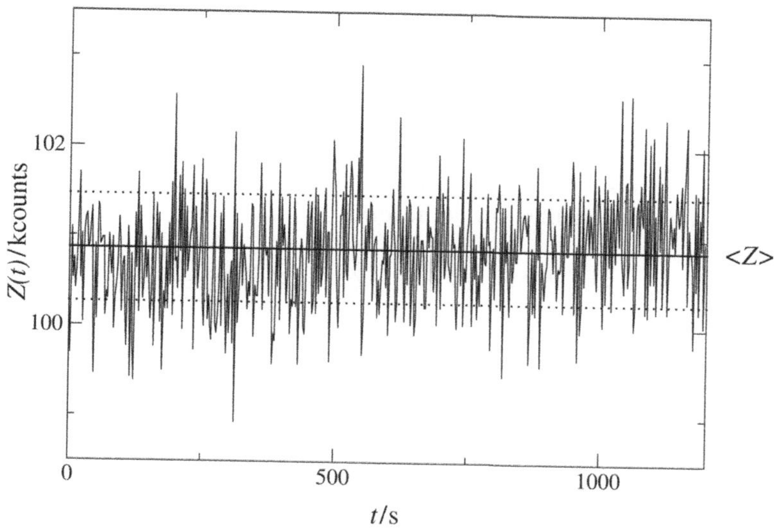

Figure 2.17 Intensity Z versus time t for a typical dynamic light scattering experiment. The signal fluctuates due to density fluctuations in the sample [80].

2.5.2.2 Time correlations

Correlation spectroscopy

In correlation spectroscopy (e.g. photon correlation spectroscopy) the intensity scattered by a sample is measured using counters [80, 86, 95, 96]. A typical example for the obtained signal is shown in Fig. 2.17.

$$\langle Z(0)Z(\tau)\rangle = \lim_{T\to\infty}\frac{1}{T}\int_0^T Z(t)Z(t+\tau)dt \tag{2.45}$$

$$\langle Z(0)Z(\tau)\rangle = \lim_{N\to\infty}\frac{1}{N}\sum_{j=1}^{N} Z_j Z_{j+n} \tag{2.46}$$

$$\tau = n\Delta t \tag{2.47}$$

$$t = j\Delta t \tag{2.48}$$

From Eq. (2.46) follows for the case $\langle Z(0)Z(0)\rangle$

$$\langle Z(0)Z(0)\rangle = \lim_{N\to\infty}\frac{1}{N}\sum Z_j^2. \tag{2.49}$$

Using the Schwartz inequality it can be shown that [80]

$$\sum_{j=1}^{N} Z_j^2 \geq \sum_{j=1}^{N} Z_j Z_{j+n} \Rightarrow \langle Z(0)^2\rangle \geq \langle Z(0)Z(\tau)\rangle. \tag{2.50}$$

A similar approach for $\langle Z(0)Z(\tau)\rangle$ in the limit $\tau \to \infty$ yields

$$\lim_{\tau\to\infty} \langle Z(0)Z(\tau)\rangle = \langle Z(0)\rangle \cdot \langle Z(\tau)\rangle = \langle Z\rangle^2. \tag{2.51}$$

That means for large τ the correlation function decays towards the static average of the quantity Z. For systems exhibiting simple dynamics the decay from $\langle Z^2\rangle$ to $\langle Z\rangle^2$ can be described by a single exponential law of the form

$$\langle Z(0)Z(t)\rangle = \langle Z\rangle^2 + (\langle Z^2\rangle - \langle Z\rangle^2)\exp\frac{-t}{\tau_r}. \tag{2.52}$$

However, in real systems the decay is more complex due to polydispersity effects and different types of motions which can contribute. A different way to formulate the above function is based on the fluctuation of Z given by $\delta Z(t) = Z(t) - \langle Z\rangle$. Using this expression leads to

$$\langle \delta Z(0)\delta Z(t)\rangle = \langle Z(0)Z(t)\rangle - \langle Z\rangle^2 = \langle \delta Z^2\rangle \exp\frac{-t}{\tau_r} \tag{2.53}$$

Dynamic light scattering

The intensity I of the light scattered from a dilute macromolecular or supra-molecular solution is a fluctuating quantity due to the Brownian motion of the scattering particles. These fluctuations can be analysed in terms of the normalised autocorrelation function $g^1(\tau)$ of the scattered electrical field E_s, which contains information about the structure and the dynamics of the scattering particles [80].

$$g^1(\tau) = \frac{\langle E_s^*(t)E_s(t+\tau)\rangle}{\langle I\rangle}. \tag{2.54}$$

Here, E_s^* is the complex conjugated of E_s. Experimentally, the intensity correlation function $g^2(\tau)$ is determined,

$$g^2(\tau) = \frac{\langle E_s^*(t)E_s(t)E_s^*(t+\tau)E_s(t+\tau)\rangle}{\langle I^2\rangle} \tag{2.55}$$

which is related to $g^1(\tau)$ by the Siegert relation [80]

$$g^2(\tau) = 1 + C|g^1(\tau)|^2. \tag{2.56}$$

C is a coherence factor and depends on the experimental conditions. For an ideal solution of mono-disperse particles the function $g^1(\tau)$ is represented by a single exponential

$$g^1(\tau) = \exp(-\Gamma\tau). \tag{2.57}$$

The relaxation rate Γ is connected with the translational diffusion coefficient D according to

$$\Gamma = Dq^2, \tag{2.58}$$

with the scattering vector $q = (4\pi n_0/\lambda)\sin(\theta/2)$. The scattering vector q depends on the wavelength λ of the incident light and the scattering angle θ. For polydisperse samples the function $g^1(\tau)$ is given by a weighted sum of exponentials.

$$g^{-1}(\tau) = \int_0^\infty G(\Gamma)\exp(-\Gamma\tau)d\Gamma. \tag{2.59}$$

The function $g^1(\tau)$ can be analysed by the method of cumulants [57] or by inverse Laplace transformation. These methods provide the mean relaxation rate $\overline{\Gamma}$ of the distribution function $G(\Gamma)$ (z-average). For the second analysis procedure mentioned above, the FORTRAN program CONTIN is available [97, 98]. It is sometimes difficult to avoid the presence of spurious amounts of dust particles or high molecular weight impurities that give small contributions to the long time tail of the experimental correlation functions. With CONTIN it is possible to discriminate these artifacts from the relevant relaxation mode contributing to $g^1(\tau)$.

The analysis of the light scattering data using CONTIN also allows for a determination of the size polydispersity of the microemulsion droplets, because all the moments $\mu_n = \int_{\Gamma_{\min}}^{\Gamma_{\max}} G(\Gamma)\Gamma^n d\Gamma$ which describe the distribution function $G(\Gamma)$ are computed (for details see Ref. [98]). The polydispersity index is obtained from

$$p^2 = \frac{\left(\frac{\mu_2}{\mu_0}\right) - \left(\frac{\mu_1}{\mu_0}\right)^2}{\left(\frac{\mu_1}{\mu_0}\right)^2}. \tag{2.60}$$

Using Eq. (2.58), from the mean relaxation rate $\overline{\Gamma}$ the average apparent translational diffusion coefficient D can be calculated. The measured apparent diffusion coefficient D depends on the concentration $[C]$ of the scattering particles. When $[C]$ is not too large ($\Phi \le 0.1$), one has

$$D = D_0(1 + k_D[C]), \tag{2.61}$$

where the diffusional virial coefficient k_D includes thermodynamic and frictional effects on D. If the interactions between the particles are negligible, k_D becomes zero and D is equal to D_0. For dilute microemulsions (small $[C]$) stabilised by uncharged surfactants D is also close to D_0 [52, 99, 100]. Knowing the value of D_0, the hydrodynamic radius of the scattering particles, R_H, can be calculated by the Stokes–Einstein equation

$$D_0 \frac{kT}{6\pi\eta R_H} \tag{2.62}$$

with η = viscosity of the solvent (the continuous phase in the case of microemulsions).

The van Hove correlation function and the dynamic structure factor

The following section is mainly based on the textbook by Egelstaff [93] and in parts on other standard texts. Again, we use the monoatomic liquid as example. In Section 2.5.2.1, only the spatial correlation of the different positions of the atoms in a liquid were discussed.

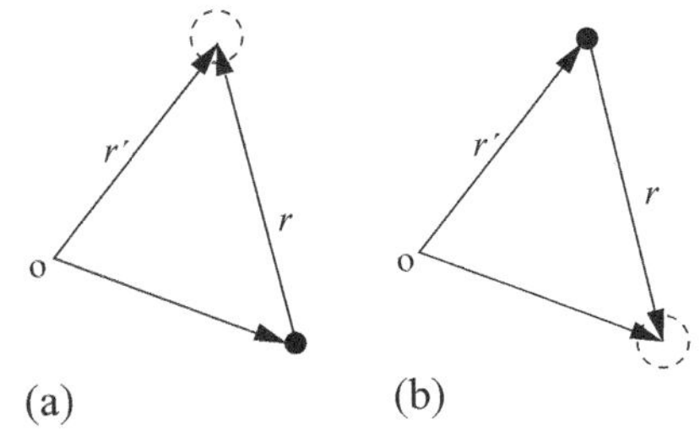

Figure 2.18 Schematic explanation of the meaning of the van Hove correlation function. Here, O indicates the origin. (a) At $t = 0$, there is a scattering centre at the position indicated by the full circle. The dashed circle indicates a position where there may or may not be such a scattering centre. (b) At time t, there is a scattering centre at the position marked by the full circle, while again there may be or may not be a scattering centre at the position of the dashed circle.

However, since the particles in a liquid are moving permanently for a complete treatment, also a discussion of the particle momenta $\vec{p}$ is required. This can be done using the general pair distribution function $F(\vec{r}_1, \vec{p}_1, t_1; \vec{r}_2, \vec{p}_2, t_2)$. F is the probability of finding and a particle at $\vec{r}_2$ with momentum $\vec{p}_2$ at time t_2 when there was an atom at $\vec{r}_1$ at time t_1 which had the momentum $\vec{p}_1$. A simplification of this general pair distribution function is obtained by integrating over the momenta

$$G(\vec{r}_1 - \vec{r}_2, t_2 - t_1) = \int\int F(\vec{r}_1, \vec{p}_1, t_1; \vec{r}_2, \vec{p}_2, t_2) d\vec{p}_1 d\vec{p}_2. \tag{2.63}$$

Hence, G is proportional to the probability of finding a scattering centre at r_2 at time t_2 if there was such a centre at (r_1, t_1). Because of the uniformity and stationarity of a liquid in equilibrium the quantity G only depends on the differences $\vec{r}_1 - \vec{r}_2 = \vec{r}$ and $t_2 - t_1 = t.G(\vec{r}, t)$ is the so-called van Hove correlation function. Figure 2.18 tries to illustrate the meaning of $G(\vec{r}, t)$. The situation in Fig. 2.18(a) can be described by the delta function $\delta[\vec{r} + r_i(0) - \vec{r}']$. In Fig. 2.18(b) the mathematical description is $\delta[\vec{r}' - r_j(t)]$. In other words, Fig. 2.18(a) shows the position of the ith scattering centre at $\vec{r}' - \vec{r} = \vec{r}_i(0)$ and Fig. 2.18(b) the position of the jth at $\vec{r}' = \vec{r}_j(t)$ at time t. Combining both delta functions leads to

$$\delta[\vec{r}' - r_j(t)] \cdot \delta[\vec{r}' - r_j(t)] \ . \tag{2.64}$$

Since the position of the origin O can be arbitrarily chosen in the volume V of the liquid this expression has to be integrated over $\vec{r}'$

$$\int_V d\vec{r}' \delta[\vec{r}' - r_j(t)] \cdot \delta[\vec{r}' - r_j(t)]. \tag{2.65}$$

The next step in the derivation of $G(\vec{r}, t)$ is the summation about all possible scattering centres i, j

$$\frac{1}{N} \sum_{ij} \int_V d\vec{r}' \delta[\vec{r}' - r_j(t)] \delta[\vec{r}' - r_j(t)]. \tag{2.66}$$

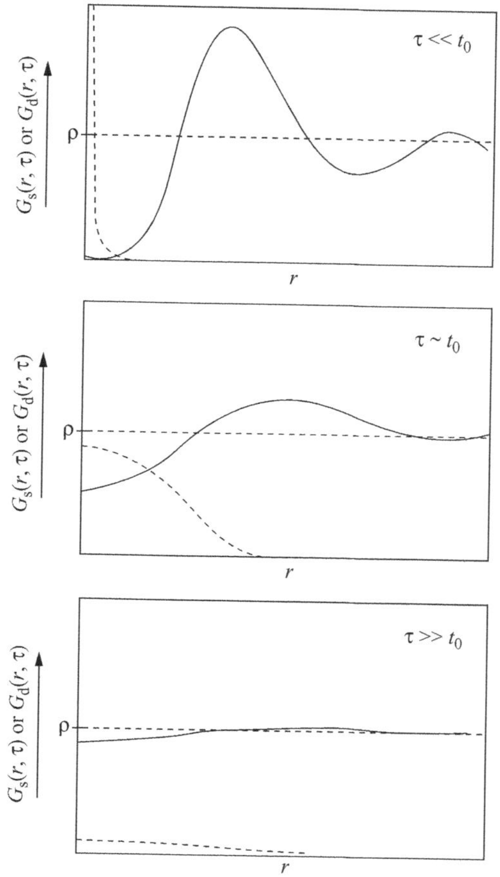

Figure 2.19 Behaviour of the self (dashed line) and the distinct part (solid line) of the van Hove pair correlation function for different values of t, where t_0 is the characteristic relaxation time of the respective liquid under investigation.

Taking now the thermal average the definition of $G(\vec{r}, t)$ reads as

$$G(\vec{r}, t) = \frac{1}{N}\left\langle \sum_{ij} \int_V d\vec{r}'\delta[\vec{r}' - r_j(t)] \cdot \delta[\vec{r}' - r_j(t)] \right\rangle. \tag{2.67}$$

For $t = 0$, the van Hove pair correlation function is directly connected to the static pair distribution function $g(r)$.

$$G(\vec{r}, 0) = \delta(r) + \rho g(r) = G_s(\vec{r}, 0) + G_d(\vec{r}, 0) \tag{2.68}$$

Here, $G_s(\vec{r}, 0)$ represents the self-part of $G(\vec{r}, t)$ corresponding to the cases where $i = j$. $G_d(\vec{r}, 0)$ is the distinct part containing the contributions from $i \neq j$. $G(\vec{r}, t)$ is schematically drawn in Fig. 2.19. In some cases it is useful to write $G(\vec{r}, t)$ as a function of the density of the liquid.

$$\rho(\vec{r},t) = \sum_i \delta[\vec{r} - \vec{r}_i(t)] \tag{2.69}$$

This leads to

$$G(\vec{r},t) = \frac{1}{N}\int d\vec{r}'\langle\rho(\vec{r}-\vec{r}',0)\rho(\vec{r}',t)\rangle. \tag{2.70}$$

The van Hove correlation function can be easily related to quasi- and inelastic scattering experiments. In the following this relation will be derived for the case of neutron scattering.

In inelastic or quasi-elastic scattering experiments, the quantity of interest is the double differential scattering cross-section $\frac{\partial^2\sigma}{\partial\Omega\partial\omega}$, which is also often called dynamic structure factor, $S(\vec{q},\omega)$. The relation between $\frac{\partial^2\sigma}{\partial\Omega\partial\omega}$ and $G(\vec{r},t)$ can be made clear by calculating the Doppler broadening for the scattering of a wave by a set of moving scattering centres in a liquid.

The following part is focusing on neutron scattering. Since neutrons are scattered by the nuclei in important quantity to describe the interaction between the neutron and the sample is the Fermi pseudo-potential given by

$$V(r') = \frac{2\pi\hbar^2}{m_n}\sum_i b_i\delta(\vec{r}' - \vec{r}_i(t_1)) \tag{2.71}$$

m_n is the mass of the neutron and b_i the scattering length of the ith nucleus. The scattering process is related to an energy change of the system from the energy E' to the energy E''. This energy change of the scattering system corresponds to a change in neutron energy of

$$E_0 - E = \hbar\omega\ . \tag{2.72}$$

Since the total energy of the system and the neutron has to be conserved this leads to

$$\delta(E_0 - E - \hbar\omega) = \frac{1}{2\pi}\int \exp(-i\omega t)\exp[it(E_0 - E)\hbar]dt \tag{2.73}$$

$$= \frac{1}{2\pi}\int \exp(-i\omega t)\exp\left[\frac{it(E' - E'')}{\hbar}\right]dt \tag{2.74}$$

The factor $\exp[\frac{it(E'-E'')}{\hbar}]$ arises from the dynamics in the scattering sample, while the first factor appears in the scattering amplitude. Introducing this scattering amplitude and accounting for the neutron flux in a given energy range leads to

$$\frac{\partial^2\sigma}{\partial\Omega\partial\omega} = \frac{k}{2\pi Nk_0}\int\int \exp[i(\vec{q}\vec{r} - \omega t)dr\,dt$$
$$\times\left\langle\sum_{ij}\int_V b_i\delta[\vec{r} + \vec{r}_i(0) - \vec{r}']b_j\delta[\vec{r}' - \vec{r}_j(t)]d\vec{r}'\right\rangle. \tag{2.75}$$

Here, k_0 and k are the neutron velocities before and after the scattering process. Looking at Eq. (2.75) immediately reveals the relation to $G(\vec{r}, t)$. The double differential scattering cross-section is directly related to the Fourier transform of the van Hove pair correlation function, which is given by

$$\frac{\partial^2\sigma}{\partial\Omega\partial\omega} = S(\vec{q}, \omega) = \frac{1}{2\pi}\int\int \exp\left[-i\left(\vec{q}\vec{r} - \omega t\right)\right]\left[G(\vec{r}, t) - \rho\right]d\vec{r}\,dt. \tag{2.76}$$

G is corrected by the subtraction of a constant ρ to exclude a $\delta(q)$ term from $S(q,\omega)$. To establish the relationship in Eq. (2.75) the nuclear scattering lengths have to be expressed in terms of the coherent and the incoherent scattering lengths. For this purpose the following equation can be used

$$\langle b_i b_j\rangle = b_{\text{coh}}^2 - b_{\text{incoh}}^2\delta_{ij}. \tag{2.77}$$

Using this expression allows to re-write Eq. (2.75) in the following way

$$\begin{aligned}\frac{\partial^2\sigma}{\partial\Omega\partial\omega} &= \frac{k}{2\pi N k_0}\int\int \exp[i(\vec{q}\vec{r} - \omega t)\left[b_{\text{coh}}^2 G(\vec{r}, t) + b_{\text{incoh}}^2 G_s(\vec{r}, t)\right] dr\,dt \\ &= \frac{k}{k_0}\left[b_{\text{coh}}^2 S(q, \omega) + b_{\text{incoh}}^2 S_s(q, \omega)\right] + \delta(q)\ldots\end{aligned} \tag{2.78}$$

This equation reveals the unique possibility to directly measure the self-part of the van Hove correlation function with neutrons. Since light scattering is also used in the present book the equivalent relationship for light will be briefly mentioned here too. For light scattering only small values of the wave number are significant. In this limit it is possible to show that

$$\frac{\partial^2\sigma}{\partial\Omega\partial\omega} \approx b_L^2 S(q, \omega) \tag{2.79}$$

where

$$b_L \approx \frac{k_0^2}{4\pi}\left(\frac{d\varepsilon(\omega_L)}{\partial\rho}\right)_T. \tag{2.80}$$

In this case k_0 and ω_L are the wave number and the frequency of the initial beam of light. The relationship is only valid when the incident and the scattered beam are vertically polarised. Because of the fact that the scattering of electromagnetic radiation is coherent, the self-structure factor $S_s(q,\omega)$ cannot be measured in a light scattering experiment. The discussion given above is strictly valid for monoatomic liquids; however, the basic principles can also be applied to colloidal systems with rather large scattering centres. A derivation of the dynamic structure factor in quantum mechanical notation can be found in Ref. [92].

Neutron spin-echo spectroscopy

Neutron spin-echo is an experimental technique which allows for quasi-elastic neutron scattering experiments with an extremely high energy resolution (in the range of neV). This is achieved by exploiting the magnetic moment of the neutrons. In a neutron spin-echo spectrometer, polarised neutrons are flying in a first set of coils thereby undergoing Larmor precession. Then they are scattered by a sample, inverted in spin and passed through a second set of coils. The scattering process leads thereby to a change in the speed of the neutrons and due to this, to a difference in the number of precessions before and after the sample if the magnetic fields on both sides of the sample are exactly identical to each other. The angle $\Delta\psi$ between the spin of the incident neutrons and the neutrons finally reaching the detector is analysed using the scattered intensity.

$$P_s = P_0 \cos(\Delta\psi) \tag{2.81}$$

Here, P_0 is the detected intensity if no sample is present. The probability for a scattering process with an energy transfer ω is described by the dynamic structure factor $S(q,\omega)$. Averaging $\cos(\Delta\psi)$ with $S(q,\omega)$ leads to

$$P_s = \langle\cos(\Delta\psi)\rangle = \frac{\int S(q,\omega)\cos(t\Delta\omega)d\omega}{\int S(q,\omega)d\omega}, \tag{2.82}$$

with $\Delta\psi = t\Delta\omega$. This Fourier transform in time can be rewritten as

$$P_s(q,t) = \frac{I(q,t)}{I(q,0)}. \tag{2.83}$$

The Fourier time is given by

$$t = \frac{\gamma l_0 H_0 m^2}{2\pi h^2}\lambda^3. \tag{2.84}$$

Here, λ is the wavelength of the neutrons, γ the gyro-magnetic ratio and $l_0 H_0$ the integral of the magnetic field along the neutrons path [101–104].

References

1. Hayter, J.B. and Penfold, J. (1983) Determination of micelle structure and charge by neutron small-angle scattering. *Colloid Polym. Sci.*, **261**, 1022–1030.
2. Hoffmann, H. and Ulbricht, W. (1989) Transition of rodlike to globular micelles by the solubilization of additives. *J. Coll. Interface Sci.*, **129**, 388–405.
3. Hellweg, Th. and Eimer, W. (1998) The micro-structures formed by Ni^{2+} -AOT/cyclohexane/water microemulsions: A light scattering study. *Colloids Surf. A*, **136**, 97–107.
4. Tlusty, T., Safran, S.A. and Strey, R. (2000) Topology, phase instabilities, and wetting of microemulsion networks. *Phys. Rev. Lett.*, **84**, 1244–1247.
5. Hellweg, Th. and von Klitzing, R. (2000) Structural changes and complex dynamics in the single phase region of a dodecane/$C_{12}E_5$/water microemulsion: A dynamic light scattering study. *Physica A*, **283**, 349–358.
6. Scriven, L.E. (1976) Equilibrium bicontinuous structure. *Nature*, **263**, 123–125.
7. Israelachvili, J., Mitchell, D.J. and Ninham, B.W. (1976) Theory of self-assembly of hydrocarbon amphiphiles into micelles and bilayers. *J. Chem. Soc. Faraday Trans.*, **72**, 1525–1568.
8. Helfrich, W. (1973) Elastic properties of lipid bilayers: Theory and possible experiments. *Z. Naturforschung.*, **28c**, 693–703.

9. Huse, D.A. and Leibler, S. (1988) Phase behaviour of an ensemble of nonintersecting random fluid films. *J. Phys. France*, **49**, 605–621.
10. Safinya, C.R., Roux, D., Smith, G.S., Sinha, S.K., Dimon, P., Clark, N.A. and Bellocq, A.M. (1986) Steric interactions in a model multimembrane system: A synchrotron X-ray study. *Phys. Rev. Lett.*, **57**, 2718–2721.
11. Roux, D. and Safinya, C.R. (1988) A synchrotron X-ray study of competing undulation and electrostatic interactions in fluid multimembrane lyotropic phases. *J. Phys. France*, **49**, 307–318.
12. Warriner, H., Keller, S., Idziak, S.H.J., Slack, N.L., Davidson, P., Zasadzinski, J.A. and Safinya, C.R. (1998) The influence of polymer molecular weight in lamellar gels based on PEG-lipids. *Biophys. J.*, **75**, 272–293.
13. Faucon, F.F., Mitov, M.D., Meleard, P., Bivas, I. and Bothorel, P. (1995) Bending elasticity and thermal fluctuations of lipid-membranes – theoretical and experimental requirements. *J. Phys. II France*, **5**, 413–425.
14. Méléard, P., Gerbeaud, C., Pott, T., Fernandez-Puente, L., Bivas, I., Mitov, M.D., Dufourcq, J. and Bothorel, P. (1997) Bending elasticities of model membranes: Influences of temperature and sterol content. *Biophys. J.*, **72**, 2616–2629.
15. Nallet, F., Roux, D. and Prost, J. (1989) Dynamic light scattering study of dilute lamellar phases. *Phys. Rev. Lett.*, **62**, 276–279.
16. Freyssingeas, É., Nallet, F. and Roux, D. (1996) Measurement of the membrane flexibility in lamellar and "Sponge" phases of the $C_12\ E_5$/Hexanol/water system. *Langmuir*, **12**, 6028–6035.
17. Pfeiffer, W., König, S., Legrand, J.F., Bayerl, T., Richter, D. and Sackmann, E. (1993) Neutron spin-echo study of membrane undulations in lipid multibilayers. *Europhys. Lett.*, **23**, 457–462.
18. Farago, B., Monkenbusch, M., Goecking, K.D., Richter, D. and Huang, J.S. (1995) Dynamics of microemulsions as seen by neutron spin echo. *Physica B*, **213/214**, 712–717.
19. Binks, B.P., Meunier, J., Abillon, O. and Langevin, D. (1989) Measurement of film rigidity and interfacial tensions in several ionic surfactant–oil–water microemulsion systems. *Langmuir*, **5**, 415–421.
20. Abillon, O., Lee, L.T., Langevin, D. and Wong, K. (1991) Microemulsions: Structures, surfactant layer properties and wetting transitions. *Physica A*, **172**, 209–218.
21. Burauer, S., Belkoura, L., Stubenrauch, C. and Strey, R. (2003) Bicontinuous microemulsions revisited: A new approach to freeze fracture electron microscopy (FFEM). *Coll. Surf. A*, **228**, 159–170.
22. Regev, O., Ezrahi, S., Aserin, A., Garti, N., Wachtel, E., Kaler, E.W., Kahn, A. and Talmon, Y. (1996) A study of the microstructure of a four-component nonionic microemulsion by Cryo-TEM, NMR, SAXS, and SANS. *Langmuir*, **12**, 668–674.
23. Ponsinet, V. and Talmon, Y. (1997) Direct imaging of lamellar phases by cryo-transmission electron microscopy. *Langmuir*, **13**, 7287–7292.
24. Bernhein-Grosswasser, A., Tlusty, T., Safran, S.A. and Talmon, Y. (1999) Direct observation of phase separation in microemulsion networks. *Langmuir*, **15**, 5448–5453.
25. Bernheim-Grosswasser, A., Wachtel, E. and Talmon, Y. (2000) Micellar growth, network formation, and criticality in aqueous solutions of the nonionic surfactant $C_{12}\ E_5$. *Langmuir*, **16**, 4131–4140.
26. Belkoura, L., Stubenrauch, C. and Strey, R. (2004) Freeze fracture direct imaging: A new freeze fracture method for specimen preparation in cryo-transmission electron microscopy. *Langmuir*, **20**, 4391–4399.
27. Roux, D., Nallet, F., Freyssingeas, E., Porte, G., Bassereau, P., Skouri, M. and Marignan, J. (1992) Excess area in fluctuating-membrane systems. *Europhys. Lett.*, **17**, 575–581.
28. Niggemann, G., Kummrow, M. and Helfrich, W. (1995) The bending rigidity of phosphatidylcholine bilayers – Dependences on experimental method, sample cell sealing, and temperature. *J. Phys. II France*, **5**, 413–425.

29. Hellweg, Th. and Langevin, D. (1998) Bending elasticity of the surfactant film in droplet microemulsions: Determination by a combination of dynamic light scattering and neutron spin-echo spectroscopy. *Phys. Rev. E*, **57**, 6825–6834.
30. Strey, R., Schomäcker, R., Roux, D., Nallet, F. and Olsson, U. (1990) Dilute lamellar and L_3 phases in the binary water–C_12E_5 system. *J. Chem. Soc. Faraday Trans.*, **86**, 2253–2261.
31. Nallet, F., Laversanne, R. and Roux, D. (1993) Modelling X-ray or neutron scattering spectra of lyotropic lamellar phases: Interplay between form and structure factor. *J. Phys. II France*, **3**, 487–502.
32. Antelmi, D.A., Kekicheff, P. and Richetti, P. (1999) The ionfinement-induced sponge to lamellar phase transition. *Langmuir*, **15**, 7774–7788.
33. Deme, B., Dubois, M., Gulik-Krzywicki, T. and Zemb, Th. (2002) Giant collective fluctuations of charged membranes at the lamellar-to-vesicle unbinding transition. 1. Characterisation of a new lipid morphology by SANS, SAXS, and electron microscopy. *Langmuir*, **18**, 997–1004.
34. Guest, D. and Langevin, D. (1986) Light scattering study of a multiphase microemulsion system. *J. Coll. Inter. Sci.*, **112**, 208–220.
35. Attwood, D. and Ktistis, G. (1989) A light scattering study on oil-in-water microemulsions. *Int. J. Pharma.*, **52**, 165–171.
36. Sheu, E.Y., Chen, S.-H., Huang, J.S. and Sung, J.C. (1989) Nonexponential relaxations in dense microemulsion near the glasslike transition. *Phys. Rev. A*, **39**, 5867–5876.
37. Dunn, C.M., Robinson, B.H. and Leng, F.J. (1990) Photon-correlation spectroscopy applied to the size characterisation of water-in-oil microemulsion systems stabilised by Aerosol-OT; effect of change in counterion. *Spectrochim. Acta*, **46A**, 1017–1025.
38. Aoudia, M., Rodgers, M.A.J. and Wade, W.H. (1991) Light scattering and fluorescence studies of o/w microemulsion: The sodium 4-dodecylbenzenesulfonate–butanol–water–NaCl–Octane system. *J. Coll. Int. Sci.*, **144**, 353–362.
39. Menge, U., Lang, P. and Findenegg, G.H. (2000) Influence of temperature and oil-to-surfactant ratio on micellar growth in aqueous solutions of $C_{12}E_5$ with decane. *Colloids Surf. A*, **163**, 81–90.
40. Evilevitch, A., Lobaskin, V., Olsson, U., Linse, P. and Schurtenberger, P. (2001) Structure and transport properties of a charged spherical microemulsion system. *Langmuir*, **17**, 1043–1053.
41. Glatter, O., Orthaber, D. Stradner, A., Scherf, G. Fanun, M., Garti, N., Clement, V. and Leser, M.E. (2001) Sugar-ester nonionic microemulsion: Structural characterization. *J. Coll. Inter. Sci.*, **241**, 215–225.
42. Safran, S.A. (1983) Fluctuations of spherical microemulsions. *J. Chem. Phys.*, **78**, 2073–2076.
43. Milner, S.T. and Safran, S.A. (1987), Dynamical fluctuations of droplet microemulsions and vesicles. *Phys. Rev. A*, **36**, 4371–4379.
44. Safran, S.A. (1991) Saddle-splay modulus and the stability of spherical microemulsions. *Phys. Rev. A*, **43**, 2903–2904.
45. Huang, J.S., Milner, S.T., Farago, B. and Richter, D. (1987) Study of dynamics of microemulsion droplets by neutron spin-echo spectroscopy. *Phys. Rev. Lett.*, **59**, 2600–2603.
46. Farago, B., Richter, D., Huang, J.S., Safran, S.A. and Milner, S.T. (1990) Shape and size fluctuations of microemulsion droplets: The role of cosurfactant. *Phys. Rev. Lett.*, **65**, 3348–3351.
47. Farago, B. (1996) Spin echo studies of microemulsions. *Physica B*, **226**, 51–55.
48. Varea, C. and Robledo, A. (1995) Bending rigidities and spontaneous curvature for a spherical interface. *Physica A*, **220**, 33–47.
49. Hellweg, Th. and Langevin, D. (1999) The dynamics in dodecane/$C_{10}E_5$/water microemulsions determined by time resolved scattering techniques. *Physica A*, **264**, 370–387.
50. Hellweg, T., Gradzielski, M., Farago, B. and Langevin, D. (2001) Shape fluctuation of microemulsion droplets: A neutron spin-echo study. *Colloids Surf. A*, **183–185**, 159–169.
51. Gradzielski, M., Hoffmann, H. and Langevin, D. (1995) Solubilisation of decane into the ternary system TDMAO/1-Hexanol/Water. *J. Chem. Phys.*, **99**, 12612–12623.

52. Gradzielski, M., Langevin, D. and Farago, B. (1996) Experimental investigation of the structure of nonionic microemulsions and their relation to the bending elasticity of the amphiphilic film. *Phys. Rev. E*, **53**, 3900–3919.
53. Leitao, H. Somoza, A.M., Telo da Gama, M.M., Sottmann, T. and Strey, R. (1996) Scaling of interfacial tension of microemulsions: A phenomenological description. *J. Chem. Phys.*, **105**, 2875–2883.
54. Sottmann, T. and Strey, R. (1997) Ultralow interfacial tensions in water–*n*-alkane–surfactant systems. *J. Chem. Phys.*, **106**, 8606–8615.
55. Komura, S. and Seki, K. (1993) Dynamical fluctuations of spherically closed fluid membranes. *Physica A*, **192**, 27–46.
56. Papoular, R.J. and Livesey, A.K. (1989) Quasielastic neutron scattering data evaluation using the maximum entropy method. In J. Skilling (ed), *Maximum Entropy and Bayesian Methods*. Kluwer Academic Publishers, Dordrecht, pp. 163–173.
57. Koppel, D.E. (1972) Analysis of macromolecular polydispersity in intensity correlation spectroscopy: The method of cumulants. *J. Chem. Phys.*, **57**, 4814–4820.
58. Sorlie, S.S. and Pecora, R. (1988) A dynamic light scattering study of a 2311 base pair DNA restriction fragment. *Macromolecules*, **21**, 1437–1449.
59. Gradzielski, M., Langevin, D., Magid, L. and Strey, R. (1995) Small-angle neutron scattering from diffuse interfaces. 2. Polydisperse shells in water–*n*-alkane–$C_{10}E_4$ microemulsions. *J. Phys. Chem.*, **99**, 13232–13238.
60. Gradzielski, M., Langevin, D., Sottmann, T. and Strey, R. (1997) Droplet microemulsions at the emulsification boundary: The influence of the surfactant structure on the elastic constants of the amphiphilic film. *J. Chem. Phys.*, **106**, 8232–8238.
61. Arleth, L. and Pedersen, J.S. (2001) Droplet polydispersity and shape fluctuations in AOT [bis(2-ethylhexyl)sulfosuccinate sodium salt] microemulsions studied by contrast variation small-angle neutron scattering. *Phys. Rev. E*, **63**, 61406–61423.
62. Strey, R. (1994) Microemulsion microstructure and interfacial curvature. *Colloid Polym. Sci.*, **272**, 1005–1019.
63. Robertus, C. (1990) *Small Angle X-Ray Scattering Studies of Clustering in a Microemulsion*. Ph.D. Thesis, Rijksuniversiteit, Utrecht.
64. Bagger-Jörgensen, H., Olsson, U. and Mortensen, K. (1997) Microstructure in a ternary microemulsion studied by small angle neutron scattering. *Langmuir*, **13**, 1413–1421.
65. Agterof, W.G.M., van Zomeren, J.A.J. and Vrij, A. (1976) On the application of hard sphere fluid theory to liquid particle dispersions. *Chem. Phys. Lett.*, **43**, 363–367.
66. Carnahan, N.F. and Starling, K.E. (1969) Equation of state for noninteracting rigid spheres. *J. Chem. Phys.*, **51**, 635–636.
67. Hellweg, T., Brûlet, A. and Sottmann, T. (2000) Dynamics in an oil-continuous droplet microemulsions as seen by quasielastic scattering techniques. *Phys. Chem. Chem. Phys.*, **2**, 5168–5174.
68. Glatter, O., Strey, R., Schubert, K.-V. and Kaler, E.W. (1996) Small angle scattering applied to microemulsions. *Ber. Bunsenges. Phys. Chem.*, **100**, 323–335.
69. Wellert, S., Tiersch, B., Koetz, J., Imhof, H., Altmann, M.-J. Dolle, M. Richardt, A. and Hellweg. T. to be published.
70. Teubner, M. and Strey, R. (1987) Origin of the scattering peak in microemulsions. *J. Chem. Phys.*, **87**, 3195–3200.
71. Sottmann, T., Strey, R. and Chen, S.-H. (1997) A small-angle neutron scattering study of nonionic surfactant molecules at the water–oil interface: Area per molecule, microemulsion domain size, rigidity. *J. Chem. Phys.*, **106**, 6483–6491.
72. Frank, C., Frielinghaus, H., Allgaier, J. and Prast, H. (2007) Nonionic surfactants with linear and branched hydrocarbon tails: Compositional analysis, phase behaviour, and film properties in bicontinuous microemulsions. *Langmuir*, **23**, 6526–6535.

73. Endo, H., Mihailescu, M., Monkenbusch, M., Allgaier, J., Gompper, G., Richter, D., Jakobs, B., Strey, R. and Grillo, I. (2001) Effect of amphiphilic block copolymers on the structure and phase behavior of oil–water–surfactant mixtures. *J. Chem. Phys.*, **115**, 580–600.
74. Beaucage, G. (1996) Small-angle scattering from polymeric mass fractals of arbitrary mass-fractal dimension. *J. Appl. Cryst.*, **29**, 134–146.
75. Zilman, A.G. and Granek, R. (1996) Undulations and dynamic structure factor of membranes. *Phys. Rev. Lett.*, **77**, 4788–4791.
76. Holderer, O., Frielinghaus, H., Byelov, D., Monkenbusch, M., Allgaier, J. and Richter, D. (2005) Dynamic properties of microemulsins modified with homopolymers and diblock copolymers: The determination of bending moduli and renormalization effects. *J. Chem. Phys.*, **122**, 094908/1–8.
77. Mihailescu, M., Monkenbusch, M., Endo, H., Allgaier, J., Gompper, G., Richter, D., Jakobs, B., Sottmann, T. and Farago, B. (2001) Dynamics of bicontinuous microemulsion phases with and without amphiphilic block-copolymers. *J. Chem. Phys.*, **115**, 9563–9577.
78. Kotlarchyk, M., Chen, S., Huang, J.S. and Kim, M.W. (1984) structure of dense sodium Di-2-ethylsulfosuccinate/D O/decane microemulsions. *Phys. Rev. Lett.*, **53**, 941–944.
79. Byelov, D., Frielinghaus, H., Holderer, O., Allgaier, J. and Richter, D. (2004) Microemulsion efficiency boosting and the complementary effect. 1. Structural properties. *Langmuir*, **20**, 10433–10443.
80. Berne, B.J. and Pecora, R. (1976) *Dynamic Light Scattering.* John Wiley & Sons, New York.
81. Higgins, J.S. and Stein, R.S. (1978) Recent developments in polymer applications of small-angle neutron, X-ray and light scattering. *J. Appl. Cryst.*, **11**, 346–375.
82. Maconnachie, A. and Richards, R.W. (1978) Neutron scattering and amorphous polymers. *Polymer*, **19**, 739–762.
83. Eisenberg, H. (1981) Forward scattering of light, X-rays and neutrons. *Q. Rev. Biophys.*, **14**, 141–172.
84. Glatter, O. and Kratky, O. (1982) *Small Angle X-ray Scattering.* Academic Press, London.
85. Schmitz, K.S. (1990) *An Introduction to Dynamic Light Scattering by Macromolecules.* Academic Press, New York.
86. Brown, W. (1993) *Dynamic Light Scattering.* Clarendon Press, Oxford.
87. Baruchel, J., Hodeau, J.-L., Lehmann, M.S., Regnard, J.-R. and Schlenker, C. (eds) (1994) *Dynamics and Diffusion in Macromelcules, Colloids and Microemulsions.* Springer-Verlag, Berlin.
88. Higgins, J.S. and Benoit, H.C. (1996) *Polymers and Neutron Scattering*, 2nd edn. Clarendon Press, Oxford.
89. Ballauff, M. (2001) SAXS and SANS studies of polymer colloids. *Curr. Opin. Colloid Interface Sci.*, **6**, 132–139.
90. Champeney, D.C. (1973) *Fourier Transforms and Their Physical Applications. Techniques of Physics.* Academic Press, London and New York.
91. Squires, G.L. (1996) *Introduction to the Thoery of Thermal Neutron Scattering.* Dover Publications, Mineola, NY.
92. Zorn, R. and Richter, D. (2001) Correlation functions measured by scattering experiments. In T. Brückel, G. Heger, D. Richter and R. Zorn. (eds), *The Laboratory Course Neutron Scattering. Vol. 9: Schriften des Forschungszentrums Jülich: Materie und Material.* Forschungszentrum Jülich, Zentralbibliothek, Jülich, pp. 5.1–5.24.
93. Egelstaff, P.A. (1992) *An Introduction to the Liquid State. Vol. 7: Oxford Series on Neutron Scattering in Condensed Matter*, 2nd edn. Clarendon Press, Oxford.
94. Pedersen, J.S. (1999) Analysis of small-angle scattering data from micelles and microemulsions: Free-form approaches and model fitting. *Curr. Opin. Colloid Interface Sci.*, **4**, 190–196.
95. Zwanzig, R. (1965) Time-correlation functions and transport coefficients in statistical mechanics. *Ann. Rev. Phys. Chem.*, **16**, 67–102.

96. Chu, B. (1974) *Laser Light Scattering*. Academic Press, New York.
97. Provencher, S.W. (1982) A constrained regularization method for inverting data represented by linear algebraic or integral equations. *Comput. Phys. Com.*, **27**, 213–217.
98. Provencher, S.W. (1982) Contin: A general purpose constrained regularization program for inverting noisy linear algebraic and integral equations. *Comput. Phys. Com.*, **27**, 229–242.
99. Olsson, U. and Schurtenberger, P. (1993) Structure, interactions, and diffusion in a ternary nonionic microemulsion near emulsification failure. *Langmuir*, **9**, 3389–3394.
100. Gradzielski, M. and Hoffmann, H. (1994) Influence of charges on the structure and dynamics of an O/W microemulsion. Effect of admixing ionic surfactants. *J. Phys. Chem.*, **98**, 2613–2623.
101. Hayter, J.B. and Penfold, J. (1979) Neutron spin-echo integral transform spectroscopy. *Z. Physik B*, **35**, 199–205.
102. Hayter, J.B. (1981) Quasielastic neutron spin-echo spectroscopy. In S.-H. Chen, B. Chu and R. Nossal. (eds), *Scattering Techniques Applied To Supramolecular and Nonequilibrium Systems*. Plenum, New York, pp. 3–33.
103. Mezei, F. (ed) (1980) *Neutron Spin Echo. Vol. 124: Lecture Notes in Physics*. Proceedings from Spin-echo Meeting in Grenoble, 1979. Springer-Verlag, Berlin.
104. Mezei, F., Pappas, C. and Gutberlet, T. (eds) (2003) *Neutron Spin Echo Spectroscopy*, 1st edn. *Vol. 601: Lecture Notes in Physics*. Springer, Heidelberg.

Chapter 3
Formulation of Microemulsions

Jean-Louis Salager, Raquel Antón, Ana Forgiarini and Laura Márquez

3.1 Basic concepts

3.1.1 Microemulsions

Although oil and water are not miscible at ambient temperature, a small amount of surfactant is able to co-solubilise them. How does that happen? Surfactant molecules tend to self-associate in structures such as a lamellar liquid crystal to accommodate their polar–apolar amphiphilic duality. These structures are able to incorporate oil and water by inclusion in between the layers, hence eliminating the direct contact between the fluids. The insertion process swells the layers, which then loose their rigidity and strict flat organisation until they transform to a twisted, sponge-like structure, the so-called bicontinuous microemulsion. Other structures may also be formed, namely water droplets dispersed in the oil phase or oil droplets dispersed in the aqueous phase (see Chapter 1 for further details).[1] When starting from a pure surfactant system, it is generally easy to dissolve (rather say, to solubilise) oil (O) or water (W) or both in the microemulsion structure. However, very often large amounts of surfactants (up to 50%) are needed to solubilise equal amounts of oil and water. Thus, the challenge is to attain a single phase with a surfactant content as low as possible (see Chapter 1).

An increased performance of the surfactant in co-solubilising O and W is essentially linked with an enlarged size of the O and W microdomains of the microemulsion. Assuming spherical domains for the sake of simplicity, it is obvious that the solubilised amount of O or W is proportional to the volume of the sphere, i.e. to the third power of the domain diameter. On the other hand, the surfactant amount which creates the interface between the domains is proportional to the surface area of the sphere, i.e. to the square of the diameter. The outcome is essentially the same for bicontinuous microemulsions in spite of not exhibiting spherical domains. The point is that the solubilisation, which is the amount of O and W that a given amount of surfactant S is able to compatibilise, is proportional to the diameter of the domains. Hence, the characteristic dimension of the structure has to be large if less surfactant ought to be used. The requirement of large domains means an almost zero mean curvature and a high flexibility of the surfactant layer resulting in the already mentioned sponge-like bicontinuous structure. This is the basic requirement to match in order to formulate the surfactant layer for attaining high solubilisation in microemulsions. It is worth noting that since the O and W domain characteristic dimension has to be large, high solubilisation microemulsions do scatter light, therefore are not transparent as often claimed in the literature. The highest solubilisation capacity reported so far is a microemulsion that contains about 49% O, 49% W and 2% S and which indeed appears quite milky although it is thermodynamically stable. In other words, transparent microemulsions are generally associated with a poor solubilisation capacity.

As already mentioned, surfactant molecules tend to self-associate in solution. In water or polar phases, they generate normal micelles with a hydrophobic core which is able to solubilise apolar substances, and in apolar media they produce inverse micelles with a polar core likely to solubilise water. Because solubilisation should be quite high in practice, the amount of the dispersed and the dispersing phases should be roughly the same. Hence, in a single-phase system containing similar amounts of oil and water, the structure may be alternatively swollen normal micelles in water or swollen inverse micelles in oil, with almost touching micelles in both cases because of the volume constraint. This kind of single-phase SOW system is referred to as Winsor type IV microemulsion. However, in most cases the system would contain swollen micelles of one type, that are 'inflated' up to the point they cannot take in anymore in their core, and thus expel the excess as a separate phase. These systems are so-called Winsor type I (type II) when they contain a solution of normal micelles in water (inverse micelles in oil) in equilibrium with excess oil (water) (see Fig. 1.2 in Chapter 1). In these cases, the microemulsions may be considered as a dispersion of tiny microdroplets which are actually swollen micelles[2] [1, 2]. The type I (type II) phase behaviour may be considered as a type IV in which the surfactant is not able to solubilise all the oil (water) present in the system and thus some excess oil (water) is expelled. These two cases are related to the presence of one specific type of micelle, i.e. to a specific curvature. If the formulation is such that the natural curvature is zero, the presence of large microdomains implies a high solubilisation, but the system might not contain enough surfactant to exhibit a type IV phase behaviour and some excess phase will be expelled. Because of the zero curvature the solubilising structure is symmetrical with respect to oil and water and might expel both excess oil and excess water at the same time, thus resulting in a three-phase system, so-called Winsor type III phase behaviour (see Fig. 1.2 in Chapter 1). In such a case, the surfactant-rich phase is a bicontinuous microemulsion, while the excess phases do not contain associated surfactant and can thus be considered as essentially pure oil and pure water.

Understanding the importance of the formulation in determining the phase behaviour is often achieved by using a simplistic description of what happens at the interface, i.e. at the oil/water boundary. Even if such approach is not strictly accurate in some cases, it helps quite a lot the formulator in seeking in the right direction. The phase behaviour is linked with the dominant affinity of the surfactant for one of the phase, and a simple explanation has been that if the surfactant 'prefers' the water phase, i.e. if it is more 'water-loving' or more 'soluble' in water than in oil, then a type I phase behaviour is attained and conversely. Another rationalisation according to Langmuir's wedge theory is that a more water-loving surfactant tends to bend the interface such that normal micelles are formed and vice versa. This approach might be considered to be too simplistic when the surfactant equally 'likes' oil and water, because it would predict a two-phase system with the surfactant equally partitioning between oil and water. Indeed, this occurs for amphiphiles such as butanol, but not for surfactants, in which a type III three-phase splitting takes place instead. The term 'like' might not be the right one to describe the surfactant interaction with O and

W. Actually, the surfactant does not like to be in water nor in oil because one part of the molecule is always lyophobic, which is why micelles are formed to hide it away from the solvent. Hence, it may be said that in type I phase behaviour the surfactant 'dislikes' more oil than water, and in type II it 'dislikes' more water than oil. Then, in type III phase behaviour, the surfactant equally dislikes both phases and would seek a third alternative, e.g. forming a bicontinuous microemulsion. In thermodynamic terms, it simply means that the chemical potential of the surfactant in such a microemulsion phase is lower than when it is adsorbed at the curved interface of a drop.

In this chapter, we will focus on the formulation of systems in which the surfactant has equal affinity for both O and W phases. These formulations do form bicontinuous microemulsions of zero mean curvature and have important properties such as minimum interfacial tension and maximum solubilisation. Such condition has been called *optimum formulation* in the 1970s, because it matches the attainment of an ultralow interfacial tension that guarantees an enhanced oil recovery from petroleum reservoirs, which was the driving force behind the research effort on microemulsions (see Chapter 10, Section 10.3 of this book) [3, 4]. High solubilisation performance microemulsions which are able to co-solubilise approximately equal amount of oil and water with less than 15–20% surfactant are attainable only at an optimum formulation.

3.1.2 Why is formulation important?

Formulation is important because the properties of surfactant–oil–water systems in general and the formation of microemulsions in particular, are very sensitive to it and slight deviations from a 'proper' formulation may result in drastic changes of the properties. Consequently, formulation has to be controlled accurately, which is quite challenging because of the high number of degrees of freedom in any practical case. This is why formulation is sometimes considered as 'magic business'.

Formulation essentially relates to the content of the systems and generally not to the way it is attained if thermodynamically stable systems are considered. The simplest microemulsion system would contain an organic oil phase (O), an aqueous phase generally referred to as water (W), and a surfactant (S) at a given temperature (T) and pressure (p). This means that at least five variables are required to describe the system. In practice, the situation is much more complicated. Water always contains electrolytes. Moreover, oils as well as nearly all commercial surfactants are mixtures. In most cases, particularly with ionic surfactant systems, a co-surfactant (e.g. an alcohol (A)) is added, among other functions, to reduce the rigidity of the surfactant layer and thus to prevent the formation of gel-like mesophases.

When mixtures are dealt with, some approximations allow to decrease the number of variables. For instance, if a commercial surfactant contains substances all of which behave similarly, then a so-called pseudo-component may be used to describe it. However, this is not the general case and in many instances so-called fractionation phenomena take place and the different components behave independently of each other. In this case, the actual number of components could be (much) larger than three or four. Aside the nature of each of the components of the SOW and eventually the SOWA system, T and p also influence the properties, sometimes to quite a large extent. Note that all variables describing the nature of the components as well as T and p are intensive, i.e. they do not depend on the quantities, the reason for which they have been called *formulation variables*. On the other hand, the relative amount of the different substances present in the system are also likely to change the properties, and are often referred to as *composition variables*; they are expressed as weight or volume fraction, percentage, or proportion. If n components are included, $(n - 1)$ independent composition variables are required to quantitatively describe the contents of the system. In the simplest SOW ternary case, the two composition variables are often selected to be the surfactant concentration and the water-to-oil ratio (WOR)[3] [5].

Before going further, it is convenient to carry out some accounting on how many variables are required to describe a simple system containing a surfactant, a co-surfactant, a pure oil (n-alkane) and a water solution of sodium chloride, at T and p constants. Considering brine as a pseudo-component, there are 4 components, hence 6 formulation variables (with T and p), and 3 independent composition variables, therefore 9 variables, which may be reduced to 8 if pressure effects are neglected. In a practical case with several electrolytes in water, a natural oil, and a commercial surfactant, the actual number of variables, thus of degrees of freedom, may be around 20. If a random trial and error procedure is taken as a method to test formulation, as often the case, the number of experiments to be carried out could reach thousands or millions. This is of course quite a problem in practice, and non-random trial techniques would of course be very valuable. This is why a numerical handling of the formulation is so important in practice, to reproduce cases, to compare formulations, to compensate effects and to predict new recipes. This is also particularly noteworthy because the number of formulation cases is far larger than the variety of properties to be dealt with, hence, there are always many different formulations with similar consequences as far as properties are concerned, that have to be studied and compared in the optimisation of any practical application.

An accurate formulation handling is extremely useful not only to make microemulsion and to adjust their properties such as their solubilisation ability, or to attain a low interfacial tension to ease emulsification or to enhance oil recovery. Formulation has been shown to be also directly linked with emulsion properties such as their type, stability, viscosity, drop size [6] and with the efficiency of the emulsification protocol [7]. The existence and persistence of foams are dependent on formulation too [8]. Solid surface wetting is also linked with formulation as well as with many related applications. This is why an accurate numerical treatment of formulation issues is paramount in industrial research and development.

3.2 Representation of formulation effects

The present book dedicated to microemulsions is concerned with single-phase systems essentially containing oil and water, with a surfactant that makes them compatible, eventually in equilibrium with excess oil, excess water or both. Consequently, there are basically four cases of phase behaviour, referred to as Winsor type I, II, III and IV, but there are scores of formulation variables, and a practical problem is how to link a few cases of phase behaviour with so many independent variables. As a matter of fact, the representation by a plot on a sheet of paper is limited to two dimensions, and some property value may be plotted as a function of an independent variable, or as a map versus two independent variables, with different cases which are roughly equivalent although they do not

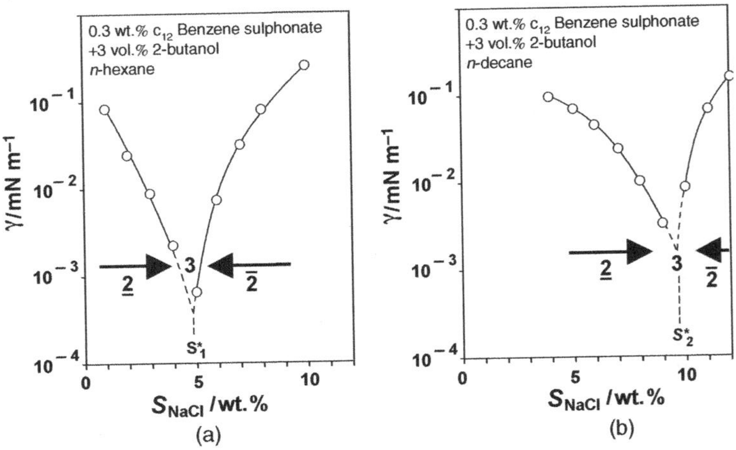

Figure 3.1 Variation of the interfacial tension (γ) as a function of the salinity (*S*) for a system containing 0.3 wt.% C_{12} benzene sulphonate, 3 vol.% 2-butanol, *n*-hexane (a) and *n*-decane (b), respectively.

supply the same information [9, 10]. In the last case, a contour map is used for properties, whereas phase behaviour may be represented as zones separated by transition boundaries. Three-dimensional representations may be carried out with a computer, but higher dimensional spaces cannot be handled in practice. Because the number of formulation and composition variables is far larger than three, something has to be done to cut down the number of dimensions. It will be shown in the following that a proper gathering of the formulation variables allows reducing the number of degrees of freedom to a manageable amount.

3.2.1 Unidimensional formulation scan representation

The first representation is to select one formulation variable to be scanned (at all others constant) and to plot the variations of a property or of the phase behaviour versus this formulation variable. Figure 3.1(a) shows such a plot of interfacial tension (γ) versus the salinity (*S*) of the aqueous phase together with the ranges in which different phases are formed for a system containing *n*-hexane as oil, an alkylbenzene sulphonate surfactant and sec-butanol as co-surfactant. The phase behaviour is indicated as $\underline{2}$ or $\overline{2}$ for two-phase systems in which the surfactant-rich phase is the lower water or the upper oil phase, corresponding to Winsor type I and II diagrams. Symbol 3 indicates the range of formulation for which a three-phase behaviour is attained.

It is seen that the minimum interfacial tension is attained for $S_1^* = 5.0$ wt.% NaCl which is called the optimum salinity of the scan, and which coincides with the centre of the three-phase region. Figure 3.1(b) represents the same kind of scan, but this time the oil phase is *n*-decane, with the same surfactant and alcohol content. The optimum salinity is now $S_2^* = 10$ wt.% NaCl. As far as the conditions for the attainment of optimum salinity are concerned the two formulations (*n*-hexane/S_1^* and *n*-decane/S_2^*) are equivalent, therefore

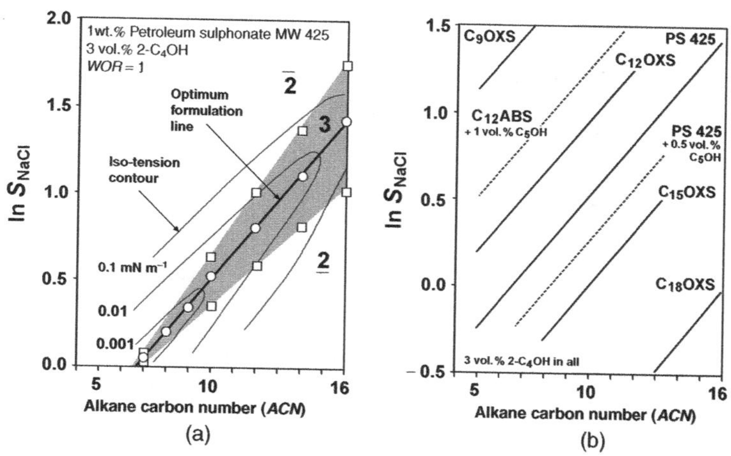

Figure 3.2 Plots of the salinity (*S*) versus the alkane carbon number (*ACN*). (a) Optimum formulation lines as the locus of the minimum interfacial tension, i.e. of the three-phase region centre. (b) Optimum formulation line as the locus in bidimensional *S-ACN* map for the same water–oil–alcohol systems containing different surfactants at constant temperature. C_nOXS stands for alkylorthoxylene sulphonates, ABS for alkyl benzene sulphonate, PS for petroleum sulphonate (the number after PS indicates the average molecular weight).

the change from hexane to decane is compensated by a change in salinity from S_1^* to S_2^*. This illustrates how two unidimensional scans generate the seed for a bidimensional scan.

3.2.2 Bidimensional map representation

3.2.2.1 Bidimensional map with two formulation variables

One step further is to plot a property as a function of two formulation variables, e.g. salinity (S) and the nature of the oil phase, referred to as alkane carbon number (ACN) when it is an n-alkane. This time the tension variation is indicated on a map with constant value contours, as in Fig. 3.2(a). The locus of the minimum tension is noted as a line which indicates the optimum formulation as a function of both S and ACN, whereas the phase behaviour is represented by regions, with the three-phase region (shaded) surrounding the optimum formulation line. For not overloading the plot, the only information shown may be this optimum formulation line, which indicates the S–ACN trade-off required to attain an optimum formulation for the system containing the given surfactant and alcohol, at constant temperature, surfactant concentration and WOR (see line PS 425 in Fig. 3.2(b)). Now, if the same map is established for a system containing other surfactants (but same alcohol, temperature and composition), other optimum lines relating S and ACN are found, as shown in Fig. 3.2(b); the shift from the previous line is a quantitative estimate of the effect of changing the surfactant, which may be expressed in terms of one of the two formulation variables (S and ACN). Hence, a bidimensional plot is able to render the trade-off or surrogate variations of three variables, e.g. the two scanned variables and

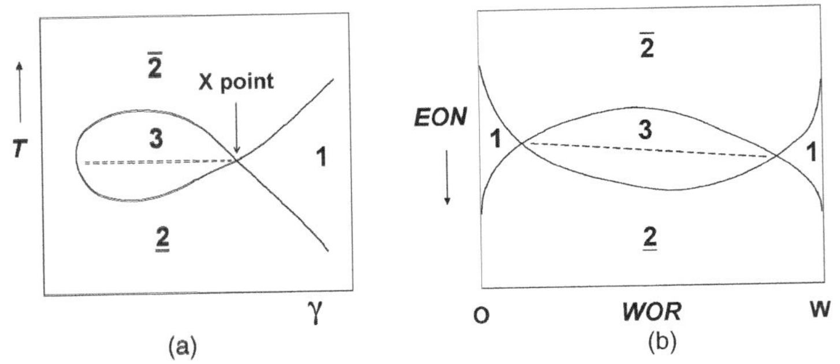

Figure 3.3 Two types of bidimensional cuts through multidimensional phase prisms. (a) Cut through a phase prism at a 1:1 water-to-oil ratio as a function of the temperature (*T*) and the surfactant concentration (γ), the so-called 'fish diagram'. (b) Cut through a phase prism at constant *T* and γ as a function of the ethoxylation degree (*EON*) and the water-to-oil ratio (*WOR*), the so-called 'χ diagram'.

a third one (like the surfactant type). This is the starting point of a multidimensional correlation.

3.2.2.2 *Bidimensional map with one formulation variable and one composition variable*

Other bidimensional maps have been used. For ethoxylated surfactant systems the temperature and the surfactant concentration are very often changed, while all other composition variables are kept constant [11]. Figure 3.3(a) shows such a map which is habitually referred to as a fish diagram [12] because of the shape of the phase boundaries (also see Fig. 1.3 in Chapter 1). This kind of diagram displays several interesting features of the system, particularly the surfactant concentration range over which a microemulsion is in equilibrium with both excess phases. This three-phase zone spans from the minimum surfactant concentration for which a three-phase behaviour is exhibited, to the minimum surfactant concentration to attain a single phase at the so-called X point. The location of the X point is the most important data of the graph because its temperature is an information on formulation, whereas the corresponding surfactant concentration is an estimate of the solubilisation performance of the microemulsion. The centre line (dashed) of the three-phase region indicates the optimum formulation, which is here the optimum temperature up to the X point. This diagram also indicates the temperature range over which three phases are formed, which is an estimate of the robustness of the formulation. The same kind of diagram is obtained when another formulation variable is selected instead of temperature, for instance the ethoxylation degree (*EON*) of the non-ionic surfactant.

In many cases, particularly when mixtures of polyethoxylated non-ionic surfactants are used, or for *WOR* different from unity, the three-phase region is distorted (see Figs. 1.8 and 1.10 in Chapter 1). This is a practical problem for the formulator because it means that the optimum formulation or temperature (dashed line at the centre of the three-phase zone) depends on the surfactant concentration, an inconvenience in many applications, since formulation is likely to be altered by the dilution of the system. This formulation–composition diagram, particularly the location of the X point, varies when any other formulation variable is changed. This allows quantifying the trade-off effects between temperature and other variables, and the variation of performance due to the change in formulation [13, 14].

Another way to plot the phase behaviour is to change the temperature and the water-to-oil ratio, while all other variables are kept constant [15, 16]. In this so-called formulation–*WOR* (or sometimes χ) diagram (Fig. 3.3(b)), one sees a three-phase region in the centre of the map at low surfactant concentration, i.e. below the one corresponding to the X point. A single-phase microemulsion wedge zone appears on each side close to pure W and pure O. These two monophasic regions extend as surfactant concentration increases, and merges as a single-phase band when it exceeds the concentration corresponding to the X point.

3.2.2.3 Bidimensional map with two composition variables

Another common case of bidimensional representation is a map of the phase behaviour versus two composition variables, while all formulation variables are kept constant. This is generally represented in a triangular diagram. If the three components are pure products, there are essentially the three Winsor's types of diagrams as in Fig. 1.2 (Chapter 1), the formation of which depends on the formulation variables. Winsor type III diagram correspond to the so-called optimum formulation situation in which there is a zone in which three phases coexist over some range of composition.

3.2.3 Other representations

Even more complex quaternary SOWA diagrams with three independent composition variables at constant formulation and T/p conditions have been proposed to be represented in an equilateral tetrahedron. Such diagrams may be useful for some peculiar cases, but they are generally not amenable to simple interpretations and in most cases they are described by a series of bidimensional cuts, i.e. cuts through the tetrahedron, which are not amenable to Winsor's types as seen in Fig. 3.4, because the four types are arranged in a different way [17, 18].

Each kind of diagram may be said to have its advantages and drawbacks. In the fish diagram, the minimum amount of surfactant to attain three-phase behaviour is readily available, as well as the surfactant concentration at the X point which is some measurement of the quality of the system or of the performance of the surfactant. However, these data vary with *WOR* and of course with other formulation variables, and the fish diagram is of no help to deal with such issues. The triangular diagram indicates the range of three-phase behaviour when surfactant concentration and *WOR* vary, but it changes readily with any formulation variable, including temperature. On the contrary, such a diagram is quite helpful to describe processes involving dilution by oil or water, as it often happens in formulation practice. The formulation–*WOR* diagram is quite useful to report emulsion properties since these variables are the most significant ones, particularly as far as inversion processes are concerned. All these diagram representations are used in this book, but essentially all fall short of representing the behaviour of any real system with nine or ten variables. It is thus imperative to cut down the number of degrees of freedom somehow,

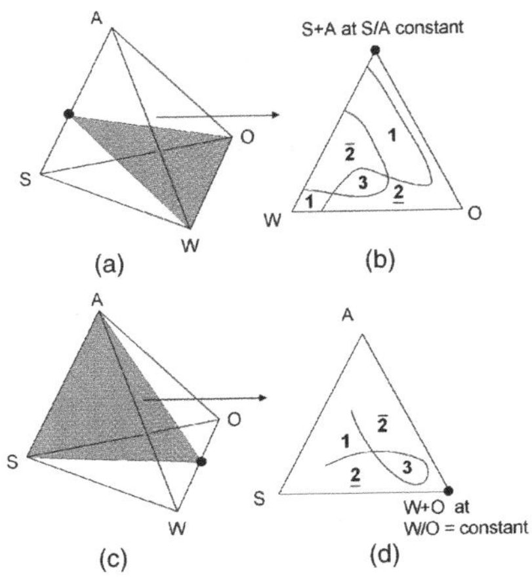

Figure 3.4 Bidimensional cuts (b, d) through a tridimensional quaternary diagram (a, c). Seen is a cut at constant surfactant-to-alcohol ratio (b) and at constant water-to-oil ratio (d), respectively

particularly those related to formulation, and this will be the focus of attention in the following sections.

3.3 Physico-chemical formulation yardsticks

3.3.1 *Early formulation concepts*

The formulation has been related with the type and properties of emulsions since Bancroft's rule of thumb (1913) and Langmuir's wedge theory (1917). The hydrophilic–lipophilic balance (HLB) was introduced by Griffin 60 years ago, probably as a selling argument for the (by the time) new non-ionic surfactants. It accounts for the relative importance of the hydrophilic and lipophilic parts of an amphiphilic molecule on a weight basis [19]. For decades there was no other numerical yardstick. The simplicity of the HLB concept was its main advantage in spite of very serious limitations, such as an inaccuracy sometimes over two units, and the fact that it does not take into account several variables which are known to alter the phase behaviour, independently of the surfactant.

The phase behaviour at equilibrium turned out to be the main property reported in Winsor's work in the late 1940s. Winsor interpreted the phase behaviour through the so-called *R* ratio of molecular interaction energies at interface. The *R* ratio was a handy theoretical concept to understand the variations of the phase behaviour of surfactant–oil–water systems and somehow of the emulsion properties. It is essentially qualitative, but for the first time the phase behaviour was linked with a condition that depended on all formulation variables, but could be expressed as a single generalised variable, i.e. the *R* ratio [1]. The original *R* ratio was

$$R = \frac{A_{\text{co}}}{A_{\text{cw}}}, \tag{3.1}$$

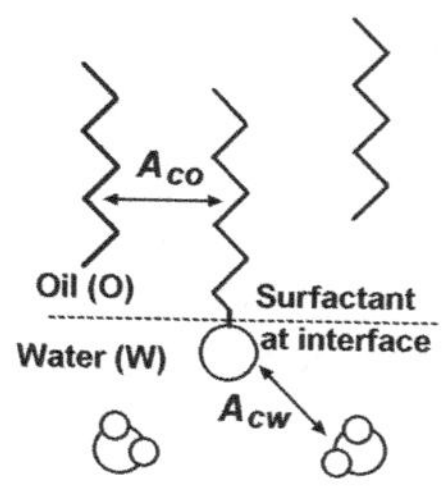

Figure 3.5 Winsor interaction energies between the adsorbed surfactant molecule and nearby oil and water molecules, respectively.

where A_{co} is the interaction between the adsorbed surfactant at the interface and the nearby oil molecules, whereas A_{cw} is the interaction with the water molecules. Both interactions are given per unit surface area (see Fig. 3.5). Later on, a more complete definition was proposed [4], namely

$$R = \frac{(A_{co} - {}^1/_2 A_{oo} - {}^1/_2 A_{LL})}{(A_{cw} - {}^1/_2 A_{ww} - {}^1/_2 A_{HH})}, \tag{3.2}$$

where the hydrophilic and lipophilic interactions at interface are referred to a reference state in which the surfactant, oil and water are apart. A_{oo} (A_{ww}) are the interactions between two oil (water) molecules, whereas H (respectively L) subscript refers to the hydrophilic (lipophilic) part of the surfactant molecule and A_{LL} (A_{HH}) is the lipophilic (hydrophilic) interaction between two surfactant molecules.

The change in the R ratio from $R < 1$ to $R > 1$ was associated with the change in phase behaviour from type I to type II, and $R = 1$ corresponded to equal interaction or neutral affinity of the surfactant for oil and water, i.e. optimum formulation. Winsor did not suggest a quantitative expression for R as a function of the different formulation variables such as the length of the tail of the surfactant or its hydrophilic group nature, the oil type, the salinity and so forth. However, the concept of energy of interaction allows some reliable guess. For instance, it is known that an increase in salinity would result in a screening effect between an ionic surfactant head group and the water molecules, hence it would result in a decrease in A_{cw}, and thus an increase in R. If the length of the tail group of the surfactant is expanded, its interaction with the oil phase is also likely to increase almost linearly, unless it becomes coiled. Some complex effects could be explained too. For instance, an increase in the length of an n-alkane would result in two opposite effects found in Eq. (3.2). First, an increase of the interaction A_{co} with the surfactant 'tail', more or less proportionally to the increase in alkane carbon number ACN. Second, an increase in the self-interaction of the alkane molecules A_{oo}, which is likely to be proportional to the square of ACN. Hence, this second term A_{oo} would generally grow faster than A_{co} with the increase in ACN, with a resulting decrease in R, and an associated transition from type II to type I phase behaviour, as it is experimentally verified. Each of the formulation variables is able to alter the R ratio one way or the other, so that the interpretation is relatively simple, may be with the exception of the effect of temperature which tends to alter all interactions. Nevertheless, it is worth noting that temperature is the unique formulation variable which could be easily reversed in experimental tests and therefore it is worth the additional difficulty in interpretation. As a matter of fact, the temperature is often used as the main formulation variable for polyethoxylated surfactants, because in this case, the main and dominant effect is the dehydration of the surfactant head group, i.e. the decrease in A_{cw} as temperature increases. Winsor's R ratio is yet the simplest way to carry out basic formulation reasoning and this is why its use is still recommended for making the first guess.

Two decades later, Beerbower's cohesive energy ratio (CER) improved over Winsor's R with a thermodynamic fundamental basis which could potentially lead to a numerical yardstick. However, it failed to do so essentially because regular solution theory was not sophisticated enough to handle a mixture of (two) different fluids-like oil and water. During this period, Shinoda proposed the temperature as a way to systematically vary the formulation, and related it to the phase behaviour and emulsion properties. Shinoda introduced the phase inversion temperature (PIT) concept [20], which is the temperature at which the emulsion type swaps from o/w to w/o or vice versa. This phase inversion essentially takes place at optimum formulation, i.e. when $R = 1$ and the phase behaviour is type III up to the X point. For such reason, it was also named HLB temperature a few years later to imply that it is related to an equilibrium thermodynamic phenomenon [21]. The PIT concept was empirical but straightforward to estimate, accurately in many cases, and it was easy to correlate it with the characteristics of all components of the system such as the surfactant hydrophilic and lipophilic groups, the salinity of the aqueous phase, or the oil nature. For the first time, it was clearly demonstrated on experimental grounds that all formulation variables were contributing to the result, i.e. the attainment of the three-phase behaviour, provided that one condition was fulfilled. It definitively showed that there was a way to express the change of formulation through the single generalised formulation variable PIT. However, the PIT concept had serious practical limitations as a universal yardstick, e.g. it could be applied only to polyethoxylated non-ionics over the 0–100°C interval.

All these attempts over a 25-year span, which are discussed elsewhere [22], ranged from empirical to theoretical. They contributed to a slow progress and were quite useful for carrying out qualitative reasoning or to get a starting guesstimate; however, there were not amenable to accurate and universal numerical accounting of the contributions of all formulation variables.

3.3.2 Correlations for the attainment of optimum formulation

It was only in the 1970s, when a huge amount of petroleum research money was dedicated to enhanced oil recovery, that surfactant–oil–water systems were scrutinised in fine tuning details as far as formulation issues were concerned. Winsor's work and his R ratio which, by this time, had been largely overlooked for 20 years, was recognised as a basic generalised formulation concept, which had to be improved upon. The requirements were to establish a precise know-how to build up models that could simulate complex interactions between surfactant, oil, water and rock, several thousand feet down hole, in non-steady state regime. A far better accuracy was definitely needed for a universal measurable formulation yardstick more wide ranging than PIT but including it. After a few years of extensive research and development, essentially aimed at quantifying Winsor's R concept, all variables could be taken into account in a numerical expression so-called correlation for the attainment of optimum formulation for minimum interfacial tension or Winsor type III three-phase behaviour. The following correlation was reported for systems containing an anionic surfactant, an n-alkane and an NaCl brine [23]; two decades later it was showed to also apply for cationic surfactants [24].

$$\ln S - k\,ACN - f(A) + \sigma - a_T(T - 25) = 0. \tag{3.3}$$

A similar correlation was reported for systems containing non-ionic surfactants of the polyethoxylated alkylphenol or alcohol type [25].

$$\alpha - EON + bS - k\,ACN - \phi(A) + c_T\,(T - 25) = 0. \tag{3.4}$$

In these expressions, S is the salinity of the aqueous phase (in wt.% NaCl) and ln S its neperian logarithm, ACN is the alkane carbon number which characterises the oil phase, σ is a characteristic parameter of the anionic or cationic surfactant which increases linearly with the length of the lipophilic tail, as well as α which is characteristic of the lipophilic group of non-ionic surfactants, EON is the average number of ethylene oxide groups per polyethoxylated surfactant molecule and T is the temperature. k, b, a_T and c_T are constants which depend on the kind of system particularly the surfactant head group and electrolyte nature [23–25]. For ethoxylated non-ionics, the characteristic parameter $\beta = \alpha - EON$ is sometimes introduced. $f(A)$ and $\phi(A)$ are function of the alcohol type and concentration which could be written in first approximation as

$$f(A) \text{ or } \phi(A) = m_A C_A, \tag{3.5}$$

where C_A is the alcohol concentration and m_A is a constant, slightly positive for short alcohols (methanol or ethanol) and negative for lipophilic alcohols such as n-butanol or n-pentanol. The longer the alcohol and the more linear it is, the higher is the absolute value of m_A. Alcohols which exhibit the same affinity for oil and water (e.g. sec-butanol or ter-pentanol) have an m_A value close to zero, which means that they do not alter the formulation according to Eqs. (3.3) and (3.4). They are appropriate candidates to formulate microemulsions as their main role is to inhibit the formation of liquid crystals.

When the correlation is satisfied (i.e. if the left term of Eqs. (3.3) and (3.4) is zero or very close to zero) an optimum formulation is attained, which means that a three-phase system is generated if the surfactant concentration is the proper one and if the performance of the surfactant and the quality of the system is suitable, as will be discussed later. If the left term of Eq. (3.3) or (3.4) is negative (respectively positive) then a type I or $\underline{2}$ (type II or $\overline{2}$) phase behaviour is found instead. A non-optimum formulation could be returned to optimum by changing any one of the variables that appears in Eqs. (3.3) and (3.4) in the proper direction and magnitude. For instance, a type I phase behaviour may be made type III by increasing salinity, decreasing ACN, increasing the length of the surfactant tail (increase in σ or α), decreasing EON, increasing temperature with polyethoxylated non-ionic surfactants or decreasing it with ionic ones, adding a more lipophilic alcohol (increasing the absolute value of a negative m_A) or increasing a lipophilic alcohol concentration C_A. The values of the constants indicate what are the equivalent changes and how compensative effects or trade-offs can be achieved. The correlations mentioned in Eqs. (3.3) and (3.4) were valid

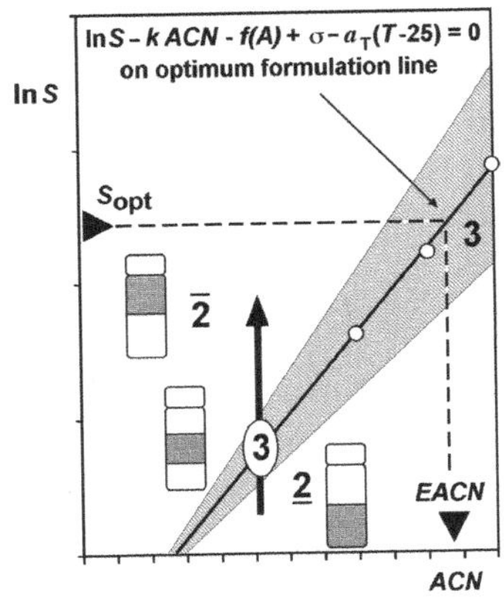

Figure 3.6 Principle of the determination of the *EACN* of unknown oil.

for relatively simple systems containing a single surfactant, NaCl brine and an n-alkane. They demonstrated the general concept of how to attain an optimum formulation, but they needed to be made extensive to any real system, particularly those including complex oils such as petroleum.

Crude oils were found to behave as an 'equivalent' alkane as far as the attainment of optimum formulation was concerned. The equivalent alkane carbon number or *EACN* was then introduced to characterise pure hydrocarbons or mixtures [26]. The *EACN* of an oil phase is defined as the *ACN* of the alkane that results in the satisfaction of the correlation in the same conditions of surfactant, salinity, alcohol and temperature. *EACN* has been experimentally determined for n-alkanes mixtures, resulting in a linear mixing rule on a molar fraction basis, namely

$$EACN = \Sigma x_i ACN_i, \tag{3.6}$$

where x_i is the molar fraction of the ith component in the oil mixture. The *EACN* of non-n-alkane oils can be estimated as follows. A base system is taken with a known surfactant, e.g. an alkylbenzene sulphonate and an alcohol. The concentration and the temperature are kept constant. Then NaCl salinity scans are carried out for a series of n-alkanes resulting in an optimum formulation straight line if ln S is plotted versus *ACN* (Fig. 3.6). With the same base system (same surfactant, alcohol and temperature) a salinity scan is then carried out with the unknown oil phase which results in a certain optimum salinity S_{opt}. The *EACN* of the unknown oil is the *ACN* of the n-alkane which exhibits optimum formulation for this same salinity S_{opt}.

The *EACN* was found to depend on the oil molecular structure. Branching was not found to alter it significantly unless it is extensive, but cyclisation tends to cut it down definitely. *EACN* was found to be 3.5 for cyclohexane and $(3.5 + n)$ for alkyl cyclohexanes with n carbon atoms in their alkyl chain; benzene *EACN* was initially found to be close to 0; alkylbenzenes were found to have an *EACN* equal to the number of carbon atoms of their alkyl chain [4]. Recent findings with extremely pure surfactants tend to indicate that these results for aromatic oils might be erroneous, or at least misleading, because these solvents induce the fractionation of different species contained in commercial surfactants [14]. It is now acknowledged that a so-called segregation [27] might take place close to interface when oils with different polarities are mixed. Consequently, and taking into account this early shortcomings, the benzene ring is likely to have a negative *EACN* contribution around –3 or –4, rather than just zero. In any case, the point is that cyclisation, in particular with double bonds, reduces the *EACN*. Limonene and pinene, which are C_{10} terpenes, respectively, with one and two cycles, were found with *EACN* around 8–9 and 6–7, respectively [5]. On the other hand, the introduction of a polar group in the oil was found to reduce considerably the *EACN*. Ethyl oleate was reported with an *EACN* around 6, and soya oil, which is essentially a C_{16-18} triglyceride, had its *EACN* estimated at about 18, a result which tends to indicate that the presence of an ester group cuts down the *EACN* by 12 units with respect to the total number of carbon atoms. These values are not accurate because the attainment of a microemulsion with these kinds of polar oils requires a new class of not commercially available surfactants, so-called extended, on which the information is limited (see Section 3.4.4). Since chlorinated hydrocarbons are quite polar oils, it is not surprising to find out that their reported *EACN* could be negative, e.g. –4 for trichloroethylene [28] and –14 for chloroform. A negative *EACN* value might sound whimsical when considering that *ACN* is a number of carbon atoms, but it is not awkward if the *EACN* scale is just taken as an oil polarity measurement.

The salinity effect of different salts, particularly divalent cation salts, is expressed through the term bS in the correlation for non-ionic surfactants of the polyethoxylated phenol or alcohol type. No information is available yet on the salinity effect on other non-ionics such as alkyl-polyglucosides. The salinity effect on ionic surfactant systems is a more complex issue because the surfactant itself is also a (more or less) dissociated electrolyte. Its degree of dissociation is paramount as far as its hydrophilicity is concerned. For instance sodium salts of alkyl sulphonic acids are essentially completely dissociated, hence they act as the sulphonate ion, and it is essentially the same with the salt of potassium or ammonium. The presence of multivalent anions produces an interference with the monovalent anionic surfactant ion, such as an alkyl benzene sulphonate, but it is essentially an ideal mixing rule.

The equivalent salinity (to replace $\ln S$ in Eq. (3.3)) of a sodium salt different from chloride has been fully established from the experimental point of view, thanks to the introduction of the valency activity factor *VAF* [29]. $\ln S$ (wt.% NaCl) has to be replaced by $1.766 + \ln S_{Ne}$ in Eq. (3.3) to take into account the change in salinity unit (from wt.% NaCl to normality of sodium ion). S_{Ne} is the equivalent salinity of the sodium salt (in mole of sodium ion per litre) defined by

$$S_{Ne} = VAF S_N = \frac{2\,S_N}{(1+Z)}, \tag{3.7}$$

where S_N is the salinity as electrolyte concentration in mole of sodium ion per litre unit, and Z is the valency of the anion of the sodium salt. The valence activity factor *VAF* is unity for monovalent anions, but it decreases for higher valence ions. It is for instance 0.5 for tri-sodium phosphate, which means that this salt-effective salinity on a sodium molar concentration basis is half the salinity produced by the same concentration of sodium chloride or fluoride. For a mixture of sodium salts, the equivalent salinity may be calculated as S_{NeM} [29].

$$S_{NeM} = \Sigma\, x_i\; VAF_i\; S_{Ni} = \frac{2\,\Sigma\, x_i\; S_{Ni}}{(1+Z_i)}, \tag{3.8}$$

where x_i is the molar faction of electrolyte i and Z_i its anion valency. On the contrary, the presence of divalent cations such as Ca^{2+} is likely to result in the formation of much less dissociated calcium sulphonate, which is not water soluble, and would precipitate or mix with completely dissociated sulphonate molecules, thus resulting in an intermediate hydrophilicity surfactant. The same phenomenon takes place with surfactants whose dissociation depends on the pH, such as fatty acid soaps or alkyl amine salts. The equilibrium constant K_a for a carboxylic acid in water may be written as

$$K_a = \frac{[H^+]\,[Ac^-]}{[AcH]}, \tag{3.9}$$

where K_a is the dissociation constant of the carboxylic acid in water (typically 10^{-5}–10^{-6} for long-chain fatty acids), $[H^+]$ is the hydrogen ion concentation, [AcH] the concentration of the undissociated acid and $[Ac^-]$ that of the dissociated carboxylate salt species. The dissociation equilibrium regulated by equilibrium equation (3.9) results in a mixture of two species, one hydrophilic (Ac^-) and the other hydrophobic (AcH). Consequently, a variation of pH has the same effect than a variation of surfactant mixture hydrophilicity and results in a phase behaviour change [30]. However, the resulting adsorption at interface, which generates the acting surfactant blend, is difficult to predict because it also depends on the equilibrium of the non-dissociated species (AcH) between oil and water which is controlled by the partitioning equilibrium constant P_a which depends on the head group and tail length

$$P_a = \frac{[AcH]_{oil}}{[AcH]_{water}}. \tag{3.10}$$

More information on how to handle such pH sensitive systems is available elsewhere [31].

The characteristic parameter of the surfactant can be estimated by the use of the corresponding correlations (Eqs. (3.3) and (3.4)). For anionic surfactants for instance, salinity scans with a given oil, alcohol type and concentration and temperature, would allow to determine the optimum salinity (S^* in wt.% NaCl) for each tested surfactant, and thus estimate the value of the surfactant characteristic parameter σ from Eq. (3.3). Another way to characterise a surfactant is by using the double-scan technique (see Fig. 3.7). A first scan, e.g. a salinity scan, is carried out with a given set of (not-to-be changed) variables such as oil phase, alcohol type and concentration and temperature. With the first (known) surfactant (subscript 1), the optimum salinity S_1^* is such that

$$\ln S_1^* - k\,ACN - f(A) + \sigma_1 - \alpha_T\,(T - 25) = 0 \tag{3.11}$$

with the second (unknown) surfactant (subscript 2), the optimum salinity S_2^* is such that

$$\ln S_2^* - k\,ACN - f(A) + \sigma_2 - \alpha_T\,(T - 25) = 0. \tag{3.12}$$

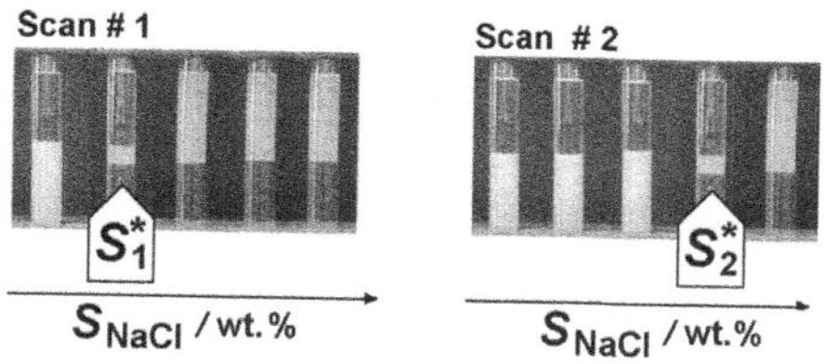

Figure 3.7 Principle of the double-scan technique to determine the characteristic parameter of an unknown surfactant.

By subtraction one obtains

$$\sigma_2 = \sigma_1 + \ln S_1^* - \ln S_2^*, \tag{3.13}$$

which allows to calculate the characteristic parameter of surfactant 2 from known surfactant 1 parameter. This double-scan technique is based on the accuracy of the correlation, particularly its linearity, which is likely to be better with small deviations. Hence, the first (known) surfactant should be relatively close to the unknown one, that is to say that it is advised that the difference between parameter values σ_2 and σ_1 should not exceed one or two units, whenever it is possible to find a proper reference surfactant 1 [32].

The principle of this double-scan technique may be used to estimate the experimental parameter for any component, not only the surfactant. One of the variables is the parameter to be estimated, whereas the other one is the scanned variable, which is in general taken as the salinity *S* (ionic surfactants) or the average *EON* (non-ionic surfactants). The temperature is also often used as the scanned variable for ethoxylated surfactants, although it is more time-consuming to carry out experimentally the experiments at different temperatures. On the other hand, the temperature range over which three-phase behaviour is exhibited is sometimes quite wide, and since the optimum temperature is not necessarily the centre of such range, it is not easy to pinpoint. The use of the emulsion phase inversion temperature is not an answer to the problem, because the emulsion inversion, although close to optimum formulation, is not necessarily coincident, particularly if the water/oil ratio is not unity. Finally, it is worth remarking that the experimental determination of the exact temperature at which the interfacial tension reaches its minimum surely allows pinpointing optimum formulation. Nevertheless, it is a tedious and uneasy task because most spinning-drop tensiometers which are used to measure ultralow interfacial tension lack of accurate temperature control.

The third alternative to determine a component characteristic parameter is to interpolate between known systems. If for instance the *EACN* of a crude oil has to be estimated, the best way is to carry out base experiments with a few *n*-alkanes for instance from heptane to tetradecane as in Fig. 3.8, in order to plot the variation of the optimum value of the scanned variable (*S*) versus *ACN*, then to carry out a scan with the unknown oil and to identify the alkane mixture that matches the optimum formulation. Figure 3.8 illustrates the determination of the *EACN* of an unknown crude oil by scanning the salinity of an anionic surfactant system. Such technique could also be used by extrapolating instead of interpolating, provided that the trend is linear and the extrapolation is not too far away, as for unknown paraffinic oil in Fig. 3.8.

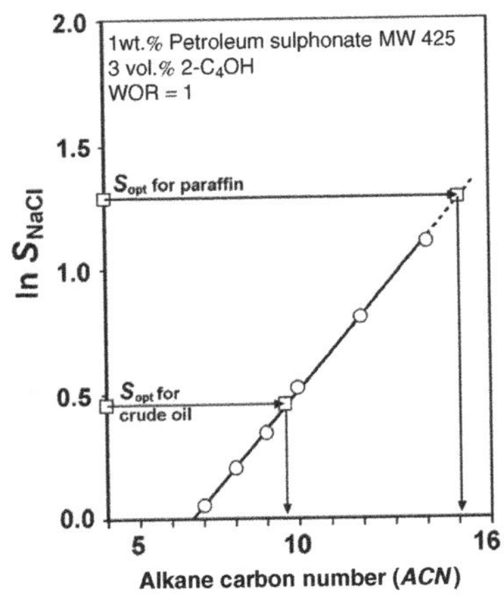

Figure 3.8 Determination of the *EACN* of unknown oil (circle data points with *n*-alkanes).

The last technique, which is the best one when there is very little information on the unknown component, is based on the mixing of a pair of known components with the unknown one, and the use of a linear mixing rule. For instance, if the characteristic parameter ($\beta = \alpha - EON$) of an unknown non-ionic surfactant is to be determined, the correlation to be used for the mixture of the two base products, such as two ethoxylated nonylphenols with different *EON*s, e.g. EON_1 and EON_2, so that the mixture that results in three-phase behaviour is EON_m is as follows:

$$\alpha - EON_{bm} + bS - k\,ACN - \phi(A) + c_T(T - 25) = 0, \tag{3.14}$$

with

$$EON_{bm} = x_{b1}EON_1 + x_{b2}EON_2 \text{ and } x_{b1} + x_{b2} = 1. \tag{3.15}$$

The *x*s are the molar fraction of the two base surfactants in the mixture, but if they are relatively close products, e.g. nonylphenols with $EON = 4$ and 6, the weight fraction, which is easier to calculate, may be used instead with insignificant error. In practice, it is preferable to use a relatively high total surfactant concentration to avoid fractionation effects, say, 2–3 wt.% at least. Note that EON_{bm} only depends on all the other formulation variables, e.g. salinity, oil, alcohol and temperature, which are fixed in all experiments. Since neither the structure nor the α value is known, the correlation to be used for the unknown non-ionic surfactant is the one including characteristic parameter β.

$$\beta + bS - k\,ACN - \phi(A) + c_T(T - 25) = 0. \tag{3.16}$$

A part of the base surfactant mixture, for instance 0.5 wt.% of the total 2 wt.%, that is 25%, is substituted by the unknown surfactant, indicated by subscript 3 in what follows. A scan is carried out by mixing the two base surfactants, i.e. by changing only a part of the three surfactant mixture, since the unknown surfactant content is constant, i.e. in the present case 0.5 wt.% ($x_3 = 0.25$) of the total mixture. Optimum formulation for three-phase behaviour is found for a certain mixture (x_{t1}, x_{t2}) of the base surfactants so that

$$EON_{tm} = x_{t1} EON_1 + x_{t2} EON_2 \text{ with } x_{t1} + x_{t2} = 0.75 \text{ and } x_{t3} = 0.25. \quad (3.17)$$

The optimum formulation for the ternary mixture is thus expressed as a combination of the three expressions of correlation (Eq. (3.16)), each with the corresponding weighting factor x.

$$(x_{t1} + x_{t2})(\alpha - EON_{tm}) + x_3\beta + bS - k\,ACN - \phi(A) + c_T(T - 25) = 0. \quad (3.18)$$

The surfactant term value is the same as in the base binary mixture, since the other variables are constant. Thus, by substituting, the following expression is obtained:

$$0.75(\alpha - x_{t1} EON_1 + x_{t2} EON_2) + 0.25\beta = x_{b1} EON_1 + x_{b2} EON_2 \quad (3.19)$$

in which the only unknown is the characteristic parameter β of tested surfactant 3. The previous calculation assumes that the mixing rule is linear, and that the deviation produced by the presence of surfactant 3 is large enough to insure accuracy. Since this is not known for sure, a more secure method is to carry out such experiments several times with different proportions (x_3) of the unknown surfactant 3, and to plot the calculated value β as a function of x_3, then if there is some variation in β with x_3, to extrapolate the trend to $x_3 = 1$. In some cases, three-phase behaviour cannot be achieved with the unknown surfactant only or with a large proportion of it. However, the method is a way to estimate a characteristic parameter value which is not directly accessible by experiment.

3.3.3 Generalised formulation as SAD and HLD

A decade after the empirical determination of the correlations for three-phase behaviour and the corroboration that the linearity and generality could not be coincidental, a simple interpretation was found through the so-called '*surfactant affinity difference*' (*SAD*) concept discussed next. When a simple ternary surfactant–oil–water system exhibits three-phase behaviour, the chemical potential μ of the surfactant is equal in the three phases (oil, water and microemulsion) at equilibrium referred to by subscripts O, W and M. It holds

$$\mu = \mu_o^* + RT \ln a_O = \mu_W^* + RT \ln a_W = \mu_M^* + RT \ln a_M, \quad (3.20)$$

where the star indicates the standard chemical potential in a reference state and 'a' is the activity of the surfactant. Since the excess phases of an optimum system do not contain much surfactant, i.e. the concentrations are below the critical micelle concentration (CMC), it may be assumed that the activities in such phases are essentially equal to the concentrations

(C). Thus

$$\mu_O^* + RT \ln C_O = \mu_W^* + RT \ln C_W \tag{3.21}$$

$$\mu_O^* - \mu_W^* = +RT \ln \frac{C_W}{C_O} = \Delta\mu_{W\to O}^* = SAD = RT \ln K_{WO}, \tag{3.22}$$

where K_{WO} is the partition coefficient of the surfactant between water and oil, which is measurable with the proper analytical technique. The variation of the standard chemical potential $\Delta\mu_{W\to 0}^*$ when a surfactant molecule passes from water to oil has been called the surfactant affinity difference (*SAD*), after the long-established definition of affinity as the negative of the standard chemical potential [33].

In ionic surfactant systems, the partition coefficient between excess phases is often found to be unity, hence an optimum formulation is defined by $SAD = 0$. With polyethoxylated surfactants, the CMC in water is often extremely low, whereas the monomer solubility in many oils is high. Consequently, the assumption of unit activity coefficient is not valid anymore and the partition coefficient between excess phases is not unity. In such cases, the partition coefficient value at optimum formulation is taken as a reference, and the deviation from this reference, the so-called hydrophilic–lipophilic deviation (*HLD*) is defined by dividing by *RT* to make the yardstick dimensionless [34].

$$HLD = \frac{(SAD - SAD_{ref})}{RT} = \ln K_{WO} - \ln K_{WOref}. \tag{3.23}$$

By definition, $HLD = 0$ at optimum formulation, and away from it, it is the expression of the relative surfactant affinity difference from it. Since the partition coefficient may be measured with a proper analytical technique, its variation with respect to formulation variables such as surfactant *EON*, oil *ACN*, water salinity *S* etc. may be determined. Such studies have been carried extensively with polyethoxylated surfactants and have shown that *HLD* dependence on the formulation variables has the same expression than the correlation for three-phase behaviour [35, 36]. It holds

$$HLD = \beta + bS - k\,ACN - \phi(A) + c_T(T - 25). \tag{3.24}$$

$$HLD = \sigma + \ln S - k\,ACN - f(A) - a_T(T - 25). \tag{3.25}$$

HLD is a generalised formulation yardstick that is some kind of extended HLB, which is function of all formulation variables (surfactant characteristics, co-surfactant type and concentration, temperature, oil nature, salinity ...) and it may be numerically estimated or measured with a much better accuracy than HLB, roughly equivalent to one-tenth of an HLB unit. From the physical chemistry point of view, it has a strong foundation, since it represents the change in standard chemical potential when a surfactant molecule passes from oil to water in the conditions of experiments.

The above expression have been extended to different families of surfactants, whose characteristic parameter (as β/k or σ/k) is expressed in *ACN* units, has been found to vary linearly with the surfactant alkyl carbon number (*SACN*), i.e. the number of atoms of carbon in the lipophilic tail group [35–37]. It holds

$$\frac{\sigma}{k} = \frac{\sigma_0}{k} + 2.25SACN \text{ and } \frac{\beta}{k} = \frac{\beta_0}{k} + 2.25SACN, \tag{3.26}$$

where the subscript 0 indicates a parameter which is characteristic of the head group of the surfactant. This is quite consistent with the fact that any additional methylene group in the alkyl chain results in an extra diminution in the variation in standard chemical potential when a surfactant molecule migrates from water to oil.

The linear temperature term in Eqs. (3.24) and (3.25) is only an approximation, which may be refined by using a Van't Hoff-type expression deduced from Eq. (3.22), particularly for the polyethoxylated non-ionic surfactant systems, for which the variation has been found to be significantly non-linear with temperature.

$$\frac{\partial \ln K_{WO}}{\partial T} = -c_T = -\frac{\Delta h^*}{RT^2}. \tag{3.27}$$

The numerical expression of c_T is seen to depend not only on temperature but also on the ethoxylation degree, for instance as follows for systems containing ethoxylated nonylphenols, *n*-heptane and water [34].

$$\frac{\partial \ln K_{WO}}{\partial T} = -\frac{(2210 + 450EON)}{T^2} (\text{in K}^{-1} \text{ units}). \tag{3.28}$$

This variation with temperature was overlooked in the early studies as it is not very significant, namely only twofold over a variation of 100°C. Another reason is that the pinpointing of optimum formulation in a temperature or *EON* scan is often inaccurate because of a wide three-phase zone with the optimum not necessarily at the centre.

The introduction of the alcohol co-surfactant effect as a formulation variable might not be completely satisfactory, since it includes the alcohol concentration, i.e. a composition variable. However, introducing the alcohol as a fourth component demands the use of a three-dimension quaternary diagram which is a very serious drawback in practice. Hence, the current use of an alcohol function is probably the best approximation for the sake of simplicity. However, it should be noted that the assumption of an amphiphilic pseudo-component made of surfactant and alcohol is not generally valid, in particular as far as the total concentration effect is concerned. It has been found that if the relative hydrophilicities of both amphiphiles are quite different, e.g. with a very hydrophilic surfactant-like dodecylsulphate and a very lipophilic alcohol-like hexanol, a severe fractionation takes place, which means that the surfactant/co-surfactant ratio is not the same in the different phases. As a consequence, a change in total amphiphile concentration or in water-to-oil ratio, at surfactant/co-surfactant ratio constant, produces a variation of the proportions of the different species at interface, therefore a variation of formulation. This is why it is advisable, unless otherwise required in some imperative way, to use a surfactant and an alcohol co-surfactant which are neither too hydrophilic nor too lipophilic, so that the pseudo-component approximation is roughly valid. This also tends to improve the role of the alcohol as a disorder providing component, whose molecules mingle with the surfactant molecules in the amphiphilic structure. The alcohols that are most likely to adsorb at the interface are the ones with an intermediate hydrophilicity with no strong overriding tendency to migrate to the oil or to the water. Other roles of alcohols will be discussed later on.

The *HLD* concept has been recently related to the so-called net-average curvature which indicates the size of the oil and water domains in the microemulsion. For marginal microemulsions, i.e. of the WI or WII type at some distance from optimum, the inverse of the swollen micelle Sauter diameter is proportional to *HLD*. The zero net curvature at optimum does not result from infinite radius but rather from the coexistence of finite curvatures of opposite signs. For bicontinuous microemulsions, it is the inverse of the characteristic length ξ which is maximised at $HLD = 0$. As discussed elsewhere [38], its value at optimum formulation ξ^* is the maximum distance that a molecule of oil or water can be separated from the surfactant layer and still interacts with it. In other words, it is the length at which the molecular interaction becomes equal to the molecular entropy.

3.4 Quality of formulation

The generalised formulation expression, for instance exemplified as the *R* ratio, indicates that there are many possibilities of achieving $R = 1$ or $HLD = 0$, i.e. many ways to attain an optimum formulation. Are all these optimum formulations equal or how do they differ? At optimum formulation, the amount of surfactant to attain the X point is essentially a quantification of its ability to solubilise immiscible fluids in a single phase and thus a measure of the system's quality. Another way to look at the performance is to measure the height of the multiphase region in a triangular diagram at some *WOR*, e.g. $WOR = 1$. It has been shown that the minimum height is attained when the multiphase region is type III. Since the minimum interfacial tension has been shown to be inversely proportional to the solubilisation, a higher solubilisation capacity coincides with a lower interfacial tension [39]. The concept of quality is hence important in practice, which is why it has attracted the attention of applied researchers in the past decades. The following sections outline the main achievements so far; more information is available elsewhere [40].

3.4.1 *Winsor's basic premise*

The first hint about boosting the quality of a microemulsion formulation was proposed by Winsor 50 years ago. In a formulation scan, the best solubilisation is attained when $R = \mathrm{N/D} = 1$, i.e. when the numerator (N) and denominator (D) are equal. Consequently, the comparison should always be carried out between two optimum formulations, i.e. two cases in which $R = 1$. But $R = 1$ may be attained in different ways, for instance as ratios such as 2/2 or 5/5, i.e. with equal interactions of the surfactant for both phases, but with different magnitudes of interactions. In order to compare a case $R = 1 = 2/2$ and a case $R = 1 = 5/5$, two compensating changes are required. For instance, a salinity change would alter the denominator, whereas an *ACN* change would alter the numerator.

The way to carry out quality testing experiments is thus to select two variables, one as the scan variable and the other as the perturbation variable, for instance salinity *S* and oil *ACN*. For a given ACN_1 oil, a salinity scan is carried out and an optimum salinity S_1^* is found. Then, the oil nature is changed to ACN_2, and a new salinity scan is carried out and optimum salinity is now found to be S_2^*. The change in ACN from ACN_1 to ACN_2 is compensated by the change in salinity from S_1^* to S_2^*. If $ACN_2 > ACN_1$ it is found that $S_2^* > S_1^*$ and also that the solubilisation has decreased, i.e. more surfactant is needed to reach the X point in case 2, hence the system quality has declined. The same perturbation–compensation dual change may be carried out by changing the ACN and the temperature. Here, the most significant effect of the increase in temperature is to reduce the interaction of the surfactant with the water phase because of the dehydration of the polyethyleneoxide chain, and the result is essentially the same than increasing salinity. If ACN is increased, a higher temperature is needed to attain optimum formulation and both N and D decrease equally which results in a lower quality.

In all previous cases, an increase in ACN, whatever the compensation variable to keep an optimum formulation, results in a decline in quality. However, the change in quality cannot be attributed to the change in a single variable, since two changes (at least) are required. If the increase in ACN is compensated by an increase in the length of the surfactant tail instead, then D is not affected by any of the changes, hence N is also invariant in the dual change. The experimental evidence indicates that these two compensating changes do not alter the solubilisation, i.e. the quality remains the same. It may be said that the quality diminution due to the ACN increase has been compensated by the enhancement due to the longer tail of the surfactant. This remark is important, because it indicates that the effect of an increase in ACN is not necessarily irremediably adverse as far as quality is concerned, but could be taken care of somehow with the proper change. The trick is to realise that when a detrimental change takes place, its effect should be neutralised on the same side of the interface, so that an optimum formulation is kept without reducing solubilisation. If the detrimental change is 'overneutralised', so that a beneficial change is required on the other side of the interface to maintain an optimum formulation, then an overall improvement could be the outcome of a three-change clever modification.

According to Winsor's premise, a general approach to increase the solubilisation in a microemulsion is thus to increase the interactions of the surfactant with both O and W at the same time, maintaining equal interactions to keep an $R = 1$ situation. For instance, if an alkyl phenol tail is increased, the head group EON has to be increased for compensating the change induced by the increasing tail length.[4]

Both changes lead to an increase in the solubilisation capacity, while the overall formulation is still the same [41]. Consequently, surfactants with bulkier groups on both sides are likely to be more efficient. For instance, in a system containing equal volumes of water and oil and a polyethoxylated alkylphenols an optimum formulation is found at $EON = 5.1$ for the nonyl species, and at $EON = 8.3$ for the dodecyl one at ambient temperature (solubilisation of 8 and 21 mL of oil and water per gram of surfactant). The value of the optimum solubility parameter changes with the alcohol content, as will be discussed next, but the trend is the same for all alcohols.

3.4.2 *Alcohol conventional effects*

Two effects of an alcohol additive have been mentioned so far. First, it contributes to the general formulation as a co-surfactant (slightly hydrophilic contribution for methanol and ethanol; lipophilic contribution for *n*-butanol and longer linear alcohols) and second, as a co-solvent. The alcohol co-adsorbs with the surfactant at the interface and thus changes the overall interaction of the amphiphilic film with the adjacent solvents. It is worth noting that the lipophilicity of the co-surfactant increases with the carbon chain length (*n*-butanol < *n*-pentanol < *n*-hexanol < *n*-heptanol). The longer the alcohol, the lower its tendency to act as co-surfactant, because it is rather solubilised in the oil phase. Consequently, the co-surfactant effect may be said to fade away, and to vanish for octanol or longer alcohols, depending of the nature of the oil phase. As the alcohol mostly partitions into the water or oil phase it behaves either as a co-solvent or a linker, as discussed later. In practice, a lipophilic co-surfactant effect as a formulation modifier is provided with alcohols such as *n*-butanol or *n*-pentanol, while a hydrophilic effect is provided with hydroxylated solvents such as ethylene glycol or butoxyethanol [41].

Alcohols such as sec-butanol and ter-pentanol, which are relatively neutral as far as their affinity for the oil and water is concerned, exhibit an *HLD* numerical contribution $f(A)$ or $\phi(A)$ close to zero, and consequently have essentially no effect modifying the formulation. These alcohols are however used because they do adsorb at the interface thus decreasing its rigidity and inhibiting the formation of lamellar liquid crystals, which are likely to be generated at optimum formulation because of the zero curvature. This second effect is often needed in formulating microemulsion with linear chain ionic surfactants, unless the temperature is high enough to provide thermal disorder. This effect is also often used with ethoxylated fatty alcohol non-ionic surfactants (unless the EON distribution is broad which automatically suppresses highly ordered phases). Branched alcohols such as sec-butanol and ter-pentanol are particularly useful to mix with linear chain surfactants because their branching increases the average area of the surfactant in the interfacial layer. Note that the driving force to adsorb at the interface is the neutral affinity for oil and water. Hence, these alcohols provide more spacing between surfactant molecules, and consequently they are the best to decrease cohesion and rigidity. However, it is worth noting that they are also the worst as far as the solubilisation performance is concerned, as will be discussed next.

The third role of the alcohol is as a co-solvent, whenever it mostly partitions into water (methanol and ethanol) or oil (*n*-hexanol or longer alcohol depending on the oil phase). Alcohol modifies the polarity of the phase in which it is dissolved and hence tends to reduce the intrinsic incompatibility between the oil and water, as indicated by a lowering of the interfacial tension. In extreme proportion, such alcohols are likely to allow the formation of a single phase, which is not to be confused with a microemulsion, because of the absence of structure. This is probably what happens with ethanol or 2-butoxyethanol for which more than 50% are required to attain the single-phase region [42, 43].

When such alcohol co-solvents are present in small proportion, they might not mix uniformly in the bulk of the oil or water phase and they could exhibit a forth effect discussed next.

3.4.3 Linker effects

3.4.3.1 Lipophilic linker

As discussed before, Winsor's premise that the quality of the system may be improved by increasing the size of both the hydrophilic and lipophilic groups of the surfactant while

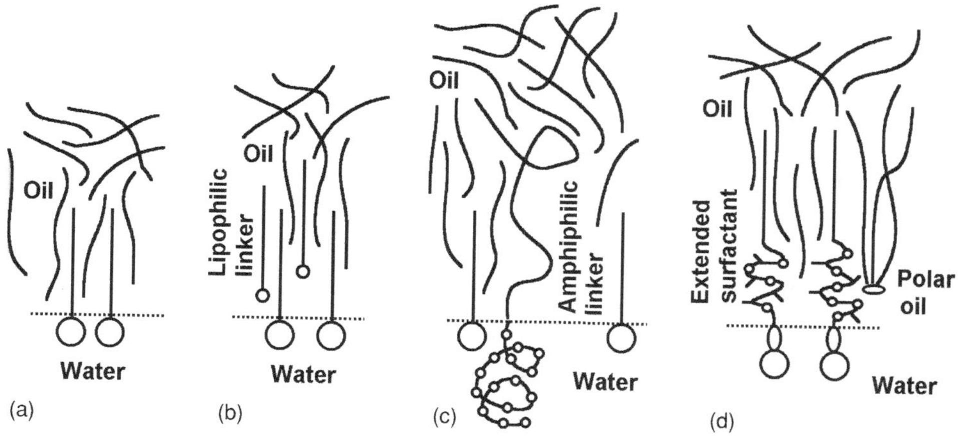

Figure 3.9 Solubilisation improvement of a conventional surfactant (a) by a lipophilic linker (b), an amphiphilic linker (c), or an extended surfactant (d).

keeping $R = 1$ has some limit, which is the solubility of the surfactant in the system, particularly in water. As a matter of fact the hydrophobic tail of the surfactant cannot be longer than 18 carbon atoms in a straight chain. On the other hand, the hydrophilic group could be made much larger than what is necessary to balance an 18 carbon atom tail. Hence, the main problem in increasing the interaction of the surfactant with both oil and water is on the oil side, particularly because of the (excessively) large hydrophobicity of a straight alkyl chain. The difficulty was overcome by a clever trick, namely by increasing the effective length of the hydrocarbon chain with the so-called *lipophilic linker* [44]. Lipophilic linkers are essentially oils with a slightly polar group, e.g. dodecanol or ethyl oleate [45]. Such polar oils tend to accumulate in the oil phase close to the interface in a mechanism called segregation [27]. Generally speaking, they may be said to be oriented perpendicular to the interface (i.e. parallel to the surfactant tail), hence contributing to a molecular organisation over distances that are longer than the chain length of the surfactant as illustrated in Fig. 3.9(b). It is essentially the extension of the chain length provided by the lipophilic linker that improves the interaction of the surfactant with the oil. The name 'linker' was chosen as these molecules create some extra link between the surfactant tail and the oil, which is the best when the lengths are matched [46]. Because the lipophilic linker is a separate molecule, the precipitation problem produced by a too long surfactant tail is avoided. As a matter of fact, the lipophilic linker presence in the oil phase close to the interface produces a slightly polar zone in the oil phase.

3.4.3.2 Hydrophilic linker

The same concept was transferred to the water side and the respective molecules were called hydrophilic linker [47, 48]. However, since the shape of the water molecules and their interactions are quite different from those of oils, hydrophilic linkers are no elongated molecules like the lipophilic linkers. Up to now the hydrophilic linker concept has been tested with alkyl naphtalene sulphonates only. Non-alkylated naphtalene sulphonate is a hydrotrope which stays in the bulk, whereas the dibutyl naphtalene sulphonate goes to the interface acting as a surfactant. On the other hand, the mono- or di-methyl naphtalene sulphonates exhibit the proper segregation in the water phase as expected from a hydrophilic linker. It is worth noting that the solubility improvement provided by a hydrophilic linker is not very high. Nevertheless, the change is significant and an even better solubilisation improvement is attained when both lipophilic and hydrophilic linker additives are used [47, 49].

3.4.3.3 Amphiphilic linker

Another improvement of the solubilisation has been accomplished by adding short amphiphilic block copolymers in low concentration (see Section 4.2 for details). Briefly, these polymers have a polyethylene–propylene hydrophobic block and a polyethylene oxide head group and are thus similar to the ethoxylated non-ionic surfactants to which these are added. The main difference is that the two blocks of the amphiphilic polymer are several times larger than the corresponding low molecular weight surfactant. The role of these polymers is to increase the reach of the amphiphilic layer such that it extends deeper into both the oil and the aqueous phase in accordance with Winsor's premise. As a consequence, they are found to notably increase solubilisation [50]. As seen in Fig. 3.9(c), these additives could be called amphiphilic linkers since they act upon both sides of the interface.

3.4.4 Extended surfactants

In Section 3.4.3, it is explained how the thickness of the transition zone can be extended by adding linkers. The resulting mixture could be an advantage because of the associated entropy disorder and flexibility, and a reduced probability of precipitation. However, the different species could also exhibit selective partitioning behaviour [32, 51] instead of a collective or ideal mixture behaviour, and some of them could move away from their proper position. This result in a loss of amphiphilic material at interface or in the microemulsion, and a decrease in performance. This is why some new composite amphiphiles, the so-called *extended surfactants*, were synthesised to achieve the same kind of behaviour with a single molecular species at the interface [52]. Extended surfactants have the conventional polar head and hydrophobic tail, but benefit from a central extension, sometimes called spacer arm, which exhibits an intermediate polarity as illustrated in Fig. 3.9(d). A typical species is dodecyl poly-propylene oxide di-ethoxy sulphate, which is abbreviated with C_{12}-PPO_{10}-EO_2-SO_3Na if 10-propylene oxide groups form the central part. The ethoxy groups are likely to have a hydrophilic linker effect, but they are generally added for another reason, namely to chemically attach the sulphate group to the polypropylene oxide chain. Their properties indicate that this kind of surfactant become more hydrophobic when the polypropylene oxide chain gets longer [52]. It is because of this finding that the PPO block has to be considered as an extension of the hydrophobic part of the surfactant, i.e. it behaves as a lipophilic linker and extends the reach of the hydrophobic tail further in the oil phase. However, its slightly polar structure allows to avoid the precipitation. These surfactants are water-soluble up to 15 PPO groups and have been shown to exhibit an excellent oil solubilisation with an X point down to a few wt.%, which is particularly outstanding with long-chain hydrocarbons like hexadecane. Moreover, they exhibit a very high solubilisation in microemulsions with polar oils such as ethyl oleate (X point below 2 wt.%) and even natural triglycerides (X point below 10 wt.%), provided that the size of the extension is matched with the size of the oil molecule. Note that this wide-ranging solubilisation capacity is most likely due to their unique molecular structure. As expected from the inverse relationship between solubilisation and interfacial tension reported 30 years ago [39], these extended surfactants are also providing a way to achieve an ultralow interfacial tension with natural and synthetic polar oils used in pharmaceuticals, cosmetics and foods. As a matter of fact, the combination of surfactants and linkers or the use of extended surfactants allows for enhanced detergency by solubilisation in microemulsions [53]. Some extended surfactants have been recently synthesised with biocompatible polar heads such as cyclic sugars, carboxylic acid or ethoxy composites, with one or two polar groups [54, 55]. Because such flexibility does not seem to alter the main features, they are likely to find scores of applications for high performance solubilisation uses such as soak-only detergency, or finely tuned functions such as toxic drug trapping in the blood, or bladder-stone in situ dissolution.

Extended surfactants behave essentially as conventional surfactants with similar head groups as far as the formulation variable effects are concerned. The polypropylene group behaves as a part of the lipophilic tail, but its presence alters the value of the k coefficient in the *HLD* expressions [52]. Extended surfactants may be mixed with conventional surfactants to attain intermediate formulations which are often closely described by a linear mixing rule. Hence, the formulation of systems containing an extended surfactant may be finely tuned by adding conventional ones.

3.4.5 Quality and transparency

As a final note of this section on quality, it is worth noting that because of the high solubilisation exhibited by systems close to optimum formulation, and independently of the fact that a conventional or extended surfactant is used, the corresponding microemulsions are not transparent but cloudy and even milky in the best solubilisation cases. If this is inconvenient for the application, i.e. if a more transparent aspect is compulsory, a decrease in solubilisation is mandatory, which may be attained essentially in two ways. First, some additional disorder could be introduced in the amphiphilic structure while keeping the optimum formulation ($HLD = 0$), for instance by using a mixture of widely different surfactants, by using branched chain amphiphiles, or by adding more alcohol co-surfactants, so that the cohesion of the amphiphile layer decreases, and its flexibility increases. On the other hand, the formulation may be shifted away from optimum, i.e. from $HLD = 0$, with two penalties (a) an increase in the required amount of surfactant and co-surfactant to attain a single-phase microemulsion and (b) a different solubilisation of oil and water in the microemulsion, which may be however a helpful feature for some applications. With the decrease in solubilisation, the characteristic length of the microemulsion, as well as the light scattering effect decreases. If the resulting microemulsion contains much more water than oil or reversely, a non-bicontinuous structure is likely to develop, which is often more robust as far as formulation effects are concerned, though much less efficient from the point of view of solubilisation.

As much as optical special effects are concerned, it is worth remarking that three decades ago microemulsions were reported to exhibit shear birefringence [56]. This phenomenon was explained at the time by the formation of a layered arrangement when the microemulsion was submitted to some shear. It was observed for bicontinuous microemulsions close to the formation of a lamellar liquid crystal, so that an extra ordering induced by the shear was assumed to generate a temporary molecular arrangement and the resulting birefringence. After the shear was terminated the disorder came back and the birefringence vanished. It is worth noting that the corresponding experiments were carried out with petroleum sulphonate surfactants which were reported to exhibit mesophases, which could be accountable for the shear birefringence as well.

3.5 Formulations for special purposes

Since the best solubilisation is attained for the Winsor type III case, the formulator always tries to keep the *HLD* generalised formulation as close to zero as possible. Nevertheless, the *HLD* expression contains many variables and hence several degrees of freedom are available to satisfy other constraints or other desirable features. The increase in solubilisation which has been discussed in the previous section is important, but not necessarily the most important in practice. On the other hand, it is known that changes may be brought to the system without altering the solubilisation, as for instance two concomitant modifications in the numerator or in the denominator of Winsor's *R* ratio. This is particularly the case when surfactant mixtures are used.

3.5.1 *Surfactant mixing rules*

Surfactant mixtures are used essentially for two reasons. The first one is that they cannot be avoided in some cases, which is very often related to the manufacturing process. The alkyl chains of many surfactants come either from the polymerisation of an olefin and thus exhibit a distribution of molecular weight, or are extracted from some natural substance such as vegetable or animal triglyceride oil, or from some petroleum distillation cut, which are all mixtures. The polycondensation of ethylene oxide also results in a distribution of the number of ethylene oxide groups per molecule, sometimes spanning over a considerable range. The addition of sugar groups coming from the hydrolysis of starch also results in mixed species, namely alkyl polyglucosides. Reducing the variety of substances in a commercial product implies some narrow cut distillation or some purification process that is often too costly. Hence, most commercial surfactants contain different species which are each likely to exhibit an independent behaviour whenever the conditions are met [32, 51]. On the other hand, mixtures are often made on purpose to attain some intermediate property or some synergetic effect. The intermediate property comes directly from the fact that in many cases the surfactant characteristic parameter in the *HLD* expression (σ or β) approximately follows a linear mixing rule, as presumed in the calculation of the HLB of a surfactant mixture 60 years ago.

Because the value of the parameter k in Eqs. (3.24) and (3.25) depends on the polar group, it has been found convenient to compare surfactants according to their characteristic

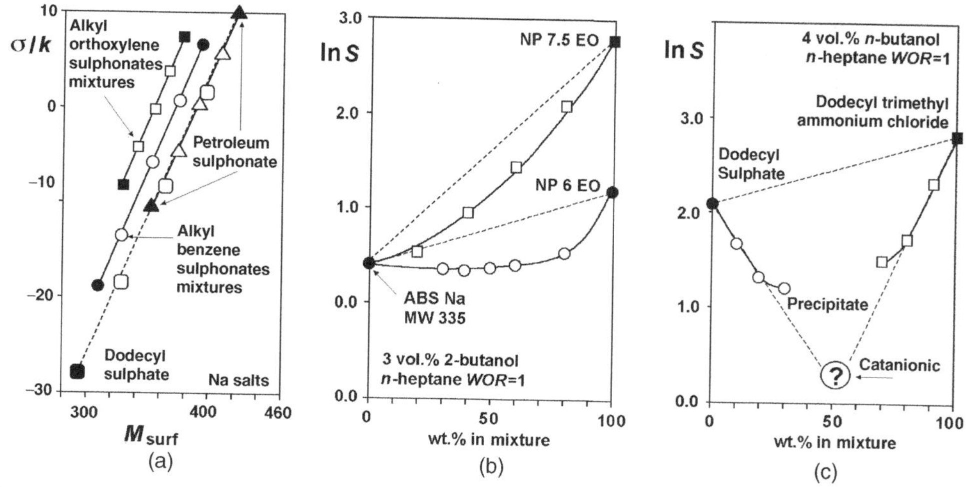

Figure 3.10 Mixing rules for surfactant mixtures. (a) σ/k as a function of the average surfactant molecular weight (M_{surf}) for a mixture of two anionic surfactants. (b) Salinity (S) versus wt.% of non-ionic surfactant in a mixture of an anionic and a non-ionic surfactant. (c) Salinity (S) versus wt.% of cationic surfactant in a mixture of an anionic and a cationic surfactant.

parameter divided by k, i.e. as σ/k and β/k. These values are expressed in *ACN* units which have exactly the same physico-chemical meaning for both *HLD* expressions. As reported previously [37], surfactant characteristic parameters σ/k and β/k vary linearly with the length of the n-alkyl lipophilic group of the surfactant. When two surfactants with same head groups are mixed, the characteristic parameter of the mixture may be expressed according to a linear mixing rule in practically all cases (see Fig. 3.10(a)). It holds

$$\frac{\sigma_M}{k} = \frac{\Sigma x_i \sigma_i}{k} \text{ and } \frac{\beta_M}{k} = \frac{\Sigma x_i \beta_i}{k}, \tag{3.29}$$

where x_i is the molar fraction of species 'i' in the interfacial mixture, which is generally unknown, but is assumed to be the same than the molar fraction in the system, provided there is no severe fractionation phenomenon [32]. If the two surfactants of the mixture have different head groups, the corresponding values of parameter k might be different and the linear mixing rule is more generally written as

$$\frac{\sigma_M}{k_M} = \frac{\Sigma x_i \sigma_i}{k_i} \text{ and } \frac{\beta_M}{k_M} = \frac{\Sigma x_i \beta_i}{k_i} \text{ with } k_M = \Sigma x_i k_i. \tag{3.30}$$

When two ethoxylated non-ionic surfactants with same hydrophobic groups are mixed, the linear mixing rule on parameter β becomes a linear mixing rule on the degree of ethoxylation *EON* and may be written as

$$EON_M = \Sigma x_i EON_i, \tag{3.31}$$

where EON_M is the average degree of ethoxylation of the mixture and EON_i is the degree of ethoxylation either of the ith oligomer or the ith component of the mixture, which is often already a mixture. For instance, if commercial ethoxylated nonylphenols with average ethoxylation degrees 4 and 6 are mixed in a 25/75 molar ratio, the average *EON* of the mixture will be 4.5; the practice indicates that this kind of linear mixing rule applies quite well, provided that the mixed products are close enough, as far as their hydrophilicity is concerned, for instance with a difference in average *EON* lower than 2 units. A handy rule of thumb is that the more complex the mixture is, the better the mixing rule applies. For instance, a mixture of four commercial products with $EON = 4$, 6, 8 and 10 would probably result in a more linear rule than a mixture of products with $EON = 4$ and 10. Mixtures of widely different products might exhibit a fractionation of some oligomer species, particularly with a large proportion of the compounds of the $EON = 4$ commercial product selectively partitioning into the oil phase, thus leaving a more hydrophilic mixture at interface and in the water phase [32].

3.5.2 Reduction in hydrophilicity with ionic–non-ionic surfactant mixtures

Establishing the characteristic parameter of an ionic–non-ionic mixture is not so easy, because many of the formulation variables do not have the same meaning in the two *HLD* expressions (see Eqs. (3.24) and (3.25)). In any case, it is recommended to use the (logarithm of) salinity as an indication of the hydrophilicity of the mixture, and to plot the optimum salinity for three-phase behaviour as a function of the mixture composition. In many cases, the rule would exhibit an optimum salinity slightly lower than the expected from a linear mixing rule in the log scale. This is illustrated in Fig. 3.10(b) by two cases of mixtures of an alkyl benzene sulphonate with polyethoxylated nonylphenols. This 'hamoc' deviation is most likely due to a shielding effect of the polyethoxylated chain of the non-ionic surfactant that wraps around the ionic head group of the other surfactant and thus reduces the interaction with water.

3.5.3 Synergy with anionic–cationic surfactant mixtures

Mixing anionic and cationic surfactants results in the formation of an equimolar catanionic species, which is likely to precipitate even at very low concentration, because it is more hydrophobic (two tails) and less ionic (the charges cancel out at least partially). It was shown, however, that this equimolar catanionic surfactant tends to behave as a hydrophobic amphoteric, i.e. ionic surfactant, which is able to exhibit a linear mixing rule with either of the ionic species provided its proportion remains small, say, less than 20% [57]. For instance, if 5 wt.% of a cationic surfactant is added to 95 wt.% of anionic surfactant, the actual mixture behaves as if it were a mixture of 90 wt.% anionic and 10 wt.% catanionic surfactant. In practice, the pure catanionic species precipitates and hence does not exist as a soluble substance in the microemulsion. Hence, its characteristic parameter has to be estimated by extrapolating the linear trends of the 1:1 mixture, as seen in Fig. 3.10(c).

Because the interaction between opposite charges is extremely strong, the expected structure when mixing anionic and cationic surfactants is a bilayered crystal precipitate, which may be a liquid crystal. Consequently, the generation of a microemulsion with such a mixture often demands the input of extra geometrical disorder in the formulation. This may be attained in different ways such as a high concentration of secondary or tertiary alcohol, a mixture with an intermediate non-ionic surfactant, different alkyl chain lengths, branched alkyl chains and a high temperature [57–59].

3.5.4 Temperature-insensitivity with anionic–non-ionic surfactant mixtures

It has been known for almost three decades that mixing anionic and ethoxylated non-ionic surfactants allows to produce microemulsions which are insensitive to temperature changes [60]. The expressions of the *HLD* for the two kind of surfactants (see Eqs. (3.24) and (3.25)) exhibit a different sign before the a_T and c_T temperature coefficients. The signs express the fact that the affinity of water for an ionic surfactant increases when the temperature increases, whereas the reverse takes place with a polyethoxylated non-ionic surfactant. Coefficient a_T is about 0.01 for alkyl benzene sulphonates and 0.02 for alkyl trimethyl quaternary ammoniums, while c_T is in the 0.05–0.1 range for ethoxylated alcohols and phenols, with a tendency to increase with the ethoxylation degree and to decrease with increased temperature. The fact is that the effect of the temperature is several times stronger with non-ionics, hence a mixture insensitive to temperature should contain more ionic than non-ionic, so that the effects could cancel out [60–62].

The calculation cannot be carried out in an accurate way because, as mentioned before, the mixing rule between anionic and non-ionic surfactants is not actually linear due to a shielding of the ionic group by the polyethylene oxide chain. However, the use of a linear approximation often leads to a fairly good estimate in some cases such as a mixture of alkylbenzene sulphonates and ethoxylated nonylphenols to be considered as an example next [61].

Assuming a linear approximation and using an *ACN* scan technique to find the optimum formulation, then the correlation for a three-phase behaviour ($HLD = 0$) could be written as

$$PACN_{NI} = \left(\frac{1}{k}\right)[\beta + bS - \phi(A) + c_T(T - 25)] \qquad (3.32)$$

for an ethoxylated non-ionic system and as

$$PACN_{AI} = \left(\frac{1}{k}\right)[\sigma + \ln S - f(A) - a_T(T - 25)] \qquad (3.33)$$

for an anionic system, where *PACN* (preferred *ACN*) indicates the measured or extrapolated *ACN* value corresponding to optimum formulation of the *ACN* scan, at which the minimum interfacial tension or three-phase behaviour is exhibited for the system in some reference state, that is at a given salinity, alcohol content and temperature. It is worth remarking that *PACN* is a parameter characteristic of the surfactant with essentially the same information than β/k or σ/k. Parameter σ/k was called EPACNUS (for extrapolated preferred *ACN* at unit salinity, no alcohol, and ambient temperature, for which $\text{Ln}S = 0$,

$f(A) = 0$, $T = 25°C$). The usual reference for ionic surfactant would be the same for all but $S = 0$.

It is seen that $PACN_{NI}$ increases as c_T/k (about 0.24 for ethoxylated nonylphenols) and $PACN_{AI}$ decreases as a_T/k (0.06 for alkylbenzene sulphonates), i.e. about four times slower. According to *HLD* expressions (Eqs. (3.24) and (3.25)), the relationship between *PACN* and *T* is a linear relation. It holds

$$PACN_{NI} = Cst_{NI} + 0.24T \text{ hence } \frac{\partial PACN^*_{NI}}{\partial T} = 0.24 \tag{3.34}$$

and

$$PACN_{AI} = Cst_{AI} - 0.06T \text{ hence } \frac{\partial PACN^*_{M}}{\partial T} = -0.06. \tag{3.35}$$

If a mixture of these anionic and non-ionic surfactants is prepared with molar fractions x_{NI} and x_{AI} at the interface (assumed to be the same than the molar fractions in the system), and if the mixing rule is assumed to be linear, then the optimum formulation of the mixture $PACN_M$ may be estimated as

$$PACN_M = x_{NI} PACN_{NI} + x_{AI} PACN_{AI}. \tag{3.36}$$

By differentiation with respect to temperature and substitution of the value of derivatives

$$\frac{\partial PACN_M}{\partial T} = x_{NI}\frac{\partial PACN_{NI}}{\partial T} + x_{AI}\frac{\partial PACN_{AI}}{\partial T} = 0.24x_{NI} - 0.06x_{AI} \tag{3.37}$$

since $x_{NI} + x_{AI} = 1$ the solution is $x_{NI} = 0.2$ and $x_{AI} = 0.8$. Because of the approximations, the actual result will be probably somehow different from these values, and some trial and error will be necessary to pinpoint the mixture which is exactly insensitive to temperature. As a matter of fact it is not easy to use *ACN* or *EACN* as a scan variable, and the salinity (as its logarithm) is probably a better choice to carry out the final trial and error. The fact that the salinity appears as *S* and ln *S* in the two *HLD* relationships (Eqs. (3.24) and (3.25)) is not critical since the trial and error is carried out close to the case in which the optimum salinity does not change with temperature, so it would show a constant value on any scale. Figure 3.11 indicates the variation of the optimum salinity for different anionic to non-ionic ratios in the mixture. It is seen in Fig. 3.11 that the variation is very close to linear over a wide range of temperature, and that one of the mixtures would produce a complete insensitivity, i.e. the one with about 36 wt.% non-ionic surfactant.

When selecting an anionic/non-ionic mixture, three choices have to be made: the anionic surfactant, the non-ionic surfactant and the proportion of the two in the mixture. When the proportion is selected such that the microemulsion is insensitive to temperature, there are still two available degrees of freedom which may be expressed as $PACN_{NI}$ and $PACN_{AI}$ since the other parameters are to be kept the same for all mixtures. If $PACN_{NI} > PACN_{AI}$ then the non-ionic surfactant is less hydrophilic than the anionic one in the reference conditions, and vice versa. The optimum formulation of the mixture would turn more hydrophilic or more lipophilic depending on the requirement to attain insensitivity to temperature. On the other hand, if $PACN_{NI} = PACN_{AI}$ at the

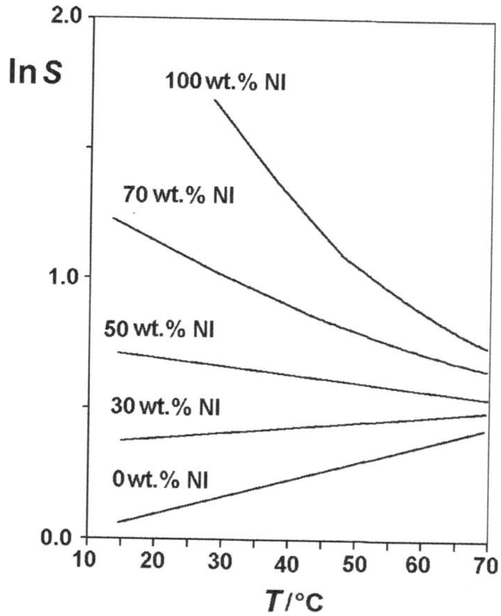

Figure 3.11 Variations of optimum salinity (S) as a function of the temperature (T) for an anionic surfactant (0 wt.% NI), a non-ionic (100 wt.% NI) surfactant and their mixtures. The systems contain equal amounts of water and n-heptane, 3 wt.% 2-butanol and 1 wt.% surfactant mixture (NI + AI). NI, polyethoxylated nonylphenol with an average of 6.5 ethyleneoxide units; AI, petroleum sulphonate sodium salt with an average molecular weight of 420 g/mol.

temperature of interest, then both surfactants have exactly the same hydrophilicity and the mixing rule between the *PACN*s (Eq. (3.36)) indicates that $PACN_M$ will be constant whatever the mixture composition, particularly for the mixture which satisfies the insensitivity to temperature condition. It has been shown [60, 61] that this peculiar situation takes place when the temperature (symbolised as preferred temperature PT_{NI}) at which the non-ionic surfactant passes from hydrophilic to lipophilic is the same than the preferred temperature PT_{AI} at which the anionic surfactant passes from lipophilic to hydrophilic, for a given system containing a given reference state for brine (S_{ref}), alcohol (A_{ref}) and oil (ACN_{ref}). It is worth remarking that PT_{NI} is essentially the same as Shinoda's PIT. It holds

$$PT_{NI} - 25 = \left(\frac{1}{c_T}\right)[-\beta - bS_{ref} + k\,ACN_{ref} + \phi(A_{ref})] \qquad (3.38)$$

$$PT_{AI} - 25 = \left(\frac{1}{a_T}\right)[\sigma + \ln S_{ref} - k\,ACN_{ref} - f(A_{ref})]. \qquad (3.39)$$

Accordingly, if the two surfactants are selected such that both their preferred temperatures coincide with the temperature of the experiment and if the mixture is insensitive to temperature changes, then the system is extremely robust, since it is insensitive to both temperature and composition, at least over an extremely wide range (which would be the whole range if the relationship were perfectly linear).

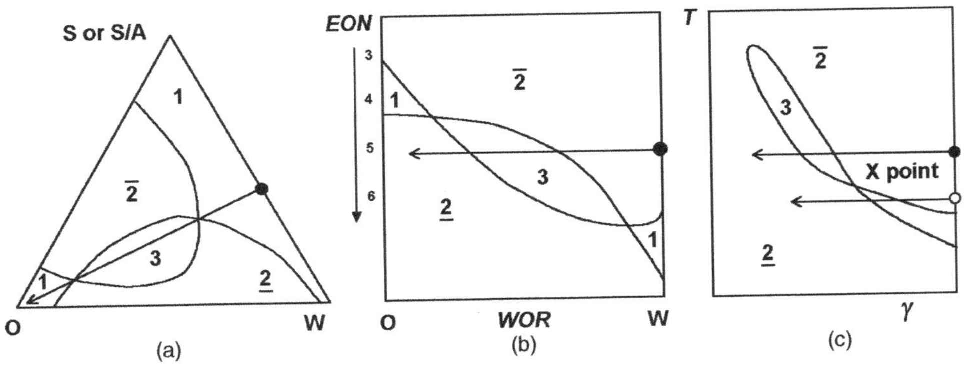

Figure 3.12 Changes in formulation (and phase behaviour) induced by a change in composition illustrated in the three bidimensional diagrams.

3.5.5 *Effect of composition variables and fractionation problems*

Changing the composition of a surfactant–oil–water system could modify the phase behaviour as indicated in Fig. 3.12 along the paths indicated by arrows. In many cases, the dilution by water or by oil results in the appearance or disappearance of a microemulsion. In the latter case, the microemulsion can be in equilibrium with excess water, excess oil or both. The problem is easily solved whenever a good phase diagram is at hand, which is not often the case as a matter of fact [63–65].

When surfactant mixtures or surfactant/co-surfactant mixtures are used, the situation may even be more difficult as was outlined above. Difficulties mainly arise if the various species are partitioned differently between the bulk phases and the interface (see also Chapter 1). Polyethoxylated alkylphenols used in commercial surfactant mixtures have been studied in detail and several rules of thumb could be extracted from these studies [32, 66]. Basically, the species with a low *EON* (i.e. the most hydrophobic) tend to partition preferentially in oil, while the remaining species are used to form the interface (note that the amount of surfactant in the aqueous phase can be neglected because of the very low CMC value). Thus, a commercial surfactant with an average *EON* of 4 may result in an interfacial formulation of $EON = 5$. The phenomenon is boosted by a reduction of the total surfactant concentration and by a decrease in *WOR* and results in slanted optimum formulation lines in phase behaviour diagram where one or two composition variables are plotted (see Fig. 3.12). A result of this partitioning is a non-Winsor III phase behaviour in a triangular diagram (see Fig. 3.12(a)), a slanted band in an *EON-WOR* map (see Fig. 3.12(b)), and a distorted fish diagram (see Fig. 3.12(c)) [32, 51].

Cheap commercial anionic surfactants of the petroleum sulphonate type might contain disulphonates that are likely to partition in water, thus resulting in a similar fractionation phenomenon. However, this time the candidates to partition in water are the very hydrophilic disulphonates and thus the remaining more lipophilic species are more likely to adsorb at interface. As a consequence the interfacial or microemulsion formulation is more lipophilic [33]. Since this is just the opposite of the previously discussed case of polyethoxylated nonyl phenols, the two phenomena are able to cancel out provided that the proper mixture is found [67]. This remark indicates that these exceptions to the simple and linear mixing rules might be used to some advantage by the clever formulator who has the know-how to harness the intricacies of multidimensional problems.

3.6 Final comment

The complexity of real systems makes the control of their formulation a very tough issue, but the number of degrees of freedom and its variety also provide to the expert formulator many possibilities to tackle the most intricate problems. The experience indicates that the few months of training needed to gain an advanced know-how and a practical expertise in formulating microemulsions are a most profitable education that returns to the formulator a time-saving capacity worth thousand times the investment.

Acknowledgements

The authors would like to acknowledge the financial support of their University Research Council CDCHT-ULA, particularly through grants I-834-05-08-AA and I-815-05-08-A.

Notes

1. Note that in Chapter 1, water is abbreviated with A, oil with B and the surfactant with C, while it is W, O and S in this chapter.
2. Note that the surfactant-rich phase that contains the swollen micelles (i.e. the microemulsion droplets) is sometimes abbreviated with the symbol Wm (Om) to indicate that it is water (oil) continuous.
3. In Chapter 1, the total mass fraction of surfactant is abbreviated with γ, while the water-to-oil ratio is abbreviated with α (equal mass fractions) or ϕ (equal volume fractions).
4. The quality cannot be boosted indefinitely by just increasing the size or length of both the head and tail groups of the surfactant molecule. In effect, there is generally some limit due to the solubility of the surfactant in the phases, the so-called Krafft temperature effect, because of the incompatibility of long straight alkyl groups with water.

References

1. Winsor, P. (1954) *Solvent Properties of Amphiphilic Compounds.* Butterworth, London.
2. Salager, J.L. (1999) Microemulsions. In G. Broze (ed), *Handbook of Detergents, Part A: Properties.* Marcel Dekker, New York, pp. 253–302.
3. Reed, R.L. and Healy, R.N. (1977) Some physicochemical aspects of microemulsion flooding: A review. In D.O. Shah and R.S. Schechter (eds), *Improved Oil Recovery by Surfactant and Polymer Flooding.* Academic Press, New York, pp. 383–437.

4. Bourrel, M. and Schechter, R.S. (1988) *Microemulsions and Related Systems*. Marcel Dekker, New York.
5. Salager, J.L. (2000) Formulation concepts for the emulsion maker. In F. Nielloud and G. Marti-Mestres (eds), *Pharmaceutical Emulsions and Suspensions*. Marcel Dekker, New York, pp. 19–72.
6. Salager, J.L. (2000) Emulsion properties and related know-how to attain them. In F. Nielloud and G. Marti-Mestres (eds), *Pharmaceutical Emulsions and Suspensions*. Marcel Dekker, New York, pp. 73–125.
7. Pérez, M., Zambrano, N., Ramirez, M., Tyrode, E. and Salager, J.L. (2002) Surfactant–oil–water systems near the affinity inversion. Part XII: Emulsion drop size versus formulation and composition. *J. Dispersion Sci. Technol.*, **23**, 55–63.
8. Lachaise, J., Breul, T., Graciaa, A., Marion, G., Monsalve, A. and Salager, J.L. (1990) Foaming properties of surfactant–oil–water systems in the neighbourhood of optimum formulation. *J. Dispersion Sci. Technol.*, **11**, 443–453.
9. Kahlweit, M., Lessner, E. and Strey, R. (1983) Influence of the properties of the oil and the surfactant on the phase behavior of systems of the type H_2O–oil–nonionic surfactant. *J. Phys. Chem.*, **87**, 5032–5040.
10. Kahlweit, M., Strey, R. and Haase, D. (1985) Phase behavior of multicomponent systems water–oil–amphiphile–electrolyte. *J. Phys. Chem.*, **89**, 163–171.
11. Bourrel, M., Chambu, C., Schechter, R.S. and Wade, W.H. (1982) The topology of phase boundaries for oil/brine/surfactant systems and its relationship to oil recovery. *Soc. Petrol. Eng. J.*, **22**, 28–36.
12. Burauer, S., Sachert, T., Sottmann, T. and Strey, R. (1999) On microemulsion phase behavior and the monomeric solubility of surfactant. *J. Phys. Chem.*, **1**, 4299–4306.
13. Kahlweit, M., Strey, R., and Busse, G. (1993) Weakly to strongly structured mixtures. *Phys. Rev. E.*, **47** (6), 4197–4209.
14. Queste, S., Salager, J.L., Strey, R. and Aubry, J.M. (2007) The EACN scale for oil classification revisited thanks to fish diagrams. *J. Colloid Interface Sci.*, **312**, 98–107.
15. Shinoda, K. and Saito, H. (1968) The effect of temperature on the phase equilibria and the types of dispersion of the ternary system composed of water, cyclohexane, and nonionic surfactant. *J. Colloid Interface Sci.*, **26**, 70–74.
16. Salager, J.L., Miñana-Pérez, M., Pérez-Sánchez, M., Ramirez-Gouveia, M. and Rojas, C.I. (1983) Surfactant–oil–water systems near the affinity inversion – Part III: The two kinds of emulsion inversion. *J. Dispersion Sci. Technol.*, **4**, 313–329.
17. Antón, R.E. (1981) Phase behavior of a quaternary surfactant–alcohol–water–oil system. Thesis, Universidad de Oriente, Pto La Cruz, Venezuela.
18. Buzier, M. and Ravey, J.C. (1983) Solubilization properties of nonionic surfactants 1. Evolution of ternary phase diagrams with temperature, salinity, HLB, and ACN. *J. Colloid Interface Sci.*, **91**, 20–33.
19. ICI Americas (1976) *The HLB System – A Time Saving Guide to Emulsifier Selection*. Atlas Division, Wilmington.
20. Shinoda, K. and Arai, H. (1964) The correlation between phase inversion temperature in emulsion and cloud point in solution of nonionic emulsifier. *J. Phys. Chem.*, **68**, 3485–3490.
21. Kunieda, H. and Shinoda, K. (1982) Phase behavior in systems of nonionic surfactant–water–oil around the hydrophile–lipopile balance temperature (HLB-temperature). *J. Dispersion Sci. Technol.*, **3**, 233–244.
22. Salager, J.L. (1996) Quantifying the concept of physico-chemical formulation in surfactant–oil–water systems. *Prog. Colloid Polym. Sci.*, **100**, 137–142.
23. Salager, J.L., Morgan, J.C., Schechter, R.S., Wade, W.H. and Vásquez, E. (1979) Optimum formulation of surfactant/water/oil systems for minimum interfacial tension or phase behavior. *Soc. Petrol. Eng. J.*, **19**, 107–115.

24. Antón, R.E., Garcés, N. and Yajure, A. (1997) A correlation for three-phase behavior of cationic surfactant–oil–water systems. *J. Dispersion Sci. Technol.*, **18**, 539–555.
25. Bourrel, M., Salager, J.L., Schechter, R.S. and Wade, W.H. (1980) A correlation for phase behavior of nonionic surfactants. *J. Colloid Interface Sci.*, **75**, 451–461.
26. Cayias, J.L., Schechter, R.S. and Wade, W.H. (1976) Modeling crude oils for low interfacial tension. *Soc. Petrol. Eng. J.*, **16**, 351–357.
27. Graciaa, A., Lachaise, J., Cucuphat, C., Bourrel, M. and Salager, J.L. (1993) Interfacial segregation of ethyl oleate/hexadecane oil mixture in microemulsion systems. *Langmuir*, **9**, 1473–1478.
28. Acosta, E., Le, M.A., Harwell, J.H. and Sabatini, D.A. (2003) Coalescence and solubilization kinetics in linker-modified microemulsions and related systems. *Langmuir*, **19**, 566–574.
29. Antón, R.E. and Salager, J.L. (1990) Effect of the electrolyte anion on the salinity contribution to optimum formulation of anionic surfactant microemulsions. *J. Colloid Interface Sci.*, **140**, 75–81.
30. Qutubuddin, S., Miller, C.A. and Fort, T. (1984). Phase behavior of pH-dependent microemulsions. *J. Colloid Interface Sci.*, **101**, 46–58.
31. Salager, J.L. and Antón, R.E. (1999) Ionic microemulsions. In P. Kumar and K.L. Mittal (eds), *Handbook of Microemulsion Science and Technology*. Marcel Dekker, New York, pp. 247–280.
32. Antón, R.E., Andérez, J.M., Bracho, C., Vejar, F. and Salager, J.L. (2008) Practical surfactant mixing rules based on the attainment of microemulsion–oil–water three-phase behavior systems. In R. Narayanan (ed), *Interfacial Processes, Phenomena and Molecular Aggregation*. Springer-Verlag, Heidelberg, Berlin.
33. Wade, W.H., Morgan, J., Schechter, R.S., Jacobson, J.K. and Salager, J.L. (1978) Interfacial tension and phase behavior of surfactant systems. *Soc. Petrol. Eng. J.*, **18**, 242–252.
34. Salager, J.L., Márquez, N., Graciaa, A. and Lachaise, J. (2000) Partitioning of ethoxylated octylphenol surfactants in microemulsion–oil–water systems. Influence of temperature and relation between partitioning coefficient and physicochemical formulation. *Langmuir*, **16**, 5534–5539.
35. Márquez, N., Antón, R.E., Graciaa, A., Lachaise, J. and Salager, J.L. (1995) Partitioning of ethoxylated alkyl phenol surfactants in microemulsion–oil–water systems. *Colloids Surf. A.*, **100**, 225–231.
36. Márquez, N., Graciaa, A., Lachaise, J. and Salager, J.L. (2002) Partitioning of ethoxylated alkylphenol surfactants in microemulsion–oil–water systems: Influence of physicochemical formulation variables. *Langmuir*, **18** (16), 6021–6024.
37. Salager, J.L., Antón, R.E., Andérez, J.M. and Aubry, J.-M. (2001) Formulation des microémulsions par la méthode du *HLD*. Techniques de L'Ingénieur, Paris [in French]. *Vol. Génie des Procédés – Formulation,* J2, Nr. **157**, 1–20.
38. Acosta, E., Szekeres, E., Sabatini, D.A. and Harwell, J.H. (2003) Net-average curvature model for solubilization and supersolubilization in surfactant microemulsion. *Langmuir*, **19**, 186–195.
39. Huh, C. (1979) Interfacial tension and solubilizing ability of a microemulsion phase that coexists with oil and brine. *J. Colloid Interface Sci.*, **71**, 408–426.
40. Salager, J.L., Antón, R.E., Sabatini, D.A., Harwell, J.H., Acosta, E. and Tolosa, L. (2005) Enhancing solubilization in microemulsions – state of the art and current trends. *J. Surfactants Detergents*, **8**, 3–21.
41. Bourrel, M. and Chambu, C. (1983) The rules for achieving high solubilization of brine and oil by amphiphilic molecules. *Soc. Petrol. Eng. J.*, **23**, 327–338.
42. Lim, K.-H. and Smith, D.H. (1991) Experimental test of catastrophe theory in polar coordinates: Emulsion inversion for the ethanol/benzene/water system. *J. Colloid Interface Sci.*, **142**, 278–290.
43. Lee, J.-M., Lim, K.-H. and Smith, D.H. (2002) Formation of two-phase multiple emulsions by inclusion of continuous phase into dispersed phase. *Langmuir*, **18**, 7334–7340.
44. Graciaa, A., Lachaise, J., Cucuphat, C., Bourrel, M. and Salager, J.L. (1993) Improving solubilization in microemulsion with aditives – Part 1: The lipophilic linker role. *Langmuir*, **9**, 669–672.

45. Graciaa, A., Lachaise, J., Cucuphat, C., Bourrel, M. and Salager, J.L. (1993) Improving solubilization in microemulsion with additives – Part 2: Long chain alcohols as lipophilic linkers. *Langmuir*, **9**, 3371–3374.
46. Salager, J.L., Graciaa, A. and Lachaise, J. (1998) Improving solubilization in microemulsion with additives – Part III: Lipophilic linker optimization. *J. Surfactants Detergents*, **1**, 403–406.
47. Uchiyama, H., Acosta, E., Tran, S., Sabatini, D.A. and Harwell, J.H. (2000) Supersolubilization on chlorinated hydrocarbon microemulsions: Solubilization enhancement by lipophilic and hydrophilic linkers. *Ind. Eng. Chem. Res.*, **39**, 2704–2708.
48. Acosta, E., Uchiyama, H., Sabatini, D. and Harwell, J.H. (2002) The role of hydrophilic linker. *J. Surfactants Detergents*, **5**, 151–157.
49. Acosta, E., Mai, P.D., Harwell, J.H. and Sabatini, D.A. (2003) Linker-modified microemulsions for a variety of oils and surfactants. *J. Surfactants Detergents*, **6**, 353–363.
50. Jakobs, B., Sottmann, T., Strey, R., Allgaier, J., Willner, L. and Richter, D. (1999) Amphiphilic block copolymer as efficiency boosters for microemulsions. *Langmuir*, **15**, 6707–6711.
51. Graciaa, A., Andérez, J.M., Bracho, C.L., Lachaise, J., Salager, J.L., Tolosa, L. and Ysambertt, F. (2006) The selective partitioning of the oligomers of polyethoxylated surfactant mixtures between interface and oil and water bulk phases. *Adv. Colloid Interface Sci.*, **123–126**, 63–73.
52. Miñana-Pérez, M., Graciaa, A., Lachaise, J. and Salager, J.L. (1995) Solubilization of polar oils with extended surfactants. *Colloids Surf. A*, **100**, 217–224.
53. Tongcumpou, C., Acosta, E.J., Quencer, L.B., Joseph, A.F., Scamehorn, J.F., Sabatini, D.A., Yanumet, N. and Chavadej, S. (2005) Microemulsion formation and detergency with oily soils: III. Performance and mechanisms. *J. Surfactants Detergents*, **9**, 147–156.
54. Scorzza, C., Gode, P., Martin, P., Miñana, M., Salager, J.L., Villa, P. and Goethals, G. (2002) Synthesis and surfactant properties of a new "extended" glucidoamphiphile made from D-glucose. *J. Surfactants Detergents*, **5**, 331–335.
55. Fernandez, A., Scorzza, C., Usubillaga, A. and Salager, J.L. (2005) Synthesis of new extended surfactants derived from a xylitol polar group. *J. Surfactants Detergents*, **8**, 193–198.
56. Thurston, G., Salager, J.L. and Schechter, R.S. (1979) Effect of salinity on the viscosity and birefringence of microemulsion systems. *J. Colloid Interface Sci.*, **70**, 517–523.
57. Antón, R.E., Gómez, D., Graciaa, A., Lachaise, J. and Salager, J.L. (1993) Surfactant–oil–water systems near the affinity inversion – Part IX: Optimum formulation and phase behavior of mixed anionic–cationic systems. *J. Dispersion Sci. Technol.*, **14**, 401–416.
58. Doan, T., Acosta, E., Scamehorn, J.F. and Sabatini, D.A. (2003) Formulating middle-phase microemulsions using mixed anionic and cationic surfactant systems. *J. Surfactants Detergents*, **6**, 215–224.
59. Upadhyaya, A., Acosta, E.J., Scamehorn, J.F. and Sabatini, D.A. (2006) Microemulsion phase behavior of anionic–cationic surfactant mixtures: Effect of tail branching. *J. Surfactant Detergents*, **9**, 169–179.
60. Salager, J.L., Bourrel, M., Schechter, R.S. and Wade, W.H. (1979) Mixing rules for optimum phase behavior formulations of surfactant–oil–water systems. *Soc. Petrol. Eng. J.*, **19**, 271–278.
61. Antón, R.E., Salager, J.L., Graciaa, A. and Lachaise, J. (1992) Surfactant–oil–water systems near the affinity inversion – Part VIII: Optimum formulation and phase behavior of mixed anionic–nonionic systems versus temperature. *J. Dispersion Sci. Technol.*, **13**, 565–579.
62. Kunieda, H. and Solans, C. (1997) How to prepare microemulsions: Temperature-insensitive microemulsions. In C. Solans and H. Kunieda (eds), *Industrial Applications of Microemulsions*. Marcel Dekker, New York, pp. 21–45.
63. Forgiarini, A., Esquena, J., Gonzalez, C. and Solans, C. (2001) Formation of nanoemulsions by low-energy emulsification methods at constant temperature. *Langmuir*, **17**, 2076–2083.
64. Pons, R., Carrera, I., Caelles, J., Rouch, J. and Panizza, P. (2003) Formation and properties of miniemulsions formed by microemulsion dilution. *Adv. Colloid Interface Sci.*, **106**, 129–146.

65. Forgiarini, A., Esquena, J., Gonzalez, C. and Solans, C. (2002) The relation between phase behavior and formation of narrow size distribution W/O emulsions. *J. Dispersion Sci. Technol.*, **23**, 209–217.

66. Graciaa, A., Lachaise, J., Sayous, J.G., Grenier, P., Yiv, S., Schechter, R.S. and Wade, W.H. (1983) The partitioning of complex surfactant mixtures between oil–water–microemulsion phases at high surfactant concentration. *J. Colloid Interface Sci.*, **93**, 474–486.

67. Andérez, J.M., Bracho, C.L., Sereno, S. and Salager, J.L. (1993) Effect of surfactant concentration on the properties of anionic–nonionic mixed surfactant–oil–brine systems. *Colloids Surf. A*, **76**, 249–256.

Chapter 4

Effects of Polymers on the Properties of Microemulsions

Jürgen Allgaier and Henrich Frielinghaus

4.1 Introduction

Polymers are widely used in aqueous systems and frequently together with low molecular weight surfactants. Therefore, it is not surprising that polymers are examined as additives in microemulsions. In order to understand their influence on microemulsion properties, it is useful to distinguish between amphiphilic and non-amphiphilic polymers. Amphiphilic polymers contain water-soluble and water non-soluble segments, which can be arranged in different ways. Examples are given in Scheme 4.1. In the simplest case linear block copolymers or telechelic polymers are obtained. In block copolymers both the hydrophilic and hydrophobic segments are of polymeric nature, whereas telechelic polymers contain low molecular weight end groups of reverse polarity. Telechelic polymers are indicative of the smooth transition from low molecular surfactants to amphiphilic polymers and reveal that there is no strict separation between these two classes. In a simple case, these polymers contain a hydrocarbon group linked to a PEO chain. If the PEO chain is shortened to a few ethylene oxide units and the hydrocarbon chain to C_8–C_{20}, respectively, the classical low molecular weight alkyl polyethylene oxide surfactants (C_iE_j) are obtained. Comb polymers represent another class of amphiphilic polymers. In this case polymeric or oligomeric side chains are linked to a backbone of opposite polarity. Usually, the side chains are placed along the backbone in a random fashion. If the side chains are of low molecular weight a random copolymer is obtained. Non-amphiphilic polymers usually are represented by homopolymers. Mainly water-soluble homopolymers were examined in microemulsions, but to a certain extent also oil soluble homopolymers were investigated. It seems obvious that amphiphilic polymers can interact with the water–oil interface, and can have a strong influence on microemulsion properties. However, non-amphiphilic polymers can influence surfactant film properties too. On the one hand, attractive forces between homopolymer and surfactant are well known, especially if polymer and surfactant are oppositely charged or one component is ionic and the other one is non-ionic. On the other hand, systems without attractive polymer–surfactant interactions are of interest too. Most homopolymers are exclusively soluble either in the aqueous or the oil phase. Consequently, the surfactant film acts as barrier that limits the extension of the polymer coil and leads to repulsive forces between polymer and surfactant film for entropic reasons. This is especially the case if the microemulsion domains and the polymer molecules resemble in size or the polymer exceeds the size of a microemulsion domain. Then confinement effects occur

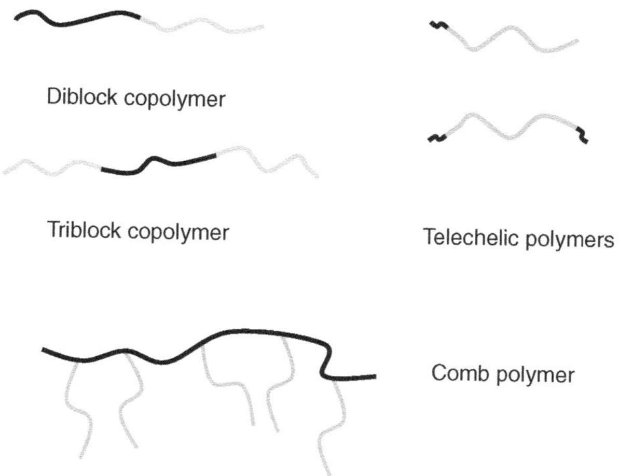

Scheme 4.1 Different architectures of amphiphilic polymers.

which strongly influence microemulsion behaviour. Typically, the size of microemulsion domains is in the order of a few nm to 10 nm. Polymer chains swollen in a good solvent and having molecular weights between 1000 and 100 000 g mol^{-1} have the same size. Therefore, the size ratio of microemulsion domains to polymer coil is of crucial importance not only for homopolymers but also for amphiphilic polymers.

4.2 Amphiphilic polymers

There are basically two topics that need to be addressed regarding the effect of amphiphilic polymers on the physical behaviour of microemulsions. The first topic is related to phase behaviour and structure formation. Amphiphilic polymers can strongly influence phase behaviour because of their impact on the bending rigidity of the surfactant film. For both droplet microemulsions and bicontinuous microemulsions such phenomena were studied. Especially in droplet microemulsions, amphiphilic polymers were used to interconnect microemulsion domains. This leads to ordering phenomena and can alter the phase behaviour. The second topic again is based on systems where microemulsion domains are connected via polymers. It covers dynamic phenomena with a focus on viscoelastic properties. Important in this area is the formation of transient or permanent networks.

4.2.1 Phase behaviour and structure formation

Amphiphilic polymers can have a strong impact on the phase behaviour of microemulsions already at very low concentrations. The most drastic consequence is that on the emulsification capacity of surfactants. A first work in this respect was carried out using hydrophobically modified ethyl hydroxyethyl cellulose [1]. This is a comb-shaped polymer, having a water-soluble backbone functionalised with low molecular weight hydrophobic

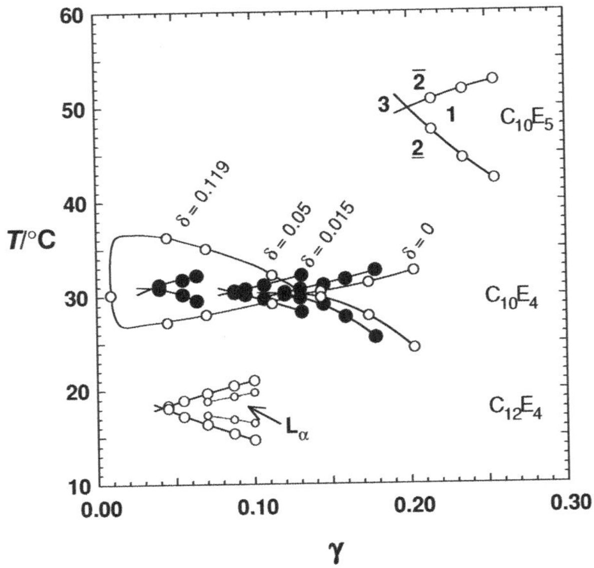

Figure 4.1 Sections through a phase prism at equal volumes of water and *n*-decane. The well-known 'fish' is shown for water–*n*-decane–$C_{10}E_4$ as hollow circles. The effect of increasing surfactant head group size ($C_{10}E_5$) and tail size ($C_{12}E_4$) is demonstrated. Note the associated temperature shifts. Adding traces of polymer PEP5-PEO5 leads to an enormous efficiency increase (full circles) at constant temperature. (From Ref. [2], reprinted with permission of the American Chemical Society.)

stickers. The polymer was added to a three-phase system, containing besides a bicontinuous phase excess water and oil. The surfactant used was $C_{12}E_5$. Up to 0.8 wt.% of polymer a weak swelling of the microemulsion phase was detected, mainly at the expense of the excess water. At higher polymer concentration the swelling decreased again.

Later this subject was studied in more detail with polyalkane–polyethylene oxide (PA-PEO) diblock copolymers [2]. Compared to the comb polymer the use of block copolymers has a much stronger effect. Using the non-ionic surfactant $C_{10}E_4$ and replacing parts of it by a polymer one obtains bicontinuous microemulsions at about 3 wt.% of amphiphile, whereas without polymer 13 wt.% are required. Interestingly, no more than 12 wt.% of surfactant was replaced by polymer in this experiment. If symmetric diblock copolymers are used the temperature behaviour is not affected. Figure 4.1 shows that with increasing polymer mass fraction of the surfactant–polymer mixture δ the one-phase region is shifted to smaller amphiphile concentrations γ. This shows that the diblock copolymer–surfactant mixture is drastically more efficient than the additive free surfactant. The polymer used for this series of experiments was a poly(ethylene-alt-propylene)-PEO diblock copolymer (PEP-PEO) having molecular weights of 5000 g mol^{-1} for both blocks. A more detailed analysis revealed that the minimum surfactant amount $\tilde{\gamma}(\delta)$, i.e. the lowest amphiphile concentration necessary in order to obtain a one-phase system, depends exponentially on the polymer content. The PA-PEO diblock copolymers are similar to the alkyl polyethylene oxide surfactants with respect to chemical structure, except that for both hydrophilic and hydrophobic moieties the molecular weights are larger by a factor of 10–300. Therefore, it is interesting to consider the consequences of surfactant chain length effects on phase behaviour. Increasing the head group size by replacing $C_{10}E_4$ by $C_{10}E_5$ reduces the efficiency, whereas the increase of the hydrocarbon tail from $C_{10}E_4$ to $C_{12}E_4$ increases the efficiency (see Fig. 4.1). However, the use of more efficient surfactants is always accompanied by a significant growth of the lamellar phase (L_α). In the case of the efficient surfactant $C_{12}E_4$ the one-phase region is almost fully superimposed by the lamellar phase. The block copolymer addition leads to a completely different scenario. The formation of lamellar phases is largely suppressed. More details will be discussed later in this chapter.

The block copolymer addition causes other changes, which are closely connected to the efficiency increase. First, the already low interfacial tension between water and oil is reduced further after polymer addition [2]. This is in agreement with the behaviour of low molecular weight surfactants, where lower interfacial tensions were found for more efficient systems. Second, the reduction of the total amphiphile content leads to turbid microemulsions. Microscopically, this means that the sizes of water and oil domains must get larger. A small-angle neutron scattering study revealed that the characteristic length scale of the water and oil domains is inversely proportional to the volume fraction of surfactant [2]. From this finding the conclusion can be drawn that the block copolymer has no influence on the overall interfacial area. This is given by the surfactant. However, the polymer allows stabilising larger structures that cannot exist without the polymeric additive.

The theoretical explanation of this effect goes back on the Helfrich free energy [3]. It assumes that the microemulsion behaviour is dominated by the elastic properties of the surfactant film, given by the bending rigidity and the saddle splay modulus. The bending rigidity is connected with a deviation of the mean from the spontaneous curvature. For systems having symmetric water to oil ratios the spontaneous curvature is zero at the phase inversion temperature, characterised by $\tilde{T}$. The saddle splay modulus is coupled to the Gaussian curvature. The bending rigidity can be measured by small-angle neutron scattering [4]. The scattering curves are described by the Teubner–Strey theory [5] with two structural parameters: the domain size and the correlation length. These parameters are connected by the Gaussian random field theory with the bending rigidity [4]. On the other hand, at the minimum amount of surfactant given by $\tilde{\gamma}$ the saddle splay modulus takes a small constant value, which allows measuring its changes with varying polymer content. It was found that the positive bending rigidity increases, and the negative saddle splay modulus decreases with increasing polymer content. This agrees well with the theory of Lipowsky [6]. The block copolymer is anchored in the film, both blocks extending in their preferred solvents in a mushroom conformation. The mushroom conformation is a consequence of the repulsive interactions between the surfactant film and the polymer coils. It results from the unique solubility of each block in either water or oil which prohibits the extension of the polymer coils beyond the surfactant film [4]. This elastic polymer deformation makes the film effectively more rigid. A more rigid film allows for the formation of larger domains with a better surface to volume ratio. Thus, the input of surfactant can be reduced. A model for droplet microemulsions describes similar effects [7].

The magnitude of the polymer effect – on the bending rigidity for instance – is proportional to the number grafting density and the projected polymer size, the square of

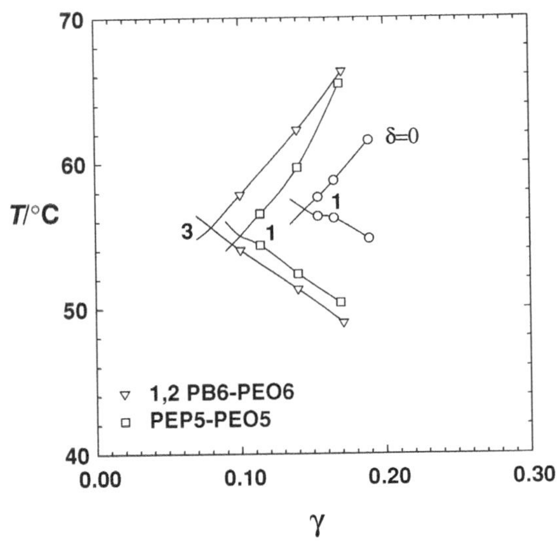

Figure 4.2 Phase behaviour of the system water–octamethyltrisiloxane–$C_4D_3E_8$ at equal volumes of water and octamethyltrisiloxane without additive ($\delta = 0$) and with the additives 1,2PB6-PEO6 and PEP5-PEO5 at $\delta = 0.05$.

the radius of gyration. Thus, either more polymer or polymer with a larger degree of polymerisation will increase this effect, but note: At a constant mass fraction of polymer δ, the variation of the degree of polymerisation influences both contributions oppositely, and the polymer effect stays nearly unchanged [4].

The increase of the emulsification capacity is not restricted to the combination of alkyl polyethylene oxide surfactants and PA-PEO diblock copolymers in bicontinuous microemulsions. It is a universal effect, independent from the nature of the surfactant and from the morphology of the microemulsion [8]. In addition, other amphiphilic diblock copolymers were tested successfully [8–13]. Most of these polymers were composed of PEO and hydrophobic polyalkylene oxides from polybutylene oxide to polydodecylene oxide. The most important property of the diblock copolymer additive is the different polarities of both blocks which force the polymer to be located at the water–oil interface. The solubilisation of each block in its preferred solvent over-compensates the loss of entropy due to the location at the interface, which, in turn, leads to the conformationally unfavoured mushroom shape of the polymer blocks. As microemulsions are usually formulated with aliphatic hydrocarbon oils the number of suitable hydrophobic polymers is limited due to insolubility in such media. This shortcoming is even more prominent for microemulsions containing non-conventional oils like silicon oils as most hydrophobic polymers are insoluble in silicon oils. Interestingly, hydrocarbon polymers with a high degree of short-chain branching like 1,2-polybutadiene (1,2PB) or hydrophobic polyalkylene oxides show an improved solubility in silicon oils. Consequently, block copolymers containing such hydrophobic blocks are useful for increasing the emulsification capacity in silicon oil microemulsions [10]. Figure 4.2 shows the influence of different diblock copolymers on the phase behaviour of mixtures containing the silicon surfactant $C_4D_3E_8$ and the

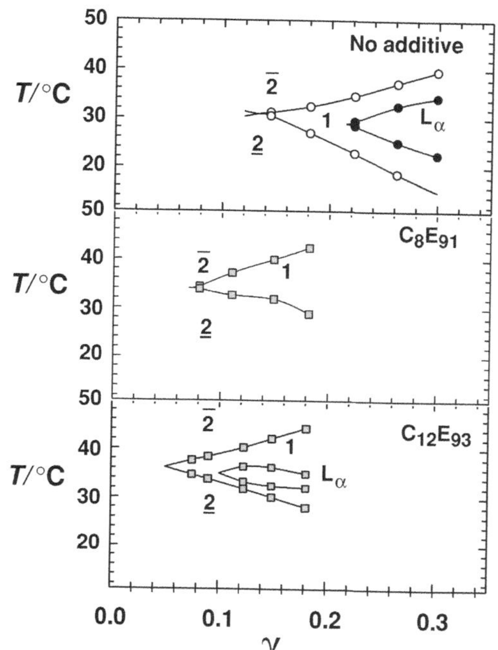

Figure 4.3 Phase diagrams of the system water–decane–$C_{10}E_4$ at equal volumes of water and decane without additive and with the hydrophilic alkyl polyethylene oxides C_8E_{91} and $C_{12}E_{93}$, respectively, at $\delta = 0.10$.

silicon oil octamethyltrisiloxane at equal volumes of water and oil. Compared to the system without polymer ($\delta = 0$) the one-phase region of the mixture containing 1,2PB6-PEO6 is extended to smaller amphiphile concentrations compared to the additive PEP5-PEO5. The different behaviour of the additives can be explained on the basis of their oil solubility. 1,2PB is soluble in octamethyltrisiloxane, while PEP is only partially soluble, leading to a much smaller volume of the hydrophobic coil. The small differences in the block molecular weights, 6000 g mol^{-1} for 1,2PB6-PEO6 and 5000 g mol^{-1} for PEP5-PEO5, do not play a major role because of the marginal molecular weight influence.

An alternative to overcome incompatibility between the oil phase and the hydrophobic polymer block is the replacement of block copolymers by telechelic polymers. It was shown that especially hydrophilic polymers equipped at one chain end with a short hydrophobic group increase the emulsification capacity similarly to PA-PEO block copolymers [14]. In the simplest case the polymer contains a short hydrocarbon group, connected to a long PEO chain. Alternatively, these polymers can be regarded as very hydrophilic alkyl polyethylene oxide surfactants. Interestingly, short hydrocarbon groups in the range of C_8–C_{12} are sufficient to strongly shift the one-phase region to lower surfactant concentrations. This is demonstrated in Fig. 4.3 which shows phase diagrams of the system water–decane–$C_{10}E_4$ at equal volumes of water and decane without additive and with the hydrophilic alkyl polyethylene oxides C_8E_{91} and $C_{12}E_{93}$. In both cases, the molecular weights of the hydrophilic parts are approximately 4000 g mol^{-1} and in the same range as the molecular

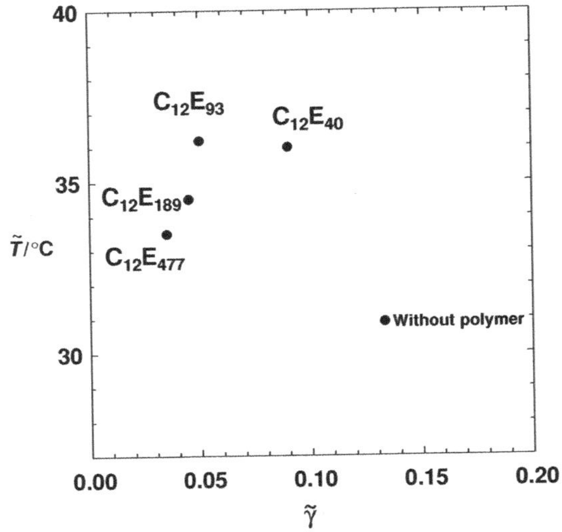

Figure 4.4 Location of the fish-tail points for the system water–decane–$C_{10}E_4$ at equal volumes of water and decane without additive and with $C_{12}E_{40}$, $C_{12}E_{93}$, $C_{12}E_{189}$ and $C_{12}E_{477}$ at $\delta = 0.10$.

weights of the PEO blocks in the diblock copolymers. Both additives strongly shift the one-phase region to lower surfactant concentrations. The effect is more pronounced for $C_{12}E_{93}$. It does not increase if longer hydrophobic units are used. This suggests that the $C_{12}E_{93}$ molecules are fully anchored at the water–oil interface and is supported by comparison with PA-PEO block copolymers. At equivalent block length and mass concentration $C_{12}E_{93}$ and PA-PEO block copolymers increase the efficiency similarly. C_8E_{91} is less effective, most likely because of a reduced interfacial activity due to the shorter hydrophobic unit. The dynamic equilibrium of anchored and non-anchored polymers and the unfavourable effect of non-anchored polymers similar to homopolymers on the efficiency makes this weaker dependence more clear (see Section 4.3.1).

The influence of the hydrophilic chain length is summarised in Fig. 4.4. In this diagram the fish-tail points in terms of $\tilde{\gamma}$ and the corresponding temperature $\tilde{T}$ are plotted for the system water–decane–$C_{10}E_4$ at equal volumes of water and decane without additive and with different hydrophilic dodecyl polyethylene oxides. In agreement with the theoretical considerations described before [15] the fish-tail points slightly move to smaller amphiphile concentrations with increasing PEO chain length. Only between $C_{12}E_{40}$ and $C_{12}E_{93}$ there is a strong efficiency increase. This behaviour is understandable considering the general behaviour of polymer molecules. The swelling degree of short chains is visibly smaller than it is the case for longer chains. Starting with oligomers it increases and gets constant usually above molecular weights of a few 1000 g mol^{-1}. Therefore, in $C_{12}E_{40}$ ($M_w \approx 1800$ g mol^{-1}) the size of the polymer coil is disproportionately small, leading to a lesser efficiency increase than for the longer chain lengths ($M_w \approx 4000$–21 000 g mol^{-1}). In those cases, it is assumed that the swelling degree basically stays constant and chain size influence and the number grafting density nearly cancel out. In addition to the increase in efficiency the temperature behaviour is interesting. For the additives $C_{12}E_{40}$ and $C_{12}E_{93}$ there is a visible increase of

$\tilde{T}$. Because of the hydrophilic nature of the polymeric additive this intuitively makes sense. However, the temperature decrease in the presence of the additives with longer chains does not fit in this scenario. It can be explained on the basis of theoretical work for polymer chains anchored at an interface [6, 15]. According to these calculations the influence on the curvature gets smaller with increasing molecular weight. Exactly this scenario is shown in Fig. 4.4 for the higher molecular weights. The reason for the deviations from the theoretical predictions at low molecular weights is again the more compact polymer coil of the short PEO chains.

Amphiphilic copolymers do not only extend the stability region of microemulsion phases, they also influence the appearance of liquid crystalline phases. Especially in systems with balanced water–oil ratios lamellar phases (L_α) exist at higher surfactant concentrations. Generally, the use of more efficient surfactants is accompanied by the formation of the L_α phase (see Fig. 4.1). Block copolymers and telechelic polymers, however, suppress the extension of L_α phases if the additives are used in small or moderate concentrations. This means that in mixtures containing the polymeric additive the location of the L_α phase in the phase diagram is similar to the system without additive. At higher additive concentrations the scenario changes. This was studied in detail for alkyl polyethylene oxide surfactants in combination with PA-PEO block copolymers [16]. It was found that above block copolymer to surfactant mass ratios of about 0.1 the additives induce liquid crystalline phases already at low amphiphile concentrations. Besides L_α phases, hexagonal and cubic phases were found. In addition to the polymer concentration its composition and especially its size can be used to influence the phase behaviour. Generally, longer polymer chains disfavour the formation of lamellar phases, possibly because of the smaller domain sizes of lamellar structures compared to those of bicontinuous structures. If the block copolymer additives are replaced by hydrophilic alkyl polyethylene oxides the scenario is partially different. The suppression of L_α phases is less pronounced than for the block copolymers but both the chain length of the hydrophobic anchor and the hydrophilic chain are important in this context. It seems that short anchors (see Fig. 4.3) and long hydrophilic chains support the suppression of L_α phases [14]. Last but not least, it has been found that the addition of an amphiphilic polymer can induce a phase separation into two lamellar phases [17–19]. The striking observation is that at certain compositions the polymer is apparently no longer incorporated into the films of the lamellar phase due to space restrictions. The polymer therefore induces a phase separation into two different lamellar phases such that it fits into one of them while the excess surfactant forms a polymer-free lamellar phase.

The strong influence of amphiphilic polymers on the phase behaviour of microemulsions is not only interesting with respect to fundamental research. It can also be exploited for applications. For economic and environmental reasons low surfactant concentrations are required in this field. The use of efficient surfactants for this purpose is disadvantageous because of the formation of liquid crystalline phases, which frequently are highly viscous and lead to phase separation if the liquid crystalline phase coexists together with a microemulsion phase. Amphiphilic polymers can help to overcome these difficulties. Especially, telechelic polymers as the hydrophilic alkyl polyethylene oxides are interesting in this context. Compared to block copolymers they are much simpler to synthesise and are advantageous because of their easy biodegradability. By contrast, polymers having larger hydrophobic segments usually show much reduced biodegradability.

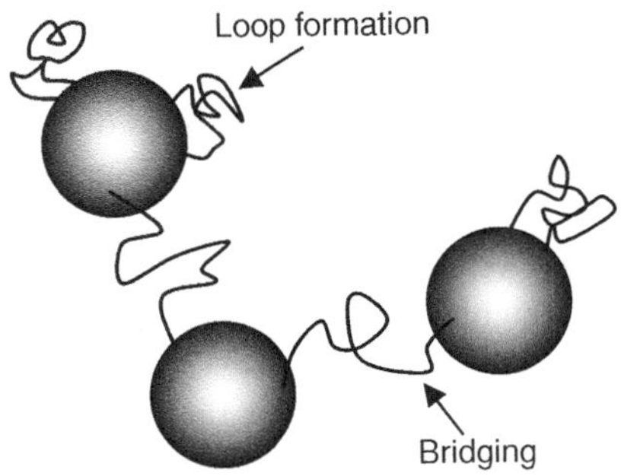

Scheme 4.2 Loop formation and bridging of difunctional polymers.

The systems described so far in this chapter have in common that the amphiphilic polymers influence the bending rigidity of the surfactant film. However, polymeric additives can alter phase behaviour also by network formation. This phenomenon was investigated using droplet microemulsions in combination with telechelic polymers (modified at both chain ends) or with ABA triblock copolymers. In contrast to monofunctional polymers or diblock copolymers, which decorate a surfactant film, for the difunctional or triblock counterparts there is a competition between loop formation and bridging of two microemulsion domains. The different scenarios are illustrated in Scheme 4.2. In order to minimise the loss of conformational entropy by stretching the middle block, solubilised in the continuous phase, generally loop formation is preferred if the inter-droplet distance and the droplet diameter are larger than the end-to-end distance of the middle block. On the other hand, the bridging event is preferred for small inter-droplet distances and small droplet diameters. For the hydrophobic end groups the situation is similar, independent of the location in the same or different droplets.

Studies were carried out with water-in-oil (w/o) droplet microemulsions and PEO-polyisoprene-PEO (PEO-PI-PEO) triblock copolymers or polyisoprene modified at both chain ends with ionic groups [20, 21]. The results showed that independent of the chemical nature of the polymers, bridges between the microemulsion droplets were formed. The microscopical changes caused by the polymers depend on the chain length and number concentration of polymer molecules and on the number concentration of droplets. Especially at inter-droplet distances similar to the end-to-end distance of the bridging polymer chains, a medium-ranged order of the droplets was induced. Without the polymeric additive the droplets were distributed irregularly. Figure 4.5 shows a freeze fracture micrograph of a w/o-droplet microemulsion containing a triblock copolymer. The ordered droplets are clearly visible. Loop formation was unfavourable in these systems because the droplet size was smaller than the polymer end-to-end distance.

The difference between PEO being functionalised at one chain end with a hydrophobic sticker and its counterpart containing the hydrophobic units at both chain ends was examined with oil-in-water (o/w) droplet microemulsions [22–24]. Cetyl pyridinium chloride/octanol or alkylphenol ethoxylate were used as surfactant. The monofunctional polymers were only capable to decorate the oil droplets whereas the difunctional polymers could also bridge them. Experiments were carried out as a function of the droplet volume

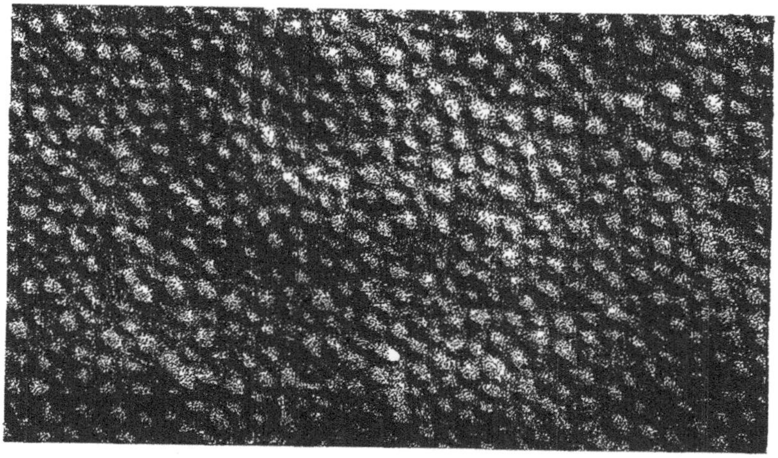

Figure 4.5 Freeze fracture micrograph of an ordered w/o-droplet microemulsion containing PEO-PI-PEO triblock copolymer. (From Ref. [20], reprinted with permission of EDP Sciences.)

fraction Φ and the number of polymer chains per droplet *r*. From scattering results it was concluded that the monofunctional polymers induce repulsive interactions between the droplets, independent from Φ. Higher values of *r* lead to phase separation where a microemulsion phase coexisted with excess oil. This is shown in the upper phase diagram of Fig. 4.6. The phase separation was attributed to smaller droplet diameters due to an increase of film curvature caused by the hydrophilic polymer, which acts as co-surfactant. For the difunctional polymer the scenario is different. The scattering investigation showed that at high Φ repulsive interactions exist whereas at low Φ attractive interactions were induced, leading to phase separation. In this case, a dilute microemulsion coexists with a more concentrated one. Therefore, the nature of this phase separation is completely different from the scenario obtained with monofunctional polymers. Besides Φ the number of polymer chains per droplet controls the phase behaviour. Phase separation only occurs if a minimum value of *r* is exceeded. This is shown in the lower phase diagram of Fig. 4.6. These results show that at low droplet concentrations or large interdroplet distances loop formation is not necessarily preferred. In the upper example the polymer end-to-end distance was large compared to the droplet diameter. Therefore, both scenarios, loop formation and bridging, would force the polymer chains in an entropically unfavourable conformation. This can be overcome by phase separation into a phase with a low droplet density and one with a high droplet density, the latter phase hosting the bridge-forming polymer.

4.2.2 Dynamic phenomena and network formation

The use of amphiphilic polymers to interconnect microemulsion domains does not only influence structural but also dynamic properties. This was investigated in droplet microemulsions of both the w/o and the o/w type. These systems allow to vary parameters

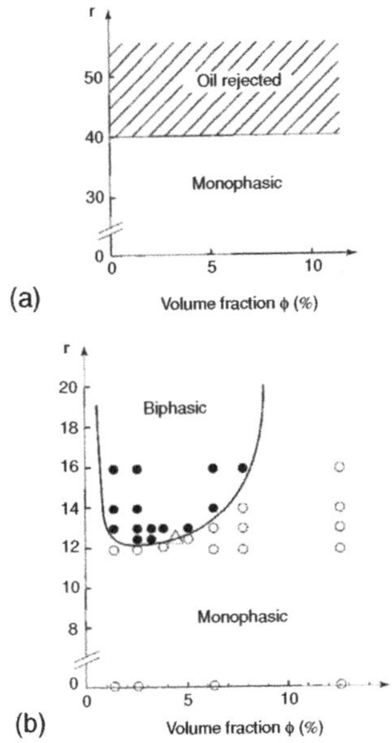

Figure 4.6 Phase behaviour of o/w-droplet microemulsions containing 0.2 M NaCl–decane–cetyl pyridinium chloride–octanol. The PEO additive is hydrophobically modified at one chain end (a) and at both chain ends (b), respectively. *r* represents the number of polymer chains per droplet. (From Ref. [23], reprinted with permission of the American Chemical Society.)

like droplet dimension, inter-droplet distance, number of polymer chains per droplet or polymer size. Most of the experiments were carried out under conditions where bridging dominates versus loop formation.

Using PEO-PI-PEO triblock copolymers and the ionic surfactant AOT self-diffusion times of these components in w/o-droplet microemulsions were investigated [25]. The system was interpreted as a transient entanglement network with the PEO blocks sticking in the water droplets that act as network junctions. Because of the insolubility of PEO in aliphatic oils it was concluded that the PEO blocks could only exchange between droplets during their collision. Their residence time in a droplet therefore must be controlled by the droplet collision rate. It was calculated that the time needed for a PEO block to move from one droplet to another during a collision event is short compared to the residence time. In another study, the same microemulsion as well as a microemulsion where AOT was replaced by an alkyl polyethylene oxide surfactant were investigated with respect to their rheological behaviour [26]. Both systems behaved qualitatively similar. In this work the polymer exchange through the oil phase was taken into consideration in addition to the exchange by droplet collision. The latter process should occur preferably at higher droplet concentrations. The first process should dominate at low droplet concentrations provided that the oil-soluble middle block is stretched. The gain in entropy by the snapping back of the stretched coil into its equilibrium state could compensate for the PEO transfer from a good solvent into a non-solvent. Figure 4.7 shows G' master curves for the ionic microemulsions. The number of polymer chains per droplet (r) is varied here between 3 and 20. For

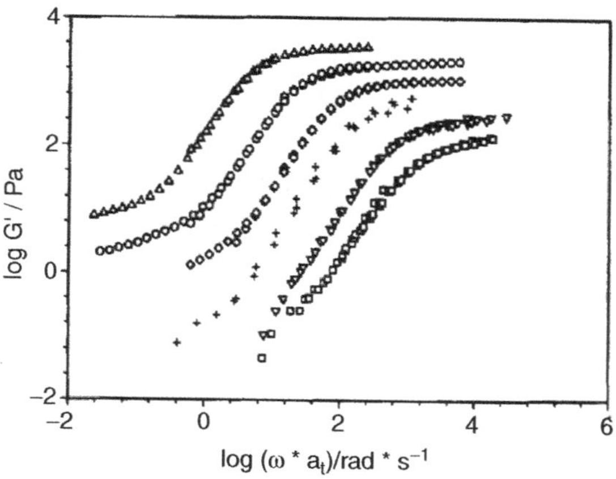

Figure 4.7 G′ master curves of the ionic microemulsion at different polymer to droplet ratios (r) at $T_{ref} = 18°C$: (□) $r = 3$, (▽) $r = 4$, (+) $r = 6$, (◇) $r = 8$, (○) $r = 12$, (Δ) $r = 20$. (From Ref. [26], reprinted with permission of the American Chemical Society.)

r-values between 8 and 20 G' is approximately proportional to $\omega^{0.5}$ in the low-frequency range. This can be interpreted as gel formation. This behaviour is missing or visible only rudimentary for the low r-values. At higher frequencies the viscoelastic properties disappear and the microemulsions show flow behaviour, indicated by the increase of slope to approximately 1.4. As expected r has a strong influence on the plateau modulus, which linearly grows with r. In addition, the curves are shifted to longer relaxation times. There is a significant difference between the microemulsions formed by the ionic and the non-ionic surfactant. In the latter case a smaller minimum value for r is required to form viscoelastic networks. This brings up another important issue, namely interactions between PEO and ionic surfactants leading to adsorption of PEO at the surfactant interface (see Section 4.3.2). In contrast, between PEO and non-ionic surfactants repulsive interactions dominate. If there is a competition between PEO forming inter-droplet bridges and interfacial adsorption it is understandable that for ionic surfactants network formation is reduced.

In order to modify the rheological behaviour not necessarily triblock copolymers are required. Low molecular weight groups of reverse polarity at both chain ends can fulfil the same task. This was studied with an o/w-droplet microemulsion stabilised with an alkyl polyethylene oxide surfactant [27]. The addition to small quantities of PEO, functionalised with C_{18} hydrocarbon groups led to a drastic decrease of the micellar self-diffusion coefficient and to an increase of the low shear viscosity by several orders of magnitude. From the rheological data the fractions of bridging and loop forming polymers were calculated. In agreement with the results presented before it was found that at high droplet concentration bridging dominates whereas at low droplet concentration loop formation is preferred. In the reverse case, where hydrophobic polymers functionalised with short PEO end groups are added to w/o microemulsions the effect is less pronounced. End capping with six ethylene oxide units on average did not influence the viscosity of the microemulsion. This first happened at a PEO polymerisation degree of about 30 [28]. This different behaviour

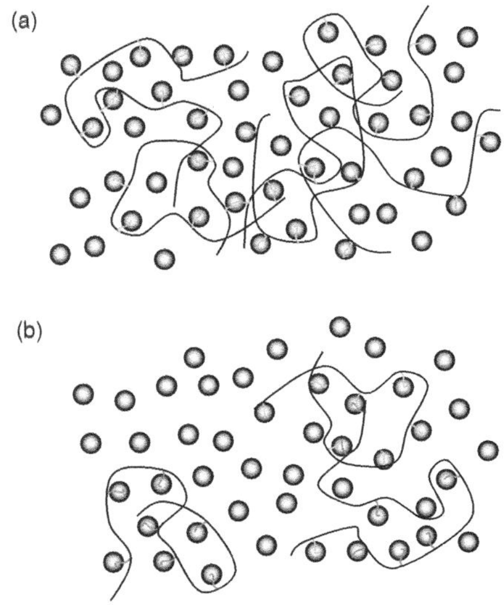

Scheme 4.3 Comb polymers interconnecting microemulsion droplets at high polymer/droplet ratio (a) and at low polymer/droplet ratio (b).

of short hydrophobic and hydrophilic end groups can be explained with their different solubility properties. Hydrocarbon chains are highly insoluble in water. Thus, chain lengths between C_{12} and C_{18} are sufficient to strongly anchor the PEO at the interface [24, 27]. On the other hand, PEO oligomers still have a measurable solubility in hydrocarbons so that hydrophobic polymers functionalised with such end groups are less interfacially active.

Besides triblock copolymers and telechelic polymers comb-shaped polymers were used as additives. The comb polymers can be advantageous because each chain can connect more than two droplets. Even if several side arms stick in the same droplet its effect is not neutralised which is illustrated in Scheme 4.3a. The ability to connect droplets certainly depends on the polymer size, the number of functionalised side chains, the droplet size and the number concentration of droplets. In a series of experiments comb polymers were used which contained an oil-soluble polydodecyl methacrylate backbone and PEO side arms [29–31]. A w/o-droplet microemulsion and the surfactant AOT were used for these investigations. The results showed that at polymer concentrations between 1 and 3 wt.% viscosities increased up to a factor of 2000. However, the efficiency of the polymer in increasing the viscosity strongly depended on the number of side arms per water droplet. At least one side arm per water droplet was needed to detect a strong effect. The number concentration of droplets influences the viscosity in different ways. At a certain droplet concentration there is a maximum viscosity. At higher droplet concentrations, viscosity drops. This behaviour was attributed to a decrease of the intermolecular connectivity because of a decrease of PEO side chains per droplet. At lower droplet concentrations the viscosity decrease was ascribed to the formation of denser clusters resulting in a less homogeneous network structure. This assumption was verified with self-diffusion measurements. From the results it was concluded that polydisperse but finite polymer-droplet aggregates coexist with free droplets. The size of the aggregates strongly depends on the polymer concentration. At low polymer concentrations isolated aggregates exist which predominantly contain only one polymer molecule (Scheme 4.3b). If the overlap concentration of the backbone in the oil is approached, more extended clusters are formed (Scheme 4.3a). However, even above the overlap concentration a large fraction of free droplets was detected, which may result from the entropy loss, associated with the incorporation of droplets into the aggregates. The self-diffusion experiments revealed another important issue. The droplet diffusion was always much faster than the polymer diffusion. In addition, for the droplets always a single diffusion coefficient was found. These results suggest a highly transient nature of the link between polymer and droplet, possibly on the basis of a rapid exchange of AOT and water between droplets or a rapid rearrangement of the aggregates. The results also show that the polymer cannot immobilise the droplets, it is only modestly efficient in slowing down droplet translation.

All the systems presented so far have in common that they form reversible networks because there is an exchange of the end groups or blocks sticking in the droplets. This exchange can be avoided by using polymerisable end groups and adding monomer to the droplet phase. After polymerisation, the polymer chains interconnecting the droplets are covalently connected and thus are permanently anchored in the droplets. This concept was applied for o/w-droplet microemulsion where PEO was used that contained methacrylate end groups [22]. In order to strengthen the connection between the end groups in the droplets, the oil contained not only monomer but also a cross-linker. This allowed to obtain mechanically stable microemulsion elastomers. By examining the rheology of the elastomers it was possible to estimate the fractions of bridging and loop-forming polymer chains. It was found that at inter-droplet distances smaller than the polymer end-to-end distance bridging dominated whereas for larger inter-droplet distances the loop-forming polymer fraction strongly increased. Microemulsion networks are interesting materials as they combine structure and phase behaviour aspects of liquid microemulsions with solid-state properties like elasticity or shape stability. As the microemulsion domains are connected via flexible polymer chains the phase behaviour of the original microemulsion is qualitatively maintained. This should allow to subsequently vary the microstructure of the polymerised elastomers. For example, temperature changes could be used to switch the droplet structure into a cylindrical one [32].

Replacing the droplet microemulsion by an L_α phase allowed to obtain liquid crystalline elastomers [22, 33]. This is shown in Scheme 4.4. The liquid crystalline phase could be macroscopically aligned in a magnetic field prior to the cross-linking process. After cross-linking the macroscopic orientation of the liquid crystalline phase could be conserved. This was even the case after extraction of the water and oil phases. Consequently, reswelling in a selective solvent like water leads to an anisotropic increase of the sample dimensions. An interesting feature of the lamellar networks is the asymmetric elasticity response to compression and elongation.

4.3 Non-amphiphilic polymers

The non-amphiphilic polymers can be widely considered as homopolymers or random copolymers. Most of the studies focussed on water-soluble polymers, but some studies on oil-soluble polymers exist as well. The water-soluble polymers can be uncharged or

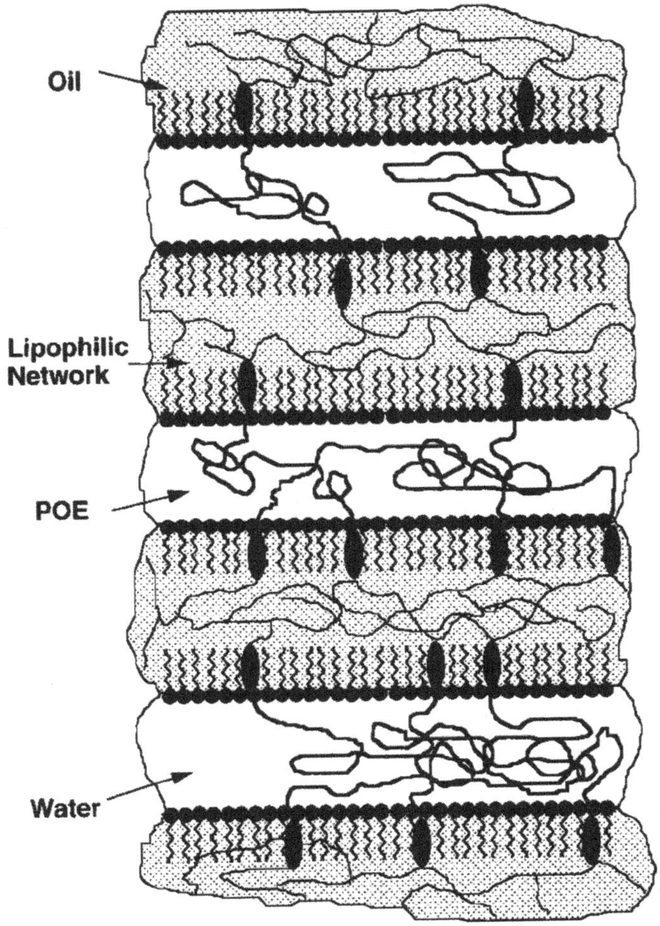

Scheme 4.4 Liquid crystalline elastomers. (From Ref. [22], reprinted with permission of the American Chemical Society.)

charged. Interactions between uncharged polymers and non-ionic surfactants are repulsive and thus the conformational entropy of the polymer dominates its behaviour. When either one component, be it the polymer or the surfactant, is charged attractive interactions lead to adsorption of the polymer on the surfactant film. If both components are charged oppositely, strongly bound complexes will form [34].

4.3.1 Repulsive interactions of polymers

The ideal case of repulsive interactions between polymers and the surfactant film was theoretically described by Eisenriegler [35]. The polymer remains dissolved either in the water or oil domain and each segment is repelled by the film surface. In this case, the

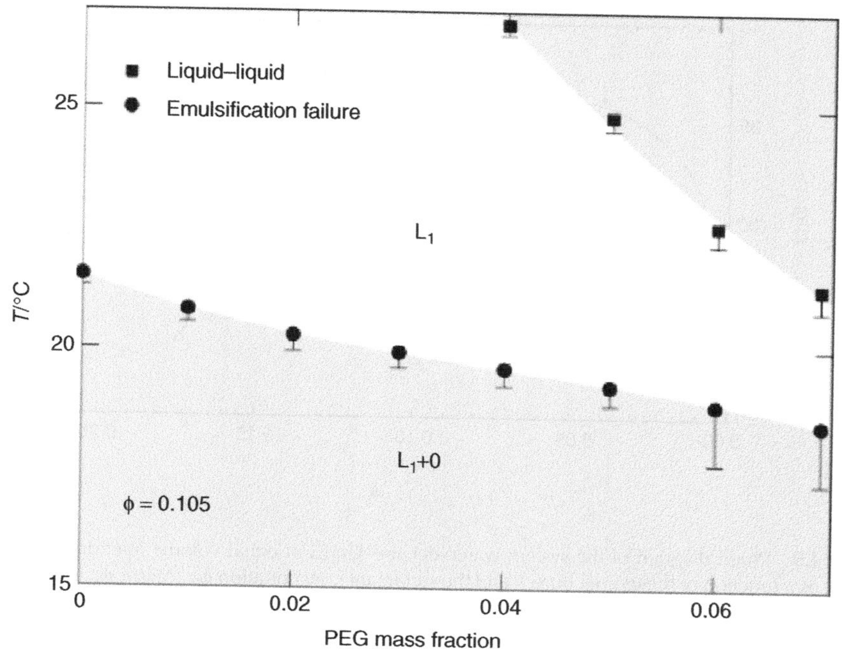

Figure 4.8 Phase diagram of the system water–decane–$C_{12}E_5$. The channel in which the o/w-droplet microemulsion (L_1) is formed closes slowly upon adding PEO polymer. (From Ref. [37], reprinted with permission of the American Chemical Society.)

conformational entropy of the polymer dominates the polymer behaviour. The depletion interaction between the polymer and large colloidal particles (oil or water droplets) leads to a depletion zone, which the polymer avoids since the conformational entropy is reduced close to the surface. The typical depletion interaction range is given by the polymer radius of gyration. The depletion interaction was already described on the basis of two interacting colloidal particles (i.e. the smaller polymer is simplified by a hard sphere) [36]. The large particles have a depletion zone, a space, where the centre of the small particles cannot be found. The width of this zone is given by the radius of the small particles. For closely packed large particles the depletion zones overlap, and effectively there is more space for the small particles. This entropically more favourable state leads to an effective attraction of the large particles. The ultimate consequence of strong depletion results in a phase separation.

The effective depletion interaction is verified plastically by reference [37]. The studied system contained water, decane, and the non-ionic $C_{12}E_5$ surfactant. The used polymer was polyethylene oxide (PEO – note: we neglect the role of the end groups of this polymer). This polymer does not adsorb on the surfactant film [38]. Upon adding more homopolymer the one-phase region (here L_1) gets narrower until it finally vanishes, which is typical for the depletion interaction (Fig. 4.8). The observed temperature decrease is connected with a decreased spontaneous curvature, i.e. the spontaneous curvature obtains a tiny tendency towards the polymer solution. The dynamic small-angle light scattering measurements are interpreted by droplets acting as hard sphere colloids and a pocket radius given by

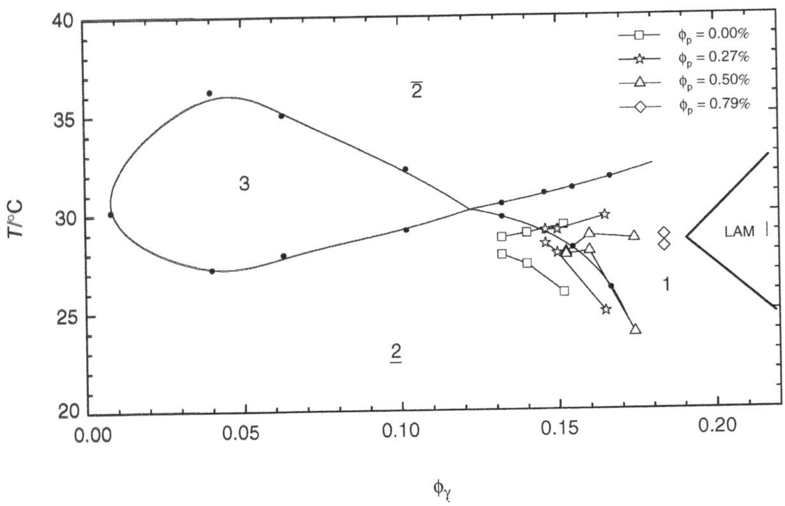

Figure 4.9 Phase diagram of the system water–decane–$C_{10}E_4$ at equal volume fractions of water and decane as a function of the temperature T and the surfactant concentration ϕ_γ. At low ϕ_γ there is a three-phase coexistence, while at moderate ϕ_γ the one-phase bicontinuous microemulsion appears. At even higher ϕ_γ the lamellar phase appears. At high and low temperatures a microemulsion phase coexists with either excess water or oil. The polymer fraction ϕ_p is raised symmetrically for the water- and oil-soluble polymers, and the one-phase microemulsion window closes continuously. The 2 K temperature shift is due to the use of heavy water. (From Ref. [40], reprinted with permission of the American Chemical Society.)

the depletion interaction. Thus, the depletion zone is consistently verified. The extreme case of phase separation is discussed in Ref. [39] for instance. The similarity between hard colloidal particles and droplet microemulsions displays that shape fluctuations are not that important for studies of depletion interactions. Only for bicontinuous microemulsions the film fluctuations can be discussed in a consistent way.

Reference [40] discusses the effect of both water- and oil-soluble polymers in a symmetric bicontinuous microemulsion in the context of the elastic moduli of the Helfrich expression. A direct consequence of the two polymers is the reduction of the one-phase region (Fig. 4.9), which is also verified in Ref. [41]. These experiments compare the bending rigidity measured by small-angle neutron scattering with the saddle splay modulus obtained from the emulsification failure boundary. Experimentally, the absolute values of both moduli decrease similarly and, thus, confirm the theory [35]. The homopolymers allow for stronger fluctuations of the surfactant film, which, in turn, destabilises large domains. Smaller domains have a larger surface to volume ratio, and thus more surfactant is needed.

In Ref. [40], the onset of polymer confinement is discussed, which leads to a higher sensitivity of the polymer effect. Although this effect is real, we do not want to discuss details in the context of bicontinuous microemulsions. For our purpose it should be mentioned that if the polymer and the domains resemble in size the polymer effects are enforced.

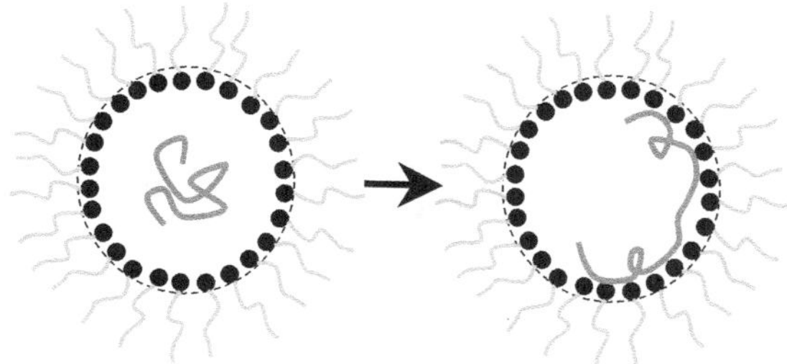

Scheme 4.5 Polymer inside a droplet microemulsion. Attractive interactions can lead to polymer adsorption.

4.3.2 *Transition to adsorbing polymers and two adsorption cases*

The transition from non-adsorbing to adsorbing polymers can be achieved simply by changing the surfactant and thus increasing the attraction between polymer and surfactant. Hydrophilic polymers in w/o-droplet microemulsions lead to polymers incorporated in the droplets (Scheme 4.5). Attractive interactions lead to adsorption at the inside of the surfactant film. With increasing chain length confinement effects eventually occur (Scheme 4.6). In this case, the polymer is incorporated in more droplets and the droplets form clusters. Polymers adsorbing on the outside can also lead to droplet clusters.

In Ref. [42], PEO was embedded in a w/o-droplet microemulsion and studied by small-angle neutron scattering. The authors state that this polymer does not adsorb considerably at the SDS monolayer. The important statement is that both the size polydispersity and the shape fluctuations are increased compared to the reference system without polymer. Larger shape fluctuations are also found for gelatine embedded in w/o-droplet microemulsions (see Fig. 4.10 in [43]). Here, by strong confinement, the elongated shapes

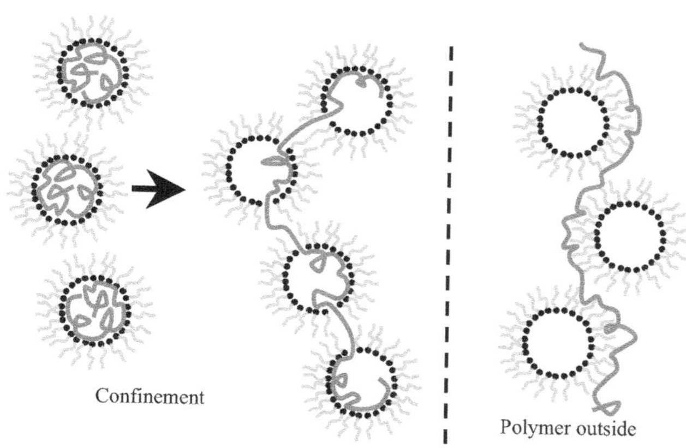

Scheme 4.6 Increasing the polymer size of an adsorbed confined polymer inside a droplet microemulsion. Large polymers lead to droplet clusters. Polymers adsorbed on the outside can also lead to droplet clusters.

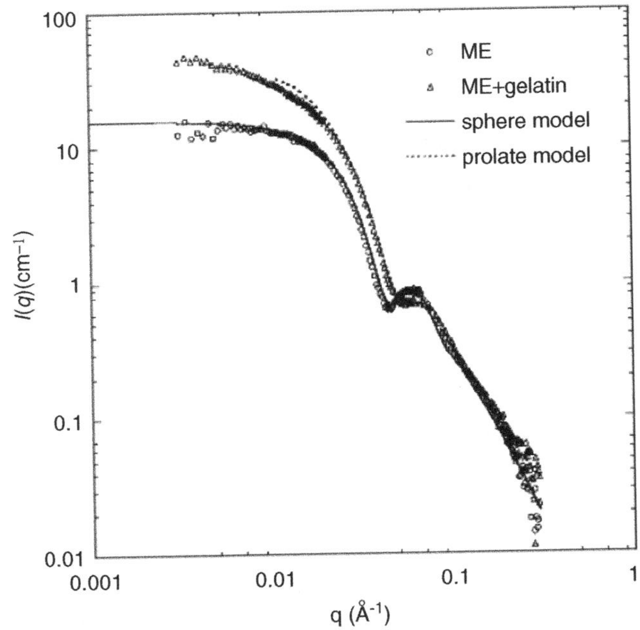

Figure 4.10 SANS profiles of w/o-droplet microemulsions consisting of water–isooctane–AOT with (Δ) and without (o) gelatin. The lines are obtained from analytical expressions for prolate (...) and spherical (—) droplets. (From Ref. [43], reprinted with permission of EDP Sciences.)

could be even driven to cylinders that lead to an isotropic–nematic phase transition. The theory [44] is in line with the observed stronger shape fluctuations and a larger size polydispersity is predicted. In this way, droplet microemulsions and bicontinuous microemulsions behave qualitatively the same, since the fluctuations become stronger in both cases.

The adsorbed polymers shall be discussed now. While the systems PEO/SDS, and gelatine/AOT did not show any or only weak adsorption, the system PEO/AOT shows clear indications of adsorption. By optical Kerr measurements [45] the prolate deformations were measured directly and the fluctuations were found to be reduced. This observation was expressed in changes of the bending rigidity, the dependence of which was explained with respect to polymer concentration and the polymer degree of polymerisation. This dependence was independently confirmed [46] by measuring the droplet–bicontinuous phase transition, which was also interpreted in terms of the bending rigidity. Furthermore, an enlarged bicontinuous region is observed. Thus, free polymers increase the fluctuations while adsorbed polymers reduce the fluctuations. These observations are connected with the bending rigidity which is described by theoretical concepts for free [35] and adsorbed [47] polymers. In this sense adsorbed non-amphiphilic polymers resemble amphiphilic polymers which are attached to the interface with a finite number of segments. We call this case of adsorption 'case 1'.

Reference [42] also reported on the hydrophilic polymer PNIPAM which is adsorbed at the outer interface of o/w-droplet microemulsions. Here, the fluctuations are increased contrary to case 1. Consequently, this can be interpreted by a lowering of the bending rigidity, which was described theoretically in [48]. We call this case of adsorption 'case 2'. The principal cases of adsorption at the outer or inner interface (Scheme 4.6) should not be regarded as completely different. The formation of droplet clusters is driven by the polymer entropy in both cases, while the name 'confinement' is usually connected to incorporated polymers.

Interestingly, opposite trends were observed in the following two cases: In Ref. [49], the size of the one-phase o/w-droplet microemulsion region is decreased upon small amounts of the hydrophilic polymer PEI. This decrease in efficiency is – in principle – connected with a decreased bending rigidity and an increased saddle splay modulus and therefore should be connected to 'case 2'. At higher polymer concentrations the one-phase region opens up and even forms a bicontinuous microemulsion channel. This opposite behaviour should be connected to 'case 1'. In another Ref. [50], a hydrophobically modified polyelectrolyte is studied in a water/oil/non-ionic surfactant system. Here, the fish-tail point is moved to higher surfactant concentrations, before it moves to lower surfactant concentrations with increasing polymer content. Again, one would address the two opposite trends to 'case 2' and 'case 1', respectively. Astonishingly, the phase inversion temperature is slightly increased once the polymer is added. This single trend means that the curvature is bent towards the oil, which would agree with the concept of 'case 1' (since similar to anchored polymers).

The theory of case 2 [48] seems to be the more general description since it includes (i) the surface energy of the adsorbed chain segments, (ii) the osmotic contributions from the enrichment of polymer near the surface and (iii) the polymer stretching of a Gaussian chain. For weak adsorption a considerable amount of polymer stays dissolved in the bulk solvent and a perturbation theory (be it mean field or scaling theory) leads to a slightly decreased bending rigidity and a slightly increased saddle splay modulus. For strong adsorption most of the polymer is adsorbed, and the scaling theory leads to a considerable decrease of the bending rigidity and an increase of the saddle splay modulus. The semi-analytical results are obtained by fitting numerical solutions to a limit where the osmotic terms should dominate, which lead to a logarithmic decrease/increase of the elastic moduli. A similar limit of dominating osmotic contributions is considered in the theory of DeGennes [47], and even the pure osmotic contributions of Brooks' theory [48] yield the opposite behaviour which we called 'case 1'. Thus, for 'case 1' the osmotic contributions dominate, whereas for 'case 2' all three contributions (i–iii) are important. It remains an open question, if the finite high adsorption of Brooks' theory yields already the infinite adsorption limit. Experimentally, it is not clear how to prepare the condition of dominating osmotic contribution.

Another approach of changing tendencies is discussed in Ref. [49] in the context of spontaneous curvatures. To summarise the above-mentioned theories [47, 48], DeGennes found that the interface bends away from the polymer ('case 1'), while Brooks finds the opposite ('case 2'). The group of Lipowsky [51] considers anchored polymers but in the limit of strong adsorption the additional fixed anchor should not matter. The additional parameter introduced is an anchoring distance, which describes a distance where the polymer starts to be flexible. In the usual limit of negligible anchoring distance, the interface bends away from the polymer (as in 'case 1'). As the anchoring distance takes finite values (in [51] quite large anchoring distances are considered) the sign of the spontaneous curvature

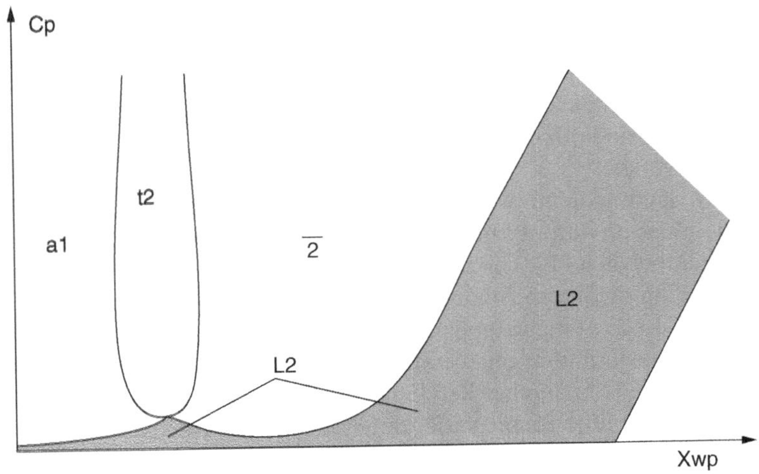

Figure 4.11 Scheme of a phase diagram of the system water–isooctane–AOT–PEO with the polymer content C_P as a function of the aqueous phase water content X_{WP}. L_2 is the w/o-droplet phase; in the region $\overline{2}$ a microemulsion coexists with an aqueous phase; a1 specifies a solid polymer coexistence with a microemulsion; in the region t2 solid polymer, aqueous polymer solution and a microemulsion are found. The real phase diagram with more phase regions is found in [52]. (From Ref. [52], reprinted with permission of the American Chemical Society.)

eventually changes. Also, this example shows that the depth of the adsorbed polymer units could be responsible for different responses of the interface on adsorbed polymers.

The general uncertainty of trends of adsorbed polymers is also supported by works of Bellocq [52, 53]. In Ref. [52], the molar isooctane/AOT ratio is fixed to 4 and polymer containing water is added consequently up 60 wt.%. The initially large w/o-droplet microemulsion phase extends to higher water contents with increasing polymer content (Fig. 4.11). In this sense the efficiency is increased, which can be interpreted as a 'case 1' system similar to the water–isooctane–AOT–PEO system [45] at really low polymer concentrations. One year later the same system [53] was studied with a fixed oil content of 40 wt.%. The one-phase regions were pushed to higher surfactant contents (Fig. 4.12), which were consequently interpreted by a decreased bending rigidity and an increased saddle splay modulus. The mean curvature was pushed towards the water domains. These observations can be interpreted as the 'case 2', even though it is the same system. In parallel, salt always caused the same changes as the polymer.

Another study [54] considers the molar mass dependence of droplet size fluctuations in a water–*n*-octane–AOT–PEO system by small-angle neutron scattering. The polydispersity increases upon increasing the molar mass and the polymer content. Usually [44], this is accompanied by larger shape fluctuations, which would make this a 'case 2' system. Unfortunately, shape fluctuations were not considered here.

The droplet size is considered in several references [55–58]. The works of Suarez considered water–decane–AOT–1-propanol(1-butanol)–PEO and water–cyclohexane–SDS–1-pentanol–PEO systems and found that the droplet size decreases as a function of the polymer content. The latter works of the Brown group considered water–cyclohexane–

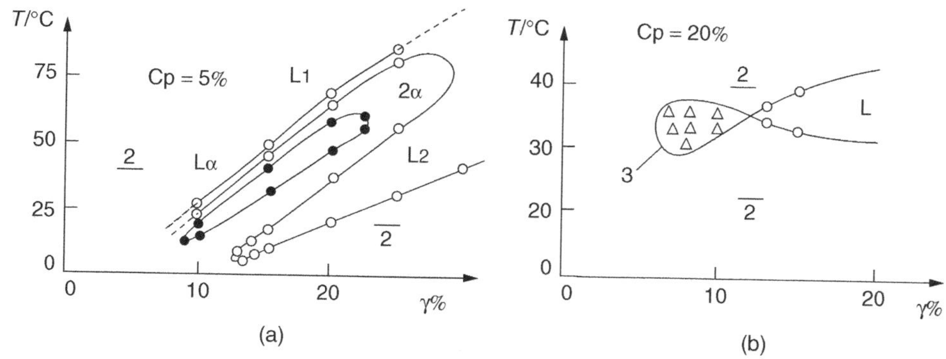

Figure 4.12 Phase diagram of the system water–isooctane–AOT–PEO as a function of the temperature T and the surfactant concentration γ. The isooctane content was fixed at 40 wt.% and the polymer content in the aqueous phase was either $C_P = 5$ or 20 wt.%. The phase regions are $\overline{2}$ for microemulsion coexisting with excess water, $\underline{2}$ for microemulsion coexisting with excess oil, 3 for three-phase coexistence of a microemulsion with two excess phases. Inside the 'fish-tail' there are one-phase regions 'L': L_1 for o/w-droplet microemulsions, L_2 for w/o-droplet microemulsions, L_α for the lamellar phase and 2α for a coexistence of a lamellar with a microemulsion phase. Note that the whole 'fish-tail' is shifted to higher surfactant concentrations upon polymer addition. (From Ref. [53], reprinted with permission of the American Chemical Society.)

SDS–1-pentanol–PEO and water–toluene–SDS–1-pentanol–PEO systems and they found an increasing droplet size as a function of the polymer content. Controversial discussions followed [59, 60]. If the elastic moduli were the only affected magnitudes one would argue: With increasing bending rigidity ('case 1') the fluctuations would be diminished, and the radius would decrease due to the opposite surface-to-volume effects. The opposite trends would support 'case 2'. However, the size could also be changed by polymer incorporation inside the surfactant layer or by polymer–surfactant complexes inside the water domains. These effects would influence the droplet size differently.

4.3.3 Cluster formation and polymer–colloid interactions

Adsorbing water-soluble polymers can cause attractive interactions between droplets. In this context, w/o- and o/w-droplet microemulsions are distinguished by the effect of confinement. At a first instance, attractive interactions are stated for w/o-droplet microemulsions with adsorbing hydrophilic polymers by FT-IR [61]. For w/o-droplet microemulsions polymer confinement is important [46]. A minimum degree of polymerisation is needed, before the polymer influences the surfactant film of a single droplet only. At higher degrees of polymerisation the polymer interconnects several droplets and so clusters are formed. The droplet–bicontinuous phase transition does not depend on the degree of polymerisation anymore. By dynamic light scattering [62], it is confirmed that these clusters do not change the size with varying degree of polymerisation, but unaffected droplets coexist. For o/w-droplet microemulsions attractive interactions are found by small-angle neutron scattering experiments [42]. These measurements directly yield a structure factor, which can be interpreted by the formation of transient clusters. The cluster formation seems to be an ultimate realisation of long adsorbing polymers independent on which side they are adsorbed. A threshold degree of polymerisation is only reported for incorporated polymers. The cluster formation of droplets resembles the network formation (see Section 4.2.2) even though the droplet spacing is minimised for the adsorbing polymers.

Adsorbing polymers also become important when the microemulsion is used as a template to synthesise ionic nanoparticles. On the one hand, the phase behaviour can be tailored for the application [63]. On the other hand, the polymer gives the product a certain stability against drying/dissolving cycles [64]. Polyampholytes embedded in the microemulsion can enlarge the one-phase region. More important is the stabilisation of the obtained colloidal particles. While the particles have a typical size of 2–4 nm as long as they are dispersed in the microemulsion, they can be as large as 108 nm after drying and redispersing if no polymer is added. However, their size nearly stays unchanged if polymer was added to the system. A similar problem is addressed using polycarboxylates which can keep $CaCO_3$ in small crystallites [65], which could be interesting in washing processes with hard water. The general issue of nanoparticle synthesis is treated in Chapter 6.

References

1. Kabalnov, A., Olsson, U., Thuresson, K. and Wennerstroem, H. (1994) Polymer effects on the phase equilibrium of a balanced microemulsion: Adsorbing versus nonadsorbing polymers. *Langmuir*, **10**, 4509–4513.
2. Jakobs, B., Sottmann, T., Strey, R., Allgaier, J., Willner, L. and Richter, D. (1999) Amphiphilic block copolymers as efficiency boosters for microemulsions. *Langmuir*, **15**, 6707–6711.
3. Helfrich, W. (1973) Elastic properties of lipid bilayers – theory and possible experiments. *Z. Naturforsch.*, **28**, 693–703.
4. Endo, H., Mihailescu, M., Monkenbusch, M., Allgaier, J., Gompper, G., Richter, D., Jakobs, B., Sottmann, T., Strey, R. and Grillo, I. (2001) Effect of amphiphilic block copolymers on the structure and phase behavior of oil–water–surfactant mixtures. *J. Chem. Phys.*, **115**, 580–600.
5. Teubner, M. and Strey, R. (1987) Origin of the scattering peak in microemulsions. *J. Chem. Phys.*, **87**, 3195–3200.
6. Hiergeist, C. and Lipowsky, R. (1996) Elastic properties of polymer decorated membranes. *J. Phys. II France*, **6**, 1465–1481.
7. Appell, J., Ligoure, C. and Porte, G. (2004) Bending elasticity of a curved amphiphilic film decorated with anchored copolymers: A small angle neutron scattering study. *J. Stat. Mech. Theor. Exp.*, **P08002**, 1–4.
8. Allgaier, J., Willner, L., Richter, D., Jakobs, B., Sottmann, T. and Strey, R. EP1109883, US6677293, Forschungszentrum Jülich, Invs.
9. Allgaier, J., Frielinghaus, H., Frank, C., Sottmannn, T. and Strey, R. WO 2006060993, Forschungszentrum Jülich, Invs.
10. Allgaier, J., Frielinghaus, H. and Frank, C. WO 2006122530, Forschungszentrum Jülich, Invs.
11. Nilsson, M., Södermann, O. and Johansson, I. (2006) The effect of polymers on the phase behavior of balanced microemulsions: Diblock-copolymers and comb-polymers. *Colloid Polym. Sci.*, **284**, 1229–1241.
12. Mock-Knoblauch, C., Wagner, N., Oetter, G., Völkel, L., Petrovic, S., Lange, A., Mijolovic, D. and Hüffer, S. WO 2005077513, BASF AG; Invs.
13. Mock-Knoblauch, C., Wagner, N., Oetter, G., Völkel, L., Steinmetz, B., Dyllick-Brenzinger, R., Schröder, J. and Petrovic, S. WO 2004103542, BASF AG; Invs.

14. Frank, C., Frielinghaus, H., Allgaier, J. and Richter, D. (2008) Hydrophilic alcohol ethoxylates as efficiency boosters for microemulsions. *Langmuir*, **24**, 6036–6043.
15. Gompper, G., Richter, D. and Strey, R. (2001) Amphiphilic block copolymers in oil–water–surfactant mixtures: Efficiency boosting, structure, phase behavior and mechanism. *J. Phys. Condens. Matter*, **13**, 9055–9074.
16. Frank, C., Sottmann, T., Stubenrauch, C., Allgaier, J. and Strey, R. (2005) Influence of amphiphilic block copolymers on lyotropic liquid crystals in water–oil–surfactant systems. *Langmuir*, **21**, 9058–9067.
17. Bagger Jorgensen, H., Olsson, U., Ilioupolos, I. and Mortensen, K. (1997) A nonionic microemulsion with adsorbing polyelectrolyte. *Langmuir*, **13**, 5820–5829.
18. Rong, G., Yang, J., Friberg, S.E., Aikens, P.A. and Greenshields, J.N. (1996) Complex lamellar structures of polyoxyethylene 20 sorbitan oleate and a fatty acid/lecithin lamellar liquid crystal. *Langmuir*, **12**, 4286–4291.
19. Frank, C., Strey, R., Schmidt, C. and Stubenrauch, C. (2007) Coexisting lamellar phases in water–oil–surfactant systems induced by the addition of an amphiphilic block copolymer. *J. Colloid Interface Sci.*, **312**, 76–86.
20. Vollmer, D., Hofmeier, U. and Eicke, H.-F. (1992) Mesoscopic structure of polymer mediated microemulsion networks. *J. Phys. II France*, **2**, 1677–1681.
21. Stieber, F. and Eicke, H.-F. (1996) Solution of telechelic ionomers in a water/AOT/oil (w/o) microemulsion: A static and dynamic light scattering study. *Colloid Polym. Sci.*, **274**, 826–835.
22. Meier, W. (1996) Structured polymer networks from o/w-microemulsions and liquid crystalline phases. *Langmuir*, **12**, 6341–6345.
23. Filali, M., Aznar, R., Svenson, M., Porte, G. and Appell, J. (1999) Swollen micelles plus hydrophobically modified hydrosoluble polymers in aqueous solutions: Decoration versus bridging. A small angle neutron study. *J. Phys. Chem. B.*, **103**, 7293–7301.
24. Filali, M., Ouazzani, M.J., Michel, E., Aznar, R., Porte, G. and Appell, J. (2001) Robust phase behavior of model transient networks. *J. Phys. Chem. B*, **105**, 10528–10535.
25. Fleischer, G., Stieber, F., Hofmeier, U. and Eicke, F.-H. (1994) On the dynamics of equilibrium networks: POE-b-PI-b-POE copolymers in H_2O/AOT/Isooctane water in oil microemulsions. *Langmuir*, **10**, 1780–1785.
26. Odenwald, M., Eicke, H.-F. and Meier, W. (1995) Transient networks by ABA triblock copolymers and microemulsions: A rheological study. *Macromolecules*, **28**, 5069–5074.
27. Bagger-Jörgensen, H., Coppola, L., Thuresson, K., Olsson, U. and Mortensen, K. (1997) Phase behavior, microstructure, and dynamics in a nonionic microemulsion on addition of hydrophobically end-capped poly(ethylene oxide). *Langmuir*, **13**, 4204–4218.
28. Lynch, I. and Piculell, L. (2004) Size, concentration, and solvency effects on the viscosifying behavior of PEO-PS-PEO triblock copolymers in AOT oil-continuous microemulsions. *J. Phys. Chem. B*, **108**, 7515–7522.
29. Holmberg, A., Piculell, L. and Wesslen, B. (1996) Viscosity effects of a graft copolymer with a hydrophobic backbone and hydrophilic side chains in a water/AOT/cyclohexane oil-continuous microemulsion. *J. Phys. Chem.*, **100**, 462–464.
30. Holmberg, A., Hansson, B., Piculell, L. and Linse, P. (1999) Effects of amphiphilic graft copolymer on an oil-continuous microemulsion. Viscosity, droplet size, and phase behavior. *J. Phys. Chem. B*, **103**, 10807–10815.
31. Holmberg, A., Piculell, L. and Nyden, M. (2002) Effects of amphiphilic graft copolymer on an oil-continuous microemulsion. Self-diffusion and viscosity. *J. Phys. Chem. B*, **106**, 2533–2544.
32. Meier, W., Falk, A., Odenwald, M. and Stieber, F. (1996) Microemulsion elastomers. *Colloid Polym. Sci.*, **274**, 218–226.
33. Meier, W. (1998) Polymer networks with lamellar structure. *Macromolecules*, **31**, 2212–2217.

34. Koetz, J., Beitz, T. and Tiersch, B. (1999) Self assembled polymer–surfactant systems. *J. Disp. Sci. Technol.*, **20**, 139–163.
35. Hanke, A., Eisenriegler, E. and Dietrich, S. (1999) Polymer depletion effects near mesoscopic particles. *Phys. Rev. E*, **59**, 6853–6878.
36. Herring, A.R. and Henderson, J.R. (2007) Hard-sphere fluid adsorbed in an annular wedge: The depletion force of hard-body colloidal physics. *Phys. Rev. E*, **75**, 011402.
37. Zackrisson, M., Anderson, R. and Bergenholtz, J. (2004) Depletion interactions in model microemulsions. *Langmuir*, **20**, 3080–3089.
38. Kabalnov, A., Olsson, U. and Wennerström, H. (1994) Polymer effects on the phase equilibrium of a balanced microemulsion. *Langmuir*, **10**, 2159–2169.
39. Pham, K.N., Egelhaaf, S.U., Pusey, P.N. and Poon W.C.K. (2004) Glasses in hard spheres with short-range attraction. *Phys. Rev. E*, **69**, 011503.
40. Byelov, D., Frielinghaus, H., Holderer, O., Allgaier, J. and Richter, D. (2004) Microemulsion efficiency boosting and the complementary effect. 1. Structural properties. *Langmuir*, **20**, 10433–10443.
41. John, A.C., Uchiyama, H., Nakamura, K. and Kunieda, H. (1997) Phase behaviour of a water/nonionic surfactant/oil ternary system in the presence of polymer oil. *J. Colloid Interface Sci.*, **186**, 294–299.
42. Lal, J. and Auvray, L. (1994) Perturbations of microemulsion droplets by confinement and adsorption of polymer. *J. Phys. II France*, **4**, 2119–2125.
43. Nakaya, K., Imai, M., Komura, S., Kawakatsu, T. and Urakami, N. (2005) Polymer-confinement-induced nematic transition of microemulsion droplets. *Europhys. Lett.*, **71**, 494–500.
44. Milner, S.T. and Safran, S.A. (1987) Dynamical fluctuations of droplet microemulsions and vesicles. *Phys. Rev. A*, **36**, 4371–4379.
45. Meier, W. (1997) Kerr effect measurements on a poly(oxyethylene) containing water-in-oil microemulsion. *J. Phys. Chem. B*, **101**, 919–921.
46. Meier, W. (1996) Poly(oxyethylene) adsorption in water/oil microemulsions: A conductivity study. *Langmuir*, **12**, 1188–1192.
47. DeGennes, P.G. (1990) Interactions between polymers and surfactants. *J. Phys. Chem.*, **94**, 8407–8413.
48. Brooks, J.T., Marques, C.M. and Cates, M.E. (1991) The effect of adsorbed polymer on the elastic moduli of surfactant bilayers. *J. Phys. II*, **1**, 673–690.
49. Note, C., Koetz, J. and Kosmella, S. (2006) Structural changes in poly(ethyleneimine) modified microemulsion. *J. Colloid Interface Sci.*, **302**, 662–668.
50. Rajagopalan, V., Olsson, U. and Iliopoulos, I. (1996) Effect of adsorbing polyectrolytes on a balanced nonionic surfactant–water–oil system. *Langmuir*, **12**, 4378–4384.
51. Breidenich, M., Netz, R.R. and Lipowsky, R. (2001) Adsorption of polymers anchored to membranes. *Eur. Phys. J. E*, **5**, 403–414.
52. Bellocq, A.M. (1998) Phase equilibria of polymer-containing microemulsions. *Langmuir*, **14**, 3730–3739.
53. Maugey, M. and Bellocq, A.M. (1999) Effect of added salt and poly(ethylene glycol) on the phase behaviour of a balanced AOT–water–oil system. *Langmuir*, **15**, 8602–8608.
54. Schübel, D., Bedford, O.D., Ilgenfritz, G., Eastoe, J. and Heenan R.K. (1999) Oligo- and polyethylene glycols in water-in-oil microemulsions. A SANS study. *Phys. Chem. Chem. Phys.*, **1**, 2521–2525.
55. Suarez, M.J. and Lang, J. (1995) Effect of addition of water-soluble polymers in water-in-oil microemulsions made with anionic and cationic surfactants. *J. Phys. Chem.*, **99**, 4626–4631.
56. Suarez, M.J., Lévy, H. and Lang, J. (1993) Effect of addition of polymer to water-in-oil microemulsions on droplet size and exchange of material between droplets. *J. Phys. Chem.*, **97**, 9808–9816.

57. Papoutsi, D., Lianos, P. and Brown, W. (1993) Interaction of a-hydro-w-hydroxypoly(oxy-1,2-ethanediyl) with water-in-oil microemulsions. 2. Medium-size polymer chain. Cyclohexane and toluene microemulsions. *Langmuir*, **9**, 663–668.
58. Lianos, P., Modes, S., Staikos, G. and Brown, W. (1992) Interaction of poly(oxyethylene glycol) with cyclohexane-pentanol-sodium dodecyl sulfate water-in-oil microemulsions. *Langmuir*, **8**, 1054–1059.
59. Lianos, P. (1996) Comment on the effect of addition of water-soluble polymers in water-in-oil microemulsions made with anionic and cationic Surfactants. *J. Phys. Chem.*, **100**, 5155–5155.
60. Lang, J. (1996) Reply to comment on the effect of addition of water-soluble polymers in water-in-oil microemulsions made with anionic and cationic surfactants. *J. Phys. Chem.*, **100**, 5156–5156.
61. González-Blanco, C., Rodríguez, L.J. and Velázquez, M.M. (1997) Effect on the addition of water-soluble polymers on the structure of aerosol OT water-in-oil microemulsions: A Fourier transform infrared spectroscopy study. *Langmuir*, **13**, 1938–1945.
62. Papoutsi, D., Lianos, P. and Brown, W. (1994) Interaction of polyethylene glycol with water-in-oil microemulsions. 3. Effect of polymer size and polymer concentration. *Langmuir*, **10**, 3402–3405.
63. Koetz, J., Andres, S., Kosmella, S. and Tiersch, B. (2006) $BaSO_4$ nanorods produced in polymer-modified bicontinuous microemulsions. *Comp. Interface*, **13**, 461–475.
64. Note, C., Ruffin, J., Tiersch, B. and Koetz, J. (2007) The influence of polyampholytes on the phase behaviour of microemulsion used as template for the nanoparticle formation. *J. Disp. Sci. Tech.*, **28**, 155–164.
65. Rieger, J., Thieme, J. and Schmidt, C. (2000) Study of precipitation reactions by X-ray microscopy: $CaCO_3$ precipitation and the effect of polycarboxylates. *Langmuir*, **16**, 8300–8305.

Chapter 5
Reactions in Organised Surfactant Systems

Reinhard Schomäcker and Krister Holmberg

5.1 Introduction

Organic reactions are usually performed in homogeneous solutions. The reaction kinetics in such systems is usually well understood and can be described numerically with a high degree of accuracy. However, in many instances reagent incompatibility requires that the reaction is carried out in a two- or multi-phase reaction system. Whereas one component may only be soluble in a non-polar solvent, the other component may require a polar environment. A typical example where this is the case is a reaction between a lipophilic organic compound and an inorganic salt. There are numerous examples of such reactions in the domain of organic synthesis. Pertinent examples are hydrolysis of esters with alkali, oxidative cleavage of olefins with permanganate–periodate, addition of hydrogen sulphite to aldehydes and to terminal olefins, preparation of alkyl sulphonates by treatment of alkyl chloride with sulphite or by addition of hydrogen sulphite to α-olefin oxides. The list can be extended further. In all examples given, there is a compatibility problem to be solved if the organic component is a large non-polar molecule.

There are various options to overcome the reagent incompatibility problem. One way is to use a solvent or a solvent combination capable of dissolving both the organic compound and the inorganic salt. Polar, aprotic solvents, such as dimethylsulphoxide (DMSO), dimethylformamide (DMF) and tetrahydrofuran (THF), are sometimes useful for this purpose but many of these are unsuitable for large-scale work due to toxicity and/or difficulties in removing them by low vacuum evaporation.

Alternatively, the reaction may be carried out in a mixture of two immiscible solvents. The contact area between the phases may be increased by agitation. Phase transfer reagents, most commonly tetraalkylammonium salts based on two or three long alkyl chains, are useful aids in many two-phase reactions. Also, crown ethers are very efficient in overcoming phase contact problems; however, their usefulness is limited by high price. Open-chain polyoxyethylene compounds often give a 'crown ether effect' and may therefore constitute a practically interesting alternative to the use of normal phase transfer reagents. (This effect is often seen with common non-ionic surfactants of the alcohol ethoxylate type. The polyoxyethylene chain of the non-ionic surfactant folds around the cation in the complex thus building up a charge and increased hydrophilicity.)

Microemulsions are excellent solvents both for hydrophobic organic compounds and for inorganic salts. Being macroscopically homogeneous, yet microscopically dispersed, they can be regarded as something between a solvent-based one-phase system and a true two-phase system. In this context microemulsions should be seen as an alternative to two-phase systems with added phase transfer catalyst. Both systems rely on an auxiliary agent – a surfactant or a phase transfer catalyst – which is not consumed in the reaction. In both cases this component needs to be separated from the product after completed reaction and the procedure should preferably be such that the surfactant or the phase transfer catalyst can be reused. Various approaches exist for such operations and work-up procedures relevant to the microemulsion approach will be treated in some detail in this chapter.

So, the disadvantages of working with polar aprotic solvents or with phase transfer catalysts can be avoided by the use of surfactant-based systems. Work-up procedures at high boiling temperatures or low vacuum, as in the case of many polar aprotic solvents, can be avoided by the selection of a suitable hydrophobic component. Since there are surfactants available which allow dispersion of all kinds of hydrophobic liquids in water, low boiling aliphatic or aromatic solvent can be used for formulation of the reaction medium. This allows a work-up without high-energy demand.

The most recent motivation for the use of surfactant-based reaction media relates to the 'Green Chemistry' discussion. It is generally recommended to use water instead of an organic solvent that could be harmful to the environment. Since water is non-toxic and inflammable, it is regarded as the ideal solvent. Methods for water purification after industrial use are well established and available in industrial scale.

5.2 Motivation for surfactant systems as reaction media

Surfactants by definition self-organise in water giving rise to micelles of varying size and shape. The core of micelles is non-polar and can solubilise reactants that are insoluble in water. Thus, a simple surfactant–water system at a surfactant concentration well above the critical micelle concentration can be used to overcome the problem of reactant incompatibility: the polar reagent will be situated in the bulk aqueous domain, the non-polar reagent will be present in the micelles, and the reaction will occur at the micelle boundary. Organic reactions in micellar systems have been reported more than 40 years ago [1, 2].

These colloidal aggregates of amphiphilic molecules have been discussed as biological membrane or bio-mimicking systems. In many publications, micelles have been described as simple model system for enzymatic catalysis. Indeed, for a variety of reactions an acceleration of the reaction rate in the presence of surfactant micelles is observed. Other publications discuss 'micellar catalysis'. A detailed kinetic analysis of the observed phenomena later showed that local accumulation of the reactants inside the colloidal dispersions is one reason for rate enhancement rather than catalysis. Another cause of 'micellar catalysis' is a higher concentration of a reactant in the vicinity of the micelle than in the bulk. The reactant, e.g. an inorganic anion, is then attracted by oppositely charged surfactant head groups. Alkaline hydrolysis of ester-containing cationic surfactants is a well-known example [3].

Thus, micellar solutions consist of three regions of distinctly different solvation properties, a continuous polar aqueous domain, non-polar cores and interfacial regions of intermediate polarity. They are all present in a single homogeneous, thermodynamically stable solution. The totality of the three regions can be treated as separate reaction regions distributed throughout the solution because the distributions of all the components are in dynamic equilibrium [4–6]. For bimolecular reactions between lipophilic and hydrophilic reactants dissolved in a micellar solution, the hydrophilic reactant partitions primarily between the aqueous and interfacial regions and the lipophilic reactant partitions primarily between the interfacial and hydrophobic regions. The two reactants meet and react in the interfacial region.

True micellar systems have low capacity for dissolving non-polar reactants, however. They are therefore of limited preparative value. Microemulsions, which contain not only surfactant and water but also an oil component, can dissolve appreciable amounts of both a polar and a non-polar reactant and are therefore much more practically useful as media for organic synthesis. There has been considerable interest in the use of microemulsions as media for organic reactions in recent years [7–11]. Not only can such a formulation be a way to overcome compatibility problems, the capability of microemulsions to compartmentalise and concentrate reactants can also lead to considerable rate enhancement compared to one-phase systems. A third aspect of interest for preparative organic synthesis is that the large oil–water interface of the system can be used as a template to induce regioselectivity. These aspects will be dealt with in this chapter.

In microemulsions, as well as in micellar solutions, the local concentrations of both reactants in the interfacial region, or reaction zone, can be much higher than their average concentrations in the whole solution, calculated with the assumption of homogeneous distribution. For example, if the surfactant is chosen with a head group charge opposite to that of an ionic water-soluble reactant, the local concentration of this ionic reactant may be one to three orders of magnitude higher than its average concentration in solution [12]. This concentration effect is a major contributor to rate enhancements in micellar solutions and it approaches a million-fold rate increase in some cases [13–15]. Conversely, if the ionic reactant is of the same charge as the surfactant, its concentration will be greatly reduced at the interface and rate inhibition is observed [15]. The observed overall rate also depends on the physical properties of the interface, which may stabilise or destabilise the transition state relative to the ground state in comparison to a homogeneous medium [6, 15]. For instance, it has been reported that a bimolecular nucleophilic substitution reaction (S_N2) was faster in a microemulsion based on an alcohol ethoxylate than in a microemulsion stabilised by a sugar surfactant [16, 17]. Both are non-ionic surfactants, so the effect cannot be related to charges of the interface. It was proposed that the effect was due to a difference in the degree of hydration of the head group layer of the surfactants. Sugar is a more polar head group than an oligooxyethylene chain. It is also known that micelles of sugar-based surfactants have a higher dielectric constant than micelles of alcohol ethoxylates [18]. The chemical potential of water in the reaction zone of a microemulsion based on a sugar surfactant is therefore likely to be higher than in the reaction zone of a microemulsion based on an alcohol ethoxylate.

At higher surfactant concentration liquid crystalline phases may be formed. Surfactant liquid crystals can also solubilise appreciable amounts of oil into the non-polar regions made up of the surfactant tails. Thus, both binary surfactant–water systems and ternary systems with oil included can be formulated into liquid crystals. Such systems can also be used as media for organic synthesis. In fact, a reaction in a surfactant liquid crystal often runs very rapidly, considerably faster than in a microemulsion based on the same surfactant [19]. Figure 5.1 shows the reaction profiles of a typical substitution reaction of

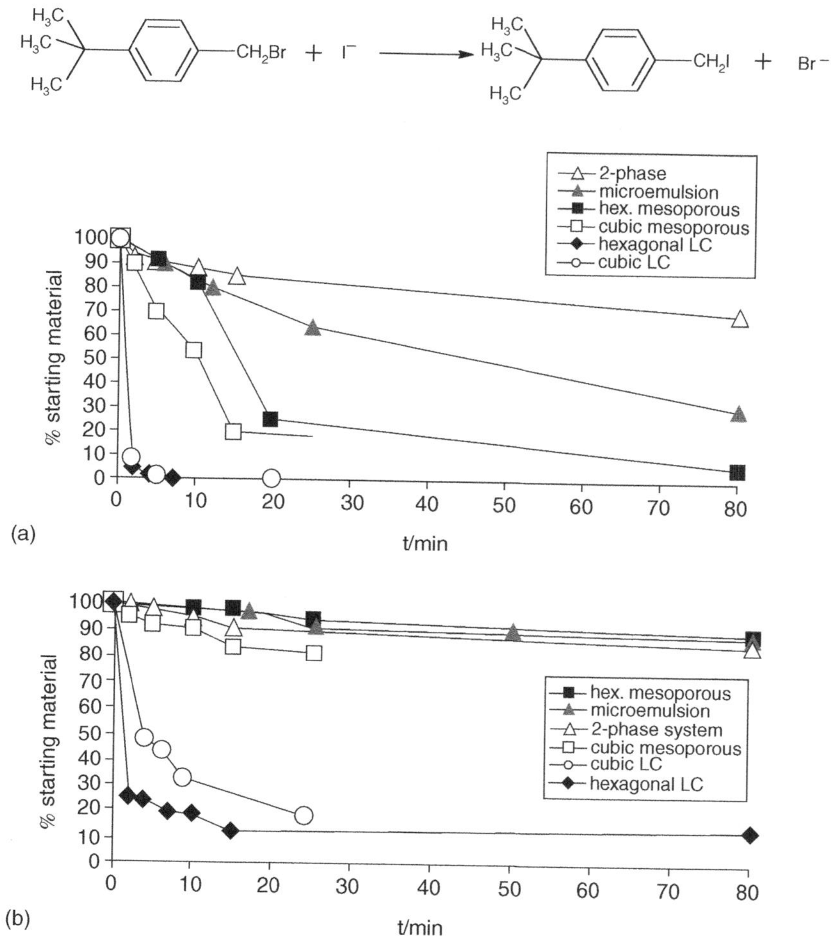

Figure 5.1 Reaction profiles for reaction between 4-*tert*-butylbenzyl bromide and potassium iodide using a 1:10 (a) or a 1:1 (b) molar ratio of the reactants. The reactions were performed in a decane–water two-phase system, in a microemulsion, in hexagonal or cubic mesoporous materials and in hexagonal or cubic liquid crystalline phases.

S_N2 type performed in a microemulsion and in two liquid crystalline phases of different geometry. The higher overall rate obtained in a liquid crystalline system as compared to a microemulsion is most probably an effect of the higher interfacial area between the polar and the non-polar domains of the former systems. Almost all added surfactant will be located at the interface in both systems, which means that the volume of the reaction zone will be proportional to the amount of surfactant in the system. As will be discussed later in this chapter, the reaction rate for a typical bimolecular substitution reaction is proportional to the interfacial area, provided one of the reactants is only soluble in the polar domain and the other reactant is only soluble in the non-polar domain of the organised surfactant system.

However, for practical use as reaction media the liquid crystalline systems are hardly realistic. They need very high surfactant concentrations, which make the work-up procedure complex, and these systems are highly viscous, which make mixing and heat removal a problem. Surfactant liquid crystals are of more interest as templates for making mesoporous oxides and such materials, in the form of suspensions of small particles, are also of interest for overcoming compatibility problems in organic synthesis [20]. The profiles for reactions in such systems are also included in Fig. 5.1.

Microemulsions and two-phase systems with added phase transfer agent are both useful means of overcoming reactant incompatibility, but on entirely different grounds. In phase transfer catalysis, the nucleophile is carried into the organic phase where it becomes poorly solvated and highly reactive. Phase transfer catalysts, like quaternary ammonium salts or crown ethers are efficient transfer agents, especially for univalent ionic reagents. But for a variety of hydrophilic reactants no phase transfer catalyst is available at justifiable prices or without ecological drawbacks. Decomposition of phase transfer catalysts and problems with their recovery are additional barriers for large-scale applications.

In the microemulsion approach, there is no transfer of reactant from one environment to another; the success of the method relies on the very large oil–water interface at which the reaction occurs. In an attempt to combine the two methods and take advantage of both the high reactivity of a poorly solvated anion in phase transfer catalysis and the very large oil–water interface of a microemulsion, ring-opening of a lipophilic epoxide was carried out in a microemulsion composed of chlorinated hydrocarbon, water and a sugar surfactant, an alkyl glucoside, in the presence of a conventional phase transfer agent, tetrabutylammonium hydrogen sulphate [21]. Reactions were also performed in a two-phase system with and without added phase transfer agent. As shown in Fig. 5.2, a very high reactivity was obtained when the phase transfer agent was added to the microemulsion. Similar results have been obtained for another bimolecular nucleophilic substitution reaction [22] and for epoxidation of α,β-unsaturated enones with alkaline hydrogen peroxide [23].

In the two-phase systems, strong agitation is usually needed in order to provide a sufficient contact area for the reactants. Still low reaction rates are often obtained, especially at larger scales. Handling of two-phase systems in micro-mixing devices with high power input is one new approach for solving this problem. The use of surfactant-based systems is another novel approach. In addition to the capability to solubilise a broad range of substances the surfactant systems provide a large internal interface between the hydrophilic and the hydrophobic subphases of the dispersion, and the interfacial area is mainly governed by the surfactant content of the system. The formation of microemulsions and normal or reverse micelles is driven by thermodynamics. Agitation is only needed for mixing and equilibration of coexisting phases, not for the formation of the interface. The large interface enables and accelerates phase transfer reactions without specific interactions between one of the reactants and a phase transfer agent. This opens the application of surfactant-based media for a broader range of reactants than phase transfer catalysis.

In surfactant-based reaction media mass transfer limitations on the reaction rate are much suppressed in comparison to stirred two-phase systems without surfactants. The reason for this pronounced difference is the different characteristic length scales of the systems. This may be illustrated by the following example. Twenty grams per litre of a

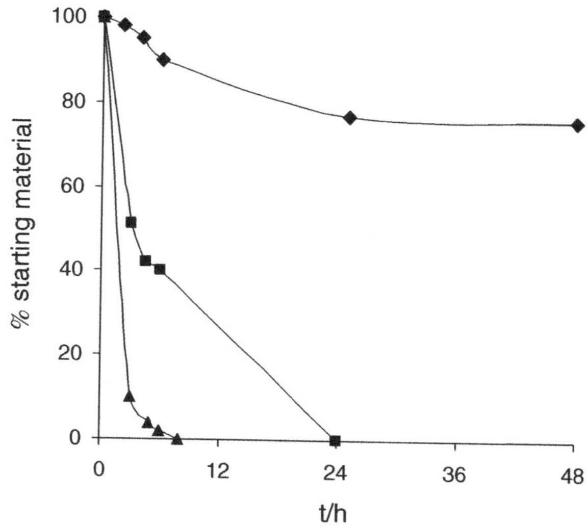

Figure 5.2 Reaction profiles for ring opening of 1,2-epoxyoctane with sodium hydrogen sulphite in a two-phase system (diamonds), a two-phase system with added phase transfer agent (squares), and a microemulsion with added phase transfer agent (triangles).

standard surfactant with a typical head group area of 0.5 nm^2 per molecule forms aggregates with a size of ~10 nm and an interface of 10 000 m^2 L^{-1}. Addition of a hydrophobic substance results in swelling of the aggregate, an increase in droplet size and volume fraction of the dispersed phase, but does not alter the interfacial area.

A typical example for a stirred two-phase system with a volume fraction of 30 vol.% organic phase dispersed in water, an interfacial tension of 25 mN m^{-1} and a specific power input of 0.5 W L^{-1} shows a droplet diameter in the range of 250 μm and a specific interface of about 10 m^2 L^{-1}. These dimensions may be estimated from simple empirical correlations between the Sauter mean diameter of the dispersed phase (d_{23}) and the characteristic Weber number (We). In case of turbulent mixing the following correlation is proposed in the literature for calculation of the mean diameter of dispersed droplets [24]

$$\frac{d_{32}}{d} = C_1\, We^{-3/5}\ (1 + C_2\ \phi_M) \tag{5.1}$$

$$\text{with } d_{32} = \frac{\sum_i n_i\ d_i^3}{\sum_i n_i\ d_i^2} \text{ and } We = \frac{d^3 N^2 \rho_C}{\sigma}.$$

The constants C_1 and C_2 depend on chemical and physical properties of the system used. Typical values for water and hydrocarbons are 0.5 and 5, respectively. This correlation must be used with caution since at larger Weber numbers deviations are reported in literature. If the factor $(1 + C_2\phi_M)$ is neglected, the following correlation between droplet diameter and specific power input by stirring (ε) is obtained

$$d_{32} = C_3\,\varepsilon^{-0.4} \tag{5.2}$$

$$\text{with } \varepsilon = \frac{P}{\rho_C V}$$

as specific power input to the system. The specific interface of the dispersion is then calculated with the assumption of spherical droplets according to

$$a = \frac{6\varphi}{d_{32}}. \tag{5.3}$$

The average mole number N_A transferred across the internal interface of the dispersion is given by

$$N_A = \beta a \Delta c \tag{5.4}$$

with a being the specific interface, Δc the concentration gradient at the place of highest mass transfer resistance and β the mass transfer coefficient. The product βa defines the time constant for mass transfer in the system. β itself is proportional to the ratio of the diffusion coefficient of the transferred molecule and the characteristic length (δ) for the mass transfer resistance D_A/δ. In a stirred two-phase system the predominant mass transfer resistance is diffusion within the droplet. Therefore, the droplet radius is the characteristic length. For very small droplets, like micelles or microemulsion droplets, the predominant mass transfer barrier is diffusion in a stagnant layer around the droplets. For high volume fractions (20–50 vol.%), the distance between the droplets is in the same range as the droplet radius. Therefore, the radius is also considered as a characteristic length. For the same molecular diffusion coefficient the ratio between the time constants for mass transfer for microemulsion and for two-phase systems can be calculated based on the data given above

$$\frac{D_A a_{ME}}{r_{mic}} : \frac{D_A a_{2\Phi}}{r_d} = 10^6. \tag{5.5}$$

Even if the uncertainties with respect to the values are high, the mass transfer in micellar systems or microemulsion is about six orders of magnitude higher than in two-phase systems. This situation is not changed with the addition of a phase transfer catalyst. The ion pair formed by the phase transfer agent and the hydrophilic reactant has a partition coefficient that is more favourable towards the organic phase. This will change the driving concentration difference, but not the mass transfer time constant βa. This value may even be decreased because of a lower diffusion coefficient of the transferred compound, the ion pair. Mass transport phenomena in two-phase systems have been discussed in detail since the 1950s for extraction processes [25]. That knowledge can be transferred to this field.

Because of the much higher mass transfer time constants in comparison to two-phase systems a pure kinetic control of the reactions is normally observed for reactions in micellar solutions and microemulsions. Even for a collision-controlled fluorescence quenching reaction with bromonaphthalene as probe and nitrite as quencher, no mass transfer limitation was observed when performed in a bicontinuous microemulsion [26].

5.3 Selected reactions

Research on the use of microemulsions as media for organic synthesis is an area where surface chemistry and organic chemistry meet. In addition, chemical engineering aspects related to mass transfer issues, work-up aspects etc. are also important. Research reports are published in surface chemistry and organic chemistry journals, as well as in journals devoted to chemical engineering. The research field has grown rapidly in recent years. A review from 2005 in Angewandte Chemie on reactions in such systems contains no less than 300 references with the majority being from 2000 or later [10]. All types of organic reactions have been studied, including oxidations, reductions, substitution reactions, concerted reactions of Diels–Alder type, various types of metal-organic reactions etc. Microemulsions have also been used for making enantioselective reactions.

The purpose of this review is not to give a new summary of all reactions that have been investigated using microemulsions as reaction media. Instead, we have chosen to put focus on two reaction types, where the microemulsion-based processes have shown to be particularly successful, nucleophilic substitutions and homogeneous catalysis. We also illustrate a generally important aspect of microemulsion-based reactions: the possibility to affect the reaction pattern and, in some instances, induce a regiospecificity that is difficult to obtain in homogeneous medium. Immediately after this section follows a section on engineering aspects, which we believe is a very important and sometimes neglected topic. Proper control of the engineering aspects can be seen as the key to success in terms of implementation of the organised surfactant solution into industrial use.

5.3.1 Nucleophilic substitution reactions

Bimolecular nucleophilic substitution reactions, S_N2 reactions, are probably the reaction type that has been investigated the most in microemulsions. Such reactions often involve one lipophilic component that is insoluble in water and one very hydrophilic component, usually the anion of a salt with virtually no solubility in hydrocarbon. This means that the components meet and react at the interface. The area of the interface is obviously of importance.

The rate of a bimolecular nucleophilic substitution reaction performed in a homogeneous medium can be expressed as

$$r = -\frac{1}{V}\frac{dn_{\mathrm{j}}}{dt} = k_{\mathrm{m}} C_{\mathrm{A}} C_{\mathrm{B}}, \tag{5.6}$$

where V is the total volume of the reaction mixture and n_{j} the amount of component j (mole), k_{m} is the rate constant and C_{A} and C_{B} are the concentrations of the two reactants.

The observed rate constant (k_m) can be calculated as the slope obtained when plotting the reverse concentration of substrate against reaction time.

Microemulsions are not homogeneous at the molecular level in that they consist of microscopic domains of water and oil separated by a surfactant film. The reaction may occur in either of the two domains, as well as at the interface. However, if the solubility of the polar reactant in hydrocarbon and of the lipophilic component in water is negligible, the reaction can be assumed to be a purely interfacial reaction, i.e. no reaction occurs in the two bulk phases.

Assuming that the reaction is entirely kinetically controlled and that the time of transfer of the reactants to the reaction zone is not rate-determining, the relationship between the rate constant for the reaction at the interface (k_A) and the rate constant based on total mass (k_m) is [27]

$$k_A = k_m a^{-1} \alpha(1-\alpha)(1-\gamma)^2. \tag{5.7}$$

In the above expression, a is the specific interfacial area, which can be calculated from the surfactant concentration combined with the value of the area occupied per surfactant molecule at the interface (all surfactant is assumed to reside at the interface; the bulk concentrations are neglected). α and γ are the mass fractions of oil and surfactant, respectively, defined as $\alpha = m_{oil}/(m_{oil} + m_{aq})$ and $\gamma = m_{surf}/(m_{surf} + m_{oil} + m_{aq})$. The concentration of the oil-soluble reactant in the oil domain is calculated from the overall concentration C and the parameters α and γ according to

$$C_{A,O} = C_A \alpha(1-\gamma). \tag{5.8}$$

The corresponding relation for the water-soluble reactant is

$$C_{B,W} = C_B(1-\alpha)(1-\gamma). \tag{5.9}$$

The rate constant at the interface (k_A) can be obtained as the slope of the straight line by plotting $k_m\alpha(1-\alpha)(1-\gamma)^2$ versus a (Fig. 5.3). This approach was used for describing the kinetics of the synthesis of 1-phenoxyoctane from sodium phenoxide and 1-bromooctane in a microemulsion based on the non-ionic surfactant Triton X-100, which is an octylphenol ethoxylate [27]. The total interfacial area was calculated from known values of the head group area of the non-ionic surfactant. As shown in Fig. 5.3, straight lines were obtained from which the rate constants could be obtained. From the values of k_A determined at the three different temperatures, an activation energy of 85 kJ mol^{-1} was calculated. This is a typical value for an S_N2 reaction, as usually determined in homogeneous reaction media.

The solubility characteristics of the substrates are important. The hydrophilic reactant must have negligible solubility in the non-polar domain and vice versa. If the lipophilic substrate is soluble in water to an appreciable extent, a bulk reaction in the water domain will accompany the reaction at the interface. This aspect has been investigated in some detail for another substitution reaction, reaction between potassium iodide and four different benzyl bromides using an oil-in-water microemulsion based on D_2O, decane and $C_{12}E_5$ as reaction medium [16]. The lipophilic components were unsubstituted benzyl bromide, 4-methylbenzyl bromide, 4-isopropylbenzyl bromide and 4-*tert*-butylbenzyl bromide. As

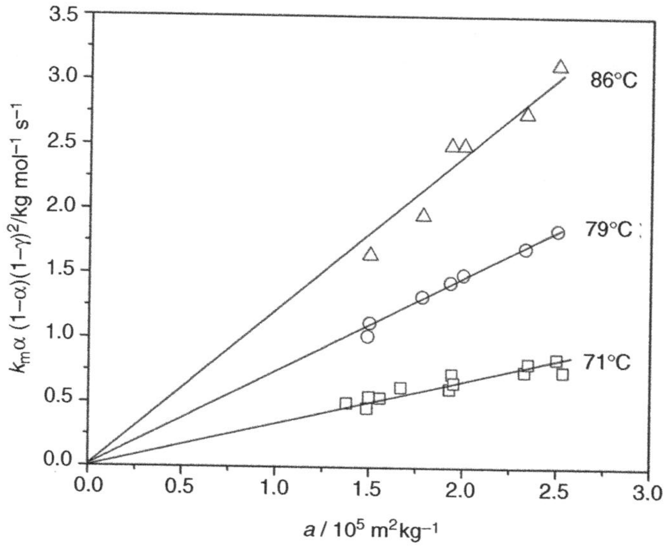

Figure 5.3 Plots of $k_m\alpha(1 - \alpha)(1 - \gamma)^2$ versus a at three different temperatures for the reaction between sodium phenoxide and 1-bromooctane. Measurements were made with both reactants as excess component.

control, reactions were also performed in deuterated methanol, d-MeOH. The relative reaction rates in d-MeOH were the expected ones with the benzyl bromides carrying a larger alkyl group in 4-position being the most reactive followed by the 4-methyl-substituted benzyl bromide. An alkyl group in *para* position will increase the electrophilicity of the benzylic methylene carbon and the inductive effect will be larger for an isopropyl and a *tert*-butyl group than for a methyl group.

The relative rates were very different when the reactions were carried out in the microemulsion. Now the methyl-substituted benzyl bromide reacted fastest followed by the unsubstituted benzyl bromide. The *tert*-butyl and the isopropyl substituted substrates reacted with the same rate. These results were interpreted as follows. The methyl substituted and, even more, the unsubstituted benzyl bromides have non-negligible water solubility. For these substrates reaction in the bulk water domain occurs in parallel to the interfacial reaction and the observed rate is the sum of the two processes. The *tert*-butyl and the isopropyl substituted benzyl bromides, on the other hand, react only at the interface. A kinetic expression has been derived for a reaction that occurs partly at the interface and partly in the aqueous domain [16]. In this treatment the microemulsion has been divided into two sub-volumes, oil and water, instead of three. The surfactant sub-volume is divided equally between the oil and water sub-volumes. Assuming again that the reactant concentration is the same in the bulk as at the interface, i.e. there is no concentration gradient for the reactants, it can be shown that the rate constant of the total reaction can be written as

$$k = k_w(1 + \Phi_{os}(K_{ow} - 1))^{-1}, \tag{5.10}$$

Table 5.1 Rate constants (k) for the reaction between 4-*tert*-butylbenzyl bromide and potassium iodide in various reaction media

Composition	$k/\mathrm{dm^3\ mol^{-1}\ s^{-1}}$
D_2O–$C_{12}E_5$–decane ($\phi = 0.1$)	0.0041
D_2O–$C_{12}E_5$–decane ($\phi = 0.2$)	0.0044
D_2O–$C_{12}E_5$–decane ($\phi = 0.3$)	0.0045
D_2O–$C_{12}E_5$–decane ($\phi = 0.4$)	0.0044
D_2O–$C_{12}E_5$–C_5OH–decane (4:96 $C_5OH/C_{12}E_5$)	0.0041
D_2O–$C_{12}E_5$–C_5OH–decane (8:92 $C_5OH/C_{12}E_5$)	0.0033
D_2O–$C_{12}E_5$–C_7OH–decane (2:98 $C_7OH/C_{12}E_5$)	0.0040
D_2O–$C_{12}E_5$–decane (10:0 $C_{12}E_4/C_{12}E_5$)	0.0038
D_2O–$C_{12}E_4$–$C_{12}E_5$–decane (1:9 $C_{12}E_4/C_{12}E_5$)	0.0037
D_2O–$C_{12}E_4$–$C_{12}E_5$–decane (1.5:8.5 $C_{12}E_4/C_{12}E_5$)	0.0035
D_2O–$C_{12}E_4$–$C_{12}E_5$–decane (2:8 $C_{12}E_4/C_{12}E_5$)	0.0034
D_2O–$C_{12}E_6$–decane (no stirring)	0.0050
D_2O–$C_{12}E_6$–decane (stirring)	0.0055
D_2O–$C_{12}E_8$–octanol–octane	0.0031
D_2O–C_8G_1–octanol–octane	0.0008
d-MeOH	0.0086
d-EtOH	0.020

where k_w is the rate constant in the aqueous domain, Φ_{os} is the combined volume fraction of oil and half the surfactant and K_{ow} is the partition coefficient for the lipophilic component (in this case the benzyl bromide derivative) between the oil and the water domains. Since K_{ow} is known to be >>1, the expression is reduced to

$$k = k_w(\Phi_{os} K_{ow})^{-1}. \tag{5.11}$$

In order to investigate the effect of aggregation size and shape on the reaction rate, the reaction between 4-*tert*-butylbenzyl bromide and potassium iodide was performed in microemulsions with varying ratio from $C_{12}E_5$ to $C_{12}E_4$. Self-diffusion NMR was used to measure the self-diffusion coefficients of the components of the system. An increase of the relative amount of $C_{12}E_4$ leads to a decrease of the observed diffusion coefficient of both surfactant and oil, which indicated an increase of the hydrodynamic radius. Since surfactant and oil diffused with the same diffusion rate one can confirm that the aggregates are discrete for the systems studied. However, the results from the kinetics experiments (Table 5.1) clearly showed that the rates were independent of the microstructure as long as the composition remained within the oil-in-water domain.

The rate constants for a variety of microemulsions based on a non-ionic surfactant and formulated with or without an alcohol as co-surfactant were determined from the slopes of the straight line obtained by plotting the reverse concentration of substrate against reaction time. The results are compiled in Table 5.1. It can be seen from the table that rather similar values were obtained for all the microemulsions based on an alcohol ethoxylate as surfactant. The reaction was more sluggish in the microemulsion based on the sugar surfactant octyl glucoside (C_8G_1). A probable reason for this difference was discussed in the section 'Motivation for surfactant systems as reaction media'. Table 5.1 also shows the rate constants for reaction in two homogeneous media, d-MeOH and d-EtOH.

Table 5.1 shows that the rate constant in EtOH is 0.020 $dm^3\ mol^{-1}\ s^{-1}$ and in MeOH 0.0086 $dm^3\ mol^{-1}\ s^{-1}$. One may expect that the rate constant would have been considerably smaller in water than in MeOH (if water would have been a possible solvent, which is not the case) because this type of S_N2 reaction runs slower the more polar the solvent, as was discussed above. The rate constant in the D_2O–decane–$C_{12}E_5$ microemulsion with $\phi = 0.1$ is 0.0041 $dm^3\ mol^{-1}\ s^{-1}$. However, since the reaction in the microemulsion is assumed to occur only inside the surfactant palisade layer, the interfacial rate constant is a more relevant parameter. The interfacial rate constant in the microemulsion was found to be 0.0071 $dm^3\ mol^{-1}\ s^{-1}$ [16]. Hence, the interfacial rate constant in the microemulsion is of the same order as in MeOH but smaller than in EtOH. The relatively large value of the interfacial rate constant for reaction in the microemulsion probably reflects the low water activity inside the surfactant palisade layer.

The above-mentioned reaction between sodium phenoxide and 1-bromooctane to synthesise 1-phenoxyoctane has been carried out in different types of microemulsion systems, all based on the same non-ionic surfactant, Triton X-100 (an octylphenol ethoxylate), the same surfactant concentration (20 wt.%), the same oil to water ratio (2:3) but different hydrocarbons as oil component [28]. This results in different phase volume ratios for the different hydrocarbons. A one-phase microemulsion is only obtained with toluene as oil component. The more hydrophobic oils, i.e. cumene, isooctane, hexadecane and paraffin oil, all give a microemulsion in equilibrium with an excess oil phase, i.e. a Winsor I system. With the more hydrophilic chlorobenzene as oil a microemulsion coexisting with an excess water phase, i.e. a Winsor II system, is obtained. As is also shown in Fig. 5.4, the reactivity is highest in the chlorobenzene- and the paraffin oil-based microemulsions, i.e. in the systems

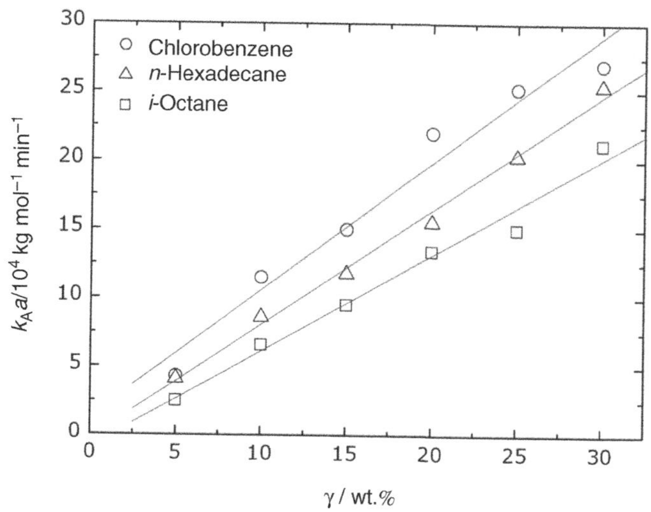

Figure 5.4 Plot of the interfacial rate constant (k_A) multiplied by the specific interface (a) against weight fraction of surfactant for different single-, two- and three-phase systems obtained by exchange of oil component.

that are at the extremes in terms of phase separation. The relative reactivity is believed to be governed by the solubility of the surfactant in the oil. The higher the solubility, the less surfactant is available to form the interface. Since the reaction occurs exclusively at the interface (sodium phenoxide has no solubility in the oil and 1-bromooctane is insoluble in water), it is important to have as large an interface as possible.

The observation that the reaction runs approximately as fast in a Winsor system as in a one-phase microemulsion has later been seen also for another bimolecular substitution reaction [16]. The fact that a Winsor system can be used instead of a one-phase microemulsion is practically important. Formulation of a one-phase microemulsion is often a problem, particularly when one wants a high loading of reactants into the oil and water domains, and one may end up with various types of two-phase or three-phase systems. Evidently, such systems may be just as useful as reaction media, as long as one of the phases is a microemulsion. The excess phase (or phases) can be regarded as reservoirs for the reactant (or reactants) while the reaction occurs at the oil–water interface of the microemulsion phase.

5.3.2 *Regioselective synthesis*

The reaction pattern can be different in microemulsions and other systems that contain an interface as compared to true homogeneous systems. The interface may influence the selectivity in at least two ways: by attracting ions that may compete with ionic reactants of same charge or by acting as a template for the reaction.

An illustrative example of the effect of counterions on selectivity is the work by Brinchi *et al.*, in which they demonstrated that reaction of sulphonate esters in the presence of equimolar amounts of bromide and hydroxyl ions took completely different paths depending on whether the reaction was performed in a micellar system based on cationic surfactant or in a homogeneous solution, see Fig. 5.5 [29]. When there is no surfactant present, attack by the hydroxyl ion dominates. In the micellar solutions, on the other hand,

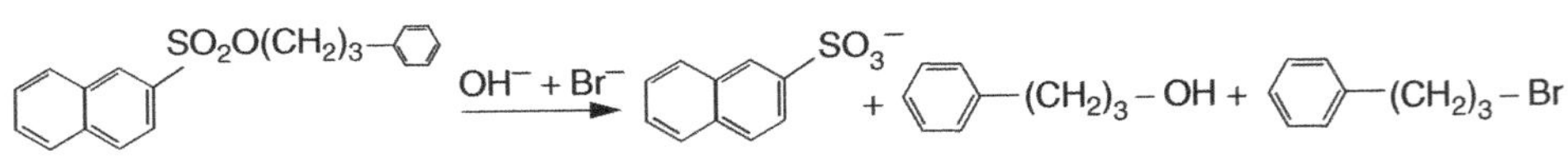

Conditions	Time for full conversion (h)	% R-OH	% R-Br
H_2O (50°C)	310	100	0
H_2O – dioxane (50°C)	162	100	0
H_2O – CTEABr (25°C)	24	9	91
H_2O – CTBABr (25°C)	67	0	100
H_2O – sulfobetaine (25°C)	136	12	88
H_2O – TBABr (25°C)	113	93	7

Figure 5.5 Effect of reaction medium on the relative reactivity of hydroxyl and bromide ions with a lipophilic sulphonate ester.

bromide is the dominating reacting species. Evidently, the larger and more polarisable bromide ion is attracted more strongly than the hydroxyl ion to the micellar interface.

It is well known that large polarisable ions bind strongly to interfaces. The ion effect is particularly pronounced for systems based on cationic surfactants. The choice of counterion for anionic surfactants seems to be of less importance. The fact that anions interact more strongly than cations with oil–water interfaces is well known and the magnitude of the interaction of different anions follows the so-called Hofmeister series [30].

It has been shown that the addition of a small amount of the anionic surfactant sodium dodecyl sulphate to a microemulsion based on non-ionic surfactant decreased the rate of decyl sulphonate formation from decyl bromide and sodium sulphite [31]. Addition of minor amounts of the cationic surfactant tetradecyltrimethylammonium salt gave either a rate increase or a rate decrease depending on the surfactant counterion. A poorly polarisable counterion, such as acetate, accelerated the reaction. A large polarisable counterion, such as bromide, gave a slight decrease in reaction rate. Evidently, the competition exerted by the bromide ion can be so powerful that the effect outweighs the favourable effect of introducing cationic charges at the interface.

The effect exerted by ions can be found also at interfaces made up solely of non-ionic surfactants, i.e. at uncharged interfaces. This has been shown for the above-mentioned reaction between 4-*tert*-butylbenzyl bromide and potassium iodide to give 4-*tert*-butylbenzyl iodide, shown in Fig. 5.1. The reaction was performed in a microemulsion based on the non-ionic surfactant penta(ethylene glycol) monododecyl ether ($C_{12}E_5$) and the temperature was varied from 23 to 29°C, which is the total temperature range of the isotropic oil-in-water region of this system [17]. It was found that the reaction rate decreased considerably when the temperature was increased from 23 to 29°C. ^{125}I-NMR showed a marked temperature effect on line broadening; the lower the temperature the broader the signals (within the temperature range 23–29°C). This indicates that the iodide ion interacts more strongly with the interface at lower temperature, which is likely to be the reason for the inverse temperature–reactivity relationship. Thus, ion binding to the microemulsion interface can have a pronounced effect on reactivity also in microemulsions based on uncharged surfactants. In another investigation of a nucleophilic substitution reaction in microemulsion, synthesis of 1-phenoxyoctane from 1-bromooctane and sodium phenoxide, no accumulation of the nucleophile at the interface could be observed based on the kinetics data [27]. This is in line with the view that only the large polarisable anions, such as iodide, become attracted to the interface due to dispersion force interactions [32].

As mentioned in the beginning of this section, the large oil–water interface of microemulsions can serve as a template for organic reactions. Organic molecules with one more polar and one less polar end will accumulate at the interface. They will orient so that the polar part of the molecule extends into the water domain and the non-polar part extends into the hydrocarbon domain. This tendency for orientation at the interface can be taken advantage of to induce regiospecificity of an organic reaction. A water-soluble reagent will react from the 'water side', i.e. attack the polar part of the amphiphilic molecule, and a reagent soluble in hydrocarbon will react at the other end of the amphiphilic molecule. The principle has previously been applied to a micellar system for controlling regioselectivity of a Diels–Alder reaction in which both reactants were surface active [33, 34]. When the reactions were run in an organic solvent, i.e. in absence of micellar orientational effects, the two regioisomers were obtained in equal amounts. When, on the other hand, the reactions were carried out in an aqueous buffer in which the reactants form mixed micelles, a regioisomer ratio of 3 was obtained.

Microemulsions can be used as a tool to differentiate between the first and the second step of a substitution of symmetrical bifunctional reactants, in this case α,ω-dibromoalkanes [35]. In a homogeneous system with good solubilising capacity for both the lipophilic reactant, the α,ω-dibromoalkane, and the hydrophilic reactant, sodium sulphite, the two substitution steps occur at the same rate. The situation proved to be different in a microemulsion. The intermediate mono-substituted species, a bromoalkanesulphonate, has one polar and one non-polar end; hence, it orients at the interface such that the sulphonate end points into the water domain, leaving the bromo end in the non-polar environment. Provided that the alignment of the intermediate is fast compared to the rate of the substitution reaction, such an orientation of the intermediate may protect it from further nucleophilic attack. This turned out to be the case for some of the α,ω-dibromoalkanes. For the species with the shortest alkane chain, 1,4-dibromobutane, there was a pronounced difference in rate between the first and the second substitution step and the intermediate, bromobutanesulphonate could be recovered in high yield. The selectivity decreased with the alkane chain length. Evidently, the second bromide is less protected for the longer derivatives. A likely explanation for the effect is illustrated in Fig. 5.6. The more limited regiochemical control for the longer derivatives is probably due to a considerable conformational freedom of these molecules, which allows the remaining bromomethyl group to be exposed at the interface. When the 'spacer group' between the sulphonate and the bromomethyl group is short, such folding is more difficult to achieve.

The regiospecificity of electrophilic aromatic substitution reactions can also be controlled by use of the oil–water interface of a microemulsion as template for the aromatic moiety. Bromination of two phenols and two anisoles has been carried out in a cationic

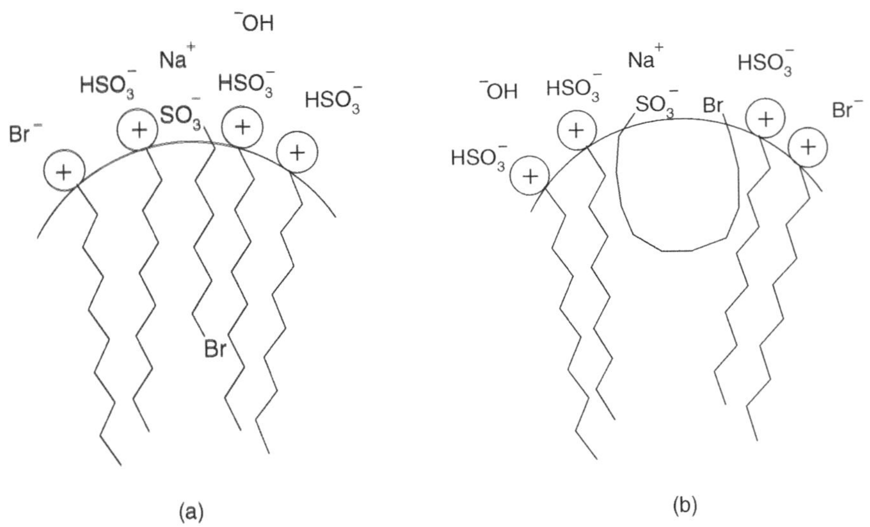

Figure 5.6 Alignment of a short-chain (a) and a long-chain (b) bromoalkanesulphonate at the oil–water interface of a microemulsion.

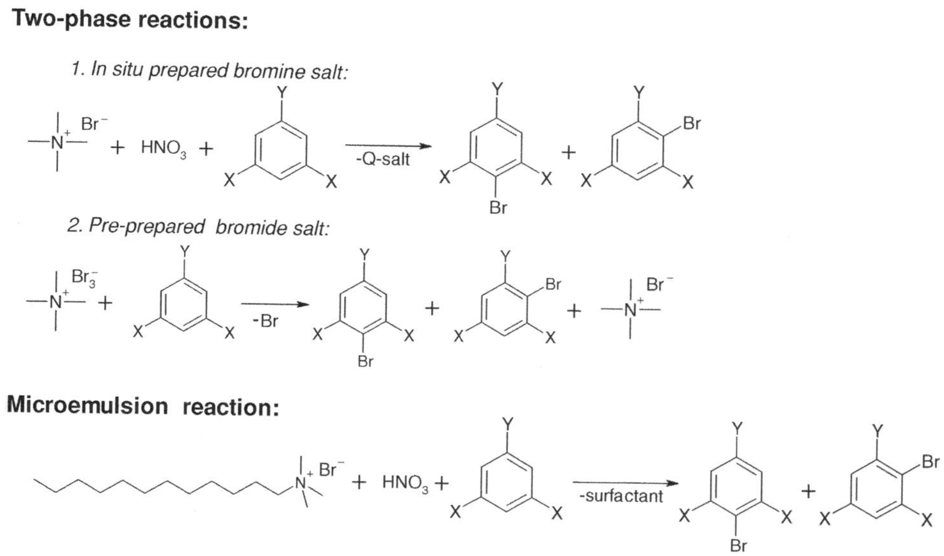

Figure 5.7 Reaction approaches used for bromination; X = H or CH_3; Y = CH_3, OH or OCH_3.

surfactant-based microemulsion using the surfactant counterion, e.g. bromide, as reagent [36]. The bromide ion was oxidised to elemental bromine by dilute nitric acid, which in turn reacted with the aromatic compound. The results have been compared with two-phase procedures using either an in situ-prepared or a pre-prepared complex between bromine and a quaternary ammonium salt as oxidising reagent and also with conventional bromination using elemental bromine. The different routes are shown in Fig. 5.7. Reaction in the microemulsion gave a more selective para-bromination than the other procedures. This is most likely due to the templating effect of the interface. The phenol or anisole derivative orients such that the para position is most susceptible for attach by the electrophile. In addition, the microemulsion-based reaction proceeded smoothly at room temperature.

5.3.3 Hydrogenation and hydroformylation reactions

Over the last decades, homogeneous catalysis has gained more and more importance in industry. Excellent catalyst activity and selectivity as well as the ability to enable mild reaction conditions are only a few advantages of homogeneous catalysts. Nonetheless, problems in recycling the mostly expensive precious metals – like rhodium, palladium and platinum – and organic ligands mark the crucial drawback. Therefore, considerable efforts have been made to overcome this disadvantage [37].

Hydroformylation of alkenes (Fig. 5.8) is an important commercial process for the production of aldehydes and alcohols from alkenes and is one of the most important examples of homogeneous catalysis in industry with a production capacity of more than 8 million tons/year [38]. Initial work by Manassen on aqueous biphasic catalysis [39]

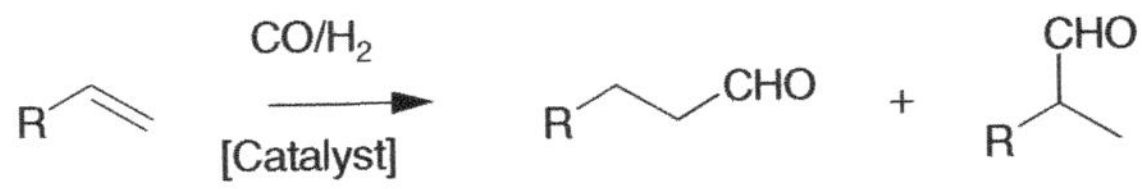

Figure 5.8 Reaction scheme for rhodium catalysed hydroformylation of an α-olefin.

resulted in the application of this technology in the Ruhrchemie/Rhône-Poulenc process in 1984 for propene hydroformylation using water soluble Rh-TPPTS (TPPTS = tri(*m*-sulphonyl)triphenyl phosphine trisodium salt) catalysts [40, 41]. The two basic problems of homogeneous catalysis, product separation and catalyst recycling, are circumvented in this process by using a simple decantation step. However, this technique is not applicable for alkenes higher than butane [42], since sufficient water solubility is required for high reaction rates. One way to overcome this solubility problem that is frequently encountered in organic reactions is performing the reaction in a microemulsion [43–45]. At high water contents the so-called reverse micelles are formed. Small water droplets in a continuous oil domain stabilised by a surfactant layer are able to host the hydrophilic Rh-TPPTS catalyst.

In a series of studies performed by different groups hydroformylation of higher alkenes in microemulsions resulted in high turn-over frequencies (TOFs) of up to 5000 per hour for 1-octene [46] and 1000 per hour for 1-decene [44]. Cationic surfactants such as cetyltrimethyl ammoniumbromide (CTAB) were used in the first studies of hydroformylation of various alkenes with the Rh-TPPTS catalyst. Both high activity and selectivity were observed in the hydroformylation of 1-octene and 1-decene using a sulphonated bisphosphine modified rhodium complex in the presence of ionic surfactants or methanol.

For unmodified rhodium catalysts, the reaction kinetics and the resting state have been extensively studied. The resting state is generally assumed to be $[HRh(CO)_3]$, which is formed by dissociation of a CO ligand from the complex $[HRh(CO)_4]$ (**4** in Fig. 5.9). Using in situ IR spectroscopy, Garland and co-workers [47] identified a rhodium acyl intermediate to be the resting state of the catalyst. The hydroformylation mechanism for phosphine modified rhodium catalysts and the coordination chemistry of several rhodium phosphine complexes that are potential intermediates in the catalytic cycle were studied by Wilkinson and co-workers [48, 49]. The principal active catalytic species was $[HRh(CO)_2(PPh_3)_2]$ (**2**) under the conditions studied. The resting state was generated prior to the catalytic cycle via several equilibria as depicted in Fig. 5.9.

Depending on the reaction conditions, the predominant catalyst species is coordinated by one or more phosphine ligands, thus influencing the selectivity of the catalyst. Starting the catalytic cycle with complex **2** will result in high *n*/iso selectivity due to the steric demand of the ligands at the metal centre. Complex **3** will result in lower selectivity, since it only contains one ligand. The lowest *n*/iso ratio will be obtained when starting with the ligand-free rhodium hydride **4**.

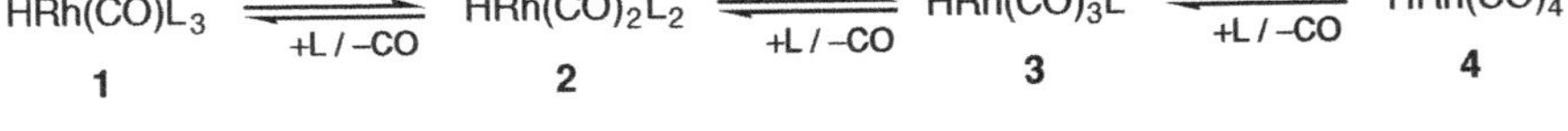

Figure 5.9 Equilibrium between different complexes of rhodium with CO and with TPP (L) as ligand.

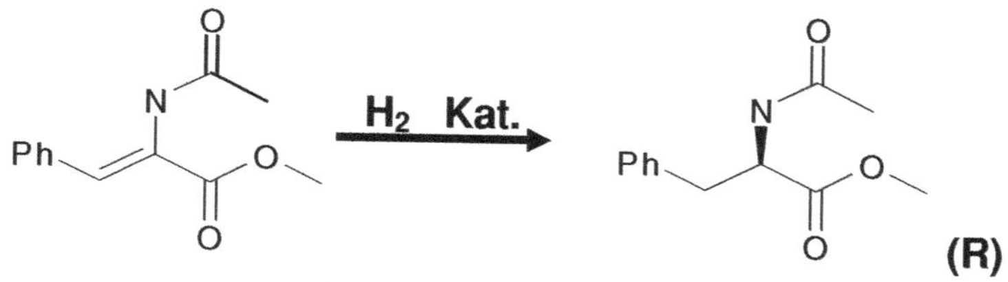

Figure 5.10 Reaction scheme for catalysed asymmetric hydrogenation of Z-a-acetamidocinnamic acid methyl ester (MAC).

Recently, it was reported that the use of non-ionic surfactants of alkylpolyglycolether-type results in high reaction rates in the hydroformylation of 1-dodecene, yielding 75% *n*-tridecanal and 25% iso-tridecanal (n/iso $= 3$) [44]. In a similar study, 7-tetradecene was converted at temperatures around 120°C and pressures of 100 bar into the corresponding pentadecanals with high regioselectivity [50]. Functionalised alkenes such as styrene, cyclohexene and 1,2-diacetoxy-2-butene were also hydroformylated with high rates in microemulsions stabilised by non-ionic surfactants [51]. In continuing investigations of hydroformulations in microemulsion systems hydroformylation of 1-octene by rhodium complexes with phosphine ligands have been carried out in various systems. The selectivity of the obtained nonanals is indicative of the active species present. Because of the high local concentration of the catalyst and excess ligand inside the microemulsion droplet the equilibrium between the different complexes is strongly shifted towards the inactive species 1. In comparison to bulk aqueous solutions a lower excess of ligand is required to establish an equilibrium with species 2 as the predominant species. Lazzaroni [52] studied the relation between the observed isomerisation and the linear to branched ratio. The activation energy for the isomerisation reaction was found to be higher than the one for hydroformylation, so higher temperatures would favour higher isomerisation rates. Furthermore, low partial pressures of CO had a positive effect on the isomerisation because isomerisation terminates via β-hydride elimination and a vacant site at the rhodium centre. At 100°C hydroformylation dominated β-hydride elimination for linear rhodium-alkyl species and vice versa for branched rhodium-alkyl species.

Homogeneous asymmetric hydrogenation reactions have been studied intensively with amino acid precursors in aqueous micellar solutions. In early work only stabilising effects of added amphiphiles were observed [53, 54]. However, for the hydrogenation of Z-a-acetamidocinnamic acid methyl ester (Fig. 5.10) with an optically active rhodium-phosphane complex (Fig. 5.11) in the presence of micelles a significant increase in activity and enantioselectivity was found in comparison to reaction in pure water [55].

This effect was observed by Oehme and co-workers for a variety of ionic and non-ionic surfactants. The surfactants were always added at a concentration of about twice the cmc but in substoichiometric ratio to the substrate. This significant effect of the amphiphiles above the cmc manifests the simultaneous solubilising effect of the micelles for the catalyst and the substrate. In kinetic studies of hydrogenation of MAC in micellar solutions of the anionic surfactant SDS and the non-ionic surfactant octa(ethylene glycol)monotridecyl ether ($C_{13}E_8$) micellar solutions the dissociation constant of the catalyst substrate complex was found to be considerably smaller than in methanol as solvent [56]. This indicates

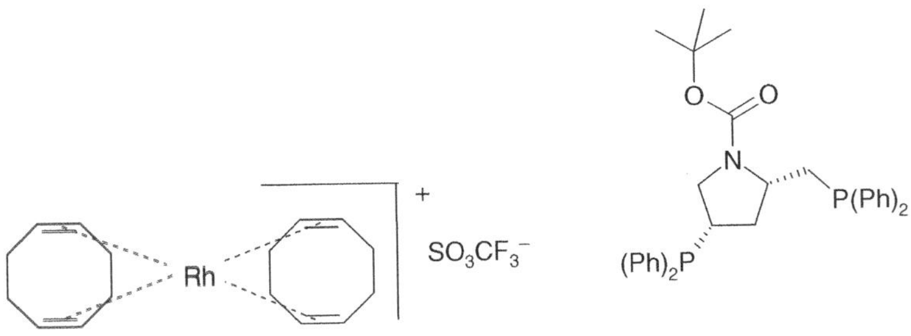

Figure 5.11 Precursor and ligand (BPPM) for catalyst synthesis.

a stabilisation of the complex because of an increased substrate concentration inside the micelles. This substrate accumulation within the micelles also causes the apparently enhanced activity. The oxidative insertion of hydrogen is again the rate-limiting step within the catalytic cycle. Yonehara *et al.* showed asymmetric hydrogenation of a series of unsaturated substrates, like itaconic acid (Fig. 5.12) and its derivatives with rhodium phosphonite catalysts [57]. However, the enantioselectivity strongly differed within this series of substrates.

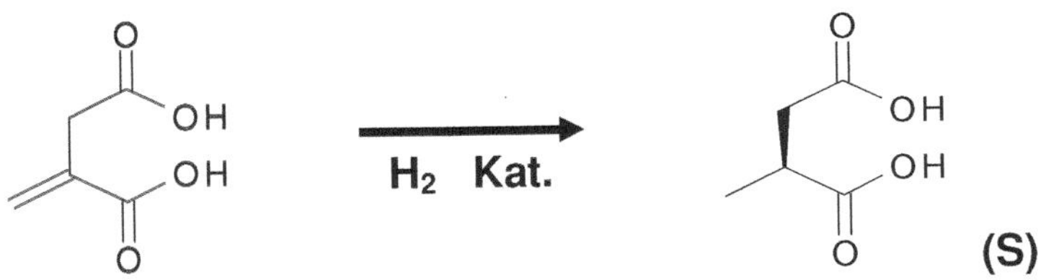

Figure 5.12 Reaction scheme for catalysed asymmetric hydrogenation of itaconic acid.

The mechanism and the kinetic model of homogeneously catalysed hydrogenation reactions are still under discussion. In most cases a rate law analogous to Michaelis–Menton kinetics for enzyme catalysis is applicable. For a precise description of the kinetics in a microheterogeneous medium the local concentrations of catalyst, hydrogen and substrate would be needed. In general only the concentration of the substrate changes with time. The concentrations of catalyst and hydrogen stay at a constant local concentration that is proportional to the overall concentration. The hydrogen concentration within the micelles or microemulsion droplets is expected to be proportional to the applied pressure, but still unknown. Because of this complex situation all kinetic studies performed so far have only considered overall concentrations of the involved species. This means that all the determined kinetic parameters are effective parameters that combine the microkinetic constants with some partition coefficients and phase volume ratio.

5.4 Engineering aspects

If promising experimental results with respect to reaction rate, selectivity and yield motivate the development of an industrial process based on a self-organised surfactant system as reaction medium, a number of additional questions have to be addressed. In addition to the synthesis as such, the work-up procedure and the recycling of solvents, catalysts and unreacted substrates have to be considered. The selected medium should allow all these operations without additional technological or economical efforts. In this context, the selection of the surfactant is of special importance. At first the surfactant has to stabilise the required thermodynamic state and microstructure at the temperature necessary for the reaction. Since the physical–chemical behaviour of surfactants is strongly dependant on temperature and physical properties of the other components of the system, the phase behaviour should be tuned to the required temperature. Many commercially available surfactants are optimised for applications at ambient temperature. For reactions that require a temperature in this range a variety of surfactants is therefore available. For reactions at elevated temperatures the assortment is somewhat restricted, especially for ionic surfactants. Within this group of products there is only little variation possible for tuning the hydrophilicity of the surfactants. Non-ionic surfactants offer a much broader range of amphiphilic properties, because the variation of both the hydrocarbon chain length and degree of ethoxylation allows a change in well-defined increments. In addition to the phase behaviour the compatibility of the surfactant and the reactants, as well as the catalyst, has to be considered. In some cases interactions of surfactant head groups with the catalyst can inhibit or deactivate homogeneous catalysts. Impurities that come with the surfactant may also be the cause of such deactivation. Finally, the chosen surfactant has to allow for a simple work-up procedure that yields a pure product. The catalysts, solvent and surfactants should preferably be reused. Dependant on the chosen unit operations for work-up, stability of the surfactant and compatibility with further tools like membranes has to be guaranteed. The surfactants and other auxiliary agents that are released from a process must have proper biodegradability.

5.4.1 Selection and tuning of surfactant systems

Many surfactants are produced for large-scale applications without recycling, especially for cleaning purposes and formulation of industrial products such as paints, polymer dispersions and agrochemicals. These applications usually require moderate prices and environmental compatibility of the surfactants. Surfactants produced for cosmetic and personal hygiene application are also available at moderate prices but with higher purity standards. If the compatibility with the components of the reaction and the tools for the work-up is given, the potential for tuning the phase behaviour guides the selection of the surfactants. Therefore, out of the range of available surfactants the group of non-ionic surfactants is often preferred for applications in reaction media. In addition to good tuning properties this group shows little sensitivity to electrolyte addition. Non-ionic surfactants of the alkylpolyglycolether type are made via ethoxylation of a synthetic or natural fatty alcohol. The hydrophilicity depends on the number of oxyethylene units in the molecule, usually ranging from 5 to 20. The phase behaviour of such a system is described in detail by the use of a Gibbs phase prism with the Gibbs triangle as base and the temperature as the ordinate (see Fig. 1.3 in Chapter 1).

As already mentioned above, the composition of the microemulsion is described by the mass fraction of oil in the mixture of water and oil $\alpha = m_{\text{oil}}/(m_{\text{oil}} + m_{\text{water}})$ and by the mass fraction of the amphiphile in the total mixture $\gamma = m_{\text{surf}}/(m_{\text{surf}} + m_{\text{oil}} + m_{\text{aq}})$ both expressed in weight percent (wt.%). The Gibbs triangle can be used to plot the compositions of microemulsion phases when temperature, pressure, type of oil and type of surfactant are fixed. If any of these variables are changed, a new Gibbs triangle must be constructed. The generic pattern of phase behaviour found upon changing variables of interest is qualitatively sketched by the three Gibbs triangles shown in Fig. 1.2 of Chapter 1. In these diagrams, regions of 1, 2 and 3 phases are shown. Liquid crystalline phases (usually lamellar L_α) typically found at higher surfactant concentrations are omitted for clarity. The '1' denotes regions where oil and water are completely mixed into a single microemulsion phase. The ends of the tie lines within the two-phase region denote the compositions of the two phases in equilibrium when a sample is mixed at the overall composition located along the line. The '$\underline{2}$' (or Winsor I system) denotes a two-phase sample where the majority of the surfactant, along with some oil, resides in the lower water-rich phase (an 'oil-in-water' microemulsion phase), and excess oil floats on top. The '$\overline{2}$' (or Winsor II system) denotes a two-phase sample where the majority of the surfactant, along with some water, resides in the upper oil-rich phase (a 'water-in-oil' microemulsion phase), with excess water lying underneath. Finally, '3' (or Winsor III system) denotes a three-phase sample, where the middle phase, rich in oil and water (a 'bicontinuous' microemulsion phase) is in equilibrium with excess water (bottom phase) and oil (top phase). The three corners of the three-phase triangle denote the compositions of the three phases in equilibrium. At very low surfactant concentrations there is a two-phase region, in which the surfactant concentration is below the 'critical microemulsion concentration' of the surfactant. In this region, there is essentially no mixing of oil and water and all surfactant is dissolved as monomers in the water and oil phases. In order to better visualise the progressions shown schematically by the three triangular phase diagrams, one may imagine a test tube, filled with equal amounts of oil and water, with enough surfactant added to achieve some mixing of oil and water (above the critical microemulsion concentration, discussed above), but not enough to completely mix oil and water into a single microemulsion phase (below the '1' region). At low temperatures non-ionic surfactants prefer water-rich phases and the test tube will contain a '$\underline{2}$' phase system (a lower 'oil-in-water' microemulsion phase in equilibrium with excess oil). At intermediate temperatures, the surfactant has affinity for both oil and water and the test tube will contain a '3' phase system (a middle 'bicontinuous' microemulsion phase, in equilibrium with excess oil and water). At high temperatures, the surfactant prefers oil-rich phases and the test tube will contain a $\overline{2}$ phase system (an upper 'water-in-oil' microemulsion phase in equilibrium with excess water). More details can be found in Chapter 1 (Fig. 1.2).

Visual observations allow determination of phase boundaries precisely and reliably. Single-phase systems are usually transparent whereas mixed multi-phase systems are turbid. Liquid crystalline phases, found at higher surfactant concentrations, are birefringent and easily identified by using crossed polarisers along with a strong light source. Differentiating between two- and three-phase regions usually requires waiting for phase separation of the samples. The relative partitioning of a non-ionic surfactant between water and oil phases strongly depends upon temperature. The polyether chains become less hydrated and, thus, less water-soluble the higher the temperature. Most non-ionic surfactants go from preferentially water-soluble to preferentially oil-soluble within a relatively narrow

Table 5.2 Qualitative effects of increasing the variables listed upon phase behaviour

Increasing parameter	$\underline{2}$	3	$\bar{2}$
Temperature		→	
Pressure			
Salt concentration		→	
Oil hydrophobicity		←	
Surfactant lipophilicity		→	
Surfactant hydrophilicity		←	

temperature interval. At an intermediate temperature range solubilisation of the hydrophilic head group by water is 'balanced' by the solubilisation of the hydrophobic tail by the oil. In this temperature range, the surfactant is most efficient in decreasing the interfacial tension between water and oil. This is also the range where the surfactant has maximum capacity of solubilising both oil and water.

The pattern of phase behaviour described above for increasing temperature (Fig. 1.2 in Chapter 1) is also found for a number of different variables of interest. Table 5.2 shows the direction in which the phase behaviour of non-ionic surfactants progresses as a function of increasing temperature, pressure, salinity, oil 'hydrophobicity' and variations of the 'hydrophilic/lipophilic' balance of the surfactant. For example, oil 'hydrophobicity' is increased by increasing the chain length of aliphatic hydrocarbons. The 'hydrophilicity' of non-ionic surfactants is increased by increasing the number of oxyethylene units of the surfactant (increasing j in C_iE_j), which is equivalent to increasing the hydrophilic–lipophilic balance (HLB) number [58]. The 'lipophilicity' increases upon increasing the length of the aliphatic chain of the amphiphile (increasing i), which is equivalent to decreasing the HLB number. Note that the 'amphiphilicity' of ethoxylated alcohols, i.e. the strength of the 'chemical dipole' between hydrophilic and lipophilic groups, is increased by increasing both i and j simultaneously.

5.4.2 *Type of organised surfactant system*

Surfactants in solutions show a broad variety of microstructures caused by molecular self-organisation. The observed structures depend essentially on the physical interactions of the involved components and the composition of the mixtures. For the selection of a suitable type of reaction medium the required composition of the reaction mixture is more important than the question of whether a micellar solution, a bicontinuous microemulsion, a w/o- or an o/w-microemulsion is formed. For synthetic purposes high concentrations of reactants are indispensable in order to avoid high-energy cost for work-up procedures. Therefore, a reaction system that allows high-reactant concentrations needs to be chosen. For stoichiometric reactions involving reactant incompatibility, like nucleophilic substitution reactions with an inorganic nucleophile, often an aqueous solution of this reactant has

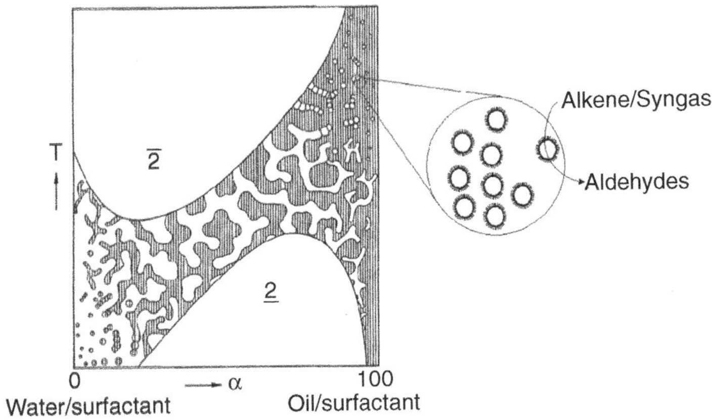

Figure 5.13 Section of the phase prism at constant surfactant concentration. Different structures within the one-phase region are indicated by hatching. In the water-rich region, swollen micelles solubilise oil. In the oil-rich region, reverse micelles of nanometre size exist. Bicontinuous structures are found in the intermediate range. (From Ref. [45], reprinted with permission of Elsevier.)

to be used. The unpolar reactant needs to be solubilised into this aqueous medium, or vice versa. The volume ratio of the polar and the unpolar subphases will be given by the stoichiometric ratio of the reactants and by their solubility characteristics. On the basis of these prerequisites the surfactant suitable for the required reaction temperature will generate a certain microstructure. For catalytic reactions a system is first chosen that solubilises a high concentration of the substrate. Addition of a surfactant that is compatible with the catalyst and enables its solubilisation will usually result in a normal or reverse micellar solution. As an example the system selected for hydroformylation of long-chain olefins is described below. Figure 5.13 shows a section of the Gibbs phase prism at constant surfactant content. A region of isotropic single-phase solutions is observed extending from the water-rich to the oil-rich side of the phase prism. At low water concentrations and higher temperatures, reverse micelles are formed with diameters in the range of nanometres resulting in a large internal interfacial area.

The small droplets act as microreactors when they contain the water-soluble catalysts. For hydroformylation reactions with water-soluble Rh/TPPTS in the droplets, the alkene, carbon monoxide and hydrogen approach the micelle surface where the reaction occurs, as is illustrated in Fig. 5.13. After the reaction is completed, phase separation can be achieved by changing the temperature of the reaction mixture. When the mixture is cooled down an aqueous bottom phase, containing most of the surfactant and the water-soluble catalyst separates from the organic upper phase, which contains the hydrophobic products and unconverted reactants. In case of incomplete catalyst recovery the micelle remaining in the product phase can be separated by means of ultrafiltration.

A first attempt at hydroformylation in a micellar system using a water-soluble rhodium catalyst (Rh-TPPTS) was made by Tinicci and Platone from Eniricerche in 1994 [59]. They converted olefins with carbon numbers up to 12 using a mixture of an anionic surfactant (SDS) and butanol (as co-surfactant). It has been shown that microemulsions made with non-ionic surfactants of the alcohol ethoxylate type are advantageous compared to ionic microemulsions. 1-Dodecene was hydroformylated in a microemulsion based on non-ionic surfactant using Rh-TPPTS at moderate temperatures and pressure [45]. Catalyst recycling is made easy by use of an ultrafiltration unit following the phase separation. Renken reported the conversion of alkenes between C_6 and C_{16} in a micellar system using SDS together with butanol [44]. Cationic surfactants were used by Fell and co-workers for the hydroformylation of unsaturated fatty acids [43]. These ionic systems provide good reaction conditions for high rates and selectivities, but work-up is more difficult than with systems based on non-ionic amphiphiles.

As was mentioned earlier in this chapter, it is not necessary to transfer every reaction mixture into a thermodynamically stable one-phase system. Often the presence of one organised surfactant phase in equilibrium with one or two excess phases is sufficient to give an appropriate reaction rate. In such two- or three-phase systems the reaction occurs in the surfactants phase while the coexisting phases act as reservoir for the reactants. This approach has been demonstrated for alkylation of phenol [28] and for rhodium catalysed hydroformylation of dodecene [50]. A major practical advantage with the multi-phase systems is that substantially less surfactant is needed. This reduces costs and simplifies the work-up.

5.4.3 Work-up procedures for product isolation

Product isolation from a surfactant-based organised reaction medium can usually not be performed by the standard work-up procedures used for reactions in conventional media. Simple distillation or extraction will result in unacceptable contamination of the product by the surfactant. This can be avoided by using procedures based on the thermodynamic properties of the microheterogeneous systems.

5.4.3.1 Use of phase transitions

Microemulsions based on non-ionic surfactants of alcohol ethoxylate type are sensitive to temperature changes and those based on ionic surfactants are sensitive to variations in the electrolyte concentration. Such variations may cause a one-phase microemulsion to form a two- or a three-phase system in which a microemulsion phase coexists with one or two excess phases. As a work-up approach the concept is particularly useful for microemulsions based on non-ionic surfactants because the transitions obtained by temperature variations are reversible.

Whereas the surfactant will always reside in the microemulsion phase, the product is likely to partition into an excess oil phase if it is an apolar substance and into an excess water phase if it is a polar compound. The principle is illustrated in Fig. 5.14 for hydrolysis of a lipophilic ester in a Winsor I system (an oil-in-water microemulsion in equilibrium with excess oil) followed by transition into a Winsor III system [60]. The ester partitions between the excess oil phase and the oil droplets and the hydroxyl ions reside in the continuous water domain of the microemulsion. The reaction takes place at the interface. After completed reaction, acid is added to protonate the alkanoate formed and the temperature is raised so that a Winsor I to Winsor III transition occurs. The lipophilic

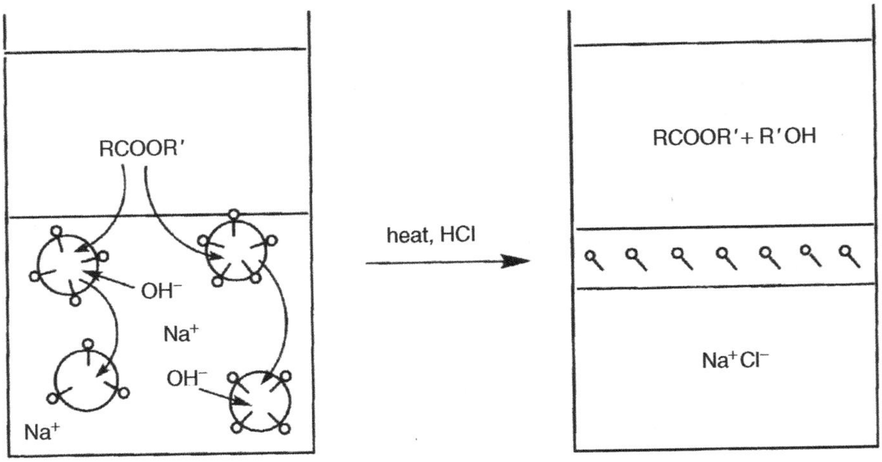

Figure 5.14 Ester hydrolysis in a Winsor I system followed by a heat-induced transition to a Winsor III system.

products, a fatty acid and an alcohol, will predominantly partition into the oil phase and the surfactant will predominantly reside in the middle phase, which is a bicontinuous microemulsion.

In order to separate the phases of a surfactant-based system for product isolation and, where necessary, for catalyst recovery the appropriate phase region of the phase diagram has to be chosen first. Within this region the phase composition and the kinetics of phase separation are essential questions. Near to the phase boundaries the composition of the phases is rather similar and separation of the components will often be incomplete. The phase separation often takes a long time because of low interfacial tension and high stability of the emulsified two-phase system. The kinetics of phase separation depends sensitively on the temperature of the system, especially on the temperature distance to phase boundaries. Figure 5.15 shows a plot of separation times for a water–oil–non-ionic surfactant system as a function of the temperature.

The separation times were measured for a mixture of 2 L of water and octane (1:1 by volume) containing 10 wt.% of C_8E_4. At each temperature, the system was mixed intensively for 5 min at a power input of 1 W L^{-1} before phase separation was observed by visual inspection. Separation time was taken at the moment when 90% of each phase was clear. Three distinct minima are recognised within this graph, one about 5° below the phase transition from two to three phases, one in the middle of the three-phase region and one around 5° above the three-phase region. This behaviour was observed as a general pattern for all non-ionic surfactant systems. The fastest phase separation is always observed within the three-phase region, because all surfactant occupies the water–oil interface within the middle phase. No surfactant is available to stabilise droplet of the excess phases within the other phases. The temperature for a fast phase separation within the two-phase region depends on the efficiency of the surfactant. For more efficient surfactants larger temperature distances to the phase boundaries are required (e.g. about 25° for $C_{12}E_5$). For isolation of hydrophobic products from an organic phase the surfactant should be transferred to the

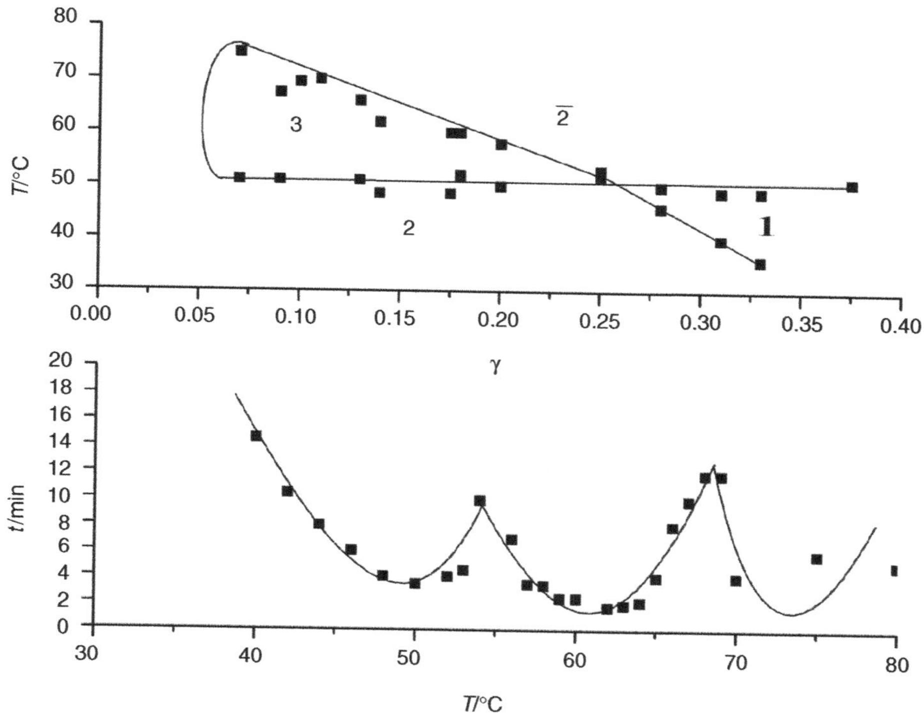

Figure 5.15 Section of phase prism and separation times of water, oil, non-ionic surfactant system as function of temperature.

aqueous phase at low temperature and vice versa. From the separated phases the products can be isolated with conventional unit operations. For reuse of the surfactant it can be re-extracted from the aqueous phase by a fresh unpolar phase [61]. The sequence of process steps is shown schematically in Fig. 5.16.

In order to avoid a pronounced shift of the phase behaviour with the progress of the reaction one reactant can be fed to the reactor in a semi-batch mode. With this concept shift in phase behaviour may be compensated and the optimal state of the system is maintained during the whole reaction time. After phase separation at the most suitable temperature conventional product isolation follows. This sequence of process steps is also possible in a continuously operating process, as illustrated in Fig. 5.17 for the synthesis of 1-phenoxyoctane in a microemulsion stabilised by Triton X-100 [28].

The product mixture that leaves the reactor (1) is cooled down so that the $\underline{2}$ region is formed. After phase separation unreacted 1-bromoctane is separated from the product by rectification (3). The aqueous phase is mixed with fresh bromooctane (4) and heated to generate phase inversion into the $\overline{2}$ region. The aqueous phase containing the by-product NaBr is released from the process (5) while the unpolar phase is fed into the reactor, where the reactant sodium phenoxide is added sequentially at different inlets of this chamber reactor.

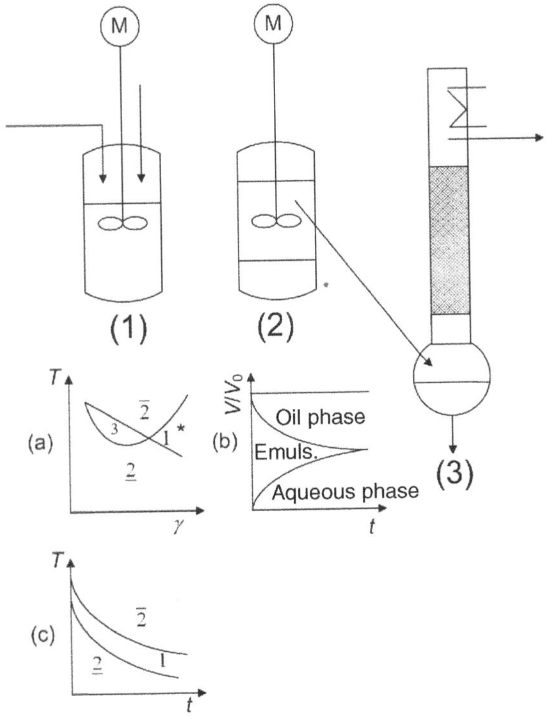

Figure 5.16 Scheme of a batch process for reaction of sodium phenoxide with bromooctane: (1) reactor, (2) phase separation into aqueous and oil phase and (3) rectification. (a) $T(\gamma)$-diagram: indication of the composition (*) of the reaction mixture. (b) $V/V_0(t)$-diagram: the evolution of the volume fractions V/V_0 as a function of time t; the phase separation finally leads to the $\underline{2}$ region. (c) $T(t)$-diagram: the evolution of phase behaviour with time and progress of reaction.

5.4.3.2 *Membrane separation processes*

In some cases, temperature-induced phase separation does not result in the required quantitative separation of all components or damages temperature sensitive catalysts. This is especially true for catalytic reactions with expensive noble metal catalysts. Such reactions often require more than 99% recovery of the catalyst in order to avoid severe economic losses due to the extremely high costs of such catalysts. Here, ultrafiltration is a suitable tool for quantitative catalyst recovery under mild conditions. In general the same conditions for reaction and ultrafiltration should be chosen in order to recycle the catalyst in its active form.

Ultrafiltration was utilised for catalyst recovery first for enzymes [62, 63] and later for polymer enlarged homogeneous catalysts [64]. The molecular weight of a homogeneous catalyst itself is usually too low in comparison to the molecular weight cut-off of standard ultrafiltration membranes of 5000 or 10 000 Da. By embedding the catalyst in a surfactant micelle the same effect is obtained as with immobilisation of the catalyst at a soluble polymer molecule or a small polymer particle. Ultrafiltration of surfactant solutions is an established technique that is often applied for complete removal of traces of small

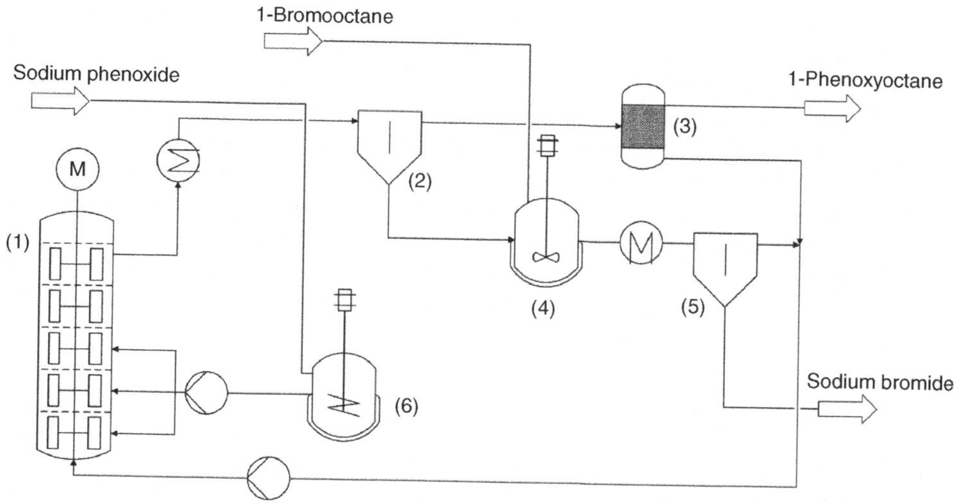

Figure 5.17 Process scheme for continuous reaction of sodium phenoxide with bromooctane: (1) reactor, (2) phase separation into the $\underline{2}$ region, (3) rectification, (4) phase inversion into the $\overline{2}$ region, (5) separation of aqueous phase and (6) reservoir for aqueous sodium phenoxide solution.

toxic or hazardous molecules from water. Ultrafiltration membranes from regenerated cellulose or polysulphone enable retention of micelles higher than 99%. Since hydrophobic or amphiphilic ligands of homogeneous catalysts can give complete embedding into the micelles also more than 99% retention of the catalyst is possible. Figure 5.18 shows a

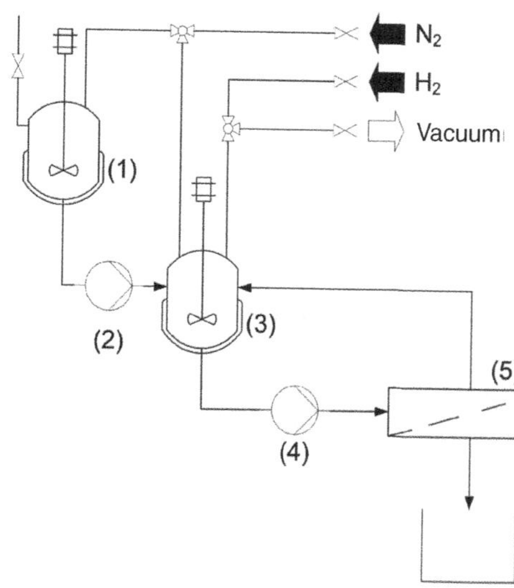

Figure 5.18 Process scheme for combination of a hydrogenation reactor with an ultrafiltration module: (1) reactant reservoir, (3) reactor, (5) ultrafiltration module, (6) product tank and (2), (4) are pumps as shown by symbols.

scheme of a laboratory scale process that was tested with the homogeneously catalysed hydrogenation of Z-a-acetamidocinnamic acid methyl ester and itaconic acid with the Rh-BPPM catalyst in micellar solutions of SDS and different non-ionic surfactants [65]. In a repetitive batch mode the catalyst was recovered more than five times and reused. In every filtration run 50% of the liquid volume was removed from the system with less than 1% of the catalyst leaching out of the process. The catalyst activity and selectivity was identical in all five runs. For an optimisation of this process the size of the filtration module has to be adjusted to the rate of the reaction in order to remove the same amount of product from the system per unit of time that is produced. With this approach of repetitive use of homogeneous catalysts the economic threshold of a turnover number of around 1000 is easily exceeded even with catalysts of medium activity.

5.4.3.3 Use of cleavable surfactants

Another approach for circumventing the often problematic separation of the surfactant from the reaction product is to use a cleavable surfactant, i.e. a surfactant that by an external stimulus breaks down to two non-surface active components, the polar head group and the tail. There exists a wide variety of cleavable surfactants with varying break-down mechanisms, such as alkali or acid sensitivity, heat or UV lability etc. [66]. After completed reaction, the conditions are changed so that the microemulsion surfactant degrades. The microemulsion will then be turned into a surfactant-free two-phase system. If the reaction product is hydrophobic, it will be contaminated by the surfactant tail residue and if it is hydrophilic it will be contaminated by the polar head group. Separating either of these non-surface active components from the reaction product is a much less complicated procedure than removing a surfactant. However, use of cleavable surfactants is a costly procedure and only feasible for high-priced products because the surfactant is consumed in the process.

5.5 Conclusion

In this chapter, we have demonstrated the potential of surfactant-based reaction media for preparative organic chemistry. We have not attempted to give a complete account of all reactions that have been investigated in microemulsions and related media. Instead, we have chosen to concentrate on three important and illustrative examples: nucleophilic substitutions, regioselective synthesis and hydrogenation and hydroformulation reactions. We believe that taken together they provide a good picture of what surfactant-based reaction media can offer to synthetic chemistry.

Little of the results have yet been transferred to industrial applications. One reason for this may be that surface chemistry is not a common tool for preparative organic chemists. There is little interaction between surface chemists and organic chemists and scientific papers on microemulsion technology are normally not read by scientists in the organic chemistry community. In this chapter, we have deliberately put emphasis on preparative and engineering aspects of reactions in microemulsions. We hope that the chapter will trigger an interest to use such system for preparative purposes. All information and tools required for process development, such as kinetics, thermodynamics, rules for selection and optimisation of media, unit operations for product isolation, etc. are in principle available. Besides technically attractive features such as high reactivity, high selectivity and good catalyst recovery, the approach of surfactant-based reaction media can be seen as a route towards the 'Green Chemistry' concept.

References

1. Motsavage, V.A. and Kostenbauder, H.B. (1963) The influence of the state of aggregation on the specific acid-catalyzed hydrolysis of sodium dodecyl sulfate. *J. Colloid Sci.*, **18**, 603–615.
2. Kurz, J. (1962) Effect of micellisation on the kinetics of the hydrolysis of monoalkylsulfates. *J. Phys. Chem.*, **66**, 2239–2246.
3. Thompson, R.A. and Allenmark, S. (1992) Factors influencing the micellar catalyzed hydrolysis of long-chain alkyl betainates. *J. Colloid Interface Sci.*, **148**, 241–246.
4. Ruasse, M-F., Blagoeva, I.B., Ciri, R., Garcia-Rio, L., Leis, J.R., Marques, A., Mejuto, J. and Monnier, E. (1997) Organic reactions in micro-organized media: Why and how? *Pure Appl. Chem.*, **69**, 1923–1932.
5. de Rocha Pereira, R., Zanette, D. and Nome, F. (1990) Application of the pseudo-phase ion-exchange model to kinetics in microemulsions of anionic detergents. *J. Phys. Chem.*, **94**, 356–361.
6. Bunton, C.A. and Romsted L.S. (1999) Organic reactivity in microemulsions. In P. Kumar and K.L. Mittal (eds), *Handbook of Microemulsion Science and Technology*. Marcel Dekker, New York, pp. 457–482.
7. Schwuger, M.J., Stickdorn, K. and Schomäcker, R. (1995) Microemulsions in technical processes. *Chem. Rev.*, **95**, 849–864.
8. Schomäcker, R. (1992) Mikroemulsionen als Medium für chemische Reaktionen. *Nachr. Chem. Tech. Lab.*, **40**, 1344–1351.
9. Holmberg, K. (2003) Organic reactions in microemulsions. *Curr Opin. Colloid Interface Sci.*, **8**, 187–196.
10. Dwars, T., Paetzold, E. and Oehme, G. (2005) Reactions in micellar systems. *Angew. Chem. Int. Ed.*, **44**, 7174–7199.
11. Holmberg, K. (2007) Organic reactions in microemulsions. *Eur. J. Org. Chem.*, 731–742.
12. Romsted, L.S. (1984) Micellar effects on reaction rates and equilibria. In K.L. Mittal and B. Lindman (eds), *Surfactants in Solution*, Vol. 2. Plenum, New York, p. 1015.
13. Bunton, C.A., Nome, F., Quina, F.H. and Romsted, L.S. (1991) Ion binding and reactivity at charged aqueous interfaces. *Acc. Chem. Res.* **24**, 357–364.
14. Romsted, L.S., Bunton, C.A. and Yao, J. (1997) Micellar catalysis, a useful misnomer. *Curr. Opin. Colloid Interface Sci.* **2**, 622–628.
15. Savelli, G., Germani, R. and Brinchi, L. (2001) Reactivity control by aqueous amphiphilic self-assembling systems. In J. Texter (ed), *Reactions and Synthesis in Surfactant Systems*, Vol. 100. Marcel Dekker, New York, pp. 175–246.
16. Häger, M., Olsson, U. and Holmberg, K. (2004) A nucleophilic substitution reaction performed in different types of self-assembly structures. *Langmuir*, **20**, 6107–6115.
17. Häger, M., Currie, F. and Holmberg, K. (2004) A nucleophilic substitution reaction in microemulsions based on either an alcohol ethoxylate or a sugar surfactant. *Colloids Surf. A*, **250**, 163–170.
18. Drummond, C.J., Grieser, F. and Healy, T.W. (1989) Effect of solvent–water mixtures on the prototropic equilibria. *J. Chem. Soc. Faraday Trans.* 1, **85**, 521–535.
19. Witula, T. and Holmberg, K. (2007) Liquid crystalline phases and other microheterogeneous systems as media for organic synthesis. *J. Disp. Sci. Technol.*, **28**, 73–79.

20. Witula, T. and Holmberg, K. (2007) Use of different types of mesoporous materials as tools for organic synthesis. *J. Colloid Interface Sci.*, **310**, 536–545.
21. Häger, M. and Holmberg, K. (2000) A substitution reaction in an oil-in-water microemulsion catalyzed by a phase transfer catalyst. *Tetrahedron Lett.*, **41**, 1245–1248.
22. Häger, M. and Holmberg, K. (2004) Phase transfer agents as catalysts for a nucleophilic substitution reaction in microemulsions. *Chem. Eur. J.*, **10** 5460–5466.
23. Wielpütz, T., Sottmann, T., Strey, R., Schmidt, F. and Berkessel, A. (2006) Dramatic enhancement of enone epoxidation rates in nonionic microemulsions. *Chem. Eur. J.*, **12**, 7565–7575.
24. Kronig, R. and Brink, C. (1950) On the theory of extraction from falling droplets. *Appl. Sci. Res.*, **2**, 142–154.
25. Oldshue, J.Y. (1983) *Fluid Mixing Technology*. McGraw-Hill Publications, New York.
26. Johannsson, R., Almgren, M. and Schomäcker, R. (1993) Diffusion-controlled reactions in bicontinuous microemulsions of AOT and C12E5 studied by triplet-state deactivation, *Langmuir*, **9**, 1269–1273.
27. Tjandra, D., Lade, M., Wagner, O. and Schomäcker, R. (1998) The kinetics of an interfacial reaction in a microemulsion. *Chem. Eng. Technol.*, **21**, 666–670.
28. Bode, G., Lade, M. and Schomäcker, R. (2000) The kinetics of an interfacial reaction in microemulsion with excess phase. *Chem. Eng. Technol.*, **23**, 405–409.
29. Brinchi, L., Di Profio, P., Germani, R., Savelli, G. and Bunton, C.A. (1998) Chemoselectivity in S_N2-E2 reactions induced by aqueous association colloids. *Colloids Surf. A*, **132**, 303–314.
30. Lyklema, J. (1995) *Fundamentals of Interface and Colloid Science*, Vol. II. Academic Press, London, pp. 3182–3186.
31. Holmberg, K., Oh, S.-G. and Kizling, J. (1996) Microemulsions as reaction medium for a substitution reaction. *Prog. Colloid Polym. Sci.*, **100**, 281–285.
32. Ninham, B.W. and Yaminsky, V. (1997) Ion binding and ion specificity: The Hofmeister effect and Onsager and Lifshitz theories. *Langmuir*, **13**, 2097–2108.
33. Jaeger, D.A. (1992) Wang, micellar effects on the regioselectivity of a Diels–Alder-reaction. *J. Tetrahedron Lett.*, **33**, 6415–6418.
34. Jaeger, D.A., Su, D., Zafar, A., Piknova, B. and Hall, S.B. (2000) Regioselectivity control in Diels-Alder reactions of surfactant 1,3-dienes with surfactant dienophiles. *J. Am. Chem. Soc.*, **122**, 2749–2757.
35. Currie, F., Jarvoll, P., Holmberg, K., Romsted, L.S. and Gunaseelan, K. (2007) Micellar induced regioselectivity in the two-step consecutive reaction of SO_3^{2-} with Br-$(CH_2CH_2)_n$-Br-($n = 2$–5). *J. Colloid Interface Sci.*, **312**, 435–459.
36. Currie, F., Holmberg, K. and Westman, G. (2003) Bromination in microemulsion. *Colloids Surf. A*, **215**, 51–54.
37. Cornils, B. and Herrmann, W.A. (2002) *Applied Homogeneous Catalysis with Organometallic Compounds*, 2nd edn. Wiley-VCH, Weinheim.
38. Frohning, C.D. and Kohlpaintner, C.W. (1996) Synthesis gas chemistry. In B. Cornils and W.A. Herrmann (eds), *Applied Homogeneous Catalysis with Organometallic Compounds*. Wiley-VCH, Weinheim.
39. Manassen, J. (ed) (1973) *Catalysis Progress in Research*. Plenum, London.
40. Wiebus, E. and Cornils, B. (1994) Die großtechnische Oxosynthese mit immobilisiertem Katalysator. *Chem. Ing. Tech.*, **66**, 916–923.
41. Cornils, B. and Kuntz, E. (1995) Introducing TPPTS and related ligands for industrial biphasic processes. *J. Organomet. Chem.*, **502**, 177–186.
42. Wachsen, O., Himmler, K. and Cornils, B. (1998) Aqueous biphasic catalysis: Where the reaction takes place. *Catal. Today*, **42**, 373–379.
43. Fell, B. and Papadogianakis, G. (1991) Rhodium-katalysierte mizellare Zweiphasenkatalyse. *J. Mol. Cat.*, **66**, 143–154.

44. Vyve, F.V. and Renken, A. (1999) Hydroformylation in reverse micellar systems. *Catal. Today*, **48**, 237–243.
45. Haumann, M., Koch, H., Hugo, P. and Schomäcker, R. (2002) Hydroformylation of 1-dodecene using Rh-TPPTS in a microemulsion. *Appl. Catal. A*, **225**, 239–249.
46. Miyagawa, C.C., Kupka, J. and Schumpe, A. (2005) Rhodium-catalyzed hydroformylation of 1-octene in microemulsions and micellar media. *J. Mol. Catal. A*, **234**, 9.
47. Garland, P. and Pino, P. (1991) Kinetics of the formation and hydrogenolysis of aclyrhodium tetracarbonyl. *Organometallics*, **10**, 1693–1704.
48. Evans, D., Osborn, J.A. and Wilkinson, G. (1968) Hydroformylation of alkenes by use of rhodium complex catalysts. *J. Chem. Soc. A*, 3133–3142.
49. Evans, D., Yagupsky, G. and Wilkinson, G. (1968) The reaction of hydridocarbonyltris(triphenylphosphine) rhodium with carbon monoxide. *J. Chem. Soc. A*, 2660–2665.
50. Haumann, M., Yildiz, H., Koch, H. and Schomäcker, R. (2002) Hydroformylation of 7-tetradecene using Rh-TPPTS in a microemulsion. *Appl. Catal. A*, **236**, 173.
51. Yildiz-Ünveren, H.H. and Schomäcker, R. (2005) Hydroformylation with rhodium phosphine-modified catalyst in a microemulsion: comparison of organic and aqueous systems for styrene, cyclohexene and 1,4-diacetoxy-2-butene. *Catal. Lett.*, **102**, 83.
52. Lazzaroni, R., Ucello-Barretta, G. and Benetti, M. (1989) Reversibility of metal-allyl intermediate formation in the rhodium catalyzed deuterioformylation of 1-hexene. *Organometallics*, **8**, 2323–2327.
53. Nuzzo, R.G., Haynie, S.L., Wilson, M.L. and Whitesides, G.M. (1981) Synthesis of functional chelating diphosphines and the use of these materials in the preparation of water soluble diphosphine complexes of transition metals. *J. Org. Chem.*, **46**, 2861–2867.
54. Ohkubo, K., Kawabe, T., Yamashita, K. and Sakaki, S. (1984) Micellar hydrogenation of atopic acid and its esters. *J. Mol. Catal.*, **24**, 83–86.
55. Oehme, G., Paetzold, E. and Selke, R. (1992) Increase in activity and enantioselectivity in asymmetric hydrogenation reactions catalyzed by chiral rhodium (I) complexes as a consequence of amphiphile action. *J. Mol. Catal.*, **71**, 11–15.
56. Weitbrecht, N., Kratzert, M., Santoso, S. and Schomäcker, R. (2003) Reaction kinetics of rhodium catalysed hydrogenations in micellar solutions. *Catal. Today*, **70–80**, 401–407.
57. Yonehara, K., Ohe, K. and Uemura, S.J. (1999) Highly enantioselective hydrogenation of enamides and itaconic acid in water in the presence of water-soluble rhodium(I) catalyst and sodium dodecyl sulfate. *Org. Chem.*, **64**, 9381.
58. Sottmann, T., Lade, M., Stolz, M. and Schomäcker, R. (2002) Phase behavior of nonionic microemulsions prepared from technical grade surfactants. *Tenside Surf. Det.*, **39**, 20–27.
59. Tinicci, L. and Platone, E. (1990) (To Eniricerche S.p.A) EP 0.380.154.
60. Lif, A. and Holmberg, K. (1997) Chemical and enzymatic ester hydrolysis in a Winsor I system. *Colloids Surf. A*, **129–130**, 273–277.
61. Wagner, O. (1994) *Reaktionsführung in Mikroemulsionen.* Dissertation, University of Clausthal.
62. Kragl, U. (1996) Enzyme membrane reactors. In T. Godfrey and S. West (eds), *Industrial Enzymology*, 2nd edn. Macmillan, Hampshire, pp. 271–283.
63. Seelbach, K. and Kragl, U. (1997) Nanofiltration membranes for cofactor retension in continuous enzymatic synthesis. *Enzyme Microb. Technol.*, **20**, 389–392.
64. Kragl, U. and Dreisbach, C. (1996) Continuous asymmetric synthesis in a membrane reactor. *Angew. Chem. Int. Ed. Engl.*, **35**, 642–644.
65. Schwarze M. and Schomäcker, R. (2006) Homogen katalysierte stereoselektive Hydrierreaktion im mizellarem Medium. *Chem. Ing. Tech.*, **78**, 931–936.
66. Stjerndahl, M., Lundberg, D. and Holmberg, K. (2003) Cleavable surfactants. In K. Holmberg (ed), *Novel Surfactants*, 2nd ed. Marcel Dekker, New York, pp. 317–345.

Chapter 6
Microemulsions as Templates for Nanomaterials

Satya P. Moulik, Animesh K. Rakshit and Ignác Capek

6.1 Introduction

The study and research of nanomaterials have emerged with a special importance in science. The subject has opened up a promise for challenging application potential. Research and development works in this field are progressing fast worldwide. The preparation of nanoparticles of homogeneous size (dimension) lower than 20 nm is of special interest. Thus, the methods of preparation of such materials are of much practical demand. Among the various methods used for the preparation of nanomaterials (viz. solid and gas phase reactions, vaporisation and condensation, sputtering, nucleation, hydrothermal reaction, precipitation etc.), chemical reactions in the presence of templates of different kinds, viz. micelles, reverse micelles (microemulsions), gels, coacervates, water-soluble polymers, subphase of insoluble monolayer etc., can produce materials under controlled conditions. These wet processes function under mild conditions and are operationally simpler and elegant. Among them, the microemulsion route is the most efficient in terms of stabilisation of the compartmentalised products and controlling their particle size.

6.1.1 *Basics of microemulsions*

6.1.1.1 *Definition*

Microemulsions are either dispersions of water in oil or oil in water stabilised by interfacially adsorbed amphiphiles. Normally, a combination of surfactant and co-surfactant is required for stable nanodispersions of fluids, while surfactants like Na-2-bis-sulfosuccinate (Aerosol Orange T or AOT) are able to efficiently stabilise the dispersion without a co-surfactant. Right kind of choice of amphiphiles and their combinations can bring down the oil/water interfacial tension to a very low value to make the dispersions thermodynamically stable. Microemulsions are thus considered as thermodynamically stable, isotropic and low viscous nanodispersions of water-in-oil (w/o) or oil-in-water (o/w).

6.1.1.2 *Water pool size*

In the synthesis of nanomaterials, w/o-type nanoreactors are used where the chemical reactions occur in the water pool whose physicochemical characteristics vary region-wise

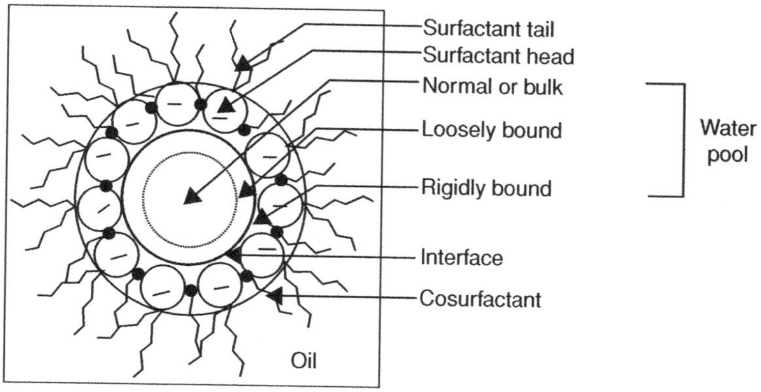

Figure 6.1 Schematic diagram of a four-component w/o microemulsion droplet.

(Fig. 6.1). The radius of the pool depends on the $[H_2O]/[\text{surfactant}]$ mole ratio or ω. At a constant [surfactant], ω increases with increasing $[H_2O]$. Thus, the water pool compartment size (or the volume) is a composition-dependant adjustable parameter. The size of the prepared materials in the water pool can, therefore, be controlled by choice. There were attempts to correlate pool radius (R_W) with ω. Pileni [1, 2] has proposed Eq. (6.1) for water/AOT/oil systems

$$R_W(\text{nm}) = 0.15\omega. \tag{6.1}$$

Also, a simple relation (Eq. (6.2)) was proposed by Marciano *et al.* [3] for AOT-derived microemulsion systems, namely

$$R_W(\text{nm}) = 0.18\omega. \tag{6.2}$$

Eastoe *et al.* [4] offered an equation of the form

$$R_W(\text{nm}) = 0.18\omega + 1.5 \tag{6.3}$$

which was almost similar to Fletcher *et al.* [5] but the coefficient was 0.175 instead of 0.18. Moulik and co-workers [6–8] have proposed equations for water/oil/AOT systems as well as for water/oil/surfactant/co-surfactant systems. It holds for the AOT-stabilised systems

$$R_W(\text{nm}) = 0.13\omega + 1.18 \tag{6.4}$$

For systems stabilised by surfactants (other than AOT) and co-surfactants, the following relation was proposed [9]

$$R_W(\text{nm}) = 0.725\omega - 2.25. \tag{6.5}$$

It has been observed that the above equations fail to correlate results at higher ω values. Besides, non-agreement between pool size and particle size formed in them has been found to depend on particle type. Quantitative accounting of correlation between the two is thus not straightforward. A rigorous study in this area is wanted. We may mention the following equations that correlate particle size with droplet size [10]. It holds

$$R_P(\text{nm}) = \left(\frac{C}{\rho}\right)^{1/3} R_W(\text{nm}), \tag{6.6}$$

where R_P is the radius of the particle, C is its concentration in the aqueous phase in terms of g/cm^3 and ρ is the theoretical density of the particle. The other recognised relation is that of Lu *et al.* [11]

$$R_W(\text{nm}) = R_P(\text{nm}) \left[\frac{\rho(1-P)}{MC}\right]^{1/3}, \tag{6.7}$$

where P is the porosity of the particle, C is the cation concentration in the aqueous solution and M is the molecular weight of the particle.

6.1.1.3 Phase formation

The mixed ternary (water/oil/surfactant) or quaternary (water/oil/surfactant/co-surfactant) systems may lead to the formation of different phases. The following are the four classes of possible phases.

1. *Biphasic*: Top oil phase and bottom o/w nanodispersion (Winsor I)
2. *Biphasic*: Top w/o nanodispersion and bottom water phase (Winsor II)
3. *Triphasic*: Top oil phase, middle bicontinuous phase (mixed regions of both w/o and o/w) and bottom water phase (Winsor III)
4. *Monophasic*: A single phase of nanodispersion of either w/o or o/w (Winsor IV)

The Winsor I and Winsor II nanodispersions are termed as L_1 and L_2 phases, respectively. Besides, the nanodispersed solutions of certain compositions may also evidence viscous- and gel-forming textures. The bicontinuous phase is considered intermediate between o/w and w/o. The microstructure of microemulsion gets affected by both temperature and additives. In case of non-ionic surfactant microemulsion, an increase in temperature changes o/w to bicontinuous to w/o microemulsion. Similar effect is observed by an increase of the NaCl concentration in an ionic microemulsion [12]. An increase of temperature and NaCl concentration both acted in the same direction for a non-ionic Brij35 microemulsion [13]. The presence of water-soluble polymers like polyacrylamide affected the area of microemulsion zone. Both increase and decrease in the microemulsion area were observed depending on which side of critical temperature the system was studied [14]. The bicontinuous structure has been attempted to be explained on the basis of two space filling models: (1) Talmon–Prager model [15, 16] in which Voroni polyhedra have been considered to fill a given space and (2) de Gennes and Taupin model [17] in which cubic lattice with size of cube as the droplet diameter was used. Elaboration of phase behaviours of microemulsion-forming systems can be found in Chapter 1 of this book.

6.1.2 Synthesis of nanoparticles

In the nanowater pools (called nanoreactors), nanosized metals and metallic salts can be conveniently synthesised. In the case of synthesis of metals, metal ions from their salts are reduced by different agents like sodium borohydride, ascorbic acid etc. Metallic salts are generally prepared by exchange reactions. The addition or mixing can be done in two ways.

1. The metal salts are taken in w/o microemulsion in a container. Concentrated solution of the reductant or the desired reacting salt is then injected into the solution to perform the reaction process. The produced nanoparticles of the metals, viz. Cu, Ag, Au, Pd, Pt etc., or their salts can be isolated by destabilisation of the microemulsion system and washing and cleaning the products and storing them in inert conditions as required.
2. The metal salts and the reductants or two reacting salts are taken in the water pools of the same microemulsion system. They are then slowly mixed with constant stirring as per stoichiometric requirements. The process of reduction or reaction takes place in situ and the desired nanoclusters of the metals or their desired salts are formed. They can be separated, washed and stored as described above.

In both the procedures, by varying ω, the dimensions of the synthesised particles can be altered. The pictorial representations of the above protocols are illustrated in Fig. 6.2. As can be seen, the internal phenomenon of droplet fusion followed by fission takes place. The materials formed during fusion by reaction get distributed among the droplets upon fission. By probability, some droplets may remain empty which is more in dilute solution of the reactants. The occurrence of the process of 'fusion and fission' has been established by the TRFQ (time-resolved fluorescence quenching method [18–20]). The internal dynamics of the disperse particles essentially guide the formation characteristics of nanoparticles.

The above-described procedures are in use in the preparation of insoluble (truly speaking, sparingly soluble) metal salts (sulphides, selenides, halides, sulphates, carbonates, oxides etc.). Isolation, cleaning, calcination (wherever required) can be performed as required. Procedural information may be found in recent literature [21–24]. According to experimental observations, the nature of the yield (particle size in particular) may depend on the sequence and rate of mixing. This aspect is not often tested in practice.

6.1.3 Characterisation and properties of nanoparticles

6.1.3.1 Techniques

By the methods of XRD (X-ray diffraction), TEM (transmission electron microscopy), SEM (scanning electron microscopy), DLS (dynamic light scattering), SANS (small-angle neutron scattering), SAXS (small-angle X-ray scattering), ultraviolet, visible, infra-red and fluorescence spectroscopy etc. the formed particles can be characterised for their shapes, sizes, clustering etc. in the states of dispersion as well as isolation. The method of AFM (atomic force microscopy) and SERS (surface-enhanced Raman spectroscopy) are also powerful tools for revealing the surface morphologies and characteristics of nanoparticles.

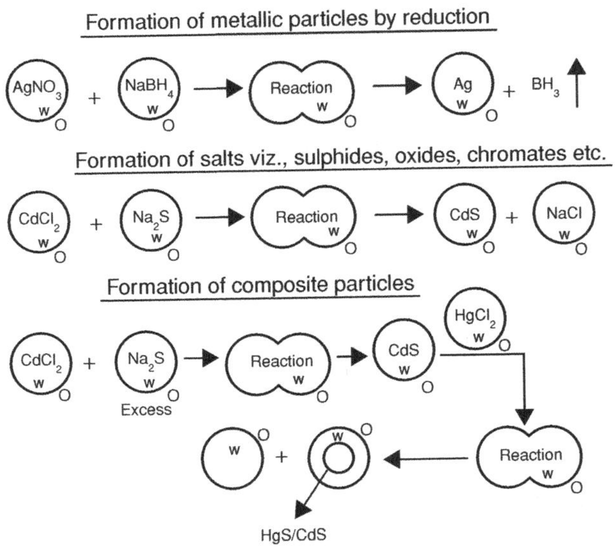

Figure 6.2 Reaction protocol and associated product formation steps.

But we have a point of concern. What research advantage is gained in simple wet chemical methods of synthesis of nanoparticles in microemulsions (or in greater perspective in compartmentalised conditions) gets greatly retarded in many laboratories for lack of modern infra-structural facilities, i.e. sophisticated (hence costly) physical techniques to physicochemically reveal the types of materials produced.

The state of aggregation of nanoparticles is very important in relation to their physical–chemical properties [23]. Their catalytic, spectral and electrochemical behaviours depend considerably on their clustering/aggregation. Tiny aggregates can show unusual physicochemical properties not manifested in the states of higher aggregation. The following are features of particle growth and dissolution. Below a critical size of nanocrystals, the particles have a tendency of dissolution but at a higher crystal size particles grow with depletion of monomers in solution, and at a much-depleted state larger particles grow at the expense of smaller ones by the process called Ostwald ripening. Such are special physicochemical features of colloid and nanodispersions [25, 26]. A detailed presentation of physicochemistry and applications of microemulsions has been recently published [27].

Monte Carlo simulation method has revealed that nanoparticle formation in microemulsion occurs by nucleation and growth by autocatalysis and ripening. Interfacial surfactant film flexibility affects the kinetics of the reaction and growth when the inter-droplet interaction becomes the rate-determining step [28]. The nucleation and growth of nanoparticles of $CuCrO_4$ and CuS (prepared in quarternary w/o microemulsion) system of water/cyclohexanone/Triton X100/*i*-propanol were found from DLS measurements to occur right from the beginning of their formation, and the growth of CuI particles (prepared in the same microemulsion) occurred after an initial induction period of nearly 10 min [29].

The particle size in microemulsion is essentially governed by two factors [30], namely (1) the number of droplets in the preparation and (2) the inter-particle interaction associated with exchange of materials. Dilute solution of the reactants in a template formulation produces tiny and weakly interacting nanoparticles whose growth by collision is also hindered by the peripherial amphiphilic film. The particle dimension in relation to water pool size has been discussed by Robinson *et al.* [31].

The formation of nanoparticles from microemulsions need not essentially follow the template shape. Pileni [32] (as quoted by Ganguli and Ganguli) has shown that with water/isooctane/$Cu(AOT)_2$ shapes like sphere to cylinder to mixed spherulites and cylinders to other polydisperse shapes were possible with increasing ω. According to Pileni [33], the presence of salt anions can control the shape while chloride ions favour formation of nanorods, nitrate ions hinder it. The surfactant content also can have a say on the shape of nanoparticles. The infrequently observed morphologies of nanoparticles, viz. wires, trigons, hexagons, cubes etc. have so far no specific and reliable reasons for formation in microemulsion templates.

6.1.3.2 Determination of band gap

The nanomaterials (viz. sulphides, selenides, chromates, iodides etc.) formed in solution may exhibit absorption in ultraviolet and visible regions [7, 29]. The concentration-dependant spectra at different ω values may obey Beer's law like normal absorbing solutions at concentrations not very high. The particles in nanodimensions behave like molecular solution, the molar extinction coefficient, however, depends on ω or the particle size. $Cu_2[Fe(CN)_6]$ preparations in H_2O/AOT/*n*-heptane w/o microemulsion medium at $\omega =$ 5, 8, 12 and 20 have exhibited good Beer's law plots with molar extinction coefficients of 350, 365, 352 and 521 $dm^3\ mol^{-1}\ cm^{-1}$, respectively, at 315 nm (310 nm at $\omega = 5$) . Interestingly, the values were reasonably smaller than true molecular solutions [7]. The visible spectral data of microemulsion encapsulated nanoparticles can be processed to evaluate the band gap of the material by the use of Tauc equation [34] which has the following form

$$(\varepsilon h v)^2 = C(hv - \varepsilon_g), \tag{6.8}$$

where ε, h and ν are the molar extinction coefficient, the Planck's constant and the frequency of light, respectively, and C is also a constant. The plot of $(\varepsilon\ h\nu)^2$ against $(h\,\nu)$ in the wavelength range on the right side of the absorption maximum [35] helps to estimate the band gap ε_g from the intercept and the slope. This is a convenient method of determination of band gap of nanosemiconductor particles [8, 36–38]. Isolated nanoparticles when dispersed in a medium in presence of suitable stabilisers may exhibit characteristic spectral behaviour depending on their type and solvent environment in which they are embedded. This area has not been studied extensively and there remains scope for further exploration.

6.2 Preparation of nanocompounds

Among the various templates employed for the synthesis of nanoparticles, microemulsion route has been found to be reasonably simple and successful to augment production of sulphides, oxides and other interesting and important nanocomposites. In the following, we present a concise account on different categories of nanomaterials produced in the domain of w/o microemulsion.

6.2.1 Sulphides

The quaternary microemulsion water/*n*-hexane/cetyltrimethylammonium bromide (CTAB)/*n*-pentanol has been used by Curri *et al.* [39] to synthesise CdS nanoclusters. In this method, $Cd(NO_3)_2$ solutions with a given *p* (molar ratio between *n*-pentanol and surfactant) and Na_2S solutions with a given ω were rapidly mixed. Fine colloidal precipitates of CdS were obtained. The authors also used AOT in preparation of microemulsion wherein the two precursor solutions of $Cd(NO_3)_2$ and Na_2S separately taken in the microemulsion were mixed together. Interestingly, they observed that in the case of AOT microemulsion there is change in colour with time due to the change in the particle size whereas in the CTAB system no such change was observed as the particle growth was instantaneous in this case. In both cases, nucleation occurred immediately. They also observed that the presence of alcohol in the microemulsion is a very important factor in controlling the size, size distribution and stability of the CdS crystallites. Dutta *et al.* [40] synthesised CdS nanoparticles in self-reproducing reverse micelles. Sodium octanoate, isooctane, octanol and water were used to make microemulsion systems with different ω values. Constant amount of cadmium perchlorate was taken as a precursor and various amounts of H_2S were rapidly added to obtain nanosized CdS particles. The sizes of the synthesised CdS were controlled by the sizes of the water pool (i.e. by ω) as well as the ratios of the concentrations of Cd^{2+} and H_2S as observed earlier [33, 41]. It was found that the addition of Cd^{2+} to H_2S or vice versa did not change the size of the CdS which was found to be 5.9 ± 0.9 nm.

Pinna *et al.* [42] synthesised triangular CdS nanocrystals by using the ternary system water/isooctane/cadmium bis(ethyl-2-hexyl)sulfosuccinate and reacting this with a gas mixture of H_2S and N_2 with a molar fraction of $\frac{1}{3}$ and bubble flow rate of 0.05 $cm^3\ s^{-1}$. TEM was used to characterise the nanoparticles. Interestingly, rod-like CdS nanoparticles was synthesised by Tong *et al.* [43] by using multilamellar vesicles stabilised by SDS and decanol. The precursors were $CdCl_2$ and Na_2S and the multilamellar vesicles were made of water/SDS/decanol (SDS:decanol = 1:1 by weight) and various amounts of poly(diallyldimethylammonium chloride). They also suggested that the sizes of the particle were affected by the temperature at which the mixing processes were done. Moreover, it was found that the nanoparticles produced a deswelling effect on the lamellar structure.

ZnS nanocrystals were synthesised in ternary w/o microemulsion stabilised by non-ionic or cationic surfactants [44]. Several morphologies, e.g. nanorods or spherical or ellipsoidal ZnS particles were obtained by varying the ω values. The product morphology was also found to be function of the absolute reactant concentration and concentration ratio of Zn^{2+} to S^{2-}, the incubation time and the ambient temperature.

CuS has been synthesised by using two different types of w/o microemulsions [29], namely (i) water/cyclohexanone/AOT and (ii) water/cyclohexanone/Triton X100/*i*-propanol. The shape of the CuS obtained from the second was found to be spherical particularly at lower ω values, while at higher values ($\omega = 14$) the shape was not completely spherical. The size of the nanoparticles increased from 38 to 95 nm with an increase in ω from 2 to 14. It can be seen from Fig. 6.3 that at $\omega = 2$, the CuS shapes are spherical.

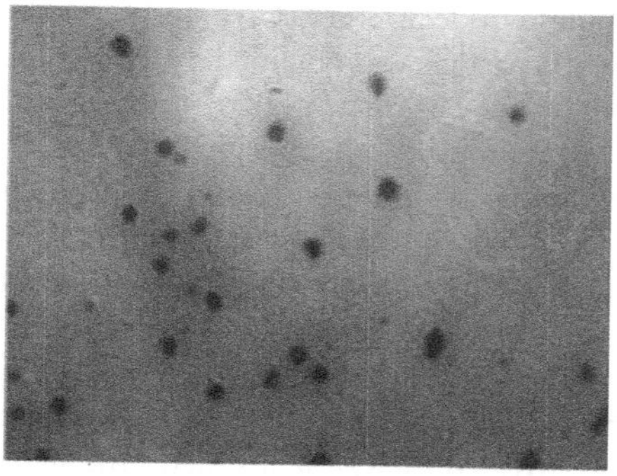

Figure 6.3 TEM image (instrument magnification 50 000×) of CuS at $\omega = 2$ from Triton X-100 microemulsion. (From Ref. [29], reprinted with permission of Taylor & Francis.)

CuS has also been prepared [36] by using Cu–ammonia complex and thiourea in alkaline pH in a water/cyclohexane/Triton X100/methylpropane-1-ol microemulsion. Other non-ionic surfactants as well as ionic sodium dodecylsulphate were also used. Biswas *et al.* [29] synthesised CuS nanoparticles in water/cyclohexanone/Triton X100/*i*-propanol w/o microemulsion. The band gap of the material and the particle growth were determined from spectral and DLS measurements along with their general characterisation.

Ward *et al.* [45] prepared PbS nanoparticles by dissolving $Pb(NO_3)_2$ or $Pb(ClO_4)_2$ in microemulsion medium and adding aqueous Na_2S solution with constant stirring into it. It was also synthesised by dissolving Na_2S in another aliquot of microemulsion and adding the same to Pb^{2+} containing aliquot. The microemulsion employed was made up of water, non-ionic dodecyltetraethylene glycol ether and hexane. The average size of the PbS particles was ~3 nm. Eastoe *et al.* [46] have also synthesised nanosized PbS particles by using (1) Na(AOT) and (2) mixed $Na(AOT)/Pb(AOT)_2$ stabilised w/o microemulsion media. Interestingly, particles synthesised with Na(AOT) were more time-stable than those prepared with $Na(AOT)/Pb(AOT)_2$ mixture.

6.2.2 Sulphates

Water-in-oil microemulsions have been conveniently used for the synthesis of sulphate nanoparticles. Mann *et al.* [47] studied the formation of $BaSO_4$ fibres which were a micrometre long. They used an unstirred reaction of $Ba(AOT)_2$ reverse micelles and NaAOT microemulsion with Na_2SO_4 at room temperature at $\omega = 10$ and $Ba^{2+}:SO_4^{2-}$ at 5:1 and 1.4:1. Nanofilaments obtained were 5 nm wide and twisted along the 010 axis. They studied $BaSO_4$ filament formation at temperatures 4, 30 and 40°C and observed that they were highly curved, cone-shaped and spindle-shaped aggregates, respectively. In other words, the shape of the nanofilaments was temperature dependant. Rees *et al.* [48] have shown that slightly irregular aggregates of $BaSO_4$ (8–10 nm in diameter) were obtained from *n*-heptane microemulsion where Na(AOT) was used. Interestingly, in presence of ammonium diethyl hexyl phosphate, the $BaSO_4$ was obtained as submicron-sized flocs (5–7 nm in diameter) from *n*-heptane microemulsion. From cyclohexane microemulsion with tetraethylene glycol monododecyl ether ($C_{12}E_4$), an almost monodisperse (8–10 nm in diameter) spherical $BaSO_4$ particle was obtained. Ivanova *et al.* [49] synthesised $BaSO_4$ of average particle size of 10 nm by using w/o microemulsion and adding Ba^{2+} containing microemulsion to $SO_4{}^{2-}$ containing microemulsion.

It was also observed by Rees *et al.* [48] that $CaSO_4$ can be prepared in w/o microemulsion medium where the surfactants might be ionic or non-ionic. For $CaSO_4$ they observed many different morphologies like nanospheres, ellipsoids, rods, nanohairs, nanowires, nanobundles etc. In presence of non-ionic $C_{12}E_4$, many different morphologies of $CaSO_4$ were obtained and they were found to be function of (i) ω, (ii) overall water content, (iii) surfactant concentration, (iv) reactant concentration etc. In contrast to this, AOT-stabilised dodecane microemulsion provided $CaSO_4$ nanospheres the shape of which was independent of the composition of the microemulsion and the reaction conditions, respectively.

Single-crystal $PbSO_4$ (anglesite) nanorods were prepared by Xiang *et al.* [50] by using a water/hexane/SDS/hexanol microemulsion as template. The effects of concentrations of reactants and surfactants and temperature variation on the formation were also studied.

6.2.3 Hydroxides

By using a microemulsion of water/cyclohexane/cetyltrimethylammonium bromide (CTAB)/*n*-pentanol, Cao *et al.* [51] synthesised three-dimensional $Ni(OH)_2$ nanoparticles. With $\omega = 40$, dandelion-like nanostructures of α-$Ni(OH)_2$ were obtained. β-$Ni(OH)_2$ phase with flower-like nanostructures was obtained from the same microemulsion at $\omega = 10$. Thus, adjustments of various microemulsion parameters led to the formation of nanoparticles of different forms and shapes. Nanosized $Al(OH)_3$ was prepared by using a supercritical fluid by Matson *et al.* [52]. In this case, $Al(NO_3)_3$, $9H_2O$ and sodium salt of AOT were charged (at $\omega = 5$) into the pressure vessel. The pressure was kept at 200 bar with the help of propane at 110°C. A clear w/o microemulsion solution was obtained on stirring for some time. Addition of dry ammonia resulted in the formation of $Al(OH)_3$ of 500 nm size. With less concentration of nitrate, the particle size was lower. Nanni *et al.* [53] synthesised $Ca(OH)_2$ nanoparticles in w/o microemulsion with cyclohexane as oil. Two non-ionic surfactants tetraethylene-glycol-monododecylether ($C_{12}E_4$) and pentaoxyethylene-glycol-nonyl phenyl ether (igepal CO520) were used at various ω values. Microemulsions containing $CaCl_2$ and NaOH were mixed together to form $Ca(OH)_2$. The product was found to be highly reactive to atmospheric CO_2 and hence it was possible to prepare ultrafine $CaCO_3$.

6.2.4 Oxides

Recently, Chen *et al.* [54] have synthesised nanoparticles of metallic Cu and also Cu_2O by radiolytic reduction of Cu^{2+}in microemulsion medium where non-ionic surfactants, e.g. Brij30, Brij56 or Triton X100 with different ω values were used. Anions and surfactants had remarkable effect on the radiolytic reduction process. They also affected the morphologies of the reduction products. Thus, in the presence of toluene with Brij56 microemulsion the radiolytic reduction product was metallic copper but replacement of toluene by naphthalene produced Cu_2O. Chen *et al.* [55] have used a CTAB-based microemulsion to synthesise crystalline tin dioxide (SnO_2). They used the typical quaternary microemulsions of water/*n*-hexane/cetyltrimethylammonium bromide (CTAB)/*n*-pentanol as space-confined microreactors for the nucleation, growth and crystallisation of SnO_2 nanoparticles under hydrothermal conditions. Tin chloride was used as the starting material. The as-prepared SnO_2 nanoparticles had large-specific surface areas (107–169 $m^2\ g^{-1}$), small particle size (~3.0 nm), high crystallinity and narrow size distribution. Indium oxide (In_2O_3) nanoparticles were prepared by chemical reaction of inorganic indium compounds and ammonia gas in a reverse microemulsion system consisting of water/*n*-octane/Triton X100/*n*-heptanol [56]. The hydroxides were precipitated in w/o microemulsion and then calcined to form In_2O_3. There was interesting temperature effect on the size and shape of the prepared In_2O_3. The product of calcination at 400°C was spherical with a narrow size distribution (average size, 7 nm) whereas the product at 800°C calcination was of irregular shape with wide size distribution (average size. 40 nm). The indium compound used had appreciable effect on the particle size. With $InCl_3$, the average size was 7 nm whereas with $In(NO_3)_3$, the average size was 15 nm. Two binary oxides, a spinel, $ZnAl_2O_4$, and a typical perovskite, $LaMnO_3$, were prepared via CTAB/1-butanol/*n*-octane/nitrate salt microemulsion both in the w/o and bicontinuous states [57]. The $ZnAl_2O_4$ obtained from both types of microemulsion showed a sponge-like structure by TEM. In case of spinels, considerable internal porosity was observed. The surface area and pore volume were 143.7 $m^2\ g^{-1}$ and 0.23 $cm^3\ g^{-1}$, respectively, for the samples obtained from reverse micelle and the corresponding values for the samples obtained from bicontinuous microemulsion were 126.7 $m^2\ g^{-1}$ and 0.21 $cm^3\ g^{-1}$. The maxima of the pore size distribution were found to be at pore diameters 4.74 nm and 4.26 nm for the reverse and bicontinuous systems, respectively. The perovskite, $LaMnO_3$, showed a peculiar doughnut-like structure when obtained from reverse microemulsion whereas the material obtained via bicontinuous microemulsion showed a uniform secondary structure. The size of the particles in the second case was almost one-tenth of the first as evidenced from SEM studies. The perovskites had low surface areas whereas the spinels had very large surface areas. The TEM photographs revealed that the particles, synthesised from w/o and bicontinuous microemulsions, were constituted of primary nanoparticles of 40–100 nm in size.

Nad *et al.* [58] synthesised titanium dioxide (TiO_2) nanoparticles of various different structures by the hydrolysis and condensation of $TiCl_4$ in the water core of water/hexane/AOT microemulsions of different ω values at 8°C. The ω was varied from 8.3 to 18 to obtain nanoparticles of sizes from 6 to 115 nm. They also observed that these were thermodynamically unstable orthorhombic crystals which on sintering at various different temperatures formed relatively stable nanorods. The variation of particle size with ω is shown in Fig. 6.4. Nad *et al.* [58] also presented TEM pictures of TiO_2 after sintering at various temperatures and showed that the particles changed from spherical (unsintered) to nanorods (Fig. 6.5).

Wu *et al.* [59] have used a combined procedure of microemulsion-mediated hydrothermal method (MMH) to prepare uniform-sized nanoparticles of TiO_2 (both rutile and anatase). Tetrabutyl titanate was dissolved in HCl or HNO_3, and the solution was then allowed to disperse in an organic phase to prepare the microemulsion. The aqueous core of the system, water/cyclohexane/Triton X100/hexanol was used as the microreactor for the controlled growth of TiO_2 particles under hydrothermal conditions. The influences

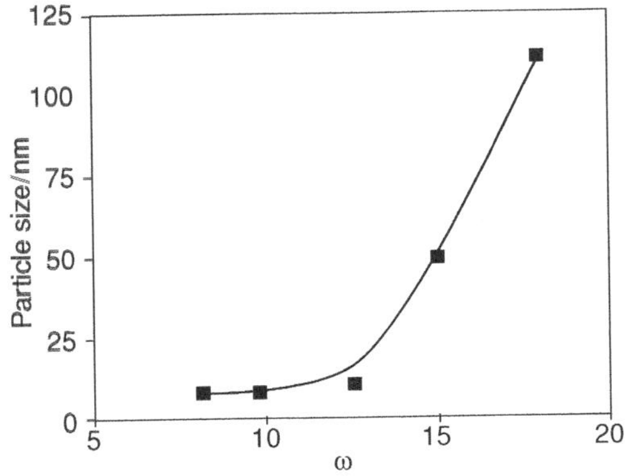

Figure 6.4 Variation of TiO_2 nanoparticle size as a function of ω. (Data are taken from Ref. [58].)

of changing variables such as different acids, their concentrations, the reaction temperature and/or the reaction time on the phase and morphology of the titania products were discussed.

Monodisperse iron oxides (magnetite, $FeO{\cdot}Fe_2O_3$) were synthesised from w/o microemulsion of AOT/isooctane or AOT/cyclohexane and aqueous solutions of $FeCl_3$ and NH_3 to which $FeCl_2$ was added with stirring [60]. Liz *et al.* [61] prepared Fe_3O_4 (~ size 4 nm) in water/*n*-heptane/AOT microemulsion using $FeCl_3{\cdot}6H_2O$ and $FeCl_2{\cdot}4H_2O$ and NH_4OH at $\omega = 10$. The other preparation protocols for different kinds of iron oxide (particularly α− and γ−Fe_2O_3) can be found in [24].

6.2.5 Core–shell products

The preparation of core–shell nanoparticles of different types from microemulsion templates has also been reported. Water-in-oil microemulsion has been used to synthesise

Figure 6.5 TEM pictures of TiO_2 after sintering at (a) 280°C, (b) 410°C, (c) 750°C and (d) 900°C. (From Ref. [58], reprinted with permission of Elsevier.)

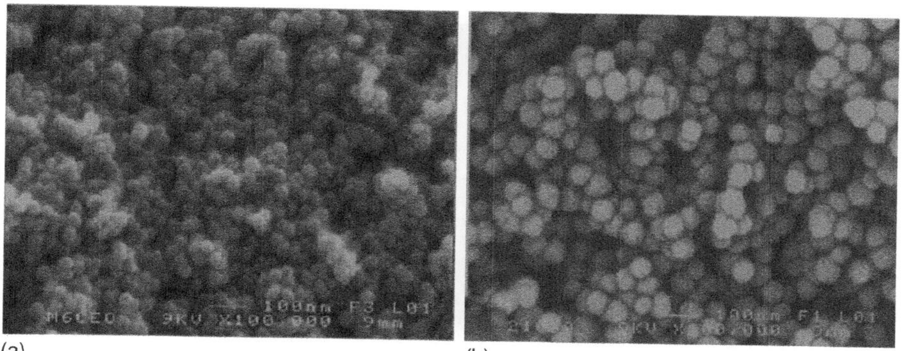

(a) (b)

Figure 6.6 SEM images of CeO_2–SiO_2 obtained by using two different microemulsions, namely (a) 63.5 wt.% *n*-heptane/27.5 wt.% surfactants/9 wt.% aqueous phase (3.2/3.7/2.1 wt.% of H_2O/TEOS/NH_3) and (b) 63.1 wt.% *n*-heptane/27.6 wt.% surfactants/9.3 wt.% aqueous phase (3.2/3.8/2.3 wt.% of H_2O/TEOS/NH_3). Note that in the later case H_2O stands for an aqueous sol of CeO_2. (From Ref. [62], reprinted with permission of Elsevier.)

silica nanoparticles [62] with a CeO_2 core and SiO_2 shell designated as CeO_2–SiO_2. In the w/o microemulsion, *n*-heptane was the oil and AOT/Brij30 mixture (1:1 w/w) was the surfactant. The mixture of CeO_2 sol, TEOS (tetraethoxysilane) and aqueous ammonia solution was the aqueous phase. The microemulsion was prepared by adding acidic CeO_2 sol into a surfactant/heptane mixture. The precursor of silica nanoparticles, TEOS, was then added to the microemulsion. Finally, ammonia solution was carefully added with stirring for the condensation of TEOS. To avoid thermally induced phase inversion, the reaction temperature was kept below 25°C. After 48 h, the synthesised samples were washed with *n*-heptane, ethanol and acetone to remove surfactant and oil, and centrifuged at 5000 rpm for 15 min. The separated materials were dried at 60°C under vacuum. TEM, SEM, DLS and other methods were used for characterisation. In Fig. 6.6, the SEM of CeO_2–SiO_2 prepared from two microemulsions are presented. Structural changes between Fig. 6.6(a) and 6.6(b) were large although the compositions of microemulsions 1 and 2 are almost the same. Grasset has reasoned (personal communication) that this is due to different sizes of CeO_2 in the two sols A and B; in (A) the size was 4 nm, while in (B) it was around 20–30 nm. Thus, the dimension of core–shell products was strongly dependant on the CeO_2 core size.

Chakraborty *et al.* [63] have synthesised CdS–HgS core–shell and composites by using a cetyltrimethylammonium bromide (CTAB) micellar solution. Here, the $CdCl_2$ (or $HgCl_2$) in CTAB solution was prepared into which Na_2S solution (more than stoichiometric requirement) was added with vigorous stirring to form CdS or HgS. $HgCl_2$ (or $CdCl_2$) solution was then slowly added with stirring to form the product HgS–CdS or CdS–HgS. The HgS–CdS core–shell particles were monodisperse, prolate particles (TEM images in Fig. 6.7(a)). The CdS–HgS particles evidenced a tendency to form needle-shaped primary particles ending in three-dimensional assemblies (Fig. 6.7(b)). These clearly indicated the controlling role of the compartmentalised systems (microemulsions or micelles) and the mode of precursor addition to influence the formation of core–shell products and their sizes.

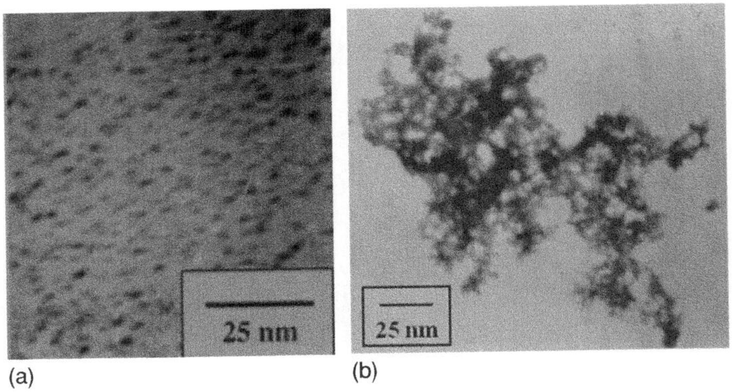

Figure 6.7 TEM pictures of (a) HgS–CdS and (b) CdS–HgS. (From Ref. [63], reprinted with permission of Springer.)

6.2.6 Miscellaneous

Besides the above-described nanomaterial systems, various other types of nanoparticles have been reported using w/o microemulsions as templates suggesting the versatility of the microheterogeneous systems in synthesising nanocompounds. Typical examples of such preparations are described below.

$PbWO_4$ of various different morphologies, e.g. nanostructures with rod-like, ellipsoid-like and sphere-like bundles have been prepared by using AOT-based microemulsion with different media conditions [64]. AOT concentration, water content and reaction temperature were responsible in controlling the morphologies of synthesised $PbWO_4$ nanostructure. The authors have discussed possible mechanism of such different nanostructures formed under various conditions. Colloidal dispersions of tungstic acid (H_2WO_4) have been also prepared in w/o microemulsion consisting of water/*n*-heptane/Triton X100/alkanol [65]. Na_2WO_4 was allowed to react with HCl in this medium. The formation of H_2WO_4 in the nanowater pool was established by FT-IR measurements. The effect of various variables on the formation of the nanoparticles were also studied.

Panda *et al.* [8] have synthesised colloidal dispersions of $PbCrO_4$ by using a w/o microemulsion system made up of water/*n*-heptane/sodium salt of AOT. The size of the lead chromates so synthesised have been found to depend on ω. These were characterised by TEM, SEM and DLS methods. $CuCrO_4$ nanoparticles were prepared in microemulsion consisting of water/cyclohexanone/AOT [29]. The ω was varied between 8 and 20 to yield particles of sizes 8–20 nm. The increase of temperature was found to produce particles of larger size. Moulik *et al.* [7] have used the microemulsion system (water/*n*-heptane/AOT) to synthesise $Cu_2[Fe(CN)_6]$ by using $CuSO_4$ and $K_4[Fe(CN)_6]$ separately in two different aliquots of the above microemulsion and mixing them together under vigorous stirring condition. The above synthesis was done at various ω and also in presence of different amounts of gelatin as a surface stabiliser.

AgCl nanoparticles have been synthesised by using a water/cyclohexane/polyoxyethylene (6) nonylphenyl ether (NP-6) microemulsion wherein $AgNO_3$ and KCl solutions were added and mixed [66]. The particle growth rate and the final particle size at a given ω were as expected from theoretical prediction. The small particle size in the microemulsion was due to low overall solubility of the solid and also the small diffusivity of the bulky droplets as silver ion carrier and not by the size of the water pool as generally stated. A novel method of preparing AgCl nanoparticles by mixing AgCl powder and a microemulsion consisting of water/isooctane/dioctyldimethylammonium chloride/*n*-decanol was introduced by Husein *et al.* [67]. In this method, powdered AgCl was mixed with the above microemulsion to obtain AgCl nanoparticles. By manipulating various operating variables, e.g. temperature, rate of mixing, concentrations of surfactant and co-surfactant, as well as ω, it was possible to understand the role of rigidity of the surface layer of surfactant on the formation of nanoparticles.

AgBr nanoparticles have been synthesised by the addition of $AgNO_3$ aqueous solution to a microemulsion consisting of water/isooctane/dioctyldimethylammonium bromide/*n*-decanol. The effect of changes in various variables was studied. Increasing the surfactant concentration at a given ω and $AgNO_3$ concentration enhanced intermicellar nucleation. This resulted in the formation of larger particles. However, when the $AgNO_3$ concentration was increased at fixed values of all the other variables, direct nucleation was enhanced resulting in the formation of smaller particles [68].

In addition to the above, preparation in w/o microemulsions of nanoparticles of various other types of compounds, viz. silica-coated iron oxide, Fe_2O_3–Ag nanocomposite, oxides of ytrium, erbium, neodymium, vanadium and cobalt, titanates of barium and lead, ferrites of barium, strontium, manganese, cobalt and zinc, oxide superconductors, aluminates, zirconium silicate, barium tungstate, phosphates of calcium, aluminium and zinc, carbonates of calcium and barium, sulphides of molybdenum and sodium, selenides of cadmium and silver etc. have been reported. Preparative sources and related elaboration can be found in [24].

6.3 Metal and metal/polymer nanoparticles

6.3.1 *General concepts*

The preparation of metal nanoparticles has received considerable attention in recent decades because nanoparticles possess unconventional physical and chemical properties [69]. The unique physical properties of nanoscale magnetic materials such as superparamagnetism have generated considerable interest for their use in a wide range of diverse applications from data information storage to *in vivo* magnetic manipulation in biomedical systems [70]. In particular, due to their large surface-to-volume ratio, the magnetic properties of nanoparticles are dominated by surface effects and particle–support interactions. They exhibit magnetic anisotropy constants that are ca. 2 orders of magnitude larger than their bulk counterparts, with correspondingly enhanced coercivities [71].

A number of techniques have been used for producing nanoparticles, including vapour phase techniques [72], sol–gel methods [73], sputtering [74], coprecipitation [75] etc. Two main methods can be employed for the preparation of metal nanoparticles: coprecipitation and chemical reduction. In both cases, the presence of surfactant is required to govern the growth process. Typically, the coprecipitation reactions involve the thermal decomposition of organometallic precursors [76]. The chemical reduction occurring in colloidal assemblies is another approach for the formation of size- and shape-controlled nanoparticles [77]. A major benefit of chemical methods is their relatively inexpensive investment of capital equipment.

Of the chemical processes, reverse micelle (microemulsion) synthesis has been recently demonstrated to be a viable method for producing a wide array of metals and metal oxide nanoparticles [78, 79] over a relatively narrow particle size distribution. Reverse micelle synthesis utilises the natural phenomenon involving the formation of spheroidal aggregates in a solution when a surfactant is introduced to an organic solvent, formed either in the presence or in the absence of water [80]. Micelle formation allows for a unique encapsulated volume of controllable size through which reactions and subsequent development of metal and metallic compounds can be produced. Aggregates containing ω (= (water)/(surfactant), see above) of less than 15 can be called as reverse micelles and have hydrodynamic diameters in the range of 4–10 nm [78], whereas ω greater than 15 constitute microemulsions, which have a hydrodynamic diameter range between 5 and 50 nm. Once the right microemulsions are obtained, the method of particle preparation consists in mixing of two microemulsions carrying the appropriate reactants in order to obtain the desired particles [81, 82]. Schematic pictures of this process are represented in Fig. 6.2.

Controlled nucleation and separation of nucleation from growth are the keys to synthesising near-monodisperse metal nanoparticles in the 1–15 nm size range [83]. This can be achieved either by providing a controlled number of preformed nanoparticles as nucleation centres in a growth medium where no secondary nucleation can occur – the seeding growth method [83] – or by varying the ratio of strong and weak reducing agents [84]. Key goals in the synthesis of metal nanostructures are that the synthesis gives nanostructures of a specific size and size distribution and that the synthesis is reproducible [85]. The simplest approaches for isotropic and anisotropic nanoparticle synthesis are various surfactant-based methods [86]. Surfactant-based anisotropic micelle templates can be easily prepared [87]. For example, the ~6 nm spherical micelles formed by a dilute (>1 mM) solution of cetyltrimethyl ammonium bromide surfactant converts to cylindrical micelles at higher concentrations (>20 mM), more elongated rod-like micelles in the presence of organic solubilizates [88], and worm-like micelle structures in the presence of salicylate [89]. Surfactant molecules can be used as 'simple' capping and stabilising agents as in the organometallic precursor decomposition reactions.

6.3.2 Anisotropic metal nanoparticles

Anisotropic nanoparticles are of considerable current interest, due to various shape-dependant properties [90, 91]. Synthesis of anisotropic nanoparticles on the 1–50 nm length scale is very challenging as they are less stable compared to spherical shapes of similar size, and the respective symmetric bulk crystal structures often create a strong barrier. Controlling the length-to-width ratio and the uniformity of the length and width distributions of metal nanoparticles, as well as synthesising particles with a width dimension below 10 nm is challenging with large-scale bench top methods [92].

Another key factor in determining the particle anisotropy was the size of the growing nanoparticle. Jana *et al.* have prepared different sizes of near-monodisperse spherical gold nanoparticles and used them as seeds for the synthesis of anisotropic nanoparticles in micellar templates [83, 93]. They observed that the yield and shape of the anisotropic nanoparticles significantly depended on the seed size. Nanorods in 90–95% yield were obtained only when the smallest seed of 1.5 nm size was used and the final nanorods had a 5–10 nm short axis and an aspect ratio between 1 and 5. When the seed size was increased to 3.5 nm, the nanoparticles produced were spheroids in 30–70% yield, and the final nanorods had a 10–20 nm short axis and an aspect ratio between 1 and 5. If the seed particle size was >5 nm, the nanorod yield was very low, regardless of the seed concentration used.

A mixture of strong and weak reducing agents can be introduced into the micellar solution of a metal salt, where the strong reducing agent initiates nucleation and the weak reducing agent helps the anisotropic nanoparticles to grow. Using this micellar template approach, nanoparticles of a wide range of shapes can be prepared for metals [93]. The anisotropic nanoparticles (the rod type of structure) generate in the microemulsion system under the influence of various cations and anions, which affects the rigidity of the interface between the hydrophilic polar head groups and the aqueous core of the micelle [94]. Another route to producing non-spherical nanoparticles is to use reverse micelles in a supersaturated regime [95].

Anisotropic noble nanoparticle dispersions are very different in colour compared to dispersions of spherical particles. This is because the surface plasmon bands are more sensitive to particle shape than size [92]. All the metal nanorods have two absorbance maxima that correspond to the longitudinal and transverse plasmon bands. The longitudinal plasmon band strongly depends on the aspect ratio. For example, platelets have additional quadrupole bands [96]. Upon transition from nanorods to platelets, as the aspect ratio decreases, the longitudinal band is blue-shifted and the transverse band becomes broad due to overlap with the quadrupole band. In cubes, all three plasmon bands merge into a single band. In contrast, transition from nanorods to nanowires increases the aspect ratio, which produces a resultant red shift of the longitudinal band and a blue shift of transverse band [83].

6.3.3 Core–shell metal nanoparticles

Of special interest are core–shell structured nanoparticles that could exhibit enhanced properties and new functionality, due to the close proximity of the two functionally different components. Such structures not only are ideal for studying proximity effects but are also suitable for structure stabilisation as the shell layer protects the core from oxidation and corrosion. Additionally, the shell layer provides a platform for surface modification and functionalisation, such as coupling the magnetic core through the shell onto organic or other surfaces, thus tuning their intrinsic magnetic properties and making them potentially biocompatible [97]. The core/shell magnetic nanoparticles, for example, can exhibit a rich variety of interesting phenomena, such as single-domain state, coercivity enhancement and quantisation of spin waves, due to their small dimensions [98]. They have exciting potential applications in magnetic recording, sensing and biological diagnosis [99].

In the case of Fe as the core, there are examples of core–shell Fe–Au [100], Fe–Fe-oxide [101], Fe-oxide/Au [102], and Fe_3O_4–polymer [103] nanoparticles. The combination of Fe core/Au shell is particularly appealing because Au is not ferromagnetic, but is noble and

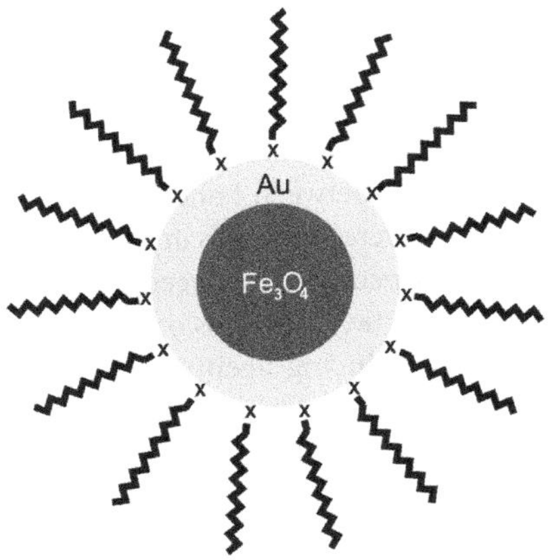

Figure 6.8 Illustration of the core–shell Fe_3O_4–Au nanoparticle with an outmost organic shell encapsulation (X = $-CO_2H$ or $-NH_2$, –SH, –OA).

relatively easy to functionalise. Initial studies on Fe–Au core–shell nanoparticles have been reported by Carpenter *et al.* [104, 105]. These Fe/Au nanoparticles were synthesised by a reverse micelle method and nanoparticles had a size distribution of 5–15 nm diameter. The Fe nanoparticle was reported not to be centred in the micelle, resulting in an asymmetric Au shell. An alternate explanation was that there may be grain boundaries in the Au shell that allow for diffusion of oxygen and oxidation of the metallic core. In the report by Kinoshita *et al.* [106], the same synthetic method was followed. Using the reverse micelle method, gold-coated iron core–shell [107] and gold (silver)-coated magnetite [108] nanoparticles have also been synthesised. The key issues are the chemical states of the core materials and whether the oxide forms during or after the synthesis process. One of the important aspects of the synthesis of the Fe_3O_4–Au nanoparticles is the formation of the gold shell at the iron oxide nanocrystal cores with high monodispersity and controllable surface capping properties (Fig. 6.8), which facilitates the subsequent control and manipulation of the inter-particle interactions and reactivities [109].

The Fe–Au nanoparticles were reported to consist of metallic cores, having an average diameter of 6.1 nm, surrounded by an oxide shell, averaging 2.7 nm in thickness, for a total average particle diameter of 11.5 nm [101]. A surfactant solution is prepared with nonylphenol poly(ethoxylate) ethers. Au-coated Fe nanoparticles were also prepared in a reverse micelle formed by cetyltrimethylammonium bromide (CTAB), 1-butanol and octane as the surfactant, the co-surfactant and the oil phase, respectively [100]. The nanoparticles were prepared in aqueous solutions of micelles by reduction of Fe(II) and Au precursors with $NaBH_4$. The typical size of the nanoparticles is about 20 nm. The existence of Fe and Au is again confirmed by energy dispersive X-ray microanalysis.

The further study of Cho *et al.* [107] has showed that the structure of Fe/Au core/shell nanomaterials is somewhat complex. Mössbauer spectra were best interpreted as Fe speciation of α-Fe, Fe^{II}, Fe^{III} and Fe–Au alloy. The Au shell was suggested to grow by nucleating from small nanoparticles on the Fe-core surface before it develops the shell structure. These nanoparticle nucleation sites form islands for the growth and coalescence of the thin Au overlayer. Specifically, Au^{3+} is reduced to Au by $NaBH_4$, which initiates minimum nanoscaled seed Au nanoparticles and they grow larger, resulting in an Au shell.

Lin *et al.* have used [110] a sequential micelle synthesis method to form a passivation layer of gold on metal and alloy nanoparticles [111]. Iron nanoparticles avoid being oxidised and maintain their magnetic properties (such as coercivity or blocking temperature) by gold coating. The Au colloid shows a red colour, while the Fe–Au colloid displays a black–blue colour. The gold colloid exhibits an absorption band with a maximum at 526 nm while the Fe–Au colloid shows an absorption band with a maximum at 555 nm. Furthermore, the latter is broader than the former. The absorption of the metallic nanoparticle colloid such as Au, Ag etc. is due to the surface plasmon absorption [112]. The red shift and broadening in the surface plasmon absorption of the Fe–Au colloid relative to the pure Au colloid reveals that the size distribution of pure Au nanoparticles is narrower than that of the Fe–Au nanoparticles and the aggregation of Fe–Au nanoparticles is more serious than the pure Au nanoparticles.

6.3.4 Core–shell metal/polymer nanoparticles

Metal(inorganic)–polymer(organic) core–shell nanoparticles have recently gathered a lot of scientific interest due to the possibility to combine different properties of core and shell in one particle [113]. In addition, many interesting technological applications are under development for this kind of materials, for example in analytical chemistry (chromatography), separation technology (ion exchange), catalysis, biochemistry and medicine etc. [114]. Magnetic composite materials comprise a new generation of multifunctional materials that combine the properties of ordinary polymer and magnetic materials (ferri- and/or ferromagnetic particles mixed or embedded in a matrix), that one could call magnetopolymeric materials. Therefore, these magnetic composite materials can be considered as a granular material consisting of small metal grains embedded in an insulating magnetic medium. These studies have been focused on the unique magnetic properties originated by dispersed nanoparticles embedded in magnetic insulating or metallic media [115]. The structural characteristics of metal/non-metal granular materials change, depending on the volume fraction of the metallic phase with respect to the non-metallic phase. Thus, Abeles *et al.* [116] classified granular microstructure based on the volume fraction of the metallic particles, the temperature coefficient of the resistivity (TCR) and the pattern of the microstructure, playing an important role in the oxygen content. When the volume fraction of the metallic particle is low, the metallic particles are surrounded and isolated by the insulating phase. The TCR becomes negative because the conduction of electrons occurs by tunnelling, which is a thermally activated process [117]. Saturation magnetisations higher than those of ferrites are widely used as soft magnetic materials [118]. Consequently, granular soft magnetic materials can show in general both properties, i.e. high resistivity and saturation magnetisation, which can be considered as advantageous features for obtaining a good frequency–permeability response [119]. On the other hand, the discovering of intrinsic conductivity in organic polymeric materials, named intrinsically conducting polymers (ICPs), opens new possibilities for the development of molecular electronics. This increasing interest has led to the combined research efforts in order to develop new generations of multifunctional materials. ICPs have been applied in different fields, e.g. electrostatic charge dissipation, electromagnetic interference shielding, metallisation of printed circuit boards, conductive fabrics and sensors.

Murillo *et al.* have presented a magnetic and microstructural study carried out in a new granular magnetic material consisting of magnetic metal grains of $CoFe_2O_4$ encapsulated in an ICPs matrix [120]. The different particle sizes of $CoFe_2O_4$ were prepared by the microemulsion method. The advantage of chemically synthesised ICPs is mainly the possibility to obtain bulk quantities of magnetopolymeric materials for industrial applications. By combining in a single material the electrical conductivity of ICPs and the magnetic properties of ferrite nanoparticles, new multifunctional materials have been developed. The value of the average size is 14.06 nm obtained at 323 K. Decreasing the reaction temperature from 323 to 303 K causes a critical reduction of the grain size. The average crystal size is 3.53 nm. This large decrease in the nanoparticle size is strongly correlated with the magnetic behaviour at room temperature.

In the covered polypyrrole samples, the amorphous broad peak is related to the amorphous polymeric shell in the nanoparticles. From the main reflection peak {311} the calculation of crystal size is 17.17 nm. The crystal growth can be due to the polymerisation process where the presence of iron oxidants influences the $CoFe_2O_4$ grains obtained in the microemulsion process. Also, the maximum of the 311 reflection peak appears with a small left displacement in the covered PPy nanoparticles that can be related to the really small compositional changes in the 1:2 (Co:Fe) ratio in $Co_{1\pm X}Fe_{2\pm X}O_4$ phase. The high electrical conductivity (ca. 160 S cm^{-1}) of the chemical synthesised PPy is due to electron fluctuations between the double bands and the charge delocalisation. The covered samples were synthesised with a ratio of 0.25/0.75 in $CoFe_2O_4$ to polypyrrole. The decrease in the electrical conductivity exhibited in the magnetopolymeric nanopowder is related to the ferrite's 25% presence in the sample and is due to the semiconducting properties of the ferrite. The magnetisation and coercive field of the sample with covered polypyrrole turned out to be larger than in the sample without polymer material. Such an increase in the magnetisation could be attributed to small stoichiometric changes taking place in the spinel cobalt ferrite nanoparticles or to grain size growth.

There has been extensive work on core/shell nanoparticles where the core is magnetic Fe_3O_4, PbS and the shell is a polymer that provides biocompatibility and long-term stability [121]. PbS particles are formed in a $Pb(AOT)_2$/polymer composite [122], according to whether this has an ordered layer structure or not, nanorods or spherical particles are obtained.

Holzinger and Kickelbick have described a general route for the synthesis of inorganic–organic particles consisting of an amorphous metal oxide core and a polymeric shell via the combination of molecular precursor design, microemulsion approach and surface grafting of polymers [123]. Pentane-2,4-dione was modified with initiating groups for atom transfer radical polymerisation (ATRP). The substituted alkoxides were used as precursors for the sol–gel process in a w/o microemulsion with cyclohexane as the continuous phase, Triton X100 as a non-ionic surfactant and 1-hexanol as a co-surfactant. The size of the obtained amorphous metal oxide particles was strongly influenced by the reaction time. TEM images of the organically surface modified amorphous nanoparticles obtained under the same reaction conditions of titania (diameter 175 nm), vanadium oxide (diameter 25 nm), and yttrium oxide (diameter 100 nm). Diameter images reveal that the growth of the nanoparticles occurred in different ways, for example in the case of titania the final surface-modified particle is formed by an agglomerate of smaller colloids, whereas other metal oxides formed more homogeneous products. This observation can be ascribed to different growing mechanisms in the micelles. The resulting surface-functionalised particles were used as multifunctional initiators for ATRP with methyl methacrylate (MMA) and styrene as monomers. ATRP allows, contrary to conventional free radical polymerisations, the exclusive grafting of the polymer chain from the particle surface. In addition, the polymerisation reaction is usually controlled, which means that the chain length and therefore the polymer layer thickness can be adjusted nicely. ATRP already showed its potential in the surface modification of nanoparticles [124]. The use of the modified nanoparticles as multifunctional initiators also proves that the initiating groups are located at the surface of the colloids because only there a polymerisation can be initiated.

Yuasa *et al.* have communicated that cobalt(II) meso-tetrakis(4-hexadecylamidophenyl) porphyrin (CoTAPP) self-assembles in ethanol/1-propanol 2/1 (v/v) to form a rod-like micelle with nanoscale dimensions [125]. Static light scattering (SLS) and spectroscopic experiments reveal that the nanorod is a face-to-face aggregate having a hydrophobic corona around a polar core and is thus characterised as a reverse micelle. Porphyrins are major building blocks for self-assembled supramolecular systems based on π–π stacking interactions [126]. In water, a variety of porphyrin amphiphiles are known to self-organise into micellar fibres [127], while in organic solutions, examples are not as frequent as those involving zinc(II) porphyrins [128]. By analogy to reverse micelles, which provide a stable dispersion of water in non-polar solvents [129], Yuasa *et al.* have anticipated that porphyrins bearing polar amido groups with which peripheral long alkyl chains are linked would self-organise even in non-aqueous media, leading to long-lived micelles stabilised through hydrogen bonding and hydrophobic interactions [125]. The length of micellar neutral nanorods is an exponential function of the end-cap energy/thermal energy ratio [128] and is usually large even at low concentrations.

PB (Prussian Blue) and related cyanometalate-based coordination polymers offer a range of compounds that exhibit unique versatility [130, 131]. PB is an important component in the study of molecular magnets because compounds with appropriate magnetic properties require further fabrication and processing if functional devices and materials are to be produced. Many attempts to synthesise PB analogue nanoparticles have recently emerged, making this a promising topic for nanomagnetic device applications [132]. Mann *et al.* first demonstrated this potential by confirming that hydrophobic PB nanoparticles with a uniform shape and size could be routinely prepared in a synthesis involving nanoscale water droplets formed in w/o microemulsions prepared from the anionic surfactant sodium bis(2-ethylhexyl) sulfosuccinate (AOT) [133]. Furthermore, Kitagawa *et al.* prepared highly dispersed PB nanoparticles that were controlled by organic polymers [134]. They could control the size of the PB nanoparticles, and they studied the size-dependence effect of the magnetic properties. Photofunctional or photoresponsive nanoparticles of PB have been developed by Taguchi *et al.* [135].

The preparation of Au and Pd nanoparticles stabilised by poly(*N*-vinyl-2-pyrrolidone) (PVP, $(C_6H_9ON)_n$) was reported [136]. Rapid injection of an aqueous solution of $NaBH_4$ into an aqueous micellar solution of the $AuCl_4^-$/PVP complexes at ca. 273 K yields the brownish Au:PVP nanoparticles with an average diameter of 1.3 nm [137]. The Au:PVP nanoparticles are allowed to grow in size by reducing $AuCl_4^-$ with Na_2SO_3, leading to the formation of reddish Au:PVP with $d_{av} = 9.5$ nm. Reduction of $PdCl_4^{2-}$/PVP by $NaBH_4$ and ethanol [138] gives the Pd:PVP nanoparticles with $D_{av} = 1.5$ nm (Pd:PVP) and 2.2 nm (Pd:PVP), respectively.

A capping agent is usually used in the synthesis of nanometre-scale particles to both control the growth of metal particles in the light of particle size and shape [139] and impart useful chemical behaviour to the final nanoscale product [140]. For example, the composite material formed by metal nanoparticles stabilised by a capping polymer show interesting physical properties and high activity in homogeneous catalytic reactions as well [141]. Hirai and co-workers reported the immobilisation of ultrafine rhodium particles on a polyacrylamide gel by forming an amide bond between the primary amino group of the support and the methyl acrylate residue in the protective polymer [142]. Liu and co-workers investigated the capture of colloidal metal particles on the surface of functionalised silica by ligand coordination [143].

6.4 Outlook

Research into nanotechnology and nanomaterials has exploded over the past several years carrying with it new ideas in both processing and utilisation of nanostructured materials for a magnitude of applications ranging from common uses to advanced technologies over all scientific and commercial fields. Numerous preparation methods for nanoscaled materials, particularly particles, have been established and documented. Nanotechnology has been receiving increased attention due to its extensive applications in the field of catalysis, electronics, high-density magnetic recording media, sensors and biotechnology. The challenge has been and remains the control of the size, the size distribution and the shape of nanoparticles. The properties of nanoscale particles have been attracting a great deal of attention because of the ways in which they differ from the atomic, molecular and bulk properties of those same materials [144, 145]. The reason for this is that nanostructured particles and materials, and the physical or chemical combination of substances at the nanometre or subnanometre scale, can lead to innovative materials with improved or even unexpected properties. However, progress in these fields will largely depend on the pace of advance of the fundamental research on nanostructured particles and materials in solid-state chemistry, solid-state physics, materials science and colloid chemistry.

Recent scientific literature demonstrates a growing interest in new methods of nanoparticle synthesis, driven primarily by an ever-increasing awareness of the unique properties and technological importance of nanostructured materials. The fabrication of nanoparticles within reverse microemulsions [40, 146] has been shown to be a convenient route to monodisperse particles of controllable size. A recognised goal of these synthetic approaches is to achieve control over the composition, size, surface species, solubility, stability, isolability and other functional properties of the nanostructures. The combination of reverse microemulsion and microwave heating has the added advantage that the oil phase in the reverse microemulsion system is transparent to microwave so that the aqueous domains are heated directly, selectively and rapidly.

Magnetic nanoparticles are of interest for a wide variety of applications: for technology, as magnetic seals, for printing, for recording [98] and for biology, as magnetic resonance imaging (MRI) agents [147, 148], and for cell tagging and sorting [149]. They have exciting potential applications in biological diagnosis [99]. Although these magnetic particles show promise for practical applications such as catalysis, magnetic recording, magnetic fluids and biomedical applications, their utility has been limited due to uncontrolled oxidation. The unique physical properties of nanoscale magnetic materials such as superparamagnetism have generated considerable interest for their use in a wide range of diverse applications from data information storage to *in vivo* magnetic manipulation in biomedical systems [70].

The ability to control the size and monodispersity in synthesising and assembling metal-coated magnetic nanoparticles is important for exploring technological applications of the nanoscale core, shell or their combinations. It is increasingly important for many applications involving magnetic nanoparticles, such as magnetic resonance imaging for medical diagnosis, high-density magnetic recording, controlled drug delivery, biological targeting or separation and catalysis [150, 151]. They have exciting potential applications in biological diagnosis [99]. While the synthesis of monolayer-capped iron oxide nanoparticles has been extensively studied [152], the synthesis of metal-coated iron oxide nanoparticles with controllable sizes and monodispersities is relatively limited.

Of special interest are core-/shell-structured metal nanoparticles that could exhibit enhanced properties and new functionality, due to the close proximity of the two functionally different components. Such structures not only are ideal for studying proximity effects but are also suitable for structure stabilisation as the shell layer protects the core from oxidation and corrosion. Additionally, the shell layer provides a platform for surface modification and functionalisation, such as coupling the magnetic core through the shell onto organic or other surfaces, thus tuning their intrinsic magnetic properties and making them potentially biocompatible [97]. Such core–shell nanostructures could find applications that explore the electronic, magnetic, catalytic, sensing and chemical or biological properties of the nanocomposite materials. To explore the magnetic properties, the formation of a gold shell with a controllable assembly allows better stability and tenability for the construction of ordered arrays [70]. While these prior studies have shown viabilities of synthesising and assembling core–shell types of nanoparticles, the precise control of the continuous nature of the metal coating, the coating thickness, the size monodispersity and the thin film assembly of the nanoparticles remain to be some of the major challenges.

Inorganic–organic core–shell nanoparticles have recently gathered a lot of scientific interest due to the possibility to combine different properties of core and shell in one particle [113]. In addition, many interesting technological applications are under development for this kind of materials, for example in analytical chemistry (chromatography), separation technology (ion exchange), catalysis, biochemistry and medicine etc. [114]. Core–shell systems with silica cores are well-established materials, but appropriate general routes for such morphologies with transition metal oxide cores are still rare. Many technological applications require magnetic nanoparticles to be embedded in a non-magnetic matrix. Over the past few years, increased attention has been focused on the preparation of various nanostructures with magnetic nanoparticulate components and on understanding the magnetic behaviour of nanoparticles due to new possible surface, inter-particle and exchange interactions in magnetic/non-magnetic matrix [113].

Anisotropic nanoparticles are of considerable current interest, due to various shape-dependant properties [90, 91]. Although synthetic methods for nanorods are well established [153], the metallic systems have very limited success due to their highly symmetric cubic crystal structure. Controlling the length-to-width ratio and the uniformity of the length and width distributions of metal nanoparticles, as well as synthesising particles with a width dimension below 10 nm is challenging with large-scale bench top methods [92].

In particular, there has been great interest in designing novel compounds whose magnetic properties can be controlled by photoillumination [154]. Photocontrollable magnetic materials are important in the development of photonic devices, such as erasable optical memory media and optical switching components. However, the number of optically switchable molecular solids that have been reported is quite small, since an appropriate strategy for achieving photo-induced switching in a solid-state system has yet to be clarified.

The outlook of magnetic nanoparticles for their use in biology is promising. One area of special interest is the development of strategies able to increase the circulation time of magnetic nanoparticles in the blood. Integration of magnetic nanoparticles in stealth liposomes [155] or artificial hollow capsules [156] seems a promising approach. Another area of recent interest is the development of nanoattractors able to concentrate magnetic nanoparticles in a desired region. Genetic engineering has brought new challenges and opportunities for medicine and biomedical research and development [157].

In the area exploring applications in biology and medicine, the use of magnetic nanoparticles and a magnetic filed *in vivo* or *in vitro* can either remotely position or selectively filter biological materials [158]. In magnetic resonance imaging, the presence of the particles at a given site can alter the contrast of certain types of cells by several orders of magnitude. The possibility of using magnetic nanoparticles to improve the effectiveness of cell manipulation and DNA sequencing could also aid to the development of pharmaceuticals, drug delivery systems and magnetic separation technologies for rapid DNA sequencing. These applications should benefit a great deal from the ability to control the surface and inter-particle spatial properties of the magnetic nanoparticles.

Acknowledgements

SPM acknowledges with thanks the Indian National Science Academy for an Honorary Scientist position during the tenure of which the article was written. AKR is very thankful to All India Council for Technical Education, New Delhi, for an Emeritus Fellowship. He also acknowledges the authorities of West Bengal University of Technology, Kolkata, for their cooperation. Thanks are due to Ms Debolina Mitra, CSS, JU, in the preparation of figures for the chapter. Thanks are also due to various copyright holders for permission to use the figures. This work is also supported by the Slovak Grand Agency (VEGA) through the grant number 2/7013/27 and Science and Technology Assistance Agency through the APVT projects (20-0173 and 0174-06).

References

1. Pileni, M.P. (ed) (1989) *Reactivity in Reverse Micelles*. Elsevier, New York.
2. Pileni, M.P., Brochette, P., Hickel, B. and Lerebours, B. (1984) Hydrated electrons in reverse micelles: 2. Quenching of hydrated electron by sodium nitrate. *J. Colloid Interface Sci.*, **98**, 549.

3. Marciano, V., Minore, A. and Liveri, V.T. (2000) A simple method to prepare solid nanoparticles of water soluble salts using water-in-oil microemulsions. *Colloid Polym. Sci.*, **278**, 250–252.
4. Eastoe, J., Robinson, B.H., Visser, A.J.W.G. and Steytler, D.C. (1991) Rotational dynamics of AOT reversed micelles in near critical and supercritical alkanes. *J. Chem. Soc. Faraday Trans.*, **87**, 1899–1903.
5. Fletcher, P.D.I., Howe, A.M. and Robinson, B.H. (1987) The kinetics of solubilizate exchange between water droplets of a water-in-oil microemulsion. *J. Chem. Soc. Faraday Trans.*, **1**, **83**, 985–1005.
6. Paul, S. and Moulik, S.P. (1995) Physicochemical studies on microemulsions. *Tenside Surfactants Detergents*, **6**, 32–38.
7. Moulik, S.P., De, G.C., Panda, A.K., Bhowmik, B.B. and Das, A.R. (1999) Dispersed molecular aggregates. 1. Synthesis and characterization of $Cu_2[Fe(CN)_6]$ in H_2O/AOT/n-heptane water-in-oil microemulsion media. *Langmuir*, **15**, 8361–8367.
8. Panda, A.K., Bhowmik, B.B., Das, A.R. and Moulik, S.P. (2001) Dispersed molecular aggregates. 3. Synthesis and characterization of colloidal lead chromate in water/sodium bis(2-ethyl hexyl) sulfosuccinate/n-heptane water-in-oil microemulsion medium. *Langmuir*, **17**, 1811–1816.
9. Qiu, S., Dong, J. and Chen, G. (1999) Preparation of Cu nanoparticles from water-in-oil microemulsions. *J. Colloid Interface Sci.*, **216**, 230–234.
10. Akinc, M. and Celikkaya, A. (1987) Advances in ceramics. In G.L. Messing, K.S. Mazdiyasni, J.W. McCauley and R.A. Haber (eds), *Ceramic Powder Science*, Vol. 21. American Ceramic Society, Westerville, OH, pp. 57–67.
11. Lu, C.H. and Yeh, C.H. (1997) Emulsion precipitation of submicron zinc oxide powder. *Mater. Lett.*, **33**, 129–132.
12. Dogra, A. and Rakshit A.K. (2004) Phase behavior and percolation studies on microemulsion system water/SDS + Myrj 45/cyclohexane in the presence of various alcohols as cosurfactants. *J. Phys. Chem. B*, **108**, 10053–10061.
13. Ajith, S., John, A.C. and Rakshit, A.K. (1994) Physicochemical studies of microemulsions. *Pure Appl. Chem.*, **66**, 509–514.
14. John, A.C. and Rakshit, A.K. (1993) Formation of microemulsion: Effect of temperature and polyacrylamide. *J. Colloid Interface Sci.*, **156**, 202–206.
15. Talmon, Y. and Prager, S. (1978) Statistical thermodynamics of phase equilibria in microemulsions. *J. Chem. Phys.*, **69**, 2984–2991.
16. Talmon, Y. and Prager, S. (1982) The statistical thermodynamics of microemulsions. II. The interfacial region. *J. Chem. Phys.*, **76**, 1535–1538.
17. de Gennes, P.G. and Taupin, C. (1982) Microemulsions and the flexibilities of oil/water interfaces. *J. Phys. Chem.*, **86**, 2294–2304.
18. Jada, A., Lang, J. and Zana, R. (1989) Relation between electrical percolation and rate constant for exchange of materials between droplets in water-in-oil microemulsions. *J. Phys. Chem.*, **93**, 10–12.
19. Lang, J., Lalem, N. and Zana. R. (1992) Droplet size and dynamics in water-in-oil microemulsions. *Colloids Surf.*, **68**, 199–206.
20. Moulik, S.P. and Paul, B.K. (1998) Structure, dynamics and transport properties of microemulsions. *Adv. Colloid Interface Sci.*, **78**, 99–195, and references therein.
21. Bongolo, G. (2003) The use of surface active agents in the preparation and assembly of quantum sized nanoparticles. *Adv. Colloid Interface Sci.*, **106**, 169–181.
22. Eastoe J., Hollamby, M.J. and Hudson, L. (2006), Recent advances in nanoparticle synthesis with reversed micelles. *Adv. Colloid Interface Sci.*, **128–130**, 5–15.
23. Moulik, S.P. and Chakraborty, I. (2006) Nanomaterials: A modern perspective. *J. Ind. Chem. Soc.*, **83**, 745–761.

24. Ganguli, D. and Ganguli, M. (2003) *Inorganic Particle Synthesis via Macro- and Microemulsions. A Micrometer to Nanometer Landscape.* Kluwer Academic/Plenum Publishers, New York.
25. Bowen Katari, J.E., Colvin, V.L. and Alivisatos, A.P. (1994) X-ray photoelectron spectroscopy of CdSe nanocrystals with applications to studies of the nanocrystal surface. *J. Phys. Chem.*, **98**, 4109–4117.
26. Peng, X., Wickham, J. and Alivisatos, A.P. (1998) Kinetics of II–VI and III–V colloidal semiconductor nanocrystal growth: "Focusing" of size distributions. *J. Am. Chem. Soc.*, **120**, 5343–5344.
27. Moulik, S.P. and Rakshit, A.K. (2006) Physicochemistry and applications of microemulsions. *J. Surf. Sci. Technol.*, **22**, 159–186.
28. Tojo, C., Blanco, M.C., Rivadulla, F. and Lopez-Quintela, M.A. (1997) Kinetics of the formation of particles in microemulsions. *Langmuir*, **13**, 1970–1977.
29. Biswas, S., Hait, S.K., Bhattacharya, S.C. and Moulik, S.P. (2004) Synthesis of Nanoparticles of CuI, $CuCrO_4$, and CuS in water/AOT/cyclohexanone and water/TritonX 100 + *i*-propanol/cyclohexanone reverse microemulsions. *J. Disp. Sci. Technol.*, **25**, 801–816.
30. Chang, C.L. and Fogler, H.S. (1997) Controlled formation of silica particles from tetraethyl orthosilicate in nonionic water-in-oil microemulsions. *Langmuir*, **13**, 3295–3307.
31. Robinson, B.H., Khan-Lodhi, A.N. and Towey, T. (1989) Structure and Reactivity. In M.P. Pileni (ed), *Reverse Micelles.* Elsevier, Amsterdam, pp. 198–220.
32. Ganguli, D. and Ganguli, M. (2003) Particle formation from microemulsions: A survey. In M.P. Pileni (ed), *Inorganic Particle Synthesis via Macro- and Microemulsions. A Micrometer to Nanometer Landscape.* Kluwer Academic/Plenum Publishers, New York, p. 97.
33. Pileni, M.P. (1997) Nanosized particles made in colloidal assemblies. *Langmuir*, **13**, 3266–3276.
34. Tauc, J. and Menth, A. (1972) States in the gap. *J. Non-cryst. Solids*, **8–10**, 569–585.
35. Thielsch, R., Bohme, T., Reiche, R., Schlafer, D., Bauer, H.-D. and Bottcher, H. (1998) Quantum size effects of PbS nanocrystallites in evaporated composite films. *Nanostructured Mater.*, **10**, 131–149.
36. Haram, S.K., Mahadeshwar, A.R. and Dixit, S.G. (1996) Synthesis and characterization of copper sulfide nanoparticles in Triton X100 water-in-oil microemulsions. *J. Phys. Chem.*, **100**, 5868–5873.
37. Chakraborty, I. and Moulik, S.P. (2004) Preparation and characterization of nanoscale semiconductor particles of ZnS, CdS and $PbCrO_4$ in polymer–surfactant gel matrix. *J. Disp. Sci. Technol.*, **25**, 849–859.
38. Zhao, X.K. and Fendler, J.H. (1991) Semiconductor particulate films on solid supports. *Chem. Mater.*, **3**, 168–174.
39. Curri, M.L., Agostiano, A., Manna, L., Monica, M.D., Catalano, M., Chiavarone, L., Spagnolo, V. and Lugara, M. (2000) Synthesis and characterization of CdS nanoclusters in a quarternary microemulsion: The role of cosurfactant. *J. Phys. Chem. B*, **104**, 8391–8397.
40. Dutta, P and Fendler, J.H. (2002) Preparation of CdS nanoparticles in self reproducing reversed micelles. *J. Colloid Interface Sci.*, **247**, 47–53.
41. Henglein, A. (1989) Small particle research: Physicochemical properties of extremely small colloidal metal and semiconductor particles. *Chem. Rev.*, **89**, 1851–1873.
42. Pinna, N., Weiss, K., Sack-Kongehl, H., Vogel, W., Urban, J. and Pileni, M.P. (2001) Triangular CdS nanocrystals: Synthesis, characterization, and stability. *Langmuir*, **17**, 7982–7987.
43. Tong, Q., Kosmella, S. and Koetz, J. (2006) Formation of rodlike CdS nanoparticles in SDS/decanol based multilamellar Vesicles. *Progr. Colloid Polym. Sci.*, **133**, 152–158.
44. Xu, J. and Li, Y. (2003) Formation of ZnS nanorods and nanoparticles in ternary w/o microemulsions. *J. Colloid Interface Sci.*, **259**, 275–281.
45. Ward, A.J.I., O'Sullivan, E.C., Rang, J.C., Nedeljkovic, J. and Patel, R.C. (1993) The synthesis of quantum size lead sulfide particles in surfactant based complex fluid media. *J. Colloid Interface Sci.*, **161**, 316–320.

46. Eastoe, J. and Cox, A.R. (1995) Formation of PbS nanoclusters using reversed micelles of lead and sodium Aerosol OT. *Colloids Surf. A*, **101**, 63–76.
47. Li, M. and Mann, S. (2000) Emergence of morphological complexity in $BaSO_4$ fibers synthesized in AOT Microemulsions. *Langmuir*, **16**, 7088–7094.
48. Rees, G.D., Evans-Gowing, R., Hammond, S.J. and Robinson, B.H. (1999) Formation and morphology of calcium sulfate nanoparticles and nanowires in water in oil microemulsions. *Langmuir*, **15**, 1993–2002.
49. Ivanova, N.I., Rudelev, D.S., Summ, B.D. and Chalykh, A.A. (2001) Barium sulfate nanoparticles in water-in-oil microemulsion systems. *Colloid J.*, **63**, 714–717.
50. Xiang, J-H., Yu, S.H., Geng, X., Liu, B.H. and Xu, Y. (2005) Growth of $PbSO_4$ single crystals nanorods and optical properties by a microemulsion approach. *Crystal Growth Design*, **5**, 1157–1161.
51. Cao, M., He, X., Chen, J. and Hu, C. (2007) Self assembled nickel hydroxide three dimensional nanostructures: A nanomaterial for alkaline rechargeable batteries. *Crystal Growth Design*, **7**, 170–174.
52. Matson, D.W., Fulton, J.L. and Smith, R.D. (1987) Formation of fine particles in supercritical fluid micelle systems. *Mater. Lett.*, **6**, 31–33.
53. Nanni, A. and Dei, L. (2003) $Ca(OH)_2$ nanoparticles from w/o microemulsions. *Langmuir*, **19**, 933–938.
54. Chen, Q., Shen, X. and Gao, H. (2007) Formation of nanoparticles in water-in-oil microemulsions controlled by the yield of hydrated electron: The controlled reduction of Cu^{2+}. *J. Colloid Interface Sci.*, **308**, 491–499.
55. Chen, D. and Gao, L. (2004) Novel synthesis of well dispersed crystalline SnO_2 nanoparticles by water-in-oil microemulsion assisted hydrothermal process. *J. Colloid Interface Sci.*, **279**, 137–142.
56. Zhan, Z., Song, W. and Jiang, D. (2004) Preparation of nanometer sized In_2O_3 particles by a reverse microemulsion Method. *J. Colloid Interface Sci.*, **271**, 366–371.
57. Giannakas, A.E., Vaimakis, T.C., Ladavos, A.K., Trikalitis, P.N. and Pomonis, P.J. (2003) Variation of surface properties and textural features of spinel $ZnAl_2O_4$ and perovskite $LaMnO_3$ nanoparticles prepared via CTAB-butanol-octane-nitrate salt microemulsions in the reverse and bicontinuous states. *J. Colloid Interface Sci.*, **259**, 244–253.
58. Nad, S., Sharma, P., Roy, I. and Maitra, A. (2003) Anomalous nanostructured titanium dioxide. *J. Colloid Interface Sci.*, **264**, 89–94.
59. Wu, M., Long, J., Huang, A., Luo, Y., Shouhua, F. and Xu, R. (1999) Microemulsion mediated hydrothermal synthesis and characterization of nanosize rutile and anatase particles. *Langmuir*, **15**, 8822–8825.
60. Gobe, M., Kon-No, K., Kandori, K. and Kitahara, A. (1983) Preparation and characterization of monodisperse magnetite sols in w/o microemulsion. *J. Colloid Interface Sci.*, **93**, 293–295.
61. Liz, L., Lopez-Quintela, M.A., Mira, J. and Rivas, J. (1994) Preparation of colloidal Fe_3O_4 ultrafine particles in microemulsions. *J. Mater. Sci.*, **29**, 3797–3801.
62. Grasset, F., Marchand, R., Marie, A.-M., Fauchadour, F. and Fajardie, F. (2006) Synthesis of CeO_2–SiO_2 core–shell nanoparticles by water-in-oil microemulsion. Preparation of functional thin film. *J. Colloid Interface Sci.*, **299**, 726–732.
63. Chakraborty, I., Mitra, D. and Moulik, S.P. (2005) Spectroscopic studies on nanodispersions of CdS, HgS, their core-shells and composites prepared in micellar medium. *J. Nanoparticle Res.*, **7**, 227–236.
64. Chen, D., Shen, G., Tang, K., Liang, Z. and Zheng, H. (2004) AOT-microemulsions-based formation and evolution of $PbWO_4$ crystals. *J. Phys. Chem. B*, **108**, 11280–11284.

65. Panda, A.K., Moulik, S.P., Bhowmik, B.B. and Das, A.R. (2001) Dispersed molecular aggregates II. Synthesis and characterization of nanoparticles of tungstic acid in $H_2O/(TX100 + \text{alkanol})/n$-heptane w/o microemulsion media. *J. Colloid Interface Sci.*, **235**, 218–226.
66. Kimijima, K. and Sugimoto, T. (2005) Effects of water content on the growth rate of AgCl nanoparticles in a reversed micellar system. *J. Colloid Interface Sci.*, **286**, 520–525.
67. Husein, M.M., Rodil, E. and Vera, J.H. (2005) A novel method for the preparation of silver chloride nanoparticles starting from their solid powder using microemulsion. *J. Colloid Interface Sci.*, **288**, 457–467.
68. Husein, M., Rodil, E. and Vera, J.H. (2004) Formation of silver bromide precipitate of nanoparticles in a single microemulsion utilizing the surfactant counterion. *J. Colloid Interface Sci.*, **273**, 426–434.
69. Service, R.F. (1996) Small clusters hit the big time. *Science*, **271**, 920–922.
70. Sun, S., Murray, C.B., Weller, D., Folks, L. and Moser, A. (2000) Monodisperse FePt nanoparticles and ferromagnetic FePt nanocrystal superlattices. *Science*, **287**, 1989–1992.
71. Feldmann, C. (2003) Polyol-mediated synthesis of nanoscale functional. *Adv. Funct. Mater.*, **13**, 101–107.
72. Siegel, R.W. and Ramasamy, S. (1998) Synthesis characterization and properties of nanophase TiO_2. *J. Mater. Rev.*, **3**, 1367–1372.
73. Fegley, B.J., White, P. and Bowen, H.K. (1985). Processing and characterization of ZrO_2 and Y-doped ZrO_2 powders. *Am. Ceram Soc. Bull.*, **64**, 1115–1120.
74. Fayet, P. and Woste, L. (1986) Experiments on size-selected metal cluster ions in a triple quadrupole arrangement. *Zeitschrift für Physik D, Atoms, Molecules and Clusters*, **3**, 177–182.
75. Tang, Z.X., Sorensen, C.M., Klabunde, K.J. and Hadjipanayis, G.C. (1991) Preparation of manganese ferrite fine particles from aqueous solution. *J. Colloid Interface Sci.*, **146**, 38–52.
76. Dumestre, F., Chaudret, B., Amiens, C., Fromen, M.C., Casanove, M.J., Renaud, P. and Zurcher, P. (2002) Shape control of thermodynamically stable cobalt nanorods through organometallic chemistry. *Angew. Chem.*, **114**, 4462
77. Pileni, M.P. (2003) The role of soft colloidal templates in controlling the size and shape of inorganic nanocrystals. *Nat. Mater.*, **2**, 145–150.
78. McLeod, M.C., McHenry, R.S., Beckman, E.J. and Roberts, C.B. (2003) Synthesis and stabilization of silver metallic nanoparticles and premetallic intermediates in perfluoropolyether/CO_2 reverse micelle systems. *J. Phys. Chem.* B, **107**, 2693–2700.
79. Capek, I. (2006) *Nanocomposite Structures and Dispersions*. Elsevier, London.
80. Misra, R.D.K., Gubbala, S., Kale, A. and Egelhoff, W.F. (2004) A comparison of the magnetic characteristics of nanocrystalline nickel, zinc, and manganese ferrites synthesized by reverse micelle technique. *J. Mater. Sci. Eng.*, **B111**, 164–174.
81. Eicke, H.F. and Zinsli, P.E. (1978) Nanosecond spectroscopic investigations of molecular processes in W/O microemulsions. *J. Colloid. Interface Sci.*, **65**, 131–140.
82. Jain, T.K., Cassin, G., Badiali, J.P. and Pileni, M.P. (1996) Relation between exchange process and structure of AOT reverse micellar system. *Langmuir*, **12**, 2408–2411.
83. Murphy, C.J. and Jana, N.R. (2002) Controlling the aspect ratio of inorganic nanorods and nanowires. *Adv. Mater.*, **14**, 80–82.
84. Jana, N.R. and Peng, X. (2003) Single-phase and gram-scale routes toward nearly monodisperse Au and other noble metal nanocrystals. *J. Am. Chem. Soc.*, **125**, 14280–14281.
85. Roucoux, A., Schultz, J. and Patin, H. (2002) Reduced transition metal colloids: A novel family of reusable catalysts? *Chem. Rev.*, **102**, 3757–3778.
86. Jana, N.R. (2004) Shape effect in nanoparticle self-assembly. *Angew. Chem.*, **116**, 1562–1566.
87. Jana, N.R. and Pal, T. (2001) Acid–base equilibria in aqueous micellar medium: A comprehensive review. *J. Surf. Sci. Technol.*, **17**, 191–212.

88. Toernblom, M. and Henriksson, U. (1997) Effect of solubilization of aliphatic hydrocarbons on size and shape of rodlike C16TABR micelles studied by 2H NMR relaxation. *J. Phys. Chem. B*, **101**, 6028–6035.
89. Lin, Z., Cai, J.J., Scriven, L.E. and Davis, H.T. (1994) Spherical-to-wormlike micelle transition in CTAB solutions. *J. Phys. Chem.*, **98**, 5984–5993.
90. Sonnichsen, C. and Alivisatos, A.P. (2005) Gold nanorods as novel nonbleaching plasmon-based orientation sensors for polarized single-particle microscopy. *Nano Lett.*, **5**, 301–304.
91. Sun, Y. and Xia, Y. (2002) Shape-controlled synthesis of gold and silver nanoparticles. *Science*, **298**, 2176–2179.
92. Jana, N.R. (2003) Nanorod shape separation using surfactant assisted self-assembly. *Chem. Commun.*, **15**, 1950–1951.
93. Jana, N.R. (2005). Gram-scale synthesis of soluble, near-monodisperse gold nanorods and other anisotropic nanoparticles. *Small*, **1**, 875–882.
94. Filankembo, A., Giorgio, S., Lisiecki, I. and Pileni, M.P. (2003) Is the anion the major parameter in the shape control of nanocrystals? *J. Phys. Chem. B*, **107**, 7492–7500.
95. Lisiecki, I. (2005) Size, shape and structural control of metallic nanocrystals. *J. Phys. Chem. B*, **109**, 12231–12244.
96. Jin, R., Cao, Y.W., Mirkin, C.A., Kelly, K.L., Schatz, G.C. and Zheng, J.G. (2001) Photo-induced conversion of silver nanospheres to nanoprisms. *Science*, **294**, 1901–1903.
97. Gerion, D., Pinaud, F., Williams, S.C., Parak, W.J., Zanchet, D., Weiss, S. and Alivisatos, A.P. (2001) Synthesis and properties of biocompatible water-soluble silica-coated CdSe/ZnS semiconductor quantum dots. *J. Phys. Chem. B*, **105**, 8861–8871.
98. Martin, J.I., Nogues, J., Liu, K., Vicent, J.L. and Schuller, I.K. (2003) Order magnetic nanostructures: Fabrication and properties. *J. Magn. Magn. Mater.*, **256**, 449–501.
99. Roos, C. (2001) Patterned magnetic recording media. *Annu. Rev. Mater. Res.*, **31**, 203–235.
100. Cho, S.J., Kauzlaricha, S.M., Olamit, J., Liu, K., Grandjean, F., Rebbouh, L. and Long, G.J. (2004) Characterization and magnetic properties of core/shell structured Fe/Au nanoparticles. *J. Appl. Phys.*, **95**, 6804–6806.
101. Carpenter, E.E., Calvin, S., Stroud, R.M. and Harris, V.G. (2003) Passivated iron as core-shell nanoparticles. *Chem. Mater.*, **15**, 3245–3246.
102. Lyon, J.L., Fleming, D.A., Stone, M.B., Schiffer, P. and Williams, M.E. (2004) Synthesis of Fe oxide core/Au shell nanoparticles by iterative hydroxylamine seeding. *Nano Lett.*, **4**, 719–723.
103. Poddar, P., Fried, T. and Marlovich, G. (2002) First-order metal-insulator transition and spin-polarized tunneling in Fe_3O_4 nanocrystals. *Phys. Rev. B*, **65**, 172405–172409.
104. Carpenter, E.E., Sangregorio, C. and O'Connor, C.J. (1999) Effects of shell thickness on blocking temperature of nanocomposites of metal particles with gold shells. *IEEE Trans. Magn.*, **35**, 3496–3498.
105. Ravel, B., Carpenter, E.E. and Harris, V.G. (2002) Oxidation of iron in iron/gold core/shell nanoparticles. *J. Appl. Phys.*, **91**, 8195–8197.
106. Kinoshita, T., Seino, S., Okitsu, K., Nakayama, T., Nakagawa, T. and Yamamoto, T.A. (2003) Magnetic evaluation of nanostructure of gold–iron composite particles synthesized by a reverse micelle method. *J. Alloys Compd.*, **359**, 46–50.
107. Cho, S.J., Idrobo, J.C., Olamit, J., Liu, K., Browning, N.D. and Kauzlarich, S.M. (2005) Growth mechanisms and oxidation resistance of gold-coated iron nanoparticles. *Chem. Mater.*, **17**, 3181–3186.
108. Mandal, M., Kundu, S., Ghosh, S.K., Panigrahi, S., Sau, T.K., Yusuf, S.M. and Pal, T. (2005) Magnetite nanoparticles with tunable gold or silver shell. *J. Colloid Interface Sci.*, **286**, 187–194.
109. Wang, L., Luo, J., Fan, Q., Suzuki, M., Suzuki, I.S., Engelhard, M.H., Lin, Y., Kim, N., Wang, J.Q. and Zhong, C.J. (2005) Monodispersed core–shell Fe_3O_4–Au nanoparticles. *J. Phys. Chem. B*, **109**, 21593–21601.

110. Lin, J., Zhou, W., Kumbhar, A., Wiemann, J., Fang, J., Carpenter, E.E. and O'Connor, C.J. (2001) Gold-coated iron (Fe–Au) nanoparticles: Synthesis, characterization, and magnetic field-induced self-assembly. *J. Solid State Chem.*, **159**, 26–31.
111. Carpenter, E.E., Seip, C.T. and O'Connor, C.J. (1999) Magnetism of nanophase metal and metal alloy particles formed in ordered phases. *J. Appl. Phys.*, **85**, 5184–5186.
112. Taleb, A., Petit, C. and Pileni, M.P. (1997) Synthesis of highly monodisperse silver nanoparticles from AOT reverse micelles: A way to 2D and 3D self-organization. *Chem. Mater.*, **9**, 950–959.
113. Caruso, F. (2001) Nanoengineering of particle surfaces. *Adv. Mater.*, **13**, 11–22.
114. Bourgeat-Lami, E. (2002) Organic–inorganic nanostructured colloids. *J. Nanosci. Nanotech.*, **2**, 1–24.
115. Chien, C.L. (1991) Granular magnetic solids (invited). *J. Appl. Phys.*, **69**, 5267–5272.
116. Abeles, B., Sheng, P., Coutts, M.D. and Arie, Y. (1975) Structural and electrical properties of granular metal films. *Adv. Phys.*, **24**, 407–461.
117. Barzilai, S, Goldstein, Y, Balberg, I. and Helman, J.S. (1981) Magnetic and transport properties of granular cobalt films. *Phys. Rev. B*, **23**, 1809–1817.
118. Fujimori, H. (1995) Structure and soft magnetic properties in multilayers and granular thin films. *Scr. Metall. Mater.*, **33**, 1625–1636.
119. Inturi, V.R. and Barnard, J.A. (1997) Studies of laminated and single layer high moment soft FeTaN films. *J. Appl. Phys.*, **81**, 4504–4506.
120. Murillo, N., Ochoteco, E., Alesanco, Y.J., Pomposo, J.A., Rodriguez, J., González, J., del Val, J.J., González, J.M., Britel, M.R., Varela-Feria, F.M. and de Arellano-López, A.R. (2004) $CoFe_2O_4$–polypyrrole (PPy) nanocomposites: New multifunctional materials. *Nanotechnology*, **15**, S322–S327.
121. Novakova, A.A., Lanchinskaya, V.Y., Volkov, A.V., Gendler, T.S., Kiseleva, T.Y., Moskvina, M.A. and Zezin, S.B. (2003) Magnetic properties of polymer nanocomposites containing iron oxide nanoparticles. *J. Magn. Magn. Mater.*, **258–259**, 354–357.
122. Wang, S. and Yang, S. (2000) Preparation and characterization of oriented PbS crystalline nanorods in polymer films. *Langmuir*, **16**, 389–397.
123. Holzinger, D. and Kickelbick, G. (2003) Preparation of amorphous metal-oxide-core polymer-shell nanoparticles via a microemulsion-based sol-gel approach. *Chem. Mater.*, **15**, 4944–4948.
124. Pyun, J., Matyjaszewski, K., Kowalewski, T., Savin, D., Patterson, G., Kickelbick, G. and Hüsing, N. (2001) Synthesis of well-defined block copolymers tethered to polysilsesquioxane nanoparticles and their nanoscale morphology on surfaces. *J. Am. Chem. Soc.*, **123**, 9445–9446.
125. Yuasa, M., Oyaizu, K., Yamaguchi, A. and Kuwakado, M. (2004) Micellar cobaltporphyrin nanorods in alcohols. *J. Am. Chem. Soc.*, **126**, 11128–11129.
126. Uno, H., Masumoto, A. and Ono, N. (2003) Hexagonal columnar porphyrin assembly by unique trimeric complexation of a porphyrin dimer with π–π stacking. *J. Am. Chem. Soc.*, **6**, 12082–12083.
127. Komatsu, T., Yanagimoto, T., Furubayashi, Y., Wu, J. and Tsuchida, E. (1999) Self-assembled fibers made of lipidporphyrinato-zinc(II) and -iron(II) complexes with an intramolecular coordinated axial imidazole, *Langmuir*, **15**, 4427–4433.
128. Terech, P., Scherer, C., Lindner, P. and Ramasseul, R. (2003) Metalloporphyrin rods in hydrocarbons: Highly shear responsive suspensions. *Langmuir*, **19**, 10648–10653.
129. Aoudia, M. and Rodgers, M.A.J. (2003) Photo-processes in AOT reverse micelles containing metalloporphyrins and oligopeptides. *J. Phys. Chem.* B, **107**, 6194–6207.
130. Ohkoshi, S., Abe, Y., Fujishima, A. and Hashimoto, K. (1999) Design and preparation of a novel magnet exhibiting two compensation temperatures based on molecular field theory. *J. Phys. Rev. Lett.*, **82**, 1285–1288.

131. Sato, O., Hayami, S., Einaga, Y. and Gu, Z.Z. (2003) Control of the magnetic and optical properties in molecular compounds by electrochemical, photochemical and chemical methods. *Bull. Chem. Soc. Jpn.*, **76**, 443–470.
132. Uemura, T., Ohba, M. and Kitagawa, S. (2004) Size and surface effects of Prussian blue nanoparticles protected by organic polymers. *Inorg. Chem.*, **43**, 7339–7345.
133. Vaucher, S., Fielden, J., Li, M., Dujardin, E. and Mann, S. (2002) Molecule-based magnetic nanoparticles: Synthesis of cobalt hexacyanoferrate, cobalt pentacyanonitrosylferrate, and chromium hexacyanochromate coordination polymers in water-in-oil emulsion. *Nano Lett.*, **2**, 225–229.
134. Uemura, T. and Kitagawa, S. (2003) Prussian blue nanoparticles protected by poly(vinylpyrrolidone). *J. Am. Chem. Soc.*, **125**, 7814–7815.
135. Taguchi, M., Yamada, K., Suzuki, O., Sato, O. and Einaga, Y. (2005) Photoswitchable magnetic nanoparticles of Prussian blue with amphiphilic azobenzene. *Chem. Mater.*, **17**, 4554–4559.
136. Tsunoyama, H., Sakurai, H., Negishi, Y. and Tsukuda, T. (2005) Size-specific catalytic activity of polymer-stabilized gold nanoclusters for aerobic alcohol oxidation in water. *J. Am. Chem. Soc.*, **127**, 9374–9375.
137. Tsunoyama, H., Sakurai, H., Ichikuni, N., Negishi, Y. and Tsukuda, T. (2004) Colloidal gold nanoparticles as catalyst for carbon–carbon bond Formation: Application to aerobic homocoupling of phenylboronic acid in water. *Langmuir*, **20**, 11293–11296.
138. Teranishi, T. and Miyake, M. (1998) Size control of palladium nanoparticles and their crystal structures. *Chem. Mater.*, **10**, 594–600.
139. Leff, D.V., Ohara, P.C., Heath, J.R. and Gelbart, W.M. (1995) Thermodynamic control of gold nanocrystal size: Experiment and theory. *J. Phys. Chem.*, **99**, 7036–7041.
140. Schmid, G., Maihack, V., Lantermann, F. and Peschel, S. (1996) Ligand stabilized metal clusters and colloids: Properties and applications. *J. Chem. Soc., Dalton Trans.*, 589–595.
141. Lewis, L.N. (1993) Chemical catalysis by colloids and clusters. *Chem. Rev.*, **93**, 2693–2730.
142. Ohtaki, M., Komiyama, M., Hirai, H. and Toshima, N. (1991) Effects of polymer support on the substrate selectivity of covalently immobilized ultrafine rhodium particles as a catalyst for olefin hydrogenation. *Macromolecules*, **24**, 5567–5572.
143. Wang, Y., Liu, H. and Jiang, Y. (1989) A new method for immobilization of polymer-protective colloidal platinum metal via co-ordination capture with anchored ligands. *J. Chem. Soc., Chem. Commun.*, 1878–1925.
144. Nickolov, Z.S., Paruchuri, V., Shah, D.O. and Miller, J.D. (2004) FTIR–ATR studies of water structure in reverse micelles during the synthesis of oxalate precursor nanoparticles. *Colloids Surf. A: Physicochem. Eng. Aspects*, **232**, 93–99.
145. Han, M., Vestal, C.R. and Zhang, Z.J. (2004) Quantum couplings and magnetic properties of $CoCrxFe_2$-xO_4 ($0 < x < 1$) spinel ferrite nanoparticles synthesized with reverse micelle method. *J. Phys. Chem. B*, **108**, 583–597.
146. Capek, I. (2004) Preparation of metal nanoparticles in water-in-oil (w/o) microemulsions. *Adv. Colloid Interface Sci.*, **110**, 49–74.
147. Calandra, P., Giordano, C., Longo, A. and Turco Liveri, V. (2006) Physicochemical investigation of surfactant-coated gold nanoparticles synthesized in the confined space of dry reversed micelles. *Mater. Chem. Phys.*, **98**, 494–499.
148. Kim, D.K., Zhang, Y., Kehr, J., Klason, T., Bjelke, B. and Muhammed, M. (2001) Characterization and MRI study of surfactant-coated superparamagnetic nanoparticles administered into the rat brain. *J. Magn. Magn. Mater.*, **225**, 256–261.
149. Niemeyer, C.M. (2001) Nanoparticles, proteins, and nucleic acids: Biotechnology meets materials science. *Angew. Chem. Int. Ed.*, **40**, 4128–4158.
150. Teng, X.W. and Yang, H. (2004) Effects of surfactants and synthetic conditions on the sizes and self-assembly of monodisperse iron oxide nanoparticles. *J. Mater. Chem.*, **14**, 774–779.

151. Wang, D., He, J., Rosenzweig, N. and Rosenzweig, Z. (2004) Superparamagnetic Fe_2O_3 beads-CdSe/ZnS quantum dots core–shell nanocomposite particles for cell separation. *Nano Lett.*, **4**, 409–413.
152. Sun, S.H., Zeng, H., Robinson, D.B., Raoux, S., Rice, P.M., Wang, S.X. and Li, G.X. (2004) Monodisperse MFe_2O_4 (M = Fe, Co, Mn) nanoparticles. *J. Am. Chem. Soc.*, **126**, 273–279.
153. Peng, X., Manna, L., Yang, W., Wickham, J., Scher, E., Kadavanich, A. and Alivisatos, A.P. (2000) Shape control of CdSe nanocrystals. *Nature*, **404**, 59–61.
154. Yamamoto, T., Umemura, Y., Sato, O. and Einaga, Y. (2004) Photoswitchable magnetic films: Prussian blue intercalated in Langmuir-Blodgett films consisting of an amphiphilic azobenzene and a clay mineral. *Chem. Mater.*, **16**, 1195–1201.
155. Kuznetsov, A.A., Filippov, V.I., Alyautdin, R.N., Torshina, N.L. and Kuznetsov, O.A. (2001) Application of magnetic liposomes for magnetically guided transport of muscle relaxants and anti-cancer photodynamic drugs. *J. Magn. Magn. Mater.*, **225**, 95–100.
156. Voigt, A., Buske, N., Sukhorukov, G.B., Antipov, A.A., Leporatti, S., Lichtenfield, H., Baumler, H., Donath, E. and Mohwald, H. (2001) Novel polyelectrolyte multilayer micro- and nanocapsules as magnetic carriers. *J. Magn. Magn. Mater.*, **225**, 59–66.
157. Anderson, L. (1999) *Genetic Engineering, Food, and Our Environment.* Chelsea Green Publishing, White River Junction, VT, USA.
158. Cui, Y.L., Wang, Y.N., Hui, W.L., Zhang, Z.F., Xin, X.F. and Chen, C. (2005) The synthesis of Gold-Mag nanoparticles and its application for antibody immobilization. *Biomedic. Microdevices*, **7**, 153–156.

Chapter 7
Non-Aqueous Microemulsions

Feng Gao and Carlos C. Co

7.1 Introduction

Self-assembled microstructures of water and surfactant with or without oil have been the subject of intense research for several decades because of their rich structural variety. Microstructures ranging from spherical micelles, rod-like micelles, bicontinuous microemulsions and liquid crystalline phases have broad commercial and scientific applications including nanomaterial synthesis, controlled delivery, coatings and detergents among many others.

Efforts towards understanding self-assembly naturally focused on aqueous systems and led to the traditional explanation for self-assembly as being driven by the directional hydrogen bonding of water molecules [1, 2]. This traditional view is challenged, however, by the work of Evans *et al.* [3, 4] on micelle formations in hydrazine and high-temperature aqueous systems. This chapter reviews more recent research demonstrating self-assembly in non-aqueous systems including polymer blends, ionic liquids, supercritical CO_2 and non-aqueous polar solvents. The discovery of self-assembly in these polymeric, gas and sometimes even glassy systems has expanded significantly the field and its applications.

7.2 Self-assembly in polymer blends

Mixtures of two homopolymers (A and B) and their corresponding diblock copolymer (A–B) are polymeric counterparts of mixtures of water, oil and surfactant. The immiscible nature between water and oil is also observed in polymer blends due to the fact that most polymers are immiscible in each other. The addition of diblock copolymers into blends of homopolymers has effects similar to adding surfactants into water–oil mixtures. The resulting reduction in interfacial tension and formation of the preferred interfacial curvature yield a variety of self-assembled structures.

Pioneering work by Bates, Lodge and co-workers [5–12] demonstrated that the addition of diblock copolymer (A–B) into the mixture of its corresponding homopolymer A and B drives self-assembly into varied structures including droplet-type microemulsions [8], bicontinuous microemulsions [5, 6, 8, 10, 12], hexagonal phases [10, 11] and lamellar phases [5, 8, 12]. Figure 7.1 shows a typical temperature–composition phase diagram of symmetric polyethylene (PE)/polyethylenepropylene (PEP)/PE–PEP mixtures [5]. To

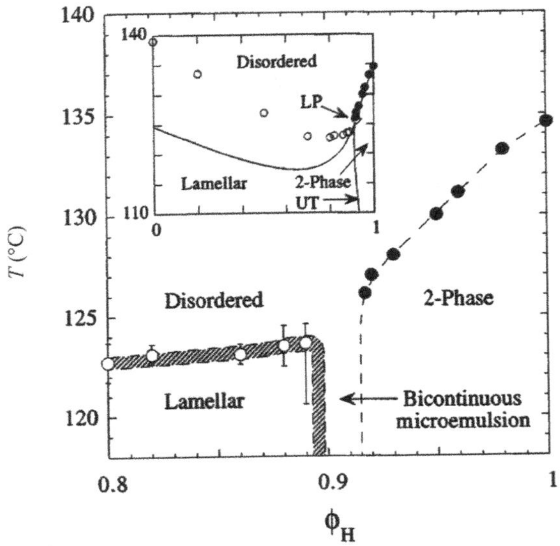

Figure 7.1 The temperature–composition phase diagram of symmetric polyethylene (PE)/ polyethylenepropylene (PEP)/PE–PEP mixtures where ϕ_H denotes the volume fraction of homopolymer. (From Ref. [5], reprinted with permission of the American Physical Society.)

reduce the number of thermodynamic variables, the symmetric mixtures contain equal volumes of PE and PEP, and the compositional phase diagram is determined by temperature and the volume fraction of homopolymer Φ_H ($\Phi_H = \Phi_{PE} + \Phi_{PEP} = 1 - \Phi_{PEP}$). Increasing the volume fraction of homopolymer Φ_H at low temperature leads to the swelling of the lamellar structures formed by PE–PEP amphiphiles. This change causes the phase transition from the lamellar phase to the polymeric bicontinuous microemulsion phase and finally to the phase separation (two phases rich in PE and PEP, respectively). The shaded portion denotes the two-phase region where the lamellar phase is in equilibrium with the polymeric bicontinuous microemulsion phase. In this polymeric system, the bicontinuous microemulsion phase is restricted to a narrow channel between the two two-phase regions.

Although the pattern of polymeric microemulsion phase behaviour, also observed in mixtures of polyisoprene and polystyrene [9], is entirely different from that observed for aqueous microemulsions, the phase behaviours of the polymeric blends are well predicted by the calculation using self-consistent mean field theory [13] (the solid curves in the inset of this illustration). The mean field theory predicts three thermodynamic regions to be present in the phase diagram. One region is a single-disordered phase which gradually transitions from a bicontinuous structure at low temperatures to a droplet structure at high temperatures. The second region consists of two equilibrated disordered liquid phases rich in PE and PEP, respectively, at high Φ_H and low temperatures. The third region is a single-ordered lamellar phase at low Φ_H and low temperatures.

Self-assembly in polymeric systems is studied and verified using small-angle scattering techniques (neutrons or X-ray) and electron microscopy. Figure 7.2 shows a typical small-angle neutron scattering (SANS) spectrum as a function of the intensity (I) and the scattering vector (q) at varying temperatures for a 40:40:20 vol.% blend of polyisobutylene

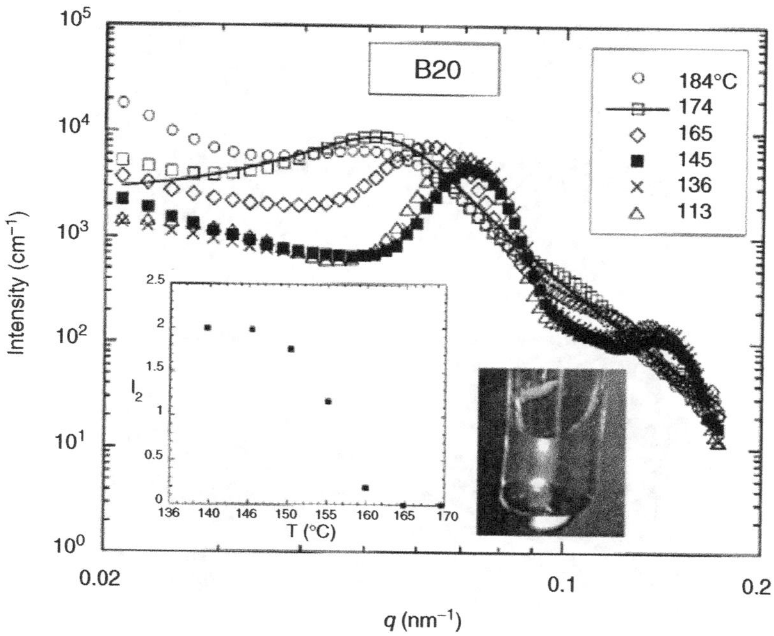

Figure 7.2 SANS spectrum of mixtures of polyisobutylene (PIB), polyethylene (PE) and polyethylene – polypropylene bi-block copolymer (PE–PP) at different temperatures. The compositions of the polymer blends are fixed at 40 vol.% of PIB, 40 vol.% of PE and 20 vol.% PE–PP. (From Ref. [12], reprinted with permission of the American Chemical Society.)

(PIB), PE and polyethylene-block-polypropylene copolymer (PE–PP) (referred as B20 in the literature). At temperatures ≤145°C, primary scattering peak and second-order peak are observed at $q_1 = 0.07$ nm^{-1} and $q_2 = 0.14$ nm^{-1}, indicative of an ordered lamellar phase with a repeat distance of 90 nm. The experimental intensity near the second peak was fit by Eq. (7.1).

$$I(q) = C \exp\left[\frac{-(q-q_{\text{peak}})^2}{\sigma^2}\right] + I_{\text{background}}(q), \tag{7.1}$$

where $1/I_{\text{background}}(q)$ is a quadratic function of q, with C, σ and the quadratic coefficients as tunable parameters. In the inset in Fig. 7.2, the area under the second-order peak I_2 ($I_2 = \sqrt{\pi}C\sigma$) is plotted as a function of temperature. I_2 is zero at temperatures ≥160°C, indicating the absence of the ordered lamellar phase. At temperatures ≥165°C, the sample is optically clear (as shown in the inset of Fig. 7.2) and the lamellar phase 'melts' into a microemulsion phase exhibiting SANS spectra with a single peak, fitted well by the Teubner–Strey model [14] (Eq. (7.2)) which was developed to explain and extract structural parameters from the scattering of bicontinuous aqueous microemulsions.

$$I(q) = \frac{1}{(a + bq^2 + cq^4) + I_{\text{background}}} \tag{7.2}$$

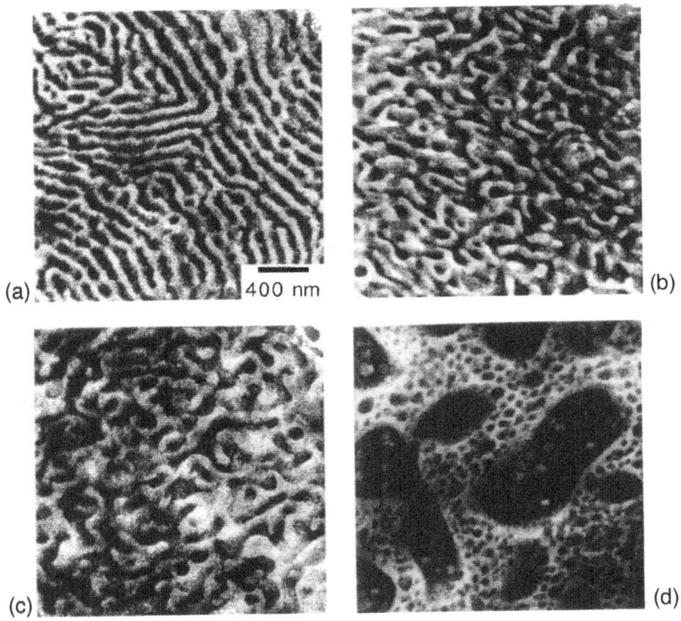

Figure 7.3 Morphology study of polymeric mixtures of PE/PEP/PE–PEP by TEM. Samples were annealed at 119°C and then frozen, sectioned and stained; volume fraction of the symmetric polymer PE–PEP: (a) 0.86, (b) 0.90, (c) 0.91 and (d) 0.92. (From Ref. [5], reprinted with permission of the American Physical Society.)

At temperatures ≥184°C, phase separation is indicated by the significant increase in low q scattering. Light scattering experiments also show that the cloud point of the polymeric mixture is located at 180 ± 5°C.

Direct visualisation by transmission (TEM) and scanning electron microscopy (SEM) yields a complementary view of self-assembled polymer structures. Care must be taken, however, that the sample preparation steps, including structure freezing [15], cryo-ultramicrotoming, staining, etching, conductive coating etc. do not alter the microstructure of the samples. In the case of PE/PEP homopolymer mixtures with PE–PEP block copolymer, the PE homopolymer and the PE block in the block copolymer crystallise below 105°C. Freezing of the hot molten self-assembled structures in liquid nitrogen, followed by gradual warming of room temperature is sufficient to preserve the microstructure. Ultramicrotoming of the samples into ~80 nm sections and selective staining of amorphous PEP with ruthenium tetraoxide vapour reveal the structures shown in Fig. 7.3. The progression from lamellar to bicontinuous structures with increasing homopolymer concentration is consistent with the phase diagram of Fig. 7.1. Thus, while the patterns of the microemulsion phase behaviour in polymeric system are different from aqueous systems, they exhibit similar bicontinuous structures.

Self-assembled polymeric structures have promising applications in nanomaterials synthesis. As demonstrated by Zhou and Lodge for bicontinuous microemulsions of polyisoprene/polystyrene [9], mesoporous polymeric networks (Fig. 7.4) can be obtained by

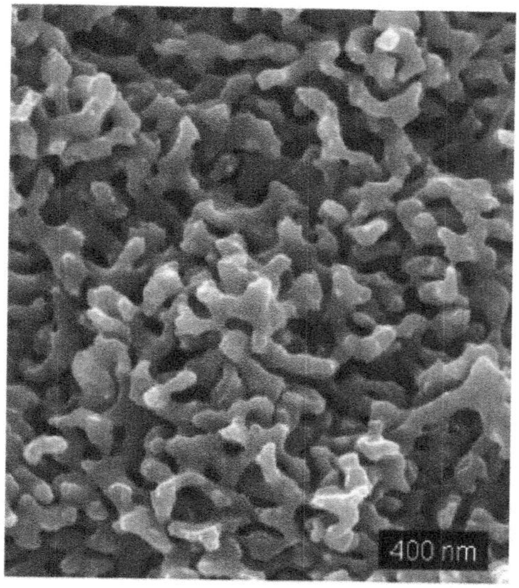

Figure 7.4 Freeze-fracture SEM micrograph mesoporous polymeric networks obtained by sulphur monochloride cross-linking of polyisoprene in bicontinuous polymeric microemulsions of polyisoprene and polystyrene. (From Ref. [9], reprinted with permission of the American Chemical Society.)

selective cross-linking of one domain (polyisoprene) subsequently followed by solvent dissolution of the other domain (polystyrene). Although the sulphur monochloride cross-linking of polyisoprene domains in this case is slow [16], it effectively fixes the bicontinuous microemulsion structure allowing the polystyrene domain to be subsequently dissolved in hexane. Moreover, the porous cross-linked polyisoprene polymer is stable up to 200°C.

7.3 Self-assembly in room temperature ionic liquids

Room temperature ionic liquids (RTILs) are molten salts whose melting points are below room temperature. RTILs are formed when the constituent ions are sterically mismatched, thereby hindering crystal formation [17]. As polar solvents, RTILs have unique applications as tunable and environmentally benign solvents with very low volatility, high fire resistance, excellent chemical and thermal stability and wide liquid temperature range and electrochemical windows [17–19]. Solvent applications of RTILs include, for example, organic synthesis [17, 20, 21], separations [22, 23], storage and transportation of hazardous chemicals [24], polymeric electrolytes [25, 26], dissolution of natural products [27] and synthesis of hollow metal oxide microspheres [28].

Because of the high polarity of RTILs, self-assembly of amphiphiles in RTILs are quite similar to that observed in aqueous systems. Alkyltrimethylammonium bromides and alkyl pyridinium bromides in the RTIL ethylammonium nitrate (EAN), for example, form micelles just like in water, albeit with critical micelle concentration (CMC) values that are five to ten times larger [4, 29]. Liquid crystalline phases are also observed in surfactant–RTIL

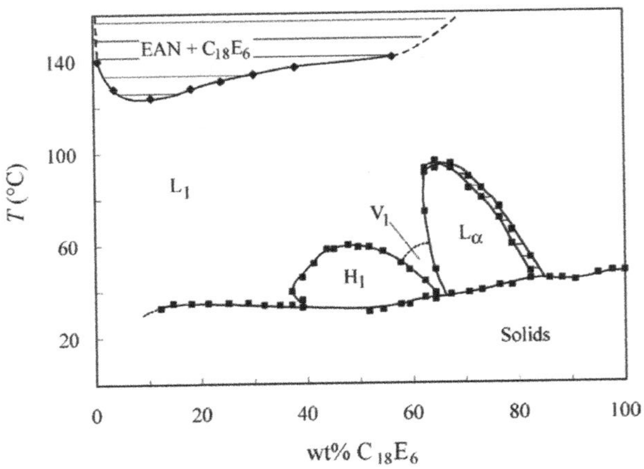

Figure 7.5 The temperature–composition binary phase diagram of $C_{18}E_6$–EAN mixtures. (From Ref. [30], reprinted with permission of the American Chemical Society.)

binary mixtures. Moreover, self-assembly in RTILs is not limited to ionic surfactants. Fig. 7.5 shows, for example, the binary phase diagram for $C_{18}E_6$ and EAN, which exhibits the Fontell sequence of hexagonal–cubic–lamellar phases. Krafft boundaries and miscibility gap were also observed for the binary system $C_{18}E_6$/EAN.

The high CMC of surfactants in RTIL parallels that observed in polar solvent wherein the hydrophobic chains of surfactant molecules exhibit decreased solvophobicity compared to that in aqueous systems. This attenuates the self-assembly of surfactants in RTILs and makes it necessary to use higher surfactant concentrations and longer surfactant tail groups to form microemulsions of oil and RTILs. Atkin and Warr [31] recently reported the microemulsions of alkanes and EAN by using non-ionic alkyl oligoethyleneoxide (C_iE_j) surfactants. Phase behaviour studies revealed that bicontinuous dodecane–EAN–C_iE_j microemulsions are strikingly similar with the corresponding aqueous systems. For example, the pattern of ternary phase diagram with equal mass of EAN and dodecane shows the fish shape in the temperature–surfactant concentration plot. Increasing the amphiphilic strength of C_iE_j yields a 'fish' body whose size initially increases and then decreases. The one-phase bicontinuous microemulsion region is also strongly structured. Small angle X-ray scattering (SAXS) spectrum of the bicontinuous dodecane–EAN–C_iE_j microemulsions are fitted well by the Teubner–Strey model [14], yielding the characteristic length scales consistent with those of aqueous microemulsion. However, the EAN–dodecane microemulsions do have differences compared to their aqueous counterparts. In general, the surfactant alkyl chain must be about four to six CH_2 groups longer in EAN to yield one-phase microemulsions and lamellar phases similar to that of aqueous systems. Other recent reports of self-assembly in RTIL include micellisation in surfactants/RTIL systems [32–34], microemulsification of RTIL in water [35], microemulsification of RTIL in oil [36] and formation of macroscopic fibres and vesicles [37].

Self-assembly in RTILs is not limited to low molar mass surfactants. For example, He *et al.* [38] have reported about spherical micelles in mixtures of poly((1,2-butadiene)-block-ethylene oxide) (PB–PEO) diblock copolymers in the RTIL 1-butyl-3-methylimidazolium

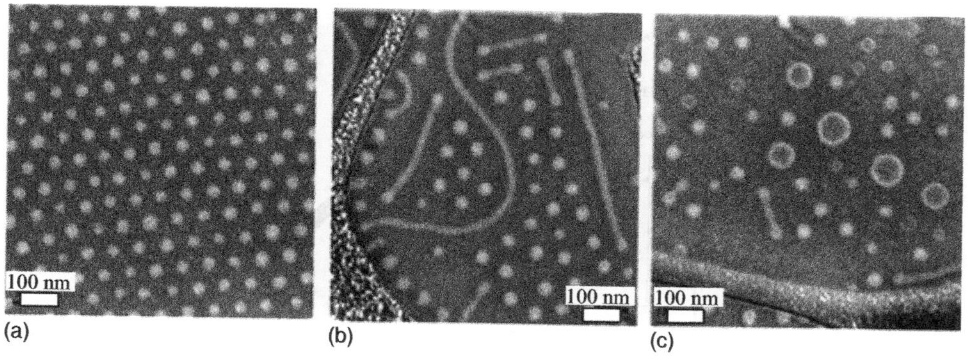

Figure 7.6 Cryo-TEM images of PB-PEO in [BMIM][PF6]. (a) micellar structure; (b) the coexistence of spherical and worm-like micelles; (c) the coexistence of spherical and wormlike micelles along with some vesicles. (From Ref. [38], reprinted with permission of the American Chemical Society.)

hexafluorophosphate ([BMIM][PF6]). A decrease of the length and the volume fraction of the PEO block results in the formation of worm-like micelles and vesicles in equilibrium with the spherical micelles (Fig. 7.6). Unlike self-assembly of PEO-based block copolymers, these self-assembled structures in an RTIL are temperature insensitive and exhibit the same structure from 25 to 100°C.

7.4 Self-assembly in supercritical CO_2

Supercritical CO_2 has been extensively studied as a solvent in many applications because of its low cost and its moderate critical conditions ($T_c = 31°C$, $p_c = 73.8$ bar). Moreover, CO_2 is non-toxic, volatile, inert, non-flammable and recyclable [39, 40]. Solvent applications of interest include polymerisation [41, 42], drying [43], cleaning of low dielectric insulators [44], nanomaterial synthesis [45], catalysis and organic synthesis [46, 47], among many others. Despite its advantages, supercritical CO_2 is, however, a poor solvent especially for polar or high molecular weight solutes. Thus, most research on CO_2 has focused on the properties of CO_2 combined with water, organics and surfactants.

Unlike water, CO_2 has no dipole moment. Therefore, hydrophilic molecules are practically insoluble even in supercritical CO_2. Van der Waals forces, arising principally from quadrupolar interactions are weaker even than in hydrocarbons. In principle, addition of surfactants would improve the solubilisation properties of compressed CO_2 and could also result in self-assembled microstructures. However, most commercially available surfactants are insoluble in supercritical CO_2 [48].

Fluorinated surfactants or graft polymers are typically necessary in CO_2 applications [48, 49]. Research into the self-assembly of these surfactants and polymers in CO_2 have focused on dilute systems where only reverse water-in-CO_2 micelles are expected [39, 50–52, 53]. Regions of the phase diagrams where CO_2-in-water micelles, bicontinuous microemulsions or liquid crystalline phases are formed remain to be investigated in detail. There is only one system known so far where all these phases have been observed. We will come back to this at the end of Section 7.4. The properties of reverse micelles in CO_2

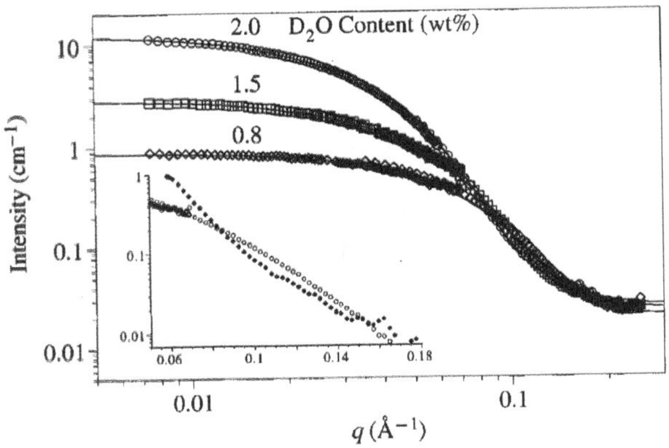

Figure 7.7 SANS spectra of 0.8 (diamonds), 1.5 (squares) and 2.0 (circles) wt.% D_2O in CO_2/PFPE mixtures at 35°C and 287 bar. Lines represent model fits to the data. (From Ref. [50], reprinted with permission of the American Chemical Society.)

are generally similar to those of reverse micelles in organic media. SANS investigations by Kaler and co-workers [50] of water-in-CO_2 micelles (Fig. 7.7), for example, confirm that the water–core radius increase from 20 to 36 Å as the D_2O concentration is increased from 0.8 to 2.0 wt.% at a fixed perfluoropolyether (PFPE) concentration of 2.1 wt.% in CO_2. The inset of Fig. 7.7 shows the high q portions of the SANS spectra where filled and open points stand for 0.8 and 2.0 wt.% D_2O, respectively. The changes in the scattering at the higher q portions also confirm the increase in droplet radius with increasing D_2O concentration.

The high cost and toxicity of fluorinated surfactants has motivated the search for CO_2-philic hydrocarbon surfactants. Eastoe and co-workers [54] first demonstrated the formation of reversed micelles in supercritical CO_2 using AOT derivatives. AOT itself is insoluble in CO_2. However, derivatives with branched trimethyl moieties are CO_2-soluble. SANS investigations confirm the formation of reverse micelles the radius of which is ~15 Å. Eastoe has also reported [55] the formation of water-in-CO_2 microemulsions with AOT derivatives that incorporate oxygen atoms in the surfactant tails (e.g. AOK and AO-vac). The structure of these water-swollen micelles, together with the corresponding scattering length density (sld or $\rho \times 10^{10}$ cm^{-2}) for SANS experiments, are shown schematically in Fig. 7.8. The core radius (r) and shell thickness (t) of the inverted micelles strongly depend on the water loading. As reported by Eastoe and co-workers, increasing the water loading ($w = [D_2O]_{added} - [D_2O]_{CO2}/[surf]$) from 8.5, 19.0 to 29.5 by adding D_2O to the inverted microemulsions, yields the increase of the core radius (r) from 15, 17 to 20 Å and the increase of the shell thickness (t) from 8, 9 to 10 Å, respectively.

Despite all difficulties mentioned above, examples for CO_2 containing microemulsion stabilised by a technical grade non-ionic surfactant have been found (see Fig. 11.3 in Chapter 11). The studied system consists of water/NaCl–n-propane/CO_2–Lutensol® XL70 with varying amounts of CO_2 in n-propane/CO_2 mixtures. All measurements were carried out at $p = 220$ bar and at equal volume fractions of the two solvents [56]. The respective phase diagrams have been studied as a function of the temperature T and the total surfactant

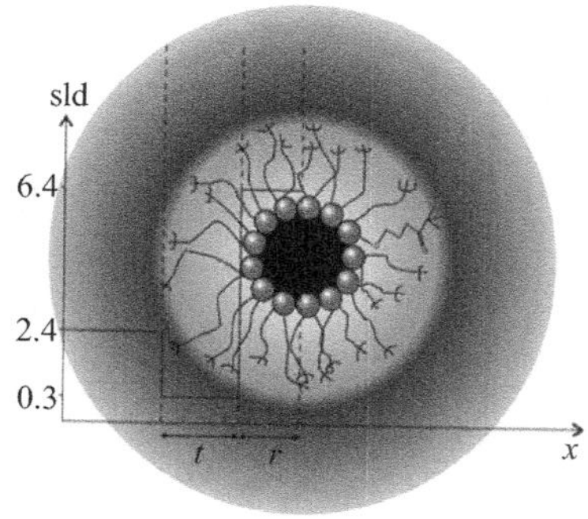

Figure 7.8 Schematic scattering length density (sld or $\rho \times 10^{10}$ cm^{-2}) profile fitted to SANS data from surfactant-stabilised D_2O-in-CO_2 microemulsion droplets. (From Ref. [55], reprinted with permission of Wiley-VCH Verlag GmbH & Co. KgaA.)

concentration γ for four different CO_2 contents, namely $\beta = 0$ (pure propane), $\beta = 0.41$, $\beta = 0.60$ and $\beta = 1$ (pure CO_2).

7.5 Self-assembly in non-aqueous polar solvents

The self-assembly of amphiphilic molecules in non-aqueous polar solvents is usually attenuated compared to that in water. The CMCs increase significantly upon the substitution of water by polar solvents [2, 57, 58]. For example, the CMC of ionic surfactants in ethylene glycol are two orders of magnitude larger than that in water [57], while the monomeric solubility of sodium dodecyl sulphate in formamide is so high that micelles do not form at all [58]. Attenuation of self-assembly in non-aqueous polar solvents is the result of the reduced free energy of repulsion between polar solvents and the solvent-phobic parts of amphiphiles compared to that in water.

Besides dampening micelle formation, the reduced free energy repulsion has a significant effect on the formation of liquid crystalline structures. Systematic phase behaviour studies of surfactant–polar solvent binary mixtures [59, 60–64] show that the liquid crystalline phase regions are reduced in size or absent in non-aqueous polar solvents. This is evident in the binary phase diagrams of alkyltrimethylammonium surfactant–polar solvent mixtures shown in Fig. 7.9. In water, these alkyltrimethylammonium surfactants exhibit liquid crystalline structures, following the Fontell sequence of isotropic (L_1), hexagonal (H), cubic (V) and lamellar (La) phases with increasing surfactant concentration. Replacing water with glycerol as solvent yields the same sequence of phases with an expansion of the L_1 region. Replacing water with ethylene glycol further expands the isotropic region at the expense of the liquid crystalline regions. In ethylene glycol mixtures of C_{12}TAB, no liquid crystalline phases are observed up to 80 wt.% surfactant concentration and the cubic phase disappears for C_{14}TAB. When water is replaced with N-methylformamide, all liquid crystalline phases are attenuated, even for the highly amphiphilic C_{16}TAB surfactant. The observed attenuation of liquid crystalline phases is related directly to the interfacial tension, which follows water >> formamide ~ glycerol > ethylene glycol > N-methylformamide

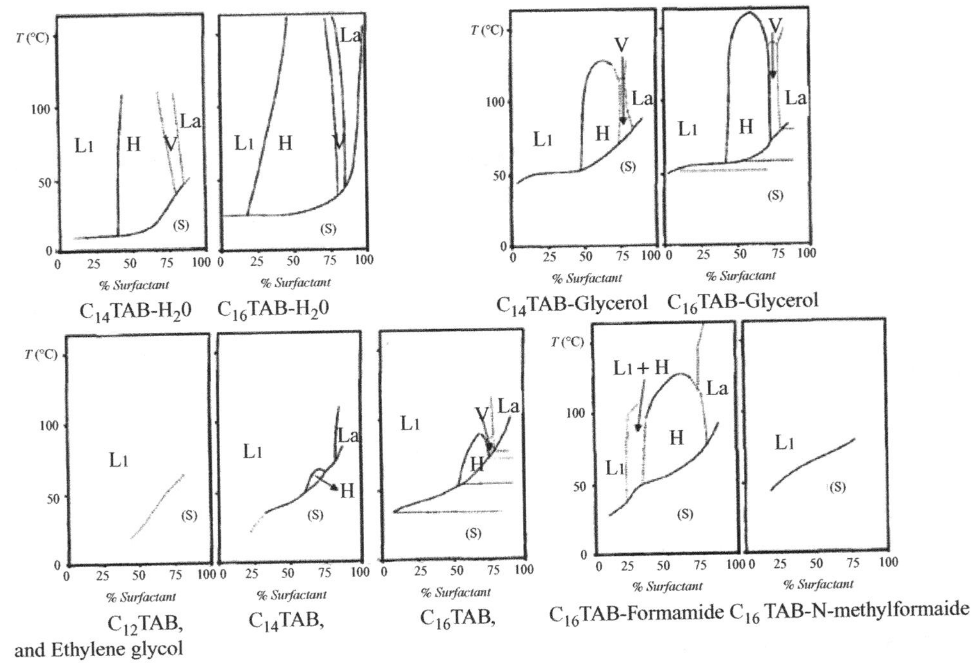

Figure 7.9 Binary phase diagrams of alkyltrimethylammonium bromide–polar solvent mixtures as a function of temperature and surfactant concentration. (From Ref. [60], reprinted with permission of Elsevier.)

[65]. Despite the attenuation, increasing the strength of amphiphiles yields the represence of certain phases which either disappear or are observed at higher surfactant concentration after replacing water with solvents. For example, when water is replaced by ethylene glycol, the cubic phase, which is found in C_{14}TAB–water binary mixtures, is not observed in the C_{14}TAB–ethylene glycol binary phase diagram. An increase of the tail length of surfactant molecules from 14 to 16 leads to the reformation of the cubic phase. In addition, the hexagonal phase forms at lower surfactant concentration when C_{16}TAB is used. The effect of increasing amphiphilic strength probed by binary phase diagram studies provides a basis for preparing solvent–oil–non-ionic surfactant microemulsions.

The decreased solvophobicity, which arises from the reduced free energy of repulsion between surfactant's hydrocarbon tails and non-aqueous polar solvents, is usually compensated by increasing the length of the hydrocarbon tail to promote self-assembly. This is necessary, for example, to form efficient microemulsions with polar solvent–oil–surfactant mixtures [66–71]. Strey and co-workers demonstrated this approach most convincingly using mixtures of formamide, octane and C_iE_j surfactants (Fig. 7.10). The presence of formamide in the ternary mixture significantly attenuates the self-assembly of C_iE_j by decreasing the free energy repulsion. As a consequence, the length of the hydrocarbon tails (i) is required to increase from 12 to 18 to offset the decreased solvophobicity and therefore form similar pattern of ternary phase diagram with other composition variables fixed. (Note the effect of increasing j by 2 is negligible compared with that of replacing water

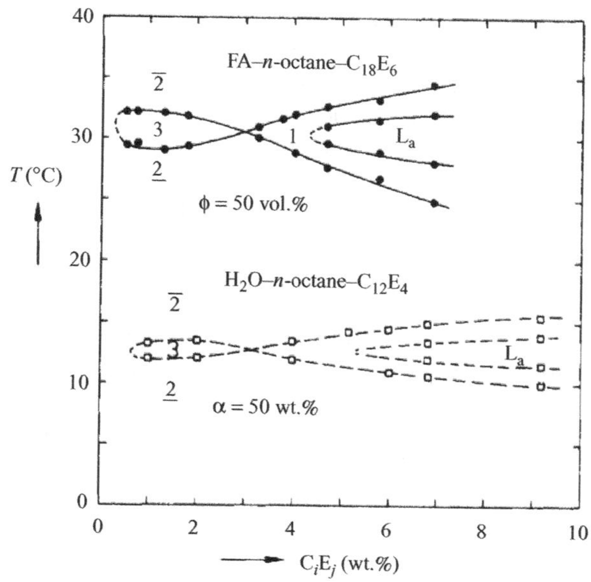

Figure 7.10 Ternary temperature–composition phase diagrams of formamide(FA)–octane–$C_{18}E_6$ mixtures (full lines) and H_2O–octane–$C_{12}E_4$ (dashed lines) mixtures, demonstrating that the effect of formamide can be compensated by increasing the carbon number i of C_iE_j by 6. (From Ref. [66], reprinted with permission of the American Chemical Society.)

by formamide.) Moreover, from SANS studies, Kaler and co-workers [68] further demonstrated that the microstructure present in these formamide microemulsions is consistent to that of corresponding aqueous microemulsions.

7.6 Self-assembly in sugar glasses

Self-assembly is a general phenomenon whose applications are not restricted to fluid-like mixtures. In our own investigations of self-assembly in concentrated sugar systems [72, 73], we discovered that microemulsions whose aqueous phase has been replaced with concentrated sugar solutions can be dehydrated to the solid glass state. These sugar-based microemulsion glasses are optically clear and contain comparable mass of oil and glassy sugar.

The principal motivation for studying these sugar-based microemulsion glasses came from the observation that water–oil–surfactant mixtures are extensively for nanomaterials synthesis with the central idea of switching dynamic self-assembly into chemically and mechanically stable supramolecular materials. Template polymerisations are classified as synergistic or transcriptive templating depending on whether the template itself participates in the reaction.

Synergistic templating involves the usage of polymerisable surfactants. The target structure is self-assembled using polymerisable surfactants in solvent (usually water) and then polymerised. After polymerisation, the nanomaterials obtained are in essence 'cured templates'. A wide-gamut of self-assembled surfactant structures, e.g. worm-like micelle [74], hexagonal and cubic micelle [75] etc. have been successfully fixed by synergistic template polymerisation to yield 'one-to-one' copies. However, synergistic templating has several disadvantages. First and foremost, the phase behaviours of polymerisable surfactants are poorly reported in the literature due to their uncommon usage. Thus, the phase behaviour of each new polymerisable surfactant, unlike regular surfactants, has to be investigated before one can proceed with the templating reactions. Secondly, the hydrophobic domain of the template is typically limited to the length of the alkyl chain present in the polymerisable surfactants. The cost of specialty polymerisable surfactants also imposes restrictions on the use of synergistic template on a commercial scale. These limitations have driven research on transcriptive templating where obtaining a 'one-to-one' copy of the template remains an elusive challenge.

Transcriptive templating involves the usage of polymerisable oil (monomer) and non-reactive surfactant. Following self-assembly of the surfactant/monomer/water mixture, the monomer is polymerised within the self-assembled template. Unlike synergistic templating, transcriptive templating usually leads to macroscopic phase separation, or replicas with much larger length scales. During polymerisation, the self-assembled surfactant template does not retain its structure, but continuously rearranges to accommodate the growing polymer chains. Reassembly and phase separation following transcriptive templating have been widely reported in the literature for droplet-like microemulsions [76–80] bicontinuous microemulsions [81–87], vesicles [88–90] and other liquid crystalline mesophases [91, 92].

As a compromise, a combination of synergistic and transcriptive templating is typically used to suppress phase separation in polymerising bicontinuous microemulsions. In this combined approach, macromonomers with slower dynamics are used as polymerisable surfactants in conjunction with conventional monomers. The slower reassembly of the macromoners and their anchoring to the oil/water interface following polymerisation, facilitates the fixation of the template structure and suppresses structural rearrangement and phase separation. Macromonomers were first used as polymerisable surfactants by Gan and his co-workers [93] to template bicontinuous microemulsions. In this initial report, the bicontinuous microemulsions, consisting of water, methyl methacrylate (MMA) and polymerisable zwitterionic surfactant acryloyloxyundecyldimethylammonium acetate (AUDMAA), were polymerised with only minimal rearrangement. Other macromonomers, such as (acryloyloxy) undecyl-trimethylammonium bromide (AUTMAB, cationic surfactant) [94], ω-methoxy poly(ethylene oxide)$_{40}$-undecyl-α-methacrylate (C1-PEO-C11-MA-40, non-ionic surfactant) [95, 96] have also been used as polymerisable surfactants to form and template bicontinuous microemulsions. Besides the additional cost of the macromonomers, it is found that macromonomers can only suppress structural rearrangements on length scales of $\sim$100 nm, just below macroscopic phase separation. Verbatim one-to-one copies patent down to smaller length scales are only rarely reported and remains the exclusive realm of synergistic templating.

Transcriptive templating of self-assembled structures is complicated by the contradictory demands of a template that self-assembles to form over a reasonable period of time and yet robust enough to retain its structure over the course of polymerisation. The thermodynamic forces faced by these templates are complex and practically unavoidable

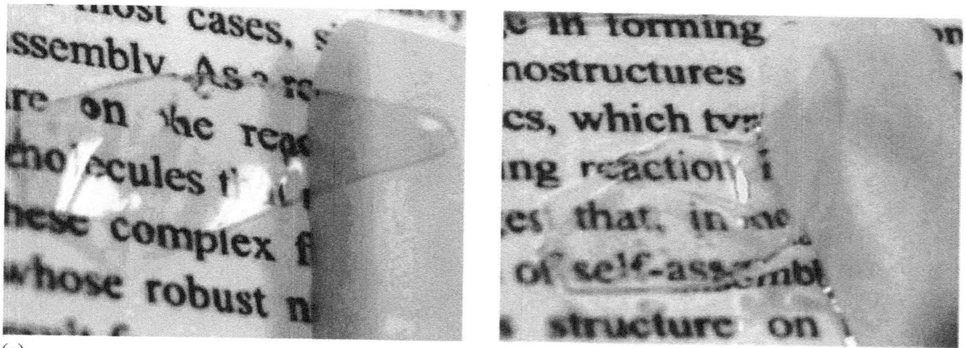

(a) (b)

Figure 7.11 (a) Optically clear microemulsion glass after UV photopolymerisation. (b) Dissolution of the sugar glass template with excess water leaves behind an optically clear and flexible poly-DVB membrane. (From Ref. [72], reprinted with permission of the American Chemical Society.)

as they arise from spatially non-uniform polymer growth, localised swelling and depletion of monomer, changes in the spontaneous curvature as monomer is converted to polymer and incompatibility of the surfactant chains with the resulting polymer [78].

We found that replacement of water typical microemulsions with glassy sugars to form solid template glasses followed by polymerisation of the hydrophobic monomer domains below the glass transition temperature of the sugar template. Structural rearrangements are effectively suppressed below the glass transition temperature and even sluggish polymerisations can therefore be polymerised to completion. The sugar glass templates are also robust and hold up well against the forces resulting from the continuously changing physico-chemical environment during polymerisation. Ultraviolet (UV) polymerisation of bicontinuous microemulsion glasses containing liquid divinylbenzene, for example, proceed with no visible alteration of the structure (Fig. 7.11(a)) and SANS confirm that the polymerised structures are indeed 'one to one' copies of the templates. Despite the significant difference in density between monomer and polymer, sugar templates do not fracture following polymerisation. The sugars and sugar surfactants are all available on a commercial scale and the method is thus amenable to scale up. Moreover, following polymerisation, the glassy sugar/surfactant templates are easily removed by dissolution in water to isolate the porous complementary polydivinylbenzene membranes (Fig. 7.11(b) [72]. If necessary, the wash liquor of sugar and surfactant may also be dried and recycled along with any residual monomer that is washed off.

The key to preparing these solid microemulsion glasses lies in detailed phase behaviour studies to identify dehydration pathways that lie exclusively within a continuous one-phase region. Following dehydration, these microemulsion glasses contain practically no water. Glass transition temperatures of sugar-based microemulsion glasses, containing up to 80 wt.% of liquid divinylbenzene oil relative to sugar, range from 64 to 75°C [72].

However, an unavoidable complication of this approach to forming microemulsion glasses is the trapping of compositional gradients during the dehydration process. This is evident sometimes in the SANS spectra of the microemulsion glasses, which do not always follow the Teubner–Strey type scattering from bicontinuous microemulsions. Such

Figure 7.12 Spontaneous formation of a microemulsion glass. Sugar and surfactant powder dried to 99.5% dryness dispersed in oil at room temperature 'dissolves' upon heating to 365 K to form a one-phase molten microemulsion glass. Gradual cooling of the molten glass to room temperature yields a solid microemulsion glass containing ~52 vol.% liquid oil with a Mohs hardness of 0.7. (Reproduced from Dave *et al.* [97].)

compositional gradients may be unacceptable in some optical and encapsulation applications. To overcome this, we have developed a direct approach to forming microemulsion glasses [97] by heating dry powder mixtures of sugar and surfactant in liquid oil at temperatures (365 K) above the glass transition of the sugar/surfactant mixture (Fig. 7.12). This approach is rapid, highly scalable, and the resulting microemulsion glasses combine liquid and solid glass properties at the nanoscale as confirmed by SANS, magnetic resonance, rheological and differential scanning calorimetry (DSC) measurements.

7.7 Conclusions

Water is not a necessary element for self-assembly, which requires only two immiscible components and a suitable amphiphile. Self-assembly in non-aqueous systems could be fluid-like as in the case of RTIL and non-aqueous polar solvents, or solid-like as in polymer blends and sugar glasses. Expanding the realm of self-assembly and complex fluids to non-aqueous systems, and in particular, to the solid state holds great promise in revolutionising several commercial encapsulation, polymerisation, membrane and optical technologies.

References

1. Franks, F. (1975) *Water: A Comprehensive Treatise.* Plenum, New York.
2. Singh, H.N., Saleem, S.M., Singh, R.P. and Birdi, K.S. (1980) Micelle formation of ionic surfactants in polar non-aqueous solvents. *J. Phys. Chem.*, **84**, 2191–2194.
3. Ramadan, M.S., Evans, D.F., Lumry, R. and Philson, S. (1985) Micelle formation in hydrazine water mixtures. *J. Phys. Chem.*, **89**, 3405–3408.
4. Evans, D.F., Yamauchi, A., Roman, R. and Casassa, E.Z. (1982) Micelle formation in ethylammonium nitrate, a low-melting fused salt. *J. Colloid Interface Sci.*, **88**, 89–96.
5. Bates, F.S., Maurer, W.W., Lipic, P.M., Hillmyer, M.A., Almdal, K., Mortensen, K., Fredrickson, G.H. and Lodge, T.P. (1997) Polymeric bicontinuous microemulsions. *Phys. Rev. Lett.*, **79**, 849–852.

6. Lee, J.H., Balsara, N.P., Chakraborty, A.K., Krishnamoorti, R. and Hammouda, B. (2002) Thermodynamics and phase behavior of block copolymer/homopolymer blends with attractive and repulsive interactions. *Macromolecules*, **35**, 7748–7757.
7. Krishnan, K., Almdal, K., Burghardt, W.R., Lodge, T.P. and Bates, F.S. (2001) Shear-induced nano-macro structural transition in a polymeric bicontinuous microemulsion. *Phys. Rev. Lett.*, **8709**, 1–4.
8. Lee, J.H., Ruegg, M.L., Balsara, N.P., Zhu, Y.Q., Gido, S.P., Krishnamoorti, R. and Kim, M.H. (2003) Phase behavior of highly immiscible polymer blends stabilized by a balanced block copolymer surfactant. *Macromolecules*, **36**, 6537–6548.
9. Zhou, N., Bates, F.S. and Lodge, T.P. (2006) Mesoporous membrane templated by a polymeric bicontinuous microemulsion. *Nano Lett.*, **6**, 2354–2357.
10. Zhou, N., Lodge, T.P. and Bates, F.S. (2006) Influence of conformational asymmetry on the phase behavior of ternary homopolymer/block copolymer blends around the bicontinuous microemulsion channel. *J. Phys. Chem. B*, **110**, 3979–3989.
11. Cavicchi, K.A. and Lodge, T.P. (2004) Anisotropic self-diffusion in block copolymer cylinders. *Macromolecules*, **37**, 6004–6012.
12. Lee, J.H., Balsara, N.P., Krishnamoorti, R., Jeon, H.S. and Hammouda, B. (2001) Designing balanced surfactants for mixtures of immiscible polymers. *Macromolecules*, **34**, 6557–6560.
13. Bates, F.S., Maurer, W., Lodge, T.P., Schulz, M.F., Matsen, M.W., Almdal, K. and Mortensen, K. (1995) Isotropic lifshitz behavior in block copolymer-homopolymer blends. *Phys. Rev. Lett.*, **75**, 4429–4432.
14. Teubner, M. and Strey, R. (1987) Origin of the scattering peak in microemulsions. *J. Chem. Phys.*, **87**, 3195–3200.
15. Khandpur, A.K., Macosko, C.W. and Bates, F.S. (1995) Transmission electron microscopy of saturated hydrocarbon block copolymers. *J. Polym. Sci. B*, **33**, 247–252.
16. Liu, G., Yan, X. and Duncan, S. (2002) Preparation and static light-scattering study of polystyrene-block-polyisoprene nanofiber fractions. *Macromolecules*, **35**, 9788–9793.
17. Welton, T. (1999) Room-temperature ionic liquids. Solvents for synthesis and catalysis. *Chem. Rev.*, **99**, 2071–2083.
18. Hoffmann, M.M., Heitz, M.P., Carr, J.B. and Tubbs, J.D. (2003) Surfactants in green solvent systems – current and future research directions. *J. Dispersion Sci. Technol.*, **24**, 155–171.
19. Huddleston, J.G., Visser, A.E., Reichert, W.M., Willauer, H.D., Broker, G.A. and Rogers, R.D. (2001) Characterization and comparison of hydrophilic and hydrophobic room temperature ionic liquids incorporating the imidazolium cation. *Green Chem.*, **3**, 156–164.
20. Zerth, H.M., Leonard, N.M., Mohan, R.S. (2003) Synthesis of homoallyl ethers via allylation of acetals in ionic liquids catalyzed by trimethylsilyl trifluoromethanesulfonate. *Org. Lett.*, **5**, 55–57.
21. Yadav, J.S., Reddy, B.V.S., Gayathri, K.U. and Prasad, A.R. (2002) Scandium triflate immobilized in ionic liquids: A novel and recyclable catalytic system for hetero-Diels-Alder reactions. *Synthesis-Stuttgart*, **17**, 2537–2541.
22. Anderson, J.L., Ding, J., Welton, T. and Armstrong, D.W. (2002) Characterizing ionic liquids on the basis of multiple solvation interactions. *J. Am. Chem. Soc.*, **124**, 14247–14254.
23. Anderson, J.L. and Armstrong, D.W. (2005) Immobilized ionic liquids as high-selectivity/high-temperature/high-stability gas chromatography stationary phases. *Anal. Chem.*, **77**, 6453–6462.
24. Freemantle, M. (2005) Catalytic nanomotors. *Chem. Eng. News*, **83**, 33–35.
25. Susan, M.A., Kaneko, T., Noda, A. and Watanabe, M. (2005) Ion gels prepared by in situ radical polymerization of vinyl monomers in an ionic liquid and their characterization as polymer electrolytes. *J. Am. Chem. Soc.*, **127**, 4976–4983.
26. Ohno, H., Yoshizawa, M., Ogihara, W. (2003) A new type of polymer gel electrolyte: zwitterionic liquid/polar polymer mixture. *Electrochim. Acta*, **48**, 2079–2083.

27. Swatloski, R.P., Spear, S.K., Holbrey, J.D. and Rogers, R.D. (2002) Dissolution of cellose with ionic liquids. *J. Am. Chem. Soc.*, **124**, 4974–4975.
28. Nakashima, T. and Kimizuka, N. (2003) Interfacial synthesis of hollow TiO_2 microspheres in ionic liquids. *J. Am. Chem. Soc.*, **125**, 6386–6387.
29. Evans, D.F., Yamauchi, A., Wei, G.J. and Bloomfield, V.A. (1983) Micelle size in ethylammonium nitrate as determined by classical and quasi-elastic light-scattering. *J. Phys. Chem.*, **87**, 3537–3541.
30. Araos, M.U. and Warr, G.G. (2005) Self-assembly of nonionic surfactants into lyotropic liquid crystals in EAN, a room-temperature ionic liquid. *J. Phys. Chem. B*, **109**, 14275–14277.
31. Atkin, R. and Warr, G.G. (2007) Phase behavior and microstructure of microemulsions with a room-temperature ionic liquid as the polar phase. *J. Phys. Chem. B*, **111**, 9309–9316.
32. Fletcher, K.A. and Pandey, S. (2004) Surfactant aggregation within room-temperature ionic liquid 1-ethyl-3-methylimidazolium bis(trifluoromethylsulfonyl)imide. *Langmuir*, **20**, 33–36.
33. Atkin, R. and Warr, G.G. (2005) Self-assembly of a nonionic surfactant at the graphite/ionic liquid interface. *J. Am. Chem. Soc.*, **127**, 11940–11941.
34. Anderson, J.L., Pino, V., Hagberg, E.C., Sheares, V.V. and Armstrong, D.W. (2003) Surfactant solvation effects and micelle formation in ionic liquids. *Chem. Commun.*, **19**, 2444–2445.
35. Seth, D., Chakraborty, A., Setua, P. and Sarkar, N. (2006) Interaction of ionic liquid with water in ternary microemulsions (triton X-100/water/1-butyl-3-methylimidazolium hexafluorophosphate) probed by solvent and rotational relaxation of coumarin 153 and coumarin 151. *Langmuir*, **22**, 7768–7775.
36. Eastoe, J., Gold, S., Rogers, S.E., Paul, A., Welton, T., Heenan, R.K. and Grillo, I. (2005) Ionic liquid-in-oil microemulsions. *J. Am. Chem. Soc.*, **127**, 7302–7303.
37. Kimizuka, N. and Nakashima, T. (2001) Spontaneous self-assembly of glycolipid bilayer membranes in sugar-philic ionic liquids and formation of ionogels. *Langmuir*, **17**, 6759–6761.
38. He, Y.Y., Li, Z.B., Simone, P. and Lodge, T.P. (2006) Self-assembly of block copolymer micelles in an ionic liquid. *J. Am. Chem. Soc.*, **128**, 2745–2750.
39. Johnston, K.P., Harrison, K.L., Clarke, M.J., Howdle, S.M., Heitz, M.P., Bright, F.V., Carlier, C. and Randolph, T.W. (1996) Water in carbon dioxide microemulsions: An environment for hydrophiles including proteins. *Science*, **271**, 624–626.
40. Goetheer, E.L.V., Vorstman, M.A.G. and Keurentjes, J.T.F. (1999) Opportunities for process intensification using reverse micelles in liquid and supercritical carbon dioxide. *Chem. Eng. Sci.*, **54**, 1589–1596.
41. Desimone, J.M., Maury, E.E., Menceloglu, Y.Z., McClain, J.B., Romack, T.J. and Combes, J.R. (1994) Dispersion polymerizations in supercritical carbon-dioxide. *Science*, **265**, 356–359.
42. Desimone, J.M., Guan, Z. and Elsbernd, C.S. (1992) Synthesis of fluoropolymers in supercritical carbon-dioxide. *Science*, **257**, 945–947.
43. Goldfarb, D.L., de Pablo, J.J., Nealey, P.F., Simons, J.P., Moreau, W.M. and Angelopoulos, M. (2000) Aqueous-based photoresist drying using supercritical carbon dioxide to prevent pattern collapse. *J. Vacuum Sci. Technol. B*, **18**, 3313–3317.
44. Zhang, X.G., Pham, J.Q., Martinez, H.J., Wolf, P.J., Green, P.F. and Johnston, K.P. (2003) Water-in-carbon dioxide microemulsions for removing post-etch residues from patterned porous low-k dielectrics. *J. Vacuum Sci. Technol. B*, **21**, 2590–2598.
45. Ohde, H., Hunt, F. and Wai, C.M. (2001) Synthesis of silver and copper nanoparticles in a water-in-supercritical-carbon dioxide microemulsion. *Chem. Mater.*, **13**, 4130–4135.
46. Jessop, P.G., Ikariya, T. and Noyori, R. (1994) Homogeneous catalytic-hydrogenation of supercritical carbon-dioxide. *Nature*, **368**, 231–233.
47. Holmes, J.D., Steytler, D.C., Rees, G.D. and Robinson, B.H. (1998) Bioconversions in a water-in-CO_2 microemulsion. *Langmuir*, **14**, 6371–6376.
48. Consani, K.A. and Smith, R.D. (1990) Observations on the solubility of surfactants and related molecules in carbon dioxide at 50°C. *J. Supercritical Fluids*, **3**, 51–65.

49. Hoefling, T.A., Enick, R.M. and Beckman, E.J. (1991) Microemulsions in near-critical and supercritical CO_2. *J. Phys. Chem.*, **95**, 7127–7129.
50. Zielinski, R.G., Kline, S.R., Kaler, E.W. and Rosov, N. (1997) A small-angle neutron scattering study of water in carbon dioxide microemulsions. *Langmuir*, **13**, 3934–3937.
51. Eastoe, J., Gold, S. and Steytler, D.C. (2006) Surfactants for CO_2. *Langmuir*, **22**, 9832–9842.
52. Eastoe, J. and Gold, S. (2005) Self-assembly in green solvents. *Phys. Chem. Chem. Phys.*, **7**, 1352–1362.
53. Fulton, J.L., Pfund, D.M., McClain, J.B., Romack, T.J., Maury, E.E., Combes, J.R., Samulski, E.T., Desimone, J.M. and Capel, M. (1995) Aggregation of amphiphilic molecules in supercritical carbon-dioxide – a small-angle X-ray-scattering study. *Langmuir*, **11**, 4241–4249.
54. Eastoe, J., Paul, A., Nave, S., Steytler, D.C., Robinson, B.H., Rumsey, E., Thorpe, M. and Heenan, R.K. (2001) Micellization of hydrocarbon surfactants in supercritical carbon dioxide. *J. Am. Chem. Soc.*, **123**, 988–989.
55. Eastoe, J., Gold, S., Rogers, S., Wyatt, P., Steytler, D.C., Gurgel, A., Heenan, R.K., Fan, X., Beckman, E.J. and Enick, R.M. (2006) Designed CO_2-philes stabilize water-in-carbon dioxide microemulsions. *Angewandte Chemie – International Edition*, **45**, 3675–3677.
56. Schwan, M. (2005) Überkritische Mikroemulsionen zur Herstellung nanozellulärer Schäume – Principle of Supercritical Microemulsion Expansion (POSME). Ph.D. Thesis, University of Cologne.
57. Ray, A. (1969) Micelle formation in pure ethylene glycol. *J. Am. Chem. Soc.*, **91**, 6511–6512.
58. Almgren, M., Swarup, S. and Lofroth, J.E. (1985) Effect of formamide and other organic polar-solvents on the micelle formation of sodium dodecyl-sulfate. *J. Phys. Chem.*, **89**, 4621–4626.
59. Friberg, S.E., Liang, P., Liang, Y.C., Greene, B. and Vangilder, R. (1986) A nonaqueous lamellar liquid-crystal with an ionic surfactant long-chain alcohol combination. *Colloids Surf.*, **19**, 249–253.
60. Warnheim, T. and Jonsson, A. (1988) Phase-diagrams of alkyltrimethylammonium surfactants in some polar-solvents. *J. Colloid Interface Sci.*, **125**, 627–633.
61. Backlund, S., Bergenstahl, B., Molander, O. and Warnheim, T. (1989) Aggregation of tetradecyltrimethylammonium bromide in water, 1,2-Ethanediol, and their mixtures. *J. Colloid Interface Sci.*, **131**, 393–401.
62. Bergenstahl, B.A. and Stenius, P. (1987) Phase-diagrams of dioleoylphosphatidylcholine with formamide, methylformamide, and dimethylformamide. *J. Phys. Chem.*, **91**, 5944–5948.
63. Belmajdoub, A., Marchal, J.P., Canet, D., Rico, I. and Lattes, A. (1987) Formamide, a water substitute. 13. Phase-behavior of CTAB in formamide. *New J. Chem.*, **11**, 415–418.
64. Wolff, T. and von Bunau, G. (1984) Light-induced viscosity changes and phase-transitions in aqueous CTAB-solutions containing aromatic-molecules. *Ber. Bunsenges. Phys. Chem.*, **88**, 1098–1101.
65. Weast, R.C. (1982) *Handbook of Chemistry and Physics*, 63rd edn, Chemical Rubber Company, Cleveland.
66. Schubert, K.V., Busse, G., Strey, R. and Kahlweit, M. (1993) Microemulsions with formamide as polar-solvent. *J. Phys. Chem.*, **97**, 248–254.
67. Martino, A. and Kaler, E.W. (1990) Phase-behavior and microstructure of nonaqueous microemulsions. *J. Phys. Chem.*, **94**, 1627–1631.
68. Martino, A. and Kaler, E.W. (1995) Phase-behavior and microstructure of nonaqueous microemulsions 2. *Langmuir*, **11**, 779–784.
69. Rananavare, S.B., Ward, A.J.I., Osborne, D.W., Friberg, S.E. and Kaiser, H. (1988) A small-angle neutron-scattering study of a nonaqueous 3-component microemulsion. *J. Phys. Chem.*, **92**, 5181–5183.
70. Schubert, K.V. and Strey, R. (1991) Small-angle neutron-scattering from microemulsions near the disorder line in water formamide octane C_iE_j systems. *J. Chem. Phys.*, **95**, 8532–8545.

71. Das, K.P., Ceglie, A. and Lindman, B. (1987) Microstructure of formamide microemulsions from NMR self-diffusion measurements. *J. Phys. Chem.*, **91**, 2938–2946.
72. Gao, F., Ho, C.C. and Co, C.C. (2006) Polymerization in bicontinuous microemulsion glasses. *Macromolecules*, **39**, 9467–9472.
73. Gao, F., Ho, C.-C. and Co, C.C. (2004) Sugar-based microemulsion glass templates. *J. Am. Chem. Soc.*, **126**, 12746–12747.
74. Liu, S.Y., Gonzalez, Y.I., Danino, D. and Kaler, E.W. (2005) Polymerization of wormlike micelles induced by hydrotropic salt. *Macromolecules*, **38**, 2482–2491.
75. Lee, Y.S., Yang, J.Z., Sisson, T.M., Frankel, D.A., Gleeson, J.T., Aksay, E., Keller, S.L., Gruner, S.M. and Obrien, D.F. (1995) Polymerization of nonlamellar lipid assemblies. *J. Am. Chem. Soc.*, **117**, 5573–5578.
76. Morgan, J.D., Lusvardi, K.M. and Kaler, E.W. (1997) Kinetics and mechanism of microemulsion polymerization of hexyl methacrylate. *Macromolecules*, **30**, 1897–1905.
77. Co, C.C. and Kaler, E.W. (1998) Particle size and monomer partitioning in microemulsion polymerization 2. Online small angle neutron scattering studies. *Macromolecules*, **31**, 3203–3210.
78. Co, C.C., de Vries, R. and Kaler, E.W. (2001) Microemulsion polymerization 1. Small-angle neutron scattering study of monomer partitioning. *Macromolecules*, **34**, 3224–3232.
79. Co, C.C., Cotts, P., Burauer, S., de Vries, R. and Kaler, E.W. (2001) Microemulsion polymerization 3. Molecular weight and particle size distributions. *Macromolecules*, **34**, 3245–3254.
80. de Vries, R., Co, C.C. and Kaler, E.W. (2001) Microemulsion polymerization 2. Influence of monomer partitioning, termination, and diffusion limitations on polymerization kinetics. *Macromolecules*, **34**, 3233–3244.
81. Burban, J.H., He, M.T. and Cussler, E.L. (1995) Organic microporous materials made by bicontinuous microemulsion polymerization. *Aiche J.*, **41**, 907–914.
82. Chieng, T.H., Gan, L.M., Teo, W.K. and Pey, K.L. (1996) Porous polymeric membranes by bicontinuous microemulsion polymerization: Effect of anionic and cationic surfactants. *Polymer*, **37**, 5917–5925.
83. Challa, V., Kuta, K., Lopina, S., Cheung, H.A. and von Meerwall, E. (2003) Microporosity of bicontinuous nanoporous polymeric materials, characterized with restricted diffusion. *Langmuir*, **19**, 4154–4161.
84. Chieng, T.H., Gan, L.M., Chew, C.H., Lee, L., Ng, S.C., Pey, K.L. and Grant, D. (1995) Microporous polymeric materials by microemulsion polymerization – effect of surfactant concentration. *Langmuir*, **11**, 3321–3326.
85. Santhanalakshmi, J. and Anandhi, K. (1996) Microporous and embedded polymeric composites of vinyltoluene from microemulsion polymerization. *Langmuir*, **12**, 3320–3325.
86. Chieng, T.H., Gan, L.M., Chew, C.H., Ng, S.C. and Pey, K.L. (1996) Microporous polymeric materials by microemulsion polymerization: Effect of the ratio of long and short alkyl chain length cationic surfactants. *Polymer*, **37**, 4823–4831.
87. Li, T.D., Gan, L.M., Chew, C.H., Teo, W.K. and Gan, L.H. (1996) Preparation of ultrafiltration membranes by direct microemulsion polymerization using polymerizable surfactants. *Langmuir*, **12**, 5863–5868.
88. Gonzalez, Y.I., Stjerndahl, M., Danino, D. and Kaler, E.W. (2004) Spontaneous vesicle formation and phase behavior in mixtures of an anionic surfactant with imidazoline compounds. *Langmuir*, **20**, 7053–7063.
89. Hubbard, F.P., Santonicola, G., Kaler, E.W. and Abbott, N.L. (2005) Small-angle neutron scattering from mixtures of sodium dodecyl sulfate and a cationic, bolaform surfactant containing azobenzene. *Langmuir*, **21**, 6131–6136.
90. McKelvey, C.A., Kaler, E.W., Zasadzinski, J.A., Coldren, B. and Jung, H.T. (2000) Templating hollow polymeric spheres from catanionic equilibrium vesicles: Synthesis and characterization. *Langmuir*, **16**, 8285–8290.

91. Jung, M., German, A.L. and Fischer, H.R. (2001) Polymerisation in lyotropic liquid-crystalline phases of dioctadecyldimethylammonium bromide. *Colloid Polym. Sci.*, **279**, 105–113.
92. Antonietti, M., Caruso, R.A., Hentze, H.P. and Goltner, C. (2000) Hydrophilic gels with new superstructures and their hybrids by nanocasting technologies. *Macromol. Symp.*, **152**, 163–172.
93. Gan, L.M., Li, T.D., Chew, C.H., Teo, W.K. and Gan, L.H. (1995) Microporous polymeric materials from polymerization of zwitterionic microemulsions. *Langmuir*, **11**, 3316–3320.
94. Chew, C.H., Li, T.D., Gan, L.H., Quek, C.H. and Gan, L.M. (1998) Bicontinuous-nanostructured polymeric materials from microemulsion polymerization. *Langmuir*, **14**, 6068–6076.
95. Liu, J., Gan, L.M., Chew, C.H., Teo, W.K. and Gan, L.H. (1997) Nanostructured polymeric materials from microemulsion polymerization using poly(ethylene oxide) macromonomer. *Langmuir*, **13**, 6421–6426.
96. Gan, L.M., Liu, J., Poon, L.P., Chew, C.H. and Gan, L.H. (1997) Microporous polymeric composites from bicontinuous microemulsion polymerization using a polymerizable nonionic surfactant. *Polymer*, **38**, 5339–5345.
97. Dave, H., Gao, F., Liberatore, M., Lee, J.H., Ho, C.C. and Co, C.C. (2007) Self-assembly in sugar-oil complex glasses. *Nat. Mater.*, **6**, 287–290.

Chapter 8

Microemulsions in Cosmetics and Detergents

Wolfgang von Rybinski, Matthias Hloucha and Ingegärd Johansson

8.1 Introduction

Products for personal care and home care have to fulfil many different requirements regarding performance, aesthetics, costs and safety. This has consequences for the use of new technologies such as microemulsions for this kind of applications. If, for example, just one of these parameters (e.g. performance) is sufficient or even excellent while the other parameters do not meet the requirements, then the new technology will not be applied. This example reflects the challenges of applying microemulsions in cosmetic products and in detergency. Although there have been numerous basic studies with microemulsions which show superiority in certain aspects of performance, broad applications of microemulsions in detergency and cosmetic products are not yet achieved. The reasons for this are manifold and depend on the specific application. From an economical point of view the high surfactant concentration is one aspect. However, the fact that the formation of a microemulsion depends on various parameters such as the type of oil, the chosen surfactant, the temperature and the electrolyte content to mention just a few is even more important.

As discussed in Chapters 1 and 3 of this book, the formation of microemulsions can be controlled by adjusting the temperature, the electrolyte content or the hydrophilic–lipophilic balance by varying the ratio of different surface active agents. All these techniques are applicable in cosmetics and detergency and have specific advantages and disadvantages. For example, most consumer products require a temperature stability which is usually not achieved with temperature-induced microemulsions. Therefore, no general rule exists for the most suitable type of microemulsion but it depends on the application. In this chapter, several examples are given which describe the status of different applications in cosmetics and detergency.

8.2 Microemulsions in cosmetics

Microemulsions are used in many cosmetic products and are in the focus of current industrial and university research activities. The reasons and motivations are manifold. The following is a summary of their utilisation in several cosmetic product categories with many references to publications and patents. Common aspects of microemulsions

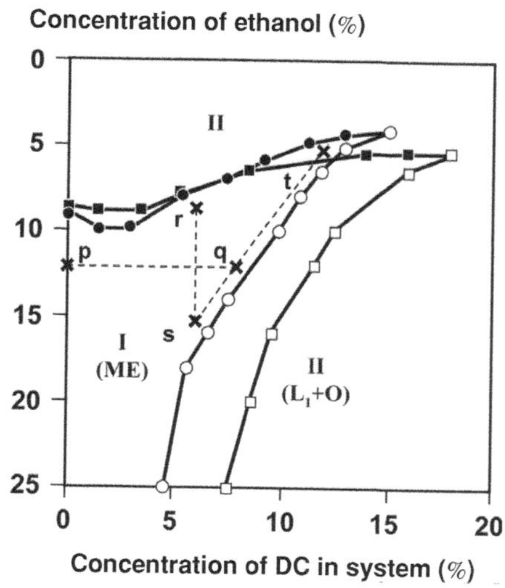

Figure 8.1 The effect of the ethanol concentration on the solubilisation of DC into a 35 wt.% PGMI aqueous solution at 25°C (circles) and 50°C (squares). Closed symbols correspond to the cloud points of the system. (From Ref. [1], reprinted with permission of JOCS.)

in cosmetics are (1) performance on oily soils, (2) product aesthetics, which includes clarity and multiphase products, (3) adjusted flow behaviour, such as shaving gels and hair waxes, (4) microemulsions as reaction media, (5) microemulsions as a carrier and protective matrix for actives and (6) the usage of intermediate microemulsion phases in the emulsification process by means of the phase inversion temperature (PIT) method.

8.2.1 *Cleanser, bath oils, sunscreens, hair treatment*

8.2.1.1 *Cleanser*

The extremely low interfacial tension of microemulsions versus oil makes them a very good candidate for the development of efficient cleanser formulations. Watanabe *et al.* have investigated silicone-based microemulsions for make-up cleanser applications [1]. They studied the phase behaviour of a system composed of the non-ionic surfactant polyoxyethylene glyceryl monoisostearate (PGMI), the silicone oil decamethyl cyclopentasiloxane (DC) and ethanol. The phase diagram for an aqueous solution of 35 wt.% PGMI is shown in Fig. 8.1 as a function of the ethanol and the oil (DC) content. A large microemulsion phase appears below an upper limit of oil and above a minimum ethanol concentration. With increasing temperature, the microemulsion region is enlarged towards higher DC concentrations. The make-up removal performance of this system was evaluated by the removal of a test soil from a textile tissue. The artificial soil consisted of a cosmetic water-in-oil foundation. Systematic trends were noticed in the comparison of detergency with formulation parameters. The performance tests for three different paths through the phase diagram are given in Fig. 8.2. The stepwise addition of oil to the microemulsion which is the denoted path

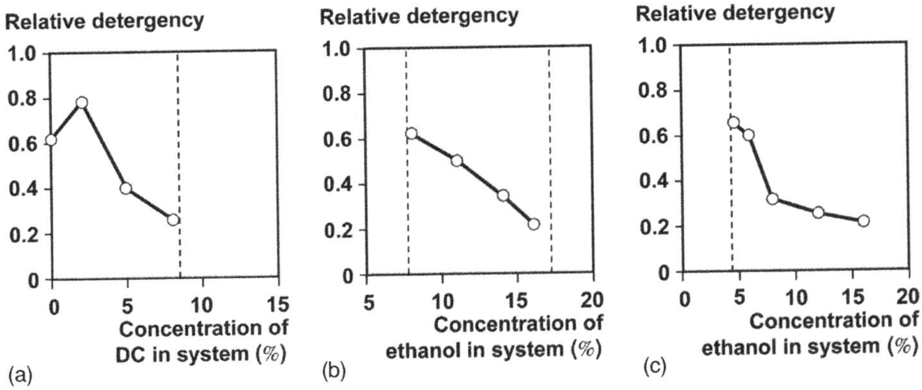

Figure 8.2 Relative detergency of cosmetic soil with 3 wt.% silicone resin in a series of microemulsions containing DC at 25°C. (a) Effect of DC concentration. The composition change corresponds to that from point p to q in Fig. 8.1. Concentration of ethanol is 12 wt.%. (b) Effect of ethanol concentration. The composition change corresponds to that from point r to s in Fig. 8.1. Concentration of DC is 6 wt.%. (c) Effect of composition change along the solubilisation limit from the point t to s in Fig. 8.1. Dashed lines represent the phase boundaries. (From Ref. [1], reprinted with permission of JOCS.)

from point p to q in Fig. 8.1, leads first to a better performance, which is decreased again if further oil is added. The addition of ethanol results in a reduction of the detergency, along two paths r to s and t to s. These three results lead to the conclusion that the detergency improves when the cloud points are approached. Another interesting aspect of this work is that a trend becomes visible when detergency and viscosity data are compared: the higher the viscosity, the better the detergency. The conclusion is that higher aggregation numbers in the microemulsion phase will lead to a better performance.

Skin cleanser products are used to remove dirt and sebum from the skin. For the development of efficient formulations for this task Komesvarakul *et al.* investigated the phase behaviour of microemulsions with artificial sebum [2]. The base formulations were composed of non-ionic and anionic surfactants, ester oil, salt and water. Subsequently, the phase behaviour as a function of added artificial sebum was investigated. It was found that adding sebum leads to the formation of microemulsions at room temperature and at low salt concentrations, which is relevant for skin cleaning applications. The single-phase microemulsion combined the desired product aesthetics with a high performance. Efficient cleanser formulation for the microemulsification of sebum can also be obtained by using non-ionic polymers [3]. A synergistic mixture of a tri-block polypropylenoxide–polyethyleneoxide ether surfactant with a block copolymer of polybutadiene and polyethylenoxide shows a significant increase in the performance on sebum and triolein.

The development of cosmetic microemulsion cleansers with alkyl polyglycosides (APG) was described by Förster *et al.* [4]. This class of non-ionic surfactants has excellent environmental and skin compatibility. Cosmetic cleanser multicomponent systems are required to have good foaming and cleansing performance. Figure 8.3 shows a pseudo-ternary phase diagram of a five-component formulation. It consists of water, the oil dioctyl cyclohexane (DOCH), the non-ionic surfactant $C_{12/14}$-APG, the anionic surfactant fatty alcohol ether sulphate (FAES) and the co-surfactant sorbitan monolaurate (SML). The phase diagram

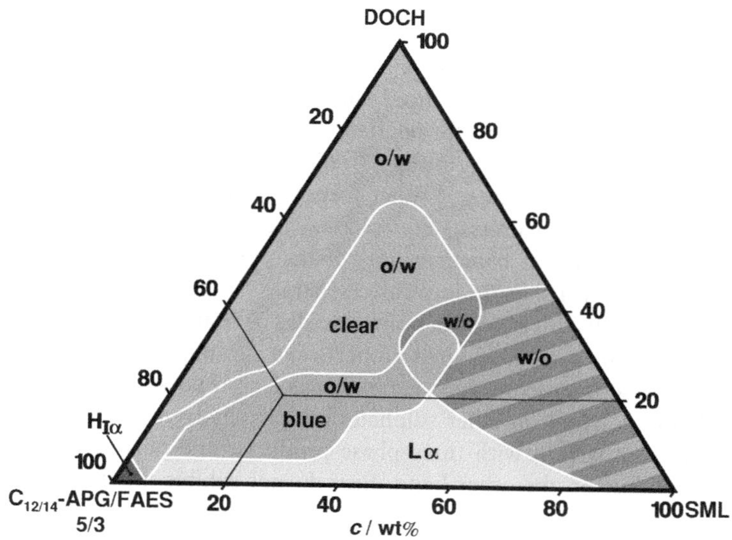

Figure 8.3 Phase behaviour of the system water – dioctyl cyclohexane (DOCH) – $C_{12/14}$-APG + FAES (5:3 mixture) – SML at room temperature and a fixed water content of 60 wt.%. (From Ref. [4], reprinted with permission of Springer.)

was measured at room temperature and at a fixed water content of 60 wt.%. Depending on the ratio between surfactant, co-surfactant and oil, different phases occur: oil-in-water emulsions (o/w, blue areas), water-in-oil emulsions (w/o, red areas), lamellar phases (L_α), hexagonal phases ($H_{1\alpha}$) and microemulsions (striped areas). A 5:3 mixture of APG and an FAES serves as hydrophilic emulsifier. The anionic surfactant (SML) is added due to its high foaming power. Starting from an oil and co-emulsifier-free system a 40% APG/FAES mixture forms a viscous hexagonal liquid crystal in water. Only a small fraction of the APG/FAES mixture has to be exchanged by the hydrophobic co-surfactant SML in order to obtain a low-viscous lamellar phase. Transparent microemulsions are obtained for higher oil and SML fractions. These formulations combine good cleaning performance and foam formation with refattening properties due to the oil compound.

For topical cleanser applications, the flow properties are also important. The product should be easily spreadable on the skin without running. Ayannidis and Ktistis investigated the rheological properties of microemulsions based on polyoxyethylensorbitanoleate, isopropylmyristate, glycerol and water [5]. The glycerol to water ratio can be used to adjust the flow behaviour. The viscosity decreases with increasing glycerol content. Another approach describes a facial wash, which combines good cleaning performance with a pleasant spreadability [6]. The formula is based on squalane, non-ionic surfactant and propanol. The resulting product performance is superior to non-microemulsion products.

Aftershave gels also require an adjusted flow behaviour. They should be easily spreadable without running off the face before the shave. Microemulsions can be used to formulate clear aftershave gels with very good sensorical properties. A promising base formula is made from a combination of a cross-linking polymer with an o/w microemulsion [7].

8.2.1.2 Bath oils

Bath oils are used to improve the well-being, cleaning and relaxing atmosphere of a bath. From a chemical point of view, they need surfactants for good foaming and cleaning properties, and also oil components for refattening the skin. Miller *et al.* have investigated the phase behaviour of bath oil formulas [8]. They found that mixtures of non-ionic surfactants, oil, polyethylenglycol and water form three-phase microemulsions with an upper excess oil and a lower excess water phase. The volume of the middle phase can be adjusted by the surfactant concentration. If the composition is shaken one gets a macroemulsion which will split up into three separate, visible phases on standing. With different dye components this effect can be increased. The result is an attractive three-phase product which gives the bath the desired properties of cleaning, foaming and refattening. The use of one-phase microemulsions for bath oils is also possible [9]. A temperature-stable formulation is based on ether sulphates and alkylpolyglucosides in combination with polyols. In comparison with three-phase products, one-phase products are more convenient to use but on the other hand, they do not exhibit the unique aesthetics of three-phase products.

8.2.1.3 Sunscreens

Modern sunscreen formulations are required to be non-sticky, waterproof and easily spreadable. Most market products are milks or macroemulsions. Carlotti *et al.* have investigated microemulsion-based sunscreen formulations [10]. These have the advantage of exhibiting new transparent aesthetics combined with good sensorical and water-resistant properties. The investigated systems consist of a complex combination of three different surfactants, solvents, lipids, sunscreen actives and water. The phase diagrams were systematically investigated in order to find the most efficient microemulsion system with the smallest amount of surfactants. The optimum, in this respect, is a combination of alkylbenzoate, lecithin, decylpolyglucose with hexanediol and water. A simpler approach is based on alkylpolyglucosides as the only surfactant type [11]. The resulting sunscreen products show a high clarity and good storage stability. Polyglycol ester sulphates are another surfactant type which can be used for the formulation of microemulsion-based sunscreens [12]. Microemulsions can also be used to create two-phase sunscreen products, which are appealing to the eye [13]. One layer is formed by an o/w microemulsion, which is based on non-ionic surfactants and organic oils. The second layer is a silicon oil phase with added sunscreen actives.

8.2.1.4 Hair treatment

Microemulsions have also been investigated for the use in chemical hair treatment [14]. Permanent wave products are based on the reduction of hair keratine cystine, which weakens the protein structure and allows a manipulation of the hair shape. Savelli *et al.* compared the cystine reduction obtained by thioglycolic acid in water with that obtained by a microemulsion. The microemulsion is based on the anionic surfactant sodium dodecylsulphate, the co-surfactant pentanol and dodecane as the unpolar oil component. The cysteine formation is evaluated over a time period of 5 min. The experimental data are

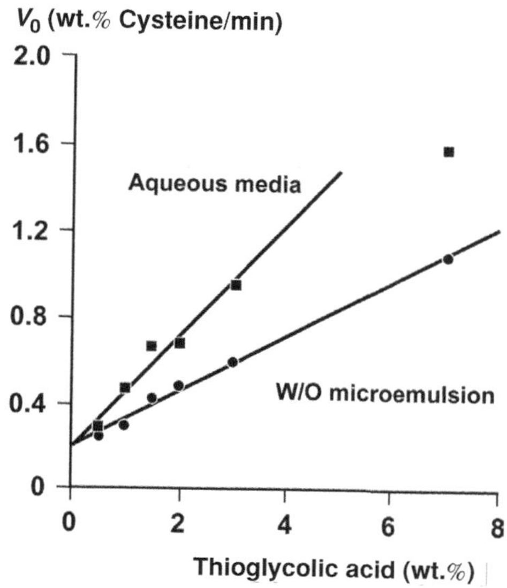

Figure 8.4 Initial cysteine formation rate (V_0) between 0 and 5 min as a function of thioglycolic acid concentration in an aqueous media and in a microemulsion, respectively. (From Ref. [14], reprinted with permission of Elsevier.)

fitted to a pseudo-first-order model (see Fig. 8.4). The cysteine formation rate is higher in the aqueous media, which leads to slower kinetics of the reduction process in the microemulsion phase. Therefore, the microemulsion is a less efficient reaction medium for permanent wave products. The activity of thioglycolic acid in the microemulsion is about one-half of the aqueous media. A second important aspect of permanent wave agents is the improvement of skin compatibility [15]. Permanent wave products, which are microemulsions based on non-ionic surfactants are less irritating to the skin than macroemulsions. Microemulsions have also been utilised for hair care products. Holloran and Hoag report on a microemulsion product with modified silicone oils and cationic surfactants. This product exhibits good long-term stability, high clarity, and shows a very good performance on the combability of treated hair [16]. Ostergaard *et al.* developed microemulsions with a quaternised silicone oil for hair care products [17]. This product improves the colour retention and the combability of hair conditioners. Moreover, clear hair conditioning formulations, which remain clear on dilution and are highly freeze stable, can be prepared from microemulsions [18]. A special example is a microemulsion that consists of a quaternary ammonium salt (which gives the hair conditioning performance), a polar solvent and an unpolar oil phase.

Hair styling waxes with a high viscosity, good spreadability, oil ingredients and visible clearity are based on microemulsions [19]. With more than 30% of non-ionic surfactants and about 10% oil, the obtained microemulsion structure results in a well performing styling product. The required amount of surfactants can be reduced by the addition of cross-linking polymers [20]. They maintain the highly viscous structure even at less than 20% of surfactants.

8.2.2 Improved skin and bio-compatibility

An important step towards more skin-friendly cosmetic microemulsions is the replacement of skin-irritating co-solvents like pentanol, which are commonly used for microemulsion formulations. Comelles and Leal [21] investigated a combination of oleic acid and different glycols as an alternative to pentanol in microemulsions. The investigated systems also contain an unpolar oil and an anionic surfactant. They proved that these systems can provide a good basis for new formulation concepts. The optimum relation of water to glycol is about one. The smaller the alkyl chain of the glycol, the larger the microemulsion phase, which makes ethylene glycol the best candidate for further developments. Comelles and Leal also investigated butyl lactate as a second alternative for pentanol [22]. They have shown that a wide microemulsion phase appears in the phase diagram for systems based on sodium dodecylsulphate, heptane and butyl lactate. Since butyl lactate is taken from renewable resources and considered to be safe and biodegradable, this approach is an interesting starting point for further product developments. Park *et al.* have extended this work and used isopropyl myristate, which is standard oil in cosmetic formulations [23]. Isopropyl myristate is a fast spreading emollient and well established for modern cosmetic products. Park *et al.* confirm that butyl lactate is a well-performing co-surfactant for microemulsions. Alcohol-free concepts for microemulsions in cosmetic use are also described [24]. An alcohol-free cleaning composition contains alkyl sulphosuccinate, non-ionic surfactants and squalane. It shows good cleaning performance on lipstick and waterproof mascara.

While the above-mentioned research work focuses on the replacement of co-surfactants, Nakamura *et al.* investigated alternative surfactants in microemulsions [25]. In their work, they investigated sucrose monododecanoate, which is a sugar-based biocompatible surfactant. In the phase diagram with hexanol and decane they found one- and three-phase microemulsion regions with a bicontinuous structure, which were examined by NMR and SAXS experiments. Sucrose monododecanoate is a promising surfactant for the formulation of microemulsions. Alkylpolyglucosides are a second class of sugar-based surfactants. Their phase behaviour has been studied in several research works. von Rybinski and Wegener studied alkylpolyglucosides with different co-surfactants: pentanol, sorbitan monolaurate, and glycerylmonolaurate [26]. Temperature insensitive microemulsions can be obtained from optimised combinations of alkylpolyglucosides and co-surfactants, which provide a basis for an extended use of microemulsions in product applications. Comelles combined alkyl polyglucosides with butyl lactate as an alternative co-surfactant [27]. This combination allows obtaining temperature insensitive microemulsions.

New biocompatible oils from renewable resources have also been investigated. Acharya *et al.* investigated the impact of the addition of ricebran oil on the phase behaviour of microemulsions [28]. In combination with isopropylmyristate as second oil a large microemulsion domain is formed in the phase diagram, which makes ricebran oil a potential oil base for microemulsions. Another approach to improve skin friendliness is a reduction of the surfactant content of microemulsions. Diec *et al.* report on optimised surfactant–co-surfactant systems in combination with a phase inversion process to reach this goal [29]. The resulting formulations are clear, stable over the long term and contain less than 10% of surfactants.

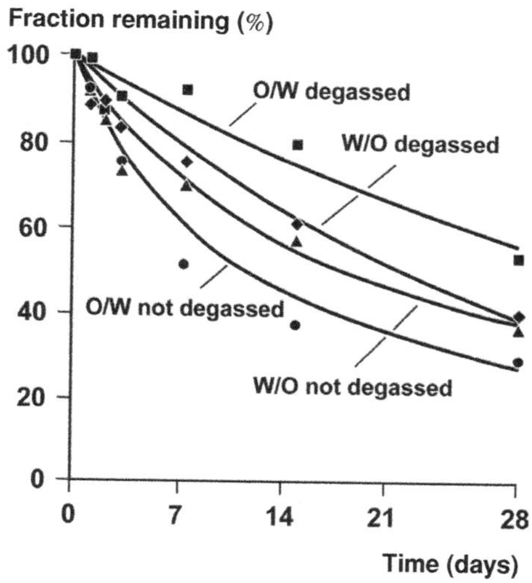

Figure 8.5 The influence of degassing on the stability of ascorbyl palmitate at an initial concentration of 2 wt.% for w/o- and o/w-microemulsions. (From Ref. [30], reprinted with permission of Elsevier.)

8.2.3 Carrier for skin actives

Microemulsions are also under investigation as potential carriers for cosmetic actives. For example, derivatives of Vitamin C have been studied in microemulsions. These substances are used to suppress skin pigmentation and to whiten the skin. Spiclin *et al.* investigated the storage stability of ascorbyl palmitate and ascorbyl phosphate in microemulsions consisting of triglyceride oil, an ethoxylated surfactant and a polyglyceryl-based surfactant [30]. They compared two different compositions, one is a w/o microemulsion and the other is an o/w microemulsion. The storage stability of ascorbyl phosphate in both microemulsions was investigated both in a non-degassed and in a degassed state. The results are summarised in Fig. 8.5. The best stability is found in the degassed o/w-microemulsion. This is in contrast to the non-degassed systems where the stability is better in the w/o-microemulsion. However, all systems exhibit a significant decomposition after 1 month's storage time.

Szymula investigated the impact of Vitamin C on the flow properties of microemulsions [31]. Microemulsions based on sodium dodecylsulphate, pentanol and water were evaluated. They exhibit Newtonian flow behaviour for o/w and w/o systems, while the bicontinuous phase is shear thinning. The addition of ascorbic acid leads to an increase in viscosity.

Jojoba oil is a widely used cosmetic oil with excellent properties. It combines good skin feel, good spreading and fast penetration properties. Shevachman *et al.* have investigated microemulsions based on jojoba oil in combination with different non-ionic surfactants [32]. The size of the microemulsion region in the phase diagram is dependant on the chain length of alcohols, which are added as co-solvents, and the choice of the non-ionic surfactant. In a second work, lycopene was added to jojoba-based microemulsions [33]. Lycopene is an antioxidant with a poor solubility in water and oil phases. Lycopene is soluble in microemulsions. The addition leads to a change in the curvature of the oil–water interface; it can thus be concluded that this compound is located at the interface. The resulting solutions are a base for further developments of clear and transparent products with this active ingredient.

There is a large consumer demand for skin moisturisation in day care products. Skin care compositions, which contain a skin moisturiser in combination with a humectant, are found to be very effective in microemulsion formulations based on non-ionic surfactant in combination with siloxanes and organic oil [34].

Aluminium chlorohydrate is a widely used active ingredient for antiperspirant compositions. This salt requires a hydrophilic solvent at low pH values to remain stable. Structured microemulsions can be used to formulate clear and viscous products with this active. These products can be used in solid applicators such as firm stick applicators [35]. The o/w microemulsions can be based on fatty alcoholethoxylates with an organic or silicon oil phase [36]. An improved temperature stability in the range of 0–70°C for such formulations can be achieved by a combination of oxyalkylene-modified siloxanes, pentacyclosiloxane and glycol [37]. The addition of alpha-hydroxylic acids results in an improved skin feel with decreased tackiness [38].

8.2.4 Perfume

The unique properties of microemulsions such as long-term stability and aesthetics make them attractive as carrier systems for perfumes in cosmetic products. von Rybinski *et al.* studied the phase behaviour of a model formulation with 20 wt.% perfume oil, 20 wt.% emulsifier mixture consisting of alkyl polyglycosides (APG) and glyceryl monooleate (GMO), an oil content of less than 1 wt.% (dicapryl ether and octyldodecanol) and water [39]. The resulting phase behaviour is shown in Fig. 8.6. The requirements for the formation of a microemulsion with 20 wt.% perfume oil are (a) a GMO concentration between 15 and 25 wt.% and (b) a suitable mixture of $C_{12/14}$-APG with $C_{8/10}$-APG. Stubenrauch *et al.* studied microemulsions with geraniol [40], which is a double unsaturated terpene alcohol and widely used as perfume compound. This study shows that geraniol acts as an efficient co-surfactant in microemulsion systems.

The consumer perception of a fragrance is related to the evaporation rate of a product. Hamdan *et al.* have investigated limonene as a model perfume compound in a microemulsion system [41]. Dioctyl sodium succinate (AOT) was used as the only surfactant without any further solvents. The evaporation rates were evaluated for different limonene to AOT ratios, and as a function of the water content. It was found that an increasing limonene to AOT ratio leads to a higher perfume evaporation rate; increasing water content will also result in a higher evaporation rate. In a second work, the impact of the environmental humidity on the evaporation rate was investigated [42]. It is shown that with increasing humidity the evaporation rate is reduced. Friberg investigated the vapour pressure of model fragrance ingredients in microemulsions [43]. The phase diagram of phenethyl alcohol, a fatty alcohol ethoxylate as the surfactant and water is given. Vapour pressure measurements show that the fragrance intensity is almost linear dependant on the mole fraction of the perfume compound in the solution.

Another important aspect is the protection of fragrance compounds in solutions against autoxidation. Carlotti *et al.* compared the stability of linalool, citral and limonene in

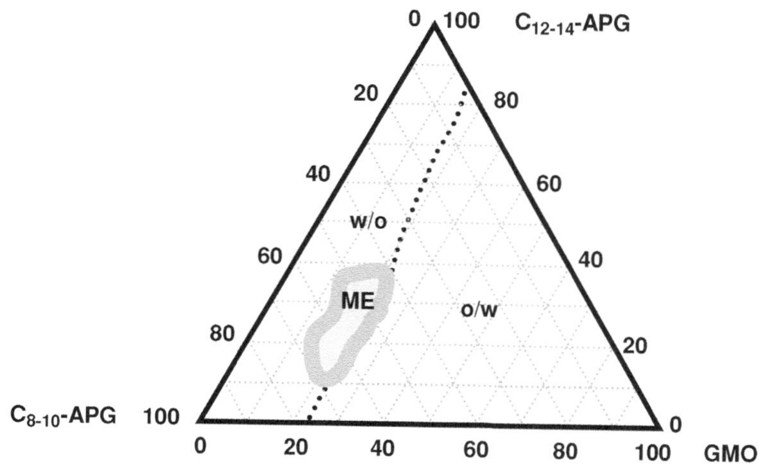

Figure 8.6 Pseudoternary phase diagram of a system containing 20 wt.% emulsifier ($C_{8/10}$-APG, $C_{12/14}$-APG and GMO), 20 wt.% perfume oil, 0.6 wt.% oil (dicapryl ether, octyldodecanol) and 59.4 wt.% water at 25°C. The formation of microemulsions was studied as a function of the emulsifier's composition. The dotted lines separate the o/w- from the w/o-region. ME indicates a one-phase microemulsion. (From Ref. [39], reprinted with permission of Elsevier.)

different formulations [44]. For example, they compared three different formulas: (A) a micellar solution of citral with decyl polyglucoside, (B) a micellar solution of citral with polyoxyethylene sorbitane monolaurate and (C) a microemulsion with citral, decyl polyglucoside, propyleneglycol and dodecanol. The results for the oxygen consumptions are summarised in Fig. 8.7. It was found that the storage stability of citral against oxidation is significantly better in the microemulsion formulation.

The transfer of well-known perfume brands into cosmetic formulations will lead to the challenge of maintaining the perfume impression despite possible interactions with cosmetic care compounds. Microemulsions have proven to be very efficient in this respect. For example, a clear aftershave microemulsion formulated with non-ionic surfactants and isoeicosane is almost non-interfering with the perfume impression [45]. Because of the trend to reduce the amount of volatile organic chemicals, ethanol-free perfume microemulsions are under further development. Non-sticky, non-fatty and ethanol-free products can be obtained through the usage of vicinal diols such as 1,2-hexanediol in microemulsion formulations [46].

8.2.5 The phase inversion temperature method

At a certain temperature some emulsions formulated with non-ionic surfactants change their structure, namely from o/w to w/o emulsions [47]. This process is reversible, i.e. that cooling below this so-called PIT leads again to the formation of an o/w emulsion. Forming emulsions via the PIT method often leads to very fine and long-term stable emulsions with particle sizes below 1 μm [48]. The main requirement that needs to be fulfilled is the

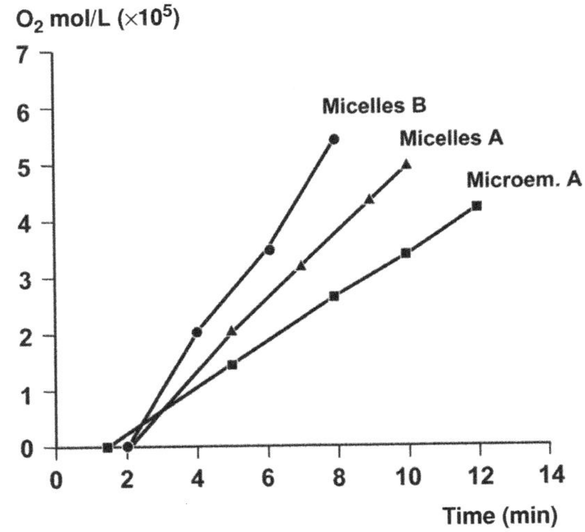

Figure 8.7 Oxygen consumption in micellar solutions and in a microemulsion by oxidation of 2 wt.% w/w citral initiated by AIBN at 45°C. (From Ref. [44], reprinted with permission of the Society of Cosmetic Chemists.)

presence of a microemulsion between the o/w- and the w/o-emulsion. It is only then that blue PIT emulsions with particles in the submicron range are formed. The principle of this process is summarised in Fig. 8.8.

In an early work with cosmetic ingredients the phase behaviour of potential base systems was investigated [49]. Different ethoxylated fatty alcohols were used in combination with

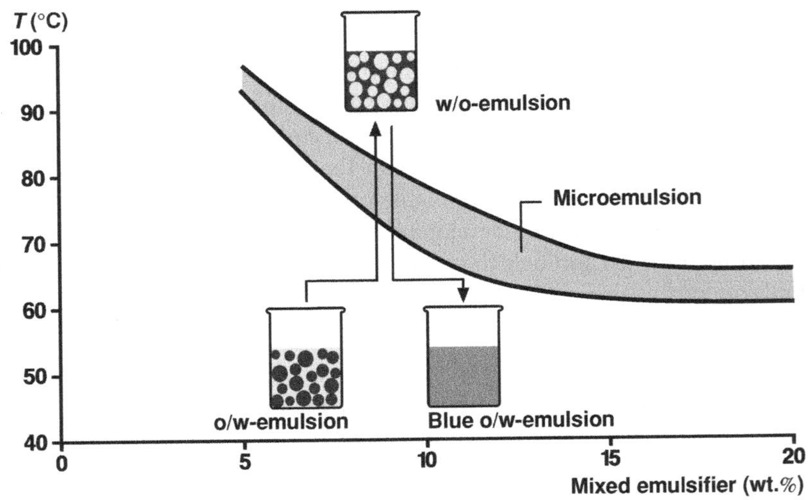

Figure 8.8 Principle of the PIT method: an o/w-emulsion changes into a w/o-emulsion above a certain temperature. In the phase inversion range a microemulsion develops, which becomes a blue o/w-emulsion after cooling down. (From Ref. [48], reprinted with permission of Elsevier.)

paraffin and further cosmetic additives. Systematic trends for the PIT were seen depending on the composition and surfactant type. The addition of polar co-surfactants such as lauryl alcohol and oleic acid leads to a reduction of the PIT, while the addition of polar oils such as castor oil and octyldodecanol results in an increase of the PIT. These results can be summarised in a simple calculation scheme to estimate the PIT of compositions with cosmetic additives [50]. This approach was extended by many works, for example by Förster *et al.* [51], who also used ethoxylated fatty alcohol with polar oils and co-emulsifiers. The long-term storage properties of fine PIT emulsions are demonstrated by particle size measurements, their rheological behaviour and the conductivity of the emulsions. An improved calculation method to predict the PIT based on oil and emulsifier parameters was developed and successfully applied to many systems [52].

An alternative surfactant combination which is free of ethoxylated molecules is based on rapeseed sorbitol ester and sodium lauroyl glutamate [53]. Here, the phase inversion from a w/o- to an o/w-emulsion can be initiated by the addition of lauroyl glutamate, which is a hydrophilic surfactant, instead of using the temperature. Penetration studies for the release of Vitamin E from PIT emulsions in comparison with other formulation concepts have been performed [54]. It has been shown that the penetration of Vitamin E into the skin is better for a w/o-cream than a PIT emulsion. The free diffusion of Vitamin E might be hindered by the oil–water interface, which acts as a barrier around the oil droplets.

Numerous cosmetic applications of PIT emulsions have been described in patent literature. In the area of skin and body care products, new concepts for sunscreen compositions with UV filters in the lipophilic phase have been proposed [55]. The application of sunscreens onto the skin is more comfortable using a spray applicator instead of using a cream. Because of their small particle size, PIT emulsions are ideal for this task, since they are sprayable and long-term stable. Another idea is to combine UV protection which acts against the harmful properties of sun exposition, with tanning actives for a healthy look [56]. This product can be used in an airbrush system.

Thin and sprayable antiperspirant formations can be obtained by the PIT method [57]. They contain aluminium chlorohydrate, which is an active antiperspirant compound. The disclosed formulations are based on ethoxylated fatty alcohols in combination with an oil mixture and propylenglycol. After phase inversion the resulting particle size is in the range of 100–300 nm, which makes the product sprayable and long-term stable. A second approach with the same deodorant active describes alcohol-free PIT emulsions with cross-linking polymers [58].

Aloe vera is a well-known active with good skin care properties. Fluid milks have been suggested for skin care products with Aloe vera [59]. They have good sensorics and no tacky or sticky feel. In the area of skin cleansers, the addition of a PIT emulsion to a cleanser formulation was described [60]. The result is a rather mild and non-irritant lotion for make-up removal. A similar approach describes the addition of a PIT emulsion with phospholipid to a body cleanser [61]. This product has two functions; it combines cleaning and refattening properties.

PIT emulsions have also been described for the use in hair care products. They can be used for an economic hair dyes production process [62]. In a first step, a PIT emulsion with the oil and the emulsifiers is prepared. After cooling, the dye components are stirred into the PIT emulsion at room temperature. Another important aspect for hair dye applications is the stability, which may be very low due to the high salt content. Thin PIT emulsions

Table 8.1 Substrates and surfaces in detergency

Hard surfaces	Glass, ceramics, metal, polymers
Fibres	Cotton, wool, polymers, glass fibres

provide an elegant solution to improve the stability of hair dyes [63]. For scalp protection a two-component kit for hair treatment with a PIT emulsion and a second component including calming actives was described [64]. The components are mixed together upon application. The treatment can be used before hair bleaching or dyeing in order to protect the scalp.

Some consumers prefer to be free of visible hair in various parts of their skin. Cosmetic products are available to reduce hair growth by actives which penetrate into the skin. It has been shown that skin penetration of difluoromethylornithine, an active to reduce hair growth from PIT emulsions, is superior to emulsions with larger particle sizes [65]. A related approach for a hair removal lotion uses thioglycolic acid as the active ingredient [66].

Nano wax dispersions which are prepared by a PIT process can be added to hair shampoo formulations [67]. The formulations improve the combability of wet hair and accelerate the hair drying after the wash. Wax dispersions can also be used to carry fragrance-active components. These additives can be used to give a long-lasting fragrance impression for cosmetic product applications.

8.3 Microemulsions in detergency

8.3.1 Introduction

The washing and cleaning of surfaces is a complex process which is influenced by many parameters [68, 69]. There are many different types of surfaces, the soils to be removed may vary significantly and the components of the detergents have different structures. Table 8.1 gives an overview of the different substrates and surfaces. The surfaces involved in cleaning processes can be very different ranging from fabrics or hair to metal surfaces, ceramics or skin. Therefore, the mechanism of the cleaning process may vary, although the basic effects are similar. The surface properties of the substrates are decisive for any cleaning process. Important surface properties are specific surface area, polarity, surface charge and porosity. Besides this the interaction of the surfaces with the components of the bulk liquid plays an important role. For example, the adsorption of ions onto the surfaces changes the surface properties. Substrates that have a high content of multivalent cations (calcium ions etc.) on the surface behave differently from surfaces that show a low adsorption of these ions. Because of these effects the different washing results of cotton (high adsorption) and synthetic fibres (low adsorption) can be explained.

The soils involved in cleaning processes can vary significantly (Table 8.2 [70]). The soils can either be solid pigments or a liquid phase like oils or fats. Usually they occur in mixtures, which may cause additional difficulties due to an interaction of the different soils. Difficult to remove soils, e.g. in the washing process of fabrics, are pigments such as carbon black or

Table 8.2 Soils in detergency

Water-soluble materials	Inorganic salts, sugar, urea, perspiration
Pigments	Metal oxides, carbonates, silicates, carbon black (soot)
Fats	Animal fat, vegetable fat, sebum, mineral oil, wax
Proteins from	Blood, egg, milk, skin residues
Bleachable dyes from	Fruit, vegetables, wine, coffee, tea
Carbohydrates	Starch

inorganic oxides and fats and waxes or denatured proteins and certain dyes. The removal of soils can be either by temperature, mechanical force, interfacial processes or by chemical degradation, e.g. by enzymes or bleaching agent.

The composition of a detergent or cleaner may be very complex, containing different types of substances. Tables 8.3 and 8.4 show the typical major components of detergents and cleansers for household and institutional applications [71]. In addition to this the components themselves are mixtures as they are usually of technical grade. This makes the description and interpretation of the interfacial processes even more complex.

Microemulsions show best performance in removing oily soils which is due to their low interfacial tension and their microstructure. Note that the systems in Figs. 8.9–8.11 are no microemulsions, but show the influence of the interfacial tension in general. The interfacial tension is one of the decisive parameters in the rolling-up process [72] and can be very different dependant on the surfactant structure and the type of the oily soil [73]. Figure 8.9 shows this general principle for two different oils and two anionic surfactants. The interfacial tension has been recorded as a function of time. For the two surfactants the interfacial tension is the same with lower values for the non-polar decane. To demonstrate the influence of the polarity of the oil on the efficiency of the surfactant, a more polar oil was chosen (Fig. 8.10). In this case the interfacial tension is significantly lower when the fatty alcohol sulphate is used instead of linear alkylbenzene sulphonate. The increase of the interfacial tension with time is probably caused by a solubility of the surfactant in the oil phase. Figure 8.11 shows the interfacial tension of different detergent formulations against mineral oil. For overall low values of the interfacial tension there are significant differences between the detergents which indicate a different performance against this non-polar oil.

As the interfacial tension should be minimised in cleaning processes, there is the need for a further decrease of the interfacial tension in formulations. One possibility is to create mixed adsorption layers of suitable surfactants [74, 75]. For example, the interfacial tension of the system water/olive oil as a function of composition for a surfactant mixture containing the anionic surfactant sodium *n*-dodecylsulphate with the non-ionic surfactant nonylphenol octaethylene glycol ether shows a pronounced minimum at a certain mixing ratio for a constant total surfactant concentration. Even small additions of one surfactant to another can lead to a significant reduction of the interfacial tension. For this specific example a minimum value of the interfacial tension is reached with a ratio of anionic surfactant to non-ionic surfactant of about 4 to 1. Kinetic effects play an important part in this process. The behaviour of the mixtures can be completely different dependant on time, showing a minimum of the interfacial tension for a certain concentration ratio of the surfactants or not [75]. This has to be taken into account for the search of an effective surfactant system. Another similar example shows the effect on cleaning efficiency by changing the ratio

Table 8.3 Powder heavy-duty detergent formulations around the world

		Composition (%)				
Ingredients	Examples	USA	South America	Europe	China, India	Japan
Surfactants	Alkylbenzene sulphonate, alcohol sulphate; alcohol ethoxylate	10–25[a]	18–25	10–25	8–18	25–40
Builders	Zeolite, sodium tri-phosphate, Na citrate, Na silicate, Na carbonate/Na bicarbonate	40–65	40–55	30–55	30–50	40–55
Co-builders	Sodium polycarboxylate	0–5	–	3–8	–	0–5
Bleaching agents	Sodium perborate; sodium percarbonate	0–10	–	8–15	–	0–6
Bleach activators	TAED, NOBS	0–3[b]	–	1–7[c]	–	0–3[b]
Antiredeposition agents	Carboxymethyl cellulose, cellulose ethers	1–2	0.5–1	0–1	0.5–1	0.5–1
Stabilisers	Phosphonates	–	–	0–1	–	–
Foam regulators	Soap, silicone oil and/or paraffins	–	–	0.1–4		
Enzymes	Protease, cellulose, amylase, lipase	0.3–2	0.3–0.8	0.3–2	0.3–0.8	0.3–1.5
Optical brighteners	Stilbene-, biphenyldistyryl derivatives	0.1–0.3	0.1–0.3	0.1–0.3	0.1–0.2	0.1–0.2
Soil repellents	Poly(ethylene glycol terephthalate) derivatives	0–1	–	0–1.5	–	–
Fillers/processing aids	Sodium sulphate	5–30	20–35	0–30	20–40	5–15
Minors	Fragrance	+/–	+	+/–	+	+
Water		5–15	5–15	5–15	5–15	5–15

[a] All figures expressed as 100% active material, except for enzymes for which figures relate to % granulate.
[b] Nonanoyloxybenzenesulphonate (NOBS).
[c] Tetraacetylethylenediamine.

between dodecyl tetraethylene glycol and sodium octyl benzene sulphatonate finding an optimal soil removal at about 45 wt.% non-ionic in the mixture [76]. Thus, the interfacial tension can be used to optimise cleaning formulations. Microemulsions offer the best way to reduce the interfacial tension to very low values, which is not possible by other ways.

In principle, there are two different possibilities for the application of microemulsions in detergency:

1. The in situ formation of microemulsions with a surfactant-containing detergent during a washing or cleaning process in which the soil acts as the oil phase. This effect has been

Table 8.4 Formulations of various types of detergents for institutional use

	Detergents		Partially built products		
Components	Base	Specialty	Surfactant boosters	Bleaching agents	Enzyme boosters
Surfactants	x	x	x		x
Sodium triphosphate or zeolite/polycarboxylate	x	x			
Alkalies (soda ash, metasilicate)	x	x	x		
Bleaching agents				x	
Fluorescent whitening agents	x		x	x	
Enzymes	x	x			
Complexing agents (phosphonates)	x			x	
Antiredeposition agents	x	x			x

described and studied in detail. The results are applied for the surfactant systems of many household detergents. A typical minimal concentration in the working solution is 3–10 times CMC of the surfactant system.

2. Direct use as (a) a detergent or cleaner (e.g. pre-treatment) or (b) a washing or cleaning liquor. This application is mainly described in basic studies; only few very specific applications are reported up to now.

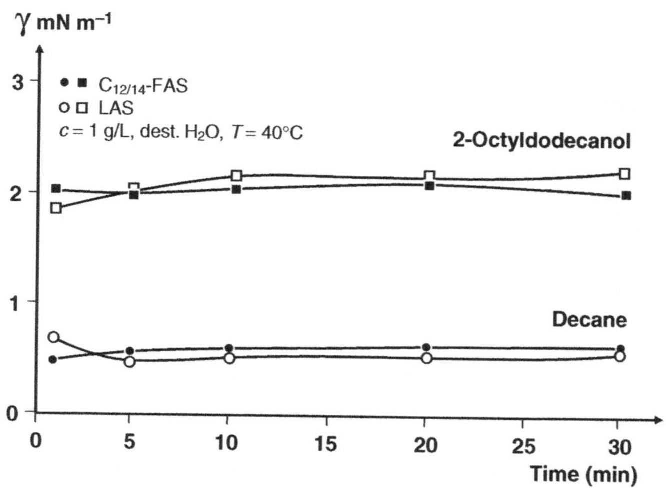

Figure 8.9 Interfacial tension between an aqueous solution of $C_{12/14}$-fatty alcohol sulphate ($C_{12/14}$-FAS) and an aqueous solution of linear alkylbenzene sulphonate (LAS), respectively, and two different oils as a function of time. (From Ref. [73], reprinted with permission of Hanser.)

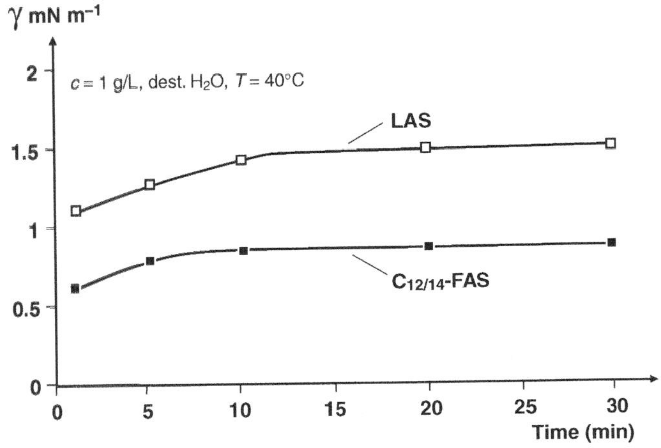

Figure 8.10 Interfacial tension between an aqueous solution of $C_{12/14}$-fatty alcohol sulphate ($C_{12/14}$-FAS) and an aqueous solution of linear alkylbenzene sulphonate (LAS), respectively, and isopropyl myristate as a function of time. (From Ref. [73], reprinted with permission of Hanser.)

8.3.2 In situ formation of microemulsions

The in situ formation of microemulsions can occur in washing processes depending on the oil type and conditions. During the oil removal from hard surfaces or fabrics ternary systems occur where two or three phases coexist in equilibrium. These systems are also referred to as Windsor I or Windsor III microemulsions. The effects were studied in detail for alkyl polyglycol ethers [77]. Depending on temperature different phases exist, having a three-phase region between the temperature T_l and T_u (see Fig. 1.3, Chapter 1). When

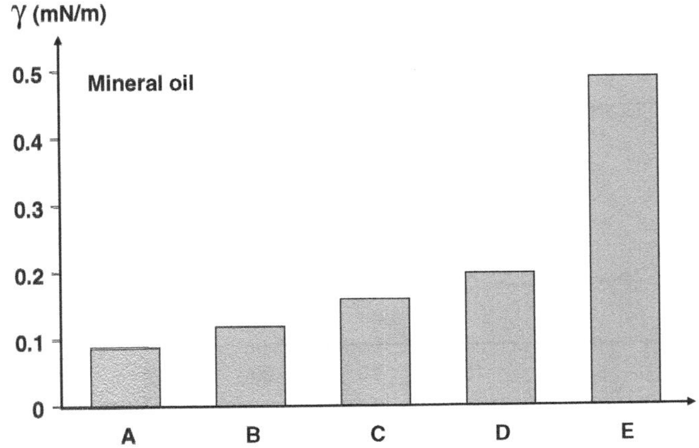

Figure 8.11 Interfacial tension between different aqueous solutions of detergents (A–E) and mineral oil. (From Ref. [73], reprinted with permission of Hanser.)

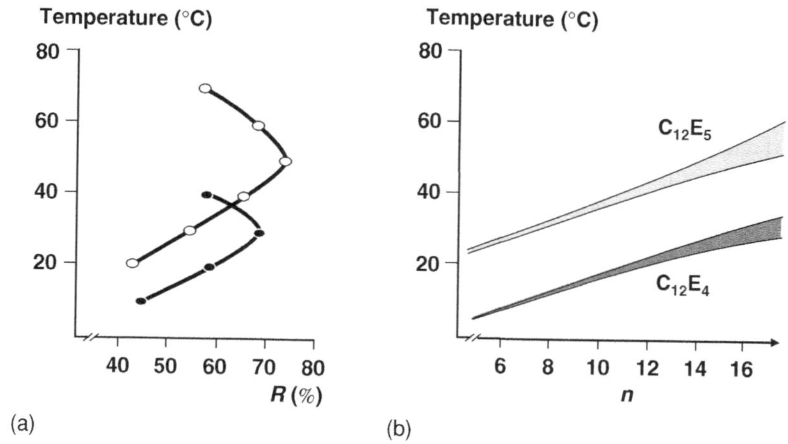

Figure 8.12 Removal (R) of n-hexadecane by $C_{12}E_4$ and $C_{12}E_5$ as a function of temperature (a) and the corresponding three-phase regions for these surfactants as a function of the number n of carbon atoms of alkanes. (From Ref. [78], reprinted with permission of Happi and Marcel Dekker.)

these three phases are formed, extremely low interfacial tensions between two phases are observed (see Fig. 1.14 in Chapter 1). Because the interfacial tension is generally the restraining force with respect to the removal of liquid soil in the cleaning process, it should be as low as possible for optimal soil removal. Other parameters such as the wetting energy and the contact angle on polyester, as well as the emulsifying ability of olive oil, also show optimum values at the same mixing ratio at which the minimum interfacial tension is observed.

Figure 8.12(b) shows the three-phase temperature intervals for $C_{12}E_4$ and $C_{12}E_5$ as a function of the number n of carbon atoms of n-alkanes. Figure 8.12(a) shows the detergency of these surfactants for hexadecane. Both parts of Fig. 8.12 indicate that the maximum oil removal is in the three-phase interval of the oil used (n-hexadecane) [78]. This means that not only the solubilisation capacity of the concentrated surfactant phase, but probably also the minimum interfacial tension existing in the range of the three-phase body are responsible for the maximum oil removal. Further details about the influence of the polarity of the oil, the type of surfactant and the addition of salt are summarised in the review of Miller and Raney [79].

Studies of diffusional phenomena have direct relevance to detergency processes. Experiments are reported which investigate the effects of changes in temperature on the dynamic phenomena, which occur when aqueous solutions of pure non-ionic surfactants contact hydrocarbons such as tetradecane and hexadecane. These oils can be considered to be models of non-polar soils such as lubricating oils. The dynamic contacting phenomena, at least immediately after contact, are representative of those which occur when a cleaner solution contacts an oily soil on a polymer surface. With $C_{12}E_5$ as non-ionic surfactant at a concentration of 1 wt.% in water, quite different phenomena were observed below, above, and well above the cloud point when tetradecane or hexadecane was carefully layered on top of the aqueous solution. Below the cloud point temperature of 31°C very slow solubilisation of oil into the one-phase micellar solution occurred. At 35°C, which is just above the cloud point, a much different behaviour was observed. The surfactant-rich L_l phase separated to the top of the aqueous phase prior to the addition of hexadecane. Upon addition of the oil, the L_l phase rapidly solubilises the hydrocarbon to form an o/w microemulsion containing an appreciable amount of the non-polar oil. After depletion of the larger surfactant-containing drops, a front developed as smaller drops were incorporated into the microemulsion phase.

Unlike the experiments carried out below the cloud point temperature, appreciable solubilisation of oil was observed in the time frame of the study, as indicated by upward movement of the oil-microemulsion interface. Similar phenomena were observed with both tetradecane and hexadecane as the oil phases. When the temperature of the system was raised to just below the PITs of the hydrocarbons with $C_{12}E_5$ (45°C for tetradecane and 50°C for hexadecane), two intermediate phases formed when the initial dispersion of L_l drops in the water contacted the oil. One was the lamellar liquid crystalline phase L_α (probably containing some dispersed water). Above it was a middle-phase microemulsion. In contrast to the studies below the cloud point temperature, there was appreciable solubilisation of hydrocarbon into the two intermediate phases. A similar progression of phases was found at 35°C using *n*-decane as the hydrocarbon. At this temperature, which is near the PIT of the water/decane/$C_{12}E_5$ system, the existence of a two-phase dispersion of L_α and water below the middle-phase microemulsion was clearly evident. These results can be utilised to optimise surfactant systems in cleaners, and in particular to improve the removal of oily soils. The formation of microemulsions is also described in the context of the pre-treatment of oil-stained textiles with a mixture of water, surfactants and co-surfactants.

8.3.3 Direct use of microemulsions

In contrast to the formation of microemulsions from aqueous surfactant systems and oily soils during the cleaning process, less basic research has been carried out on microemulsions as a direct cleaning medium [80]. Some examples will be presented in the following sections.

8.3.3.1 Textile cleaning

Initial studies of textile cleaning with microemulsions on a water base by Solans *et al.* [81] were published in 1985. At washing temperatures between 296 and 307 K homogeneous microemulsions of the system water/*n*-hexadecane/$C_{12}E_4$ and of systems with technical non-ionic surfactant mixtures removed 1.5–2 times more soil from wool, cotton and cotton–polyester blended fabrics stained with oily and particulate soils than a highly concentrated liquid detergent (Fig. 8.13). Soil removal by the microemulsions was increased by 20–25% by adding 0.05 M of the electrolytes sodium triphosphate and sodium citrate, which act as builders. The microemulsions also proved superior to the liquid detergent, in that they could be used seven times without losing any of their cleaning effectiveness.

Dörfler *et al.* [82] systematically studied the phase behaviour of quaternary systems, consisting of water, non-ionic surfactants, a co-surfactant and a hydrocarbon, with regard to possible applications in the textile-cleaning sector. As an example, Fig. 8.14 shows the

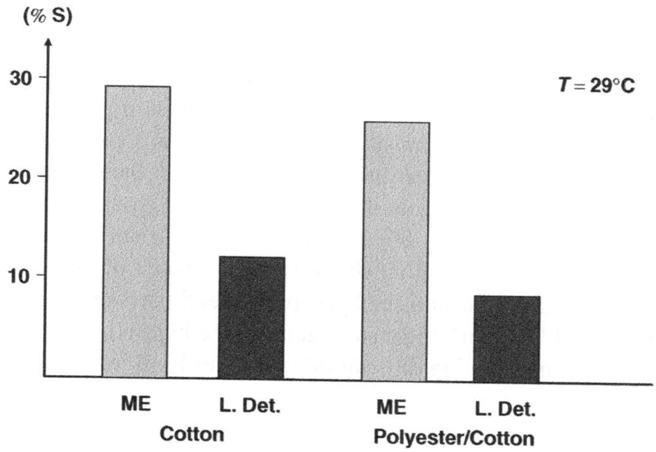

Figure 8.13 Soil removal (*S*) by a surfactant phase microemulsion (ME) and by a 1 wt.% aqueous liquid detergent solution (L. Det) from different fabrics. (From Ref. [81], reprinted with permission of Taylor & Francis.)

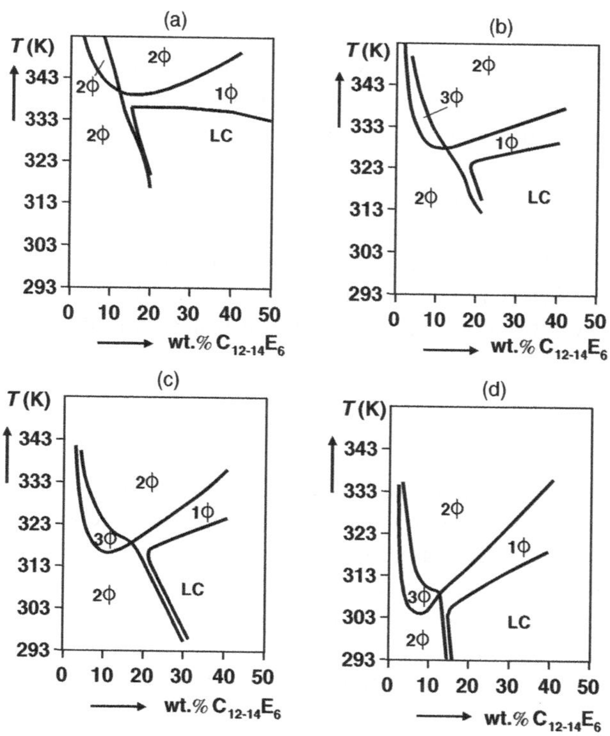

Figure 8.14 Phase behaviour of water/oil/$C_{12/14}E_6$ mixtures without co-surfactant (a), with 2 wt.% *n*-pentanol (b), with 4 wt.% *n*-pentanol (c) and with 6 wt.% *n*-pentanol (d). The water:oil ratio equals 1:1, with the oil being a mixture of *n*-alkanes with 95 wt.% undecane. (From Ref. [82], reprinted with permission of Hanser.)

influence of the co-surfactant on the phase behaviour of the water–oil–surfactant system. In this case, the PIT range decreases by an average of about 5 K per added mol.% co-surfactant. The extent of the three-phase zone is scarcely affected. The oil type has a great influence on the formation and extension of the microemulsion phases. Especially polar oils are usually more difficult to emulsify in a microemulsion system. Many surfactant systems are claimed for a better microemulsion formation with these types of oils, e.g. ester solvents [83] or amino-functional polysiloxanes [84]. For triglyceride oils the 'extended' surfactants are said to be efficient [85]. An 'extended chain surfactant' is a surfactant having an intermediate polarity linking chain, such as a block of poly-propylene oxide, inserted between the hydrophilic and the hydrophobic end parts of the surfactant. Smith and Hand claim that a blend with these surfactants and a high HLB non-ionic surfactant is particularly efficient in removing difficult oily stains and soil from a variety of surfaces [86].

Wegener *et al.* [87] propose mixtures of alkylpolyglycosides and glyceryl monooleate for the formation of microemulsions which have a broader temperature range for stability than microemulsions with alkyl ethoxylates. This makes them more suitable for the application for detergents. Microemulsions containing alkylpolyglycosides are claimed for the use as a stain pre-treatment agent [88], as a detergent for removing hydrophobic soil [89] or for cleansers with increased dynamic wetting effects for tiles and plates [90].

An overview of the cleaning properties of single-phase hydrocarbon-based microemulsions, i.e. oil-continuous systems, shows that both microemulsion structure and viscosity influence the solubilisation rate [91]. A comparison between water-continuous and oil-continuous microemulsions shows a much better degreasing effect in the oil-continuous system (Fig. 8.15) [92]. The difference is that the cleaned surface will have a thin oil coverage after the cleaning process. This can be an advantage with metal surfaces where the risk for corrosion will diminish, but in other situations like household cleaning the surface should be water-wet and feel clean afterwards. Simpson *et al.* [93] disclose a w/o-microemulsion based on terpene, C4–C5 alcohol and a surfactant which aims at efficient degreasing of metal surfaces.

8.3.3.2 Hard surface cleaning

Within certain industrial applications like gas and oil industry and ink and printing industry there is a need for cleaning when the remaining surface should be water-wet. A neutral microemulsion system based on a surfactant, a lactate ester as co-surfactant and an organic solvent like limonene is suggested by Harrison for this purpose. Butyl lactate is shown to enlarge the one-phase (Winsor IV) area in the phase diagram, for instance SDS and limonene in water [94, 95].

The high viscosity of certain microemulsions is used for the adhesion of cleaner concentrates on vertical surfaces while on dilution mobile microemulsions are formed [96]. The need for this type of behaviour is especially evident when it comes to household cleaners like toilet bowl cleaners where the formulation needs to be acidic to cope with the special dirt met there, e.g. soap scum. An example of microemulsions with high viscosity for this purpose can be found in [97] which discloses acidic thickened sprayable microemulsion composition based on a balanced mixture of anionic and non-ionic surfactants,

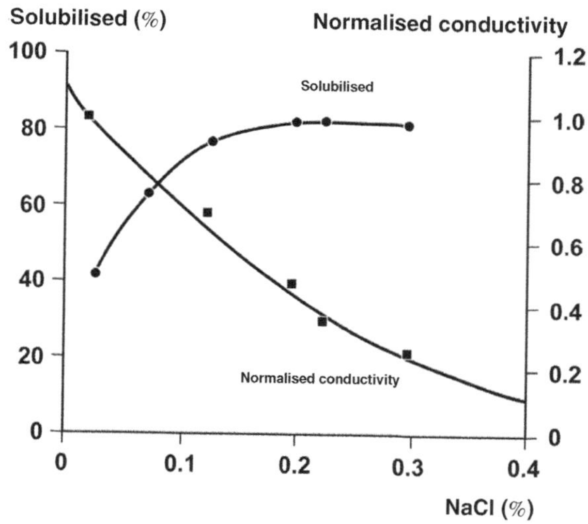

Figure 8.15 Diagram showing cleaning of petroleum jelly versus microemulsion structure containing 50 wt.% of water. Adding electrolyte drives microemulsions from water-continuous towards solvent-continuous (low conductivity) and improves cleaning performance to a value similar to that of the water-free control system. (From Ref. [92], reprinted with permission of Wiley VCH.)

dicarboxylic acid(s) and xanthan gum. Durbut *et al.* use charged surfactant–polymer complexes for microemulsions in all-purpose cleaning [98]. These systems shall be especially effective in the removal of oily or greasy soils.

Along with the general trend in all cleaning areas, multipurpose formulations are being developed also for microemulsions. Antimicrobial multipurpose formulations containing a cationic surfactant, based on a mixture of non-ionic surfactants, an amphoteric surfactant, a water-soluble solvent, hydrocarbon, essential oil or perfume and water is claimed to be effective in disinfecting and in the removal of oily and greasy soil without leaving streaks [99]. Microemulsions containing high concentrations of hydrogen peroxide are described for hard surface cleaners with bleach [100]. These emulsions have a stable viscosity in the presence of hydrogen peroxide and a ternary emulsifier mixture containing esterquats, alkyl oligoglycosides and alkylbenzenesulphonates. Microemulsions can also be used for manual dishwashing where the customer demands high and stable foam. To overcome the weak foaming of microemulsions, Pollak and Gomes [101] suggest a post-foaming microemulsion based on alkyl sulphosuccinate, alkyl ethoxylate, glycol ether, water-insoluble oil, perfume and isopentane in water. This microemulsion is said to foam after it has been sprayed onto the surface to be cleaned. Hutton *et al.* claim a dishwashing kit which comprises a container with a foam-generating dispenser and a dishwashing composition within the container whereas the dishwashing composition is a microemulsion [102]. The soil removal shall be improved by this combination.

Different steering mechanisms come into play also in this context, for instance legal or environmental demands. This might be exemplified by looking at the different oils being claimed as ingredients in the microemulsions for cleaning purposes. As is seen above, the first investigations were made with hydrocarbons [81] where the basic research had started when it came to investigating the general microemulsion behaviour. However, hydrocarbons were not efficient enough with difficult oily soil and terpenes and chlorocarbons etc. were suggested as alternative oils [103]. Both these were not desirable from environmental point of view and terpene-free alternatives were disclosed [104] based on methyl esters instead. Because of some remaining problems with evaporation and smell a late suggestion has come for low-odour ester-based microemulsions for cleaning of hard surfaces [105] where dibasic esters are put forward as an alternative.

8.3.3.3 Vehicle cleaning

An area of hard surface cleaning where especially efficient degreasing and cleaning of mixed soils is required is vehicle cleaning. To achieve this result often microemulsions are used for pre-treatment of the cars before entering the automatic high pressure or brush cleaning systems that are used both by taxi companies and private car owners. Typical ingredients in these microemulsions are hydrophobic non-ionic surfactants, cationic co-surfactants and solvents like paraffins or esters in water with electrolytes as complexing agents at a rather high pH [106]. An alternative solution uses low pH microemulsion, which is a development from the use of HF solutions that traditionally has been applied, causing corrosion problems. The new concept relies on a salt of citric acid, anionic surfactant like phosphate ester, non-ionic surfactant, hydrotrope, glycol ether, 5–25 wt.% glycolic or citric or lactic acid and an oil-phase-like limonene, pine oil, lemon oil etc. balanced with water [107].

8.3.3.4 Dry cleaning

Besides in detergency microemulsions are evaluated as aqueous-based solvents to replace organic solvents for dry cleaning, degreasing and hard surface cleaning. Acosta *et al.* [108] describe the formulation of biocompatible microemulsions using lecithin as the main surfactant and biocompatible linker molecules (hexyl polyglucoside as the hydrophilic linker and sorbitan monooleate as the lipophilic linker) as potential substitutes for chlorinated solvents in dry-cleaning applications. Formulation parameters were evaluated using isopropylmyristate as the model oil. The linker-based formulations were able to form alcohol-free microemulsions while achieving higher solubilisation capacity than similar systems reported in the literature.

The synergisms of mixtures of anionic–cationic surfactant systems can be used to form middle-phase microemulsions without adding short-chain alcohols [109, 110]. The surfactants studied were sodium dihexyl sulphosuccinate and benzethonium chloride. The amount of sodium chloride required for the middle-phase microemulsion decreased dramatically as an equimolar anionic–cationic surfactant mixture was approached. Under optimum middle-phase microemulsion conditions, mixed anionic–cationic surfactant systems solubilised more oil than the anionic surfactant alone. Upadhyaya *et al.* [109] proposed a model for the interaction of branched-tail surfactants (Fig. 8.16). According to this model the anionic–cationic pair allows oil to penetrate between surfactant tails and increases the oil solubilisation capacity of the surfactant aggregate. Detergency studies were conducted to test the capacity of these mixed surfactant systems to remove oil from

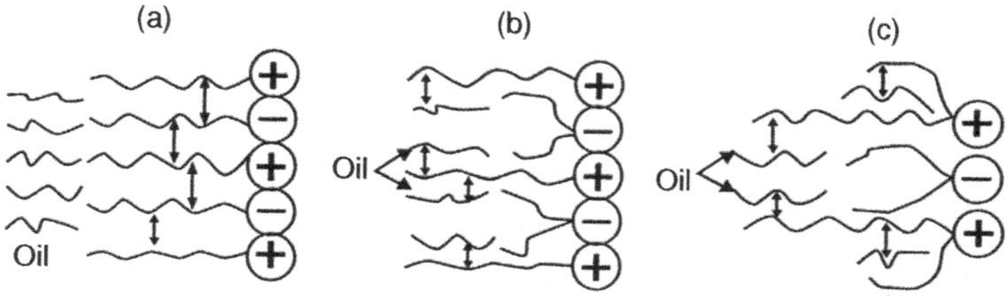

Figure 8.16 Schematic drawing of the surfactant layer and oil penetration in anionic–cationic surfactant mixtures where (a) both surfactants have linear tails, (b) one surfactant has a linear and one a branched tail and (c) both surfactants have branched tails. (From Ref. [109], reprinted with permission of AOCS.)

fabrics. It was found that anionic-rich mixed surfactant formulations yielded the largest oil removal, followed by cationic-rich systems.

8.3.3.5 *Temperature-stable microemulsions*

To overcome the narrow temperature range of stability of microemulsions with ethoxylated non-ionic surfactants the addition of polyethylene glycol was proposed [111]. The polyethylene glycol reduces the temperature sensitivity and shifts the optimum to higher ethylene oxide numbers of the surfactant, i.e. higher HLB numbers for three-phase microemulsion systems. In single-phase microemulsions, the polyethylene glycol also reduces temperature sensitivity, though the effect seems to be less than for three-phase systems. It also promotes the solubilisation of higher molecular weight oils.

8.3.3.6 *Carbon dioxide systems*

An interesting approach for the application of microemulsions is given by water/liquid carbon dioxide systems. Liquid carbon dioxide has been discussed since a long time as an alternative for dry cleaning with organic solvents [112]. It offers an important and economically viable pollution prevention solution for many of the problems facing the cleaning industries. However, cleaning performance is not adequate especially for pigment and water-soluble soil. The addition of small amounts of water improves the soil removal, but the water solubility in liquid carbon dioxide is low and the two-phase system has to be stabilised against phase separation. Water-in-carbon dioxide microemulsions seem to be a unique solution for this problem as they offer both the properties of a solvent phase and an aqueous phase. It was shown that only very few surfactants are suitable for the stabilisation of water–CO_2 microemulsions [113]. Fluorinated amphiphiles seem to be superior to carbon-based surfactants as their solubility in liquid carbon dioxide is much higher [114]. Interfacial properties of these surfactants and the structures of the resulting microemulsion phases were investigated [114]. To improve the properties of the microemulsions modified surfactant structures were synthesised [115]. Liquid carbon dioxide with the addition of specific surfactants has already been commercialised for dry cleaning of textiles, e.g. [116]. See Section 7.4 of Chapter 7 for more details regarding microemulsions containing CO_2.

References

1. Watanabe, K., Noda, A., Masuda, M. and Nakamura, K. (2004) Bicontinuous microemulsion type cleansing containing silicon oil. *J. Oleo Sci.*, **53**, 537.
2. Komesvarakul, N., Sanders, M.D., Szekeres, E., Acosta, E.J., Faller, J.F., Mentlik, T., Fisher, L.B., Nicoll, G., Sabatini, D.A. and Scamehorn, J.F (2006) Microemulsions of triglyceride-based oils: The effect of co-oil and salinity on phase diagrams. *J. Cosmet. Sci.*, **57**, 309.
3. Huang, L., Co, C. and Lips, A. (2005) (to Unilever). WO Patent 2005020939.
4. Förster, T., Guckenbiehl, B., Hensen, H. and von Rybinski, W. (1996) Physico-chemical basics of microemulsions with alkyl polyglucosides. *Prog. Colloid Polym. Sci.*, **101**, 105.
5. Ayannides, C.A. and Ktistis, G. (1999) A rheological study on microemulsion gels of isopropyl myristate, polysorbate 80, glycerol and water. *J. Cosmet. Sci.*, **50**, 1.
6. Hua, X.Y., Van, G.L., Aronson, M.P. and Zhu, Z. (2004) (to Unilever). EP Patent 1397108.
7. Diec, K.H., Meier, W. and Schreiber, J. (1999) (to Beiersdorf). EP Patent 934053.
8. Miller, D.J. and Henning, T. (2003) Three phase bath oils – an attractive application for microemulsions. *SÖFW J.*, **129**, 22.
9. Kahre, J., Engels, T., Hensen, H., Böttcher, A., Gutsche, B. and Jackwerth, B. (1999) (to Henkel). EP Patent 927553.
10. Carlotti, M.E., Gallarate, M. and Rossatto, V. (2003) O/W microemulsion as a vehicle for sunscreens. *J. Cosmet. Sci.*, **54**, 451.
11. Förster, T., Class, M., Guckenbiehl, B. and Ansmann, A. (2003) (to Henkel). EP Patent 813861.
12. Ansmann, A., Fabry, B. and Hensen, H. (2000) (to Henkel). EP Patent 1007508.
13. Schulz, J. and Göppel, A. (2004) (to Beiersdorf). EP Patent 1478326.
14. Savelli, M.P., Solans, C., Rodenas, E., Pons, R., Clausse, M. and Erra, P. (1998) Kinetic study of keratin cystine reduction in W/O microemulsion. *Colloid Surf. A*, **143**, 103.
15. Cliftan, T.W. and Cade, P.H. (1995) (to Croda). EP Patent 645998.
16. Halloran, D.J. and Hoag, C. (1998) Organofunctional silicone microemulsions for hair formulations. *J. Cosmet. Sci.*, **113**, 61.
17. Ostergaard, T., Gomes, A., Quackenbush, K. and Johnson, B. (2004) Silicone quaternary microemulsion: A multifunctional product for hair care. *Cosmet. Toiletries*, **119**, 45.
18. Dalrymple, D.M. and Manning, M. (2002) (to Goldschmidt). EP Patent 12518250.
19. Maillefer, S., Chambettaz, D. and Franzke, M. (2005) (to Wella). EP Patent 1559395.
20. Barreleiro, P., Hloucha, M., Dorn, K., Hentrich, D., Knappe, T. and Scheffler, R. (2006) (to Henkel). WO Patent 2006136331.
21. Comelles, F. and Leal, J.S. (1998) Oleic acid/glycol: An alternative to pentanol in microemulsions with anionic surfactant. *J. Disp. Sci. Technol.*, **19**, 521.
22. Comelles, F. and Leal, J.S. (1999) Butyl lactate: A useful cosurfactant to prepare O/W microemulsions with SDS. *J. Disp. Sci. Technol.*, **20**, 1777.
23. Park, S.J., Won, J.H., Lim, J.-C., Kim, J.H. and Park, S. (2005) Phase behavior and characterization of W/O microemulsion systems containing sodium dodecylsulfate/butyl lactate/isopropyl myristate/water. *J. Ind. Eng. Chem.*, **11**, 20.
24. Komesvarakul, N., Faller, J., Jones, B., Schiltz, J., Szekeres, E., Mentlik, A., Fisher, L., Nicoll, G., Sabatini, D. and Scamehorn, J. (2006) (to Mary Kay). WO Patent 2006074177.
25. Nakamura, N., Yamaguchi, Y., Häkansson, B., Olsson, U., Tagawa, T. and Kunieda, H. (1999) Formation of microemulsion and liquid crystal in biocompatible sucrose alkanoate systems. *J. Disp. Sci. Technol.*, **20**, 535.
26. von Rybinski, W. and Wegener, M. (2003) Phase behavior of microemulsion systems based on optimized nonionic surfactants. *Surf. Sci. Series*, **109**, 387.
27. Comelles, F. (1999) Alternative cosurfactants and cosolvents to prepare microemulsions suitable for practical applications. *J. Disp. Sci. Technol.*, **20**, 491.

28. Acharya, A., Sanyal, S.K. and Moulik, S.P. (2001) Formulation and characterization of a useful biological microemulsion system. *J. Disp. Sci. Technol.*, **22**, 551.
29. Diec, K.H., Eitrich, A., Schmidt, T., Sokolowski, T. and Schreiber, J. (2001) PIT microemulsions with low surfactant content. *Cosmet. Toiletries*, **116**, 61.
30. Spiclin, P., Homar, M., Zupancic-Valant, A. and Gasperlin, M. (2001) Stability of ascorbyl palmitate in topical microemulsions. *Int. J. Pharm.*, **222**, 271.
31. Szymula, M. (2005) The influence of ascorbic acid on the rheological properties of the microemulsion region of the SDS/pentanol/water system. *J. Cosmet. Sci.*, **56**, 267.
32. Shevachman, M., Shani, A. and Garti, N. (2004) Formation and investigation of microemulsions based on jojoba oil and nonionic surfactants. *J. Am. Oil Chem. Soc.*, **81**, 1143.
33. Garti, N., Shevachman, M. and Shani, A. (2004) Solubilization of lycopene in jojoba oil microemulsion. *J. Am. Oil Chem. Soc.*, **81**, 873.
34. Linn, E.E.C., West, M.P. and York, T.O. (1996) (to Eli Lilly). EP Patent 226337.
35. Brucks, R.M., Gransden, K.E. and Ma, Z. (2002) (to Unilever).WO Patent 2002083091.
36. Kux, U. and Diec, K.H. (2005) (to Beiersdorf). Patent EP 1500388.
37. Bialek, A.I., Hill, R.M., Randal, M., Kadlec, D.A. and Van Dort, H.M. (2004) (to Dow Corning). Patent EP 1408912.
38. Schulz, U., Warnke, K., Christ, G., Rohde, C., Diec, K.H., Engfeldt, L. and Miertsch, H. (2005) (to Beiersdorf). Patent EP 1753513.
39. von Rybinski, W., Guckenbiehl, B. and Tesmann, H. (1998) Influence of cosurfactants on microemulsions with alkyl polyglycosides. *Colloid Surf. A*, **142**, 333.
40. Stubenrauch, C., Paeplow, B. and Findenegg, G.H. (1997) Microemulsions supported by octyl monoglucoside and geraniol. 1. The role of the alcohol in the interfacial layer. *Langmuir*, **13**, 3652.
41. Hamdan, S., Dai, Y.Y. and Ahmad, F.B.H. (1997) Evaporation from microemulsion with perfume. *Orient. J. Chem.*, **13**, 111.
42. Hamdan, S., Ahmad, F.B.H., Dai, Y.Y., Dzulkefly, K. and Bulat, K.H. (1999) Effect of humidity on evaporation from aqueous and nonaqueous microemulsion with perfume. *J. Disp. Sci. Technol.*, **20**, 415.
43. Friberg, S.E. (1997) Vapour pressure of some fragrance ingredients in emulsion and microemulsion formulations. *Int.J. Cosmet. Sci.*, **19**, 75.
44. Carlotti, M.E., Gallarate, M., Morel, S. and Ugazio, E. (1999) Micellar solutions and microemulsions of odorous molecules. *J. Cosmet. Sci.*, **50**, 281.
45. Siciliano, M. (1996) (to Arden). Patent US 5490982.
46. Piechocki, C. and Shick, R.A. (2007) (to Dow). Patent EP 1758544.
47. Förster, T. and Tesmann, H. (1991) Phase inversion emulsification. *Cosmet. Toiletries*, **106**, 49.
48. Förster, T., von Rybinski, W. and Wadle, A. (1995) Influence of microemulsion phases on the preparation of fine-disperse emulsions. *Adv. Colloid Interface Sci.*, **58**, 119.
49. Mitsui, T., Machida, Y. and Harusawa, F. (1970) An application of the phase-inversion-temperature method to the emulsification of cosmetics. *Bull. Chem. Soc. Jpn.*, **43**, 3044.
50. Mitsui, T., Machida, Y. and Harusawa, F. (1972) Application of the phase-inversion-temperature method to the emulsification of cosmetics. *Am. Cosmet. Perfumery*, **87**, 33.
51. Förster, T., Schambil, F. and Tesmann, H. (1990) Emulsification by the phase inversion temperature method: The role of self-bodying agents and the influence of oil polarity. *Int. J. Cosmet. Sci.*, **12**, 217.
52. Wadle, A., Tesmann, H., Leonard, M. and Förster, T. (1997) Phase inversion in emulsions: CAPICO concept and application. In M.M. Rieger and L.D. Rhein (eds), *Surfactants in Cosmetics*. Marcel Dekker, New York, p. 207.
53. Miller, D.J., Henning, T. and Grünbein, W. (2001) Phase inversion of W/O emulsions by adding hydrophilic surfactant – a technique for making cosmetic products. *Colloid Surf. A*, **183**, 681.

54. von Rybinski, W. and Förster, T. (1998) Fundamentals of the development and applications of new emulsion types. *Prog. Colloid Polym. Sci.*, **109**, 126.
55. Knüppel, A. and Groteluschen, B. (2003) (to Beiersdorf). Patent DE 10134415.
56. Hamson, S. and Pirotelli, H.R. (2004) Patent US 2004228810.
57. Tesmann, H., Thomas, H., Wadle, A., Förster, T., Banowski, B., Weiler, C. and Zinken, M. (1996) (to Henkel). Patent WO 1996032923.
58. Kux, U., Cierpisz, Y., Mählmann, K., Menzel, N. and Diec, K.H. (2004) (to Beiersdorf). Patent EP 1438010.
59. Lacoutiere, S. (2005) (to L'Oreal). Patent US 2005118285.
60. Lange, R., Syed, H. and Rasato, P. (2006) (to Johnson & Johnson). Patent EP 1618925.
61. Hueglin, D., Roeding, J.F., Supersaxo, A.W. and Weder, H.G. (1999) (to Ciba). Patent EP 956851.
62. Pitfield, A., Kahre, J., Busch, P., Förster, T., Hensen, H., Tesmann, H. and Sumser, M. (2005) (to Henkel). Patent EP 981321.
63. Höffkes, H., Meingke, B., Oberkobusch, D., Förster, T. and Hollenbrock, M. (2003) (to Henkel). Patent EP 1321132.
64. Zuedel-Fernandes, N., Hoting, E. and Poppe, E. (2006) (to Henkel). Patent DE 102004061610.
65. Styczynski, P., Passi, R.K., Ahluwalia, G.S. and Shander, D. (2004) (to Gilette). Patent US 2004141935.
66. Förster, T., Heide, B., Heinen, S., Heyer, M., Neubueser, I. and Wethmar, A. (1999) (to Henkel). Patent DE 19735305.
67. Busch, P., Striepling, G.L., Thiele, K., Mueller-Kirschbaum, T., Tesmann, H., Wadle, A. and Kahre, J. (1994) (to Henkel). Patent DE 4318171.
68. Smulders, E. (2002) *Laundry Detergents, in Ullmann's Encyclopedia of Industrial Chemistry.* Wiley, Weinheim, Germany.
69. von Rybinski, W. (2007) Physical aspects of cleaning processes. In I. Johansson and P. Somasundaran (eds), *Handbook for Cleaning Decontamination*, Vol. 1. Elsevier, Amsterdam, The Netherlands.
70. Lange, R. (1994) *Detergents and Cleaners.* Hanser, Munich.
71. Hauthal, G. and Wagner, G. (2003) *Household Cleaning, Care and Maintenance Products.* Verlag f. Chemische Industrie, Augsburg, Germany.
72. Kling, C. (1949) Rolling-up process as basis for detergency. *Kolloid-Z*, **115**, 37.
73. Nickel, D., Speckmann, H.D. and von Rybinski, W. (1995) Interfacial tension and wetting – parameters for the characterization of products. *Tenside Surfactants Det.*, **32**, 470.
74. Jakobi, G. and Löhr, A. (1986) *Detergents and Textile Washing.* VCH, Weinheim.
75. Jost, F., Leiter, H. and Schwuger, M.J. (1988) Synergisms in binary surfactant mixtures. *Colloids Polym. Sci.*, **266**, 554.
76. Malmsten, M. and Lindman, B. (1989) Ellipsometric studies of cleaning of hard surface. Relation to spontaneous curvature of the surfactant monolayer. *Langmuir*, **5**, 1105.
77. Kahlweit, M., Strey, R., Haase, D., Kunieda, H., Schmeling, T., Faulhaber, B., Borkovec, M., Eicke, H.F., Busse, G., Eggers, F., Funck, T., Richmann, H., Magit, L., Soederman, O., Stilbs, P., Winkler, J., Dittrich, A. and Jahn, W. (1987) How to study microemulsions. *J. Colloid Interface Sci.*, **118**, 436.
78. Benson, L., Cox, K.R. and Zweig, J.E. (1985) Happi, 50 and Strey, R. (1987), The Phase Behavior of H_2O–Oil–Nonionic Amphiphile Ternary Systems, In H.L. Rosano (ed), *Microemulsion Systems.* Marcel Dekker, New York.
79. Miller, C.A. and Raney, K.H. (1993) Solubilization–emulsification mechanisms of detergency. *Colloids Surf.A*, **74**, 169.
80. Foerster, T. and von Rybinski, W. (1998) *Applications of Emulsions in Modern Aspects of Emulsion Science.* The Royal Society of Chemistry, Cambridge, UK, p. 418.

81. Solans, C., Garcia Dominguez, J. and Friberg, S.E. (1985) Evaluation of textile detergent efficiency of microemulsions in systems of water, nonionic surfactant and hydrocarbon at low temperature. *J. Disp. Sci. Technol.*, **6**, 523.
82. Dörfler, H.D., Grosse, A. and Krüßmann, H. (1995) Microemulsions and their use in model washing tests. *Tenside Surfactants Det.*, **32**, 484.
83. Motson, H.R. (2001) (to Imperial Chemical Industries PLC, UK).WO 2001090291.
84. Chrobaczek, H., Kvita, P., Dubini, M., Strub, R., Weiss, E., Tschida, G., Goretzki, R. and Geubtner, M. (2003) (to Ciba Speciality Chemicals Holding Inc. Switz.). WO 2003060051.
85. Acosta, E. (2007) Application of microemulsions in cleaning technologies and environmental remediation. In I. Johansson and P. Somasundaran (eds), *Handbook for Cleaning Decontamination*, Vol. 2. Elsevier, Amsterdam, The Netherlands.
86. Smith G.A. and Hand K.R. (2006) (to Huntsman Corporation). Patent application, US 2006/0211593 A1.
87. Wegener, M. and von Rybinski, W. (2001) Surfactant systems for microemulsions and their importance for applications. *Tenside Surfactants Det.*, **38**, 24.
88. Bastigkeit, T., Pfennig-Dahmen, R., Wegener, M., Rogge, B. and Riebe, H.-J. (2003) (to Henkel KGaA).WO 03000834.
89. Smith, K., Wiseth, W., Hei, R.D.P., Falbaum, D., Mattia, P. and Man, V.F. (2001) (to Ecolab Inc. USA).WO 2001059059.
90. Kuhn, H., Mueller, F., Peggau, J., Richter, N. and Thie, G. (2005) (to Goldschmidt G.m.b.H., Germany). DE 102004010152.
91. Klier, J., Suarez, R.S., Green, D.P., Kumar, A.M., Hoffmann, M., Tucker, C.J., Landes, B. and Redwine, D. (1997) Cleaning properties of single-phase hydrocarbon-based microemulsion systems. *JAOCS*, **74**, 861.
92. Klier, J., Tucker, C.J., Kalantar, T.H. and Green, D.P. (2000) Properties and applications of microemulsions. *Adv. Mater.*, **12(23)**, 1751.
93. Simpson, E.O., McFarland, J.M., Law, M.P. and Kestling, H.S. (1997) (to ARCO Chemical Technology). US5679628.
94. Harrison, J. (2004) Microemulsion technology for surfactants. *Spec. Chem. Magn.*, **24(10)**, 32.
95. Harrison, J. and Zwinderman, M. (2006) (to Surfactant Technologies Ltd). WO 2006/051255 A1.
96. Farnworth, D.M. and Alexander, M. (1996) (to Unilever Plc, UK).WO 19950629.
97. Lysy, R.O., Maurice, M.O.-M. and Blanvalet, C. (1995) (to Colgate-Palmolive Company). US 5472629.
98. Durbut, P., Broze, G. and Mathieu, F. (2003) (to Colgate-Palmolive Company). US 20030902.
99. Fonsny, P., Burke, J. and Dormal, D. (1999) (to Colgate-Palmolive Company). WO 99/31216.
100. Pi, S.R., Bogorra, J. and Bonastre, G.N. (1998) (to Henkel KGaA, Germany). DE 19801086.
101. Pollak, C. and Gomes, G. (1999) (to Colgate-Palmolive Company). US6004920.
102. Hutton, H.D., Hildebrand, R.E., Lamb, C., Turner, R.D. and Foley P.R. (2004) (to The Procter & Gamble Company). WO 2004078903.
103. Mihelic, J. and Luttinger, L.B. (1995) (to Drew Chemical Corporation). US5401326.
104. Gross, S.F., Barabash, M.J. and Hessel, J.F. (2001) (to Henkel Corporation). US6224685.
105. Gross, S., Doerr, M. and Morris, T.C. (2007) (to Cognis GmbH). EP 1780259 A1.
106. Company, M. and Karsa, D. (2007) Vehicle cleaning. In I. Johansson and P. Somasundaran (eds), *Handbook for Cleaning/Decontamination of surface*, Vol 2. Elsevier, Amsterdam, The Netherlands.
107. Chernin, V., Martens, R. and Kubala, R.W. (2004) (to Cleaning Systems Inc). US6696399 B1.
108. Acosta, E.J., Nguyen, T., Witthayapanyanon, A., Harwell, J.H. and Sabatini, D.A. (2005) Linker-based bio-compatible microemulsions. *Environ. Sci. Technol.*, **39**, 1275.

109. Upadhyaya, A., Acosta, E.J., Scamehorn, J.F. and Sabatini, D.A. (2006) Microemulsion phase behavior of anionic–cationic surfactant mixtures: Effect of tail-branching. *J. Surfactants Det.*, **9**, 169.
110. Doan, T., Acosta, E., Scamehorn, J.F. and Sabatini, D.A. (2003) Formulating middle-phase microemulsions using mixed anionic and cationic surfactant systems. *J. Surfactants Det.*, **6**, 215.
111. Miller, D.J. and Henning, T. (2005) Microemulsions containing polyethylene glycol. *Tenside Surfactants. Det.*, **42**, 34.
112. Taylor, D.K., Carbonell, R. and DeSimone, J.M. (2000) Opportunities for pollution prevention and energy efficiency enabled by the carbon dioxide technology platform. *Annu. Rev. Energy Environ.*, **25**, 115.
113. Consani, K.A. and Smith, R.D. (1990) Observations on the solubility of surfactants and related molecules in carbon dioxide at 50°C. *J. Supercrit. Fluids*, **3**, 51.
114. Eastoe, J., Downer, A., Paul, A., Steytler, D.C., Rumsey, E., Penfold, J. and Heenan, R.K. (2000) Fluoro-surfactants at air/water and water CO_2 interfaces. *Phys. Chem. Chem. Phys.*, **2**, 5235.
115. Dupont, A., Eastoe, J., Martin, L., Steytler, D.C., Heenan, R.K., Guittard, F. and Taffin de Givenchy, E. (2004) Hybrid fluorocarbon–hydrocarbon CO_2–philic surfactants. 2. Formation and properties of water-in-CO_2 microemulsions. *Langmuir*, **20**, 9960.
116. http://www.fredbutler.com.

Chapter 9

Microemulsions: Pharmaceutical Applications

Vandana B. Patravale and Abhijit A. Date

9.1 Introduction

Pharmaceutical research is aimed at delivery of the active pharmaceutical ingredient (API) to the target organ at therapeutically relevant levels, with negligible discomfort and side effects to the patient. This is being accomplished by mainly two ways, namely (1) the design of new chemical entities with more target specificity and desirable pharmacokinetic behaviour and/or (2) the development of novel delivery strategies to modulate the physicochemical and pharmacokinetic properties of established therapeutic agents to achieve optimal therapeutic efficacy with maximum patient compliance. The first approach has its own limitations due to the enormous cost involved in the drug discovery research and stringent regulatory requirements. Furthermore, up to 40% of the new chemical entities discovered by the pharmaceutical industry today are poorly soluble or lipophilic compounds and are difficult to deliver by conventional means [1]. Additionally, the manipulation of physicochemical properties of such APIs is not always possible in many cases. Due to these reasons, there has been a paradigm shift in pharmaceutical research towards the development of novel delivery strategies. According to the Biopharmaceutical Classification System [2, 3], the APIs are classified as follows:

Class I: Drugs with high solubility and high permeability (e.g. Metoprolol)
Class II: Drugs with low solubility and high permeability (e.g. Tacrolimus)
Class III: Drugs with high solubility and low permeability (e.g. Atenolol)
Class IV: Drugs with low solubility and low permeability (e.g. Paclitaxel)

Except for the Class I, the APIs belonging to the other classes often show poor therapeutic performance and require special delivery strategies to maximise their efficacy. Paradoxically, only 35% of the currently approved drugs and around 5% of the new chemical entities belong to Class I. The novel delivery strategy should be able to improve the efficacy of the APIs belonging to all the aforementioned classes and should also circumvent several other factors responsible for poor therapeutic performance such as high degree of first-pass metabolism, poor physical/chemical stability, poor *in vivo* stability (especially in case of peptides) and inter-individual variability. In addition to this, the novel delivery strategy should be amenable to manufacture and scale-up, should be cost-effective and should follow the stringent regulatory requirements in terms of safety and biocompatibility.

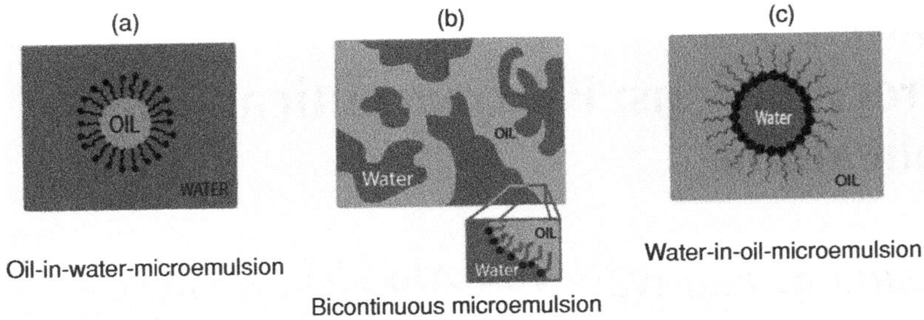

Figure 9.1 Types of microemulsions. (From Ref. [4], reprinted with permission of Elsevier.)

Microemulsions have gained a great attention as a novel delivery strategy in the past two decades to improve the therapeutic performance of an array of drugs because of their attractive features. Microemulsions are thermodynamically stable, transparent, isotropic, low-viscosity colloidal dispersions consisting of microdomains of oil and/or water stabilised by an interfacial film consisting of surfactant (and co-surfactant) molecules. Structures that can be formed are oil-in-water droplets (o/w), water-in-oil droplets (w/o), or bicontinuous phases [4] as shown schematically in Fig. 9.1. A comparison of microemulsions with other colloidal carriers such as micelles, emulsions and liposomes is shown in Table 9.1. The successful arrival and commercialisation of products such as Sandimmune® Neoral or Tropicaine® is sufficient to highlight importance of the microemulsion technology in pharmaceutical research. This chapter provides an overview of the applications of microemulsions as a drug delivery vehicle for various routes of administration.

9.2 Microemulsions

9.2.1 Overview of general advantages of microemulsions

The various advantages of microemulsions in the pharmaceutical research are as follows:

1. *Thermodynamic stability*: The thermodynamic stability of microemulsions helps in improving the shelf-life of the product making them carriers of choice.

Table 9.1 Comparison of microemulsions with other colloidal carriers

	Microemulsions	Micelles	Emulsions	Liposomes	Nanoemulsions
Spontaneity of formation	Yes	Yes	No	No	No
Thermodynamic stability	Yes	Yes	No	No	Approaching thermodynamic stability
Size range	~50 nm	<10 nm	0.5–50 μm	0.025–25 μm	20–200 nm
Surfactant content	>10%	<5%	1–20%	–	<10%
Appearance	Transparent	Transparent	Turbid	Turbid	Translucent

2. *Ease of manufacture and scale up*: The spontaneity of formation of microemulsions makes their manufacturing and scale up very easy as compared to other colloidal carriers such as emulsions or liposomes which require specialised instruments such as high pressure homogenisers for their manufacturing.
3. *Ability to entrap hydrophilic and hydrophobic therapeutic agents*: The various structures of microemulsions enable entrapment of hydrophilic and hydrophobic drugs either alone or in combination, thus making them a versatile carrier.
4. *Improved stability of the therapeutic agents*: Encapsulation of the therapeutic agents in the microemulsion structures can offer improvement in the chemical, photo-chemical and enzymatic stability of the therapeutic agents such as chloramphenicol (chemical stability) [5], arbutin (photostability) [6] and peptides (enzymatic stability) [7].
5. *Improved oral bioavailability*: Microemulsions have shown tremendous potential in improving oral bioavailability of an array of therapeutic agents such as simvastatin [8], carvedilol [9] and Cyclosporine A (Sandimmune® Neoral) [4].
6. *Improved dermal and mucosal transport*: Microemulsions are capable of penetrating the dermal and mucosal barriers leading to improvement in therapeutic efficacy of an API [10].
7. *Use as a template to fabricate nanoparticulate systems*: The inherent thermodynamic stability, large interfacial area and small droplet size of the microemulsions enable them to act as a template for facile synthesis of pharmaceutical nanoparticulates systems such as solid lipid nanoparticles [11] and nanosuspensions [12]. Additionally, microemulsions represent nanoreactors which can be tailored to fabricate pharmaceutical nanomaterials.

9.2.2 Formulation considerations

The phenomenon of microemulsification is mainly governed by factors such as (1) nature and concentration of the oil, surfactant, co-surfactant and aqueous phase, (2) oil/surfactant and surfactant/co-surfactant ratio, (3) temperature, (4) pH of the environment and (5) physicochemical properties of the API such as hydrophilicity/lipophilicity, p*K*a and polarity. Hence, all these factors should be considered while formulating microemulsions. From a pharmaceutical perspective, one of the most important factors to be considered is acceptability of the oil, surfactant and co-surfactant for the desired route of administration. This factor is very important while developing microemulsions for parenteral and ocular delivery as there is only limited number of excipients which are approved for the parenteral and ocular route. In Chapter 3 of this book a more general overview of formulating microemulsions is given and formulation considerations with respect to the components of microemulsions used in pharmaceutical applications are discussed below.

9.2.2.1 Oil phase

Selection of an appropriate oil phase is very important as it governs the selection of the other ingredients of microemulsions (mainly in the case of o/w microemulsions). Usually,

Table 9.2 List of commonly used oils

General class	Examples	Commercial name	Acceptability
'Fixed' oils	Soybean oil, Castor oil		P/O/T/Oc/M
Medium-chain triglycerides (MCTs)	Triglycerides of capric/caprylic acids	Miglyol 810, 812, Labrafac CC, Croadamol GTCC, Captex 300, 355	P/O/T/Oc/M
	Triacetin	Captex 500	
Medium-chain mono- and di-glycerides	Mono- and di-glycerides of capric/carylic acids	Capmul MCM, Imwitor 742, Akoline MCM	O/T
Long-chain mono-glycerides	Glyceryl monooleate Glyceryl monolinoleate	Peceol, Capmul-GMO Maisine-35	O/T
Propylene glycol (PG) fatty acid esters	PG monocaprylate PG monolaurate	Capryol 90, Capmul PG-8 Lauroglycol 90, Capmul PG-12	O/T O/T
	PG dicaprylate/caprate	Miglyol 840, Captex 200	O/T
Fatty acid esters	Ethyl oleate Isopropyl myristate (IPM) Isopropyl palmitate (IPP)	Crodamol EO	P/O/T/Oc/M P/T/Oc/M P/T/Oc/M
Fatty acids	Oleic acid, Caprylic acid	Crossential O94	O/T/M
Vitamins	Vitamin E		P/O/T/Oc/M

P, parenteral; O, oral; T, topical (dermal); Oc, ocular; M, mucosal.

the oil which has a maximum solubilising potential for the selected drug candidate is selected for formulation of microemulsions. This is very important to achieve maximal drug loading in the microemulsions. The ability of the selected oil to increase the region where the microemulsion is formed is equally important. It is difficult to amalgamate both these requirements with a single oil. It is known fact that oils with excessively long hydrocarbon chains (or high molecular volume) such as soybean oil are difficult to microemulsify whereas oils with shorter chain (or low molecular volume) such as medium-chain triglycerides (MCTs), medium-chain mono- and di-glycerides are easier to microemulsify. On the contrary, the capacity of solubilising lipophilic moieties usually increases with the chain length of the oil [13–15]. Therefore, the choice of the oil is often a compromise between its ability to solubilise the API and its ability to facilitate formation of microemulsions with desired characteristics. In certain cases, mixtures of oils are also used to meet both requirements. The various classes of oils which are available for pharmaceutical microemulsions as well as examples with commercial names and acceptability for oral, parenteral and dermal routes are listed in Table 9.2. Amongst the various oils described in Table 9.2, medium-chain mono- and di-glycerides are preferred for oral and dermal delivery mainly due to their ability to enhance permeation of the API across the biological membranes [16, 17] whereas for parenteral and ocular applications, MCTs and fatty acid esters are preferred [15, 18, 19].

9.2.2.2 Surfactants

Choice of the surfactant is also very critical for the formulation of microemulsions. The selected surfactant/s should microemulsify the selected oil and should also possess good solubilising potential for the selected API. In addition to this, the acceptability for the desired route of administration is an equally important criterion for the surfactant selection. It should be noted that the surfactants are not innocuous. Even pharmaceutically acceptable surfactants can show adverse effects above the recommended concentrations [20]. Surfactants, at higher concentration, may cause haemolysis on parenteral administration. Furthermore, they might cause irritation to the gastrointestinal mucosa and skin at higher concentrations. These factors must be considered while choosing a type and the concentration of surfactant. Generally, surfactants of natural origin are preferred over synthetic surfactants, e.g. phospholipids are preferred over synthetic surfactants wherever possible. Amongst various synthetic surfactants, non-ionic surfactants are preferred over cationic and anionic surfactants as the use of ionic surfactants may result in membrane perturbation and skin irritation [4]. By and large, the surfactant concentration in microemulsions should be as low as possible irrespective of the nature, origin and type [15]. The choice of the surfactant depends also on the type of microemulsion to be formulated. Low hydrophile–lipophile balance (HLB) surfactants such as Sorbitan monoesters are preferred for w/o microemulsion, whereas high HLB surfactants such as polysorbate 80 are preferred for o/w microemulsion. In several cases, a mixture of lipophilic (low HLB) and hydrophilic surfactants (high HLB) may be required to obtain a microemulsion. The various classes of surfactants that are available for the pharmaceutical microemulsions as well as examples with commercial names and acceptability for oral, parenteral and dermal routes are listed in Table 9.3. Amongst the various surfactants, phospholipids, polysorbate 80, poloxamer 188 and Solutol HS 15 have good acceptability for oral, dermal and parenteral delivery of APIs [18, 19]. Polyglycolysed glycerides are another class of surfactants commonly used in oral and dermal delivery. Most of the surfactants belonging to this class have the ability to enhance permeation of the API across the biological membranes [19].

9.2.2.3 Co-surfactants

Very often a surfactant alone cannot lower the oil–water interfacial tension sufficiently to yield a microemulsion which necessitates the addition of an amphiphilic short-chain molecule or co-surfactant. Their short-chain amphiphilic nature (with the length of the carbon chain ranging from C_2 to C_{10}) enables them to interact with surfactant monolayers at the interface thereby affecting their packing [4, 15]. In short, co-surfactants have three functions (see Chapter 1 for further details): (1) They modify the curvature of the interface. The chain length of the co-surfactants has considerable influence on the curvature of the interface and the microemulsion structure [21, 22]. In a homologous series, shorter chain alcohols swell the head region more than the tail region (positive curvature), whereas longer chain alcohols swell the tail region more than the head region (negative curvature) which, in turn, affects the extension and the type of the microemulsion formed [15]. (2) They modify the fluidity of the interfacial film. Liquid crystalline phases are formed when the surfactant film is too rigid. Co-surfactants penetrate into the surfactant monolayer providing additional fluidity to the interfacial film and thus impede the formation of

Table 9.3 List of commonly used surfactants

General class	Examples	Commercial name	Acceptability
Polysorbates	POE-20-sorbitan monooleate POE-20-sorbitan monolaurate	Tween 80, Crillet 4 Tween 20, Crillet 1	P/O/T/Oc/M P/O/T/Oc/M
Sorbitan esters	Sorbitan monooleate Sorbitan monolaurate Sorbitan monostearate	Span 80, Crill 4 Span 20, Crill 1 Span 60, Crill 3	P/O/T/Oc/M P/O/T/Oc/M O/T/M
PEO-PPO- block co-polymers	Poloxamer 188 Poloxamer 407	Pluronic/Lutrol F 68 Pluronic/Lutrol F 127	P/O/T/Oc/M O/T/Oc/M
POE alkyl ethers	POE-10-oleyl ether	Brij 96 V	T
POE castor oil	POE-35-castor oil	Cremphore EL, Etocas 35 HV	P/O/T/Oc/M
POE hydrogenated castor oil	POE-40-hydrogenated castor oil POE-60-hydrogenated castor oil	Cremophore RH 40, HCO-40, Croduret 40 LD Cremophore RH 60, HCO-60	P/O/T/Oc/M P/O/T/Oc/M
POE-stearate	PEG-660–12-hydroxystearate	Solutol HS 15	P/O/T/Oc/M
POE-vitamin E	Tocopheryl-PEG 1000-succinate	Vitamin E TPGS	T/O/Oc/M
Sucrose esters	Sucrose laurate Sucrose palmitate		O/T O/T
Alkyl polyglucosides	Caprylyl glucosides	Oramix CG 10, NS 10	T
Polyglycolysed glycerides	Linoleoyl macrogol glycerides Oleoyl macrogol glycerides Caprylocaproyl macrogol glyceride Polyglyceryl oleate Lauroyl macrogol glycerides Stearoyl macrogol glycerides	Labrafil 2125 CS Labrafil 1944 CS Labrasol Plurol oleique CC 497 Gelucire 44/14 Gelucire 50/13	O/T O/T O/T O/T O/T O/T
Polyether alcohols	Polyetetramethyl butyl phenol ether Isononyl phenyl POE 9 ether	Tyloxapol Nonoxynol 9	Oc/T/Pulm Vaginal
Phospholipids	Soybean lecithin Egg lecithin Diolelyl phosphatidyl choline Distearoyl phosphatidyl glycerol PEGylated phospholipids Dimyristoyl phosphotidyl choline		All routes
Alkyl sulphates	Dioctyl sodium sulphosuccinate	Aerosol OT (AOT)	O/T

POE, polyoxyethylene; P, parenteral; O, oral; T, topical (dermal); Oc, ocular; M, mucosal.

Table 9.4 List of commonly used co-surfactants

General class	Examples	Acceptability
Short-chain alcohols	Ethanol, Benzyl alcohol	P/O/T/Oc/M
	Isopropanol, Phenethyl alcohol	T
Alkane diols and triols	Propylene glycol (PG)	P/O/T/Oc/M
	Glycerol	P/O/T/Oc/M
	Butylene glycol	T
Polyethylene glycols (PEG)	PEG 400	P/O/T/Oc/M
Glycol ethers	Diethylene glycol monoethyl ether (Transcutol)	O/T
	Tetrahydrofurfuryl PEG ether (Tetraglycol or Glycofurol)	O/T
Pyrrolidine derivatives	*N*-methyl pyrrolidone (Pharmasolve)	T
	2-Pyrrolidone (Soluphor P)	O/T
Bile salts	Sodium deoxycholate	O/T/P
Organic acids and salts	Caprylic acid	T/O
	Sodium caprylate	O/T/M
	Potassium sorbate	O/T

P, parenteral; O, oral; T, topical (dermal); Oc, ocular; M, mucosal.

liquid crystalline phases. (3) Co-surfactants are distributed between the aqueous and the oil phase, thereby altering the chemical composition and hence the relative hydro/lipophilicity of the solvents [4, 15, 23–25]. The various classes of co-surfactants that are available for the pharmaceutical microemulsions as well as examples along with their acceptability for oral, parenteral and dermal routes are listed in Table 9.4. Amongst various co-surfactants, Transcutol® is the most widely used co-surfactant for the oral and dermal delivery due to its amphiphilic nature and permeation enhancing property and high solubilisation capacity. Short-chain alcohols like ethanol and polyhydric alcohols such as propylene glycol and Polyethyleneglycol 400 (PEG 400) are preferred for parenteral and ocular delivery [18, 19].

9.2.2.4 Aqueous phase

The nature of the aqueous phase is also important mainly in the case of w/o microemulsions. By and large, water is used as an aqueous phase in most microemulsions, especially for dermal delivery. However, depending upon the route of administration, aqueous phase compositions may have to be changed. For example, parenteral and ocular microemulsions should be isoosmotic to the blood and tear fluids, respectively. In order to achieve isotonicity, additives such as electrolytes (sodium chloride), glycerol, dextrose and sorbitol are required. These additives can affect the conditions under which the microemulsion is formed (i.e. the phase behaviour) and therefore their influence must be studied in the presence of other constituents of the microemulsion. Salinity affects the phase diagrams when ionic surfactants are used and decreases the phase inversion temperature (PIT) of non-ionic surfactants. The preparation of microemulsions is very sensitive to temperature, if the PIT is close to the operating conditions [23, 26, 27]. In certain cases, physiological fluids such as Ringer's solution may have to be used as an aqueous phase to predict the behaviour of microemulsions in *in vivo* conditions [28]. Another important factor is the pH which also has considerable influence on the phase behaviour of the microemulsions. Especially, in case of microemulsions for oral delivery, the pH of the aqueous phase is of great importance. On oral delivery, microemulsions are exposed to gastric fluids of various pH values. Hence, instead of water, simulated gastric fluid (pH 1.2) and/or simulated intestinal fluid (pH 6.8) may be used as aqueous phase. The phase behaviour of microemulsions may be different when simulated gastric and intestinal fluids are used as aqueous phase. In the case of lecithin-based microemulsions, adjustment of the initial pH at 7–8 is also important in order to minimise the hydrolysis of the phospholipids and the triglycerides to fatty acids, which can decrease the pH of the microemulsion and may affect the stability [15, 23]. Other additives in the aqueous phase such as preservatives may also affect the microemulsion phase behaviour. Preservatives like methyl paraben and propyl paraben are known to form complexes with surfactants like polysorbates. Such interactions may also affect microemulsion properties.

9.2.2.5 Drug

The therapeutic agent has a considerable influence on the phase behaviour and thus the structure of microemulsions. Various physicochemical properties of the drug such as log P, p*K*a, molecular structure and weight, presence of ionisable groups affect the microemulsion. For example, if the drug is a water-soluble electrolyte such as lidocaine hycrochloride or diclofenac sodium, it may have a significant effect on the properties of microemulsions formed by ionic surfactants like AOT and lecithin and to a lesser extent on microemulsions stabilised by non-ionic surfactants. Interestingly, certain drugs like sodium salicylate and ascorbic acid are surface active and have beneficial effect on the microemulsion formation and structure [29, 30]. These drugs have shown to increase the region where the microemulsion is formed. Furthermore, there are drugs like tricyclic amines which behave like co-surfactants obviating or diminishing the quantity of co-surfactant [31].

Lipophilic drugs may exert different kind of effects on the microemulsion properties. They might essentially act as oil or would compete with or add to the oil phase during the microemulsion formation. This may require a balancing of the amount of surfactant and co-surfactant to yield the microemulsion [10, 13]. In certain cases, drugs might affect properties of microemulsions such as conductivity and rheology. For example, Djordjevic *et al.* investigated microemulsion systems composed of water, isopropyl myristate, PEG-8 caprylic/capric glycerides (Labrasol®), and polyglyceryl-6 dioleate (Plurol Oleique®), as potential diclofenac diethylamine delivery vehicles. It was observed that diclofenac diethylamine participates in the vehicle structure due to its amphiphilic properties [32]. Furthermore, it was found that in the experimental conditions, conductivity and rheological properties were influenced significantly by the incorporation of the drug. Conductivity values for drug-loaded microemulsions were increased by a factor of 2–3 in comparison to microemulsions without the drug. On the contrary, the drug did not significantly influence either the pH value of the vehicles, or the stability and the optical texture of the examined systems [32]. Drugs like propofol which exist as liquid at room temperature can themselves be treated as an oil phase. Another important drug property to be considered is pH-dependent behaviour which depending upon the aqueous phase pH may exhibit positive or negative effect on the microemulsion behaviour. In addition to these factors, the target dose of the drug is also very important. The amount of the drug also has impact on the particle size of microemulsions. In short, it is always advisable to study the effect of the drug on the behaviour of the microemulsions in order to arrive at microemulsions with optimal characteristics.

9.2.3 Effect of temperature on microemulsions

Non-ionic surfactants are usually employed for the synthesis of pharmaceutical microemulsions. However, non-ionic surfactants show a temperature-dependent change in their physical properties (see also Chapter 1). For example, the solubility of the polyoxyethylene (POE)-based surfactants like Solutol HS 15 decreases with increasing temperature, whereas block copolymer surfactants like poloxamers 407 are more soluble at the freezing temperature and gel at $T = 37°C$. Hence, the microstructure of microemulsions prepared with such non-ionic surfactants can be very sensitive to temperature and should be critically evaluated. It is essential to study the properties of microemulsions at the physiologic temperature of 37°C, especially in the case of parenteral microemulsions. Joubran *et al.* studied the behaviour of the microemulsion system water–soybean oil–POE-40-sorbitol hexaoleate–alcohol. The w/o microemulsion region was found to be heavily dependent on the temperature [33]. Recently, Flanagan *et al.* have also demonstrated the effect of temperature on the formulation of soybean oil microemulsions formulated with POE-alkyl ethers [34]. Note that sugar surfactants and alkyl polyglucosides are relatively temperature insensitive [35] as is further described in Chapter 1.

9.2.4 Microemulsion characterisation and evaluation

Detailed discussion on the microemulsion characterisation methods is out of the scope of this chapter but discussed in Chapter 2 of this book. However, some important parameters of pharmaceutical relevance should be addressed briefly. In general, the effect of surfactant to co-surfactant ratio, oil type, temperature and the drug incorporation on the phase behaviour of the microemulsions should be characterised. The most important parameter is the particle size and polydispersity index of the microemulsion which is measured with dynamic light-scattering methods such as photon correlation spectroscopy and/or laser diffractometry [4, 35, 41]. The type of microemulsion formed can be determined by simple conductivity measurements whereas advanced techniques such as pulse-field gradient NMR can be utilised to determine the location or partitioning of the drug between the oil and water phase or even for drug and microemulsion excipient interaction [4, 23, 36]. The viscosity of the microemulsions is another important factor which should be determined. Viscosity measurements can indicate the presence of rod-like or worm-like reverse micelles [4]. The viscosity of the parenteral microemulsions is a major concern as it governs the syringeability of the microemulsion. Too viscous microemulsions are difficult to administer and could be painful. The pharmaceutical microemulsions should also be characterised for the general parameters like pH, robustness to dilution and electrolyte tolerance. Parenteral microemulsions should further be evaluated with respect to osmolarity, compatibility with intravenous dilution fluids, pain on injection and ability to cause haemolysis. The microemulsions intended for oral and mucosal applications should mainly be evaluated for their effects on the integrity of mucosal surfaces. The histopathological evaluation of the mucosal surfaces can provide information about these effects. The microemulsions for ocular and dermal delivery should mainly be evaluated for the skin and ocular irritation potential in rabbits. Recently, *in vitro* cytotoxicity tests have also gained great interest for determining the biocompatibility and tolerability of the microemulsions.

9.3 Applications in transdermal and dermal delivery

Human skin, the integument of humans, is the largest organ of the human body. It can be viewed as a natural protective barrier against penetration of toxic exogenous compounds/microbes, physical assaults, excessive loss of water and essential compounds from the human body or as a promising portal of entry of drugs for the local or systemic therapy [37–40]. Drug delivery to or via skin presents both unique opportunities and obstacles due to the particular skin structure, physiology and barrier properties. The barrier function of the skin is thus reflected in its multilayered structure, viz. stratum corneum (SC), viable epidermis, dermis and subcutaneous tissue. Each layer is known to represent different levels of cellular or epidermal differentiation. However, the main barrier function is represented by the SC which is an impediment to the delivery of many drugs at therapeutic levels. The SC is 10–15 cell layers thick over much of the body and is composed of dead cells or corneocytes. The intercellular spaces between corneocytes are filled with stacked sheets of lipid bilayer membranes (composed of ceramides, free fatty acids and cholesterol) whose organisation and unique chemical composition confer a high degree of water impermeability. It is these lipid lamellae that constitute the epidermal permeability barrier, both to water (which permits terrestrial life) and to other penetrants [37–40].

For dermal and transdermal delivery, it is necessary to overcome this barrier without disrupting normal skin functions. It should also be noted that there is a distinct difference between the requirements for the dermal and transdermal delivery and their therapeutic advantages. Dermal (topical) delivery is required for skin presenting with several dermatological conditions such as skin cancer, acne, psoriasis, eczema and microbial/fungal infections. The main aim in dermal delivery is to target the drug to the pathological sites within the skin (epidermal or deep dermal) ensuring minimal systemic absorption. However, in case of transdermal delivery, drug diffuses through the various layers of the normal skin and reaches systemic circulation to exert a desired therapeutic effect [37–40].

Transdermal delivery is preferred for the drugs presenting with problems such as high first pass metabolism (β-blockers [41], calcium channel blockers) and adverse systemic effects (such as gastrointestinal ulceration in case of anti-inflammatory agents). There are several approaches for improving dermal/transdermal delivery of therapeutic agents which are out of the scope of this chapter. Microemulsions have gained a considerable interest in deep dermal and transdermal delivery of an array of therapeutic agents [10, 42] and have succeeded in commercialisation for the delivery of a local anaesthetic lidocaine (Tropicaine®).

9.3.1 Potential mechanisms for improved dermal/transdermal transport

The transport of medication across the skin is a complex process and follows several routes: transcellular, intercellular and appendageal (through eccrine (sweat) glands or hair follicles). The basic aspects of this transport mechanism, however, are controlled by three fundamental physiochemical parameters. Primarily, these consist of the partition (K), diffusion (D) and solubility (C_s) coefficients. In order to enhance the transfer of a medication across the SC, these variables need to be manipulated and targeted. Synergism is often seen when the formulation influences more than one of these parameters. Some of the potential mechanisms by which microemulsions would improve the dermal or transdermal transport are (1) denaturation of intracellular keratin or modification of its conformation [43], (2) perturbation/fluidisation of lipid bilayers [44], (3) creation of liquid pools and extraction of SC lipids, (4) increased partitioning and solubility in stratum corneum, (5) increased thermodynamic activity of the drug [45], (6) higher concentration gradient [46] and (7) appendageal transport [47].

9.3.2 Microemulsions as smart dermal/transdermal delivery vehicles

The number of studies reported on the utility of the microemulsions is the highest for dermal delivery. Numerous studies are reported in the literature which establish the utility of the microemulsions to improve the transdermal delivery of APIs. However, only some important examples will be discussed in this chapter.

9.3.2.1 Delivery of local anaesthetic agents

Local anaesthetic agents are employed to relieve distress in children and adults during venepuncture and venous cannulation, intravenous catheterisation and in capillary heel prick blood sampling in neonates. Furthermore, a number of surgical procedures that are performed on superficial lesions, such as skin biopsies, shave and deep excisions, electrosurgical procedures, collagen injections and laser surgery also require use of local anaesthetics. A fast-acting and long-lasting local anaesthetic formulation with concomitant reduction in systemic absorption of the local anaesthetics would be of considerable clinical benefit in reducing pain associated with invasive medical procedures. However, the available formulations, EMLA® cream (a eutectic mixture of 2.5% lidocaine and 2.5% prilocaine; Astra, Lakemedel, Sweden) or Ametop® gel (4% w/w amethocaine base preparation; Smith & Nephew, Dublin, Ireland) have a number of disadvantages, particularly a long delay between application and anaesthetic effect and the need for an occlusive dressing. The instructions for usage mention that painful procedures should only be performed 60 min and 30–45 min after the application of ELMA® and Ametop®, respectively. Furthermore, both the preparations must be covered with a plastic film to be effective [48]. Microemulsions appeared to be an apt strategy due to their penetrating nature which may offer deep dermal delivery for long periods along with considerably short lag time. With this rationale, investigators have explored the potential of microemulsions in the delivery of local anaesthetic agents such as lidocaine, prilocaine, tratracaine and amethocaine.

In vitro reports

Lidocaine is a local anaesthetic which is available as a free base and hydrochloride salt. Lidocaine base is known to form eutectic mixture with prilocaine base and this eutectic mixture forms a dispersed phase of EMLA® emulsion cream. Shukla *et al.* reported that with the help of microemulsification, the drug loading can be increased from 5% (present in EMLA® cream) to 20% which may result in quick action by virtue of higher concentration gradient across the skin [49].

Kreilgaard *et al.* [50] investigated the influence of the structure and composition of microemulsions on the delivery of lipophilic lidocaine and hydrophilic prilocaine hydrochloride. A series of microemulsions with varying compositions and structures (containing 4.8 wt.% of lidocaine and 2.4 wt.% of prilocaine hydrochloride) were formulated and the transdermal flux of these drugs from microemulsions was evaluated in comparison to the conventional formulations such as 5% lidocaine cream (Xylocain®, AstraZeneca), 2% lidocaine hydrochloride gel (Xylocain® gel, AstraZeneca) and EMLA® cream. The transdermal flux of lidocaine and prilocaine hydrochloride was found to vary substantially according to both the composition of applied microemulsion vehicle, and the incorporated drug concentration relative to the saturation limit of each vehicle. The optimised microemulsions increased the transdermal flux of lidocaine up to four times, compared to a conventional oil-in-water emulsion, and increased the transdermal flux of prilocaine hydrochloride almost 10 times compared to hydrogels. The flux was dependant on drug solubility in the microemulsion as well as on drug mobility in the individual vehicle. It was found that the solubility of the drugs in the microemulsions was greater in comparison to the solubility observed with the individual components of the microemulsion. Lidocaine showed a 28–62% increase in solubility and prilocaine hydrochloride showed an increase of 24–40%. The large increase in solubility is due to surfactant film located between the oil and water phases leading to additional solubilisation sites for drugs, compared to the molecular organisation of bulk surfactants. Because of the high solubilisation property of microemulsions, which leads to high concentration gradients, the transdermal flux increases. A linear relationship between self-diffusion of the drugs in the vehicles and transdermal flux was found. It suggests that the percutaneous delivery potential of microemulsion vehicles may be decelerated because the diffusion is hindered due to the internal structure of the microemulsion which may be useful in reducing the systemic absorption.

Sintov *et al.* and Lee *et al.* have investigated the differences in the transdermal flux of lidocaine base and lidocaine hydrochloride from two different microemulsion systems [51, 52]. In both microemulsions, the transdermal flux of lidocaine base was significantly greater than that of lidocaine hydrochloride indicating the impact of the lipophilicity of the drug on the transdermal delivery. Furthermore, Lee *et al.* also observed that the permeation rate of lidocaine base and lidocaine hydrochloride was higher in o/w than in w/o microemulsions showing the influence of the microemulsion structure on drug delivery.

Last but not least, Changez *et al.* have explored the potential of lecithin-based microemulsions in improving the transdermal delivery of tetracaine hydrochloride and compared them with tetracaine hydrochloride solution [44]. It was observed that the microemulsions deliver significantly higher amounts than the aqueous solution and the rate of permeation of tetracaine was dependant on the quantity of water present in the microemulsions. Escribano *et al.* investigated the potential of microemulsions in improving topical delivery of amethocaine and the transport across the human skin was compared with that of commercial formulation (Ametop® gel). Permeability coefficient, transdermal flux and percentage permeation of amethocaine microemulsion were 1.5 times higher than those of the amethocaine commercial gel [53].

In vivo reports

Kreilgaard *et al.* extended their *in vitro* investigations to *in vivo* studies in rats to study the absorption rate and mean lag time of lidocaine and prilocaine from microemulsions, Xylocain® 5% cream, Xylocain® 2% gel and EMLA® cream in the dermis of rat skin [54]. Similar to *in vitro* investigations, large differences in absorption rate were observed depending on the composition of the microemulsion, emphasising the importance of the internal microemulsion structure. The optimised microemulsion formulation resulted in significantly higher delivery of lidocaine and prilocaine as compared to the conventional formulations. Interestingly, mean lag time for the penetration of these drugs from microemulsion was also significantly lower than that of conventional formulations indicating that microemulsions would have faster onset of action. The investigations were further extended to a small-scale human trial in which lidocaine microemulsions were compared with Xylocain® 5% cream [55]. The cutaneous bioequivalence of the formulations was evaluated in eight subjects with respect to pharmacokinetic parameters by using the microdialysis technique. As indicated by the initial *in vitro/in vivo* studies, a substantial increase in absorption of lidocaine was found, when applied in the microemulsion vehicle compared to the Xylocain® 5% cream. The pharmacokinetic study demonstrated that a four times higher amount of lidocaine was absorbed into the skin from the microemulsion relative to the o/w emulsion during a 4-h application period. Interestingly, reduction in the mean lag time was also observed.

Changez *et al.* evaluated the potential of AOT-based microemulsions in improving the delivery of tetracaine by measuring the analgesia produced in the rats. Microemulsions demonstrated significantly higher (eightfolds) topical analgesia in the rats in comparison to the tetracaine solution. It was also observed that the analgesic effect varied depending upon the concentration of water and AOT in the microemulsions. Increase in the water and AOT resulted in higher analgesia [56]. In another study, they evaluated the potential of lecithin-based microemulsions in the delivery of tetracaine and the results correlated with their *in vitro* studies described earlier [57]. Both microemulsion formulations were proved to be safe for the topical application.

Escribano *et al.* evaluated the analgesic activity of amethocaine in microemulsions in comparison to the commercial formulation in hyperalgesic or allodynic rats [48]. Antihyperalgesic activity appeared at 4.2 and 13.8 min after application of amethocaine microemulsion and gel, respectively. Furthermore, values of AUC and other analgesic parameters were significantly higher for microemulsions as compared to commercial formulation. These results suggest that microemulsion could be a valuable strategy for improving amethocaine permeation and thus offer rapid pain relief.

In summary, microemulsions exhibit a great potential in improving the delivery of local anaesthetic agents and the commercialisation of lidocaine microemulsion gel (Tropicaine®) is sufficient to establish their utility.

9.3.2.2 Delivery of non-steroidal anti-inflammatory agents

Anti-inflammatory agents are employed for local treatment of inflammation as well as for counteracting the acute and chronic rheumatoid arthritis or osteoarthritis which requires systemic treatment. However, oral delivery of most of the non-steroidal anti-inflammatory agents (NSAIDs) is severely compromised by disadvantages such as short half-life (which necessitates frequent administration), high first-pass hepatic metabolism, severe gastrointestinal side effects such as irritation and ulceration. Furthermore, oral administration of NSAIDs is very difficult in case of patients already suffering from gastric ulcers. All these factors have stimulated researchers to focus on the transdermal delivery of NSAIDs for systemic treatment and numerous investigations have been reported in the literature for transdermal delivery of NSAIDs. Transdermal patches of NSAIDs like diclofenac diethylammonium (OXA SAT®, Laboratories Beta S.A., Argentina) have already been marketed indicating the potential of this approach. Microemulsions due to their attractive advantages seemed an appropriate approach for the delivery of NSAIDs.

In vitro reports

Diclofenac has been studied by several investigators for its transdermal delivery from microemulsions. Kriwet and Muller-Goymann [58] fabricated a variety of lecithin-based colloidal structures of diclofenac diethylammonium such as liposomes, lamellar liquid crystals and microemulsions. Amongst these, microemulsions resulted in significantly higher transdermal flux as compared to the other carriers. Kweon *et al.* have recently studied the permeation of profile of diclofenac diethyl ammonium from several microemulsions with varying compositions through rat skin [59]. They observed that the diclofenac transport was dependant upon the amount of dispersed phase and the ratio of surfactant to co-surfactant. Sintov *et al.* evaluated the transdermal transport of diclofenac sodium microemulsion and solution through rat, guinea pig and porcine skin. It was observed that in all the cases, microemulsions delivered significantly high amount of diclofenac as compared to that of solutions [60].

Park *et al.* fabricated piroxicam-loaded microemulsions and evaluated the effect of oil content and surfactant to co-surfactant ratio on the permeation of piroxicam through rat skin [61]. It was observed that the permeation rate increased with increasing oil content as well as with increasing surfactant to co-surfactant ratio. Yuan *et al.* observed similar results for meloxicam microemulsions [62]. Recently, the ability of microemulsions to improve the transdermal transport of NSAIDs like ibuprofen [63], celecoxib [64] and rofecoxib [65] has also been established.

In vivo reports

Sintov *et al.* evaluated the pharmacokinetics of diclofenac from three formulations, viz. topical emulgel (Voltaren® emulgel), topical microemulsion and subcutaneous injection

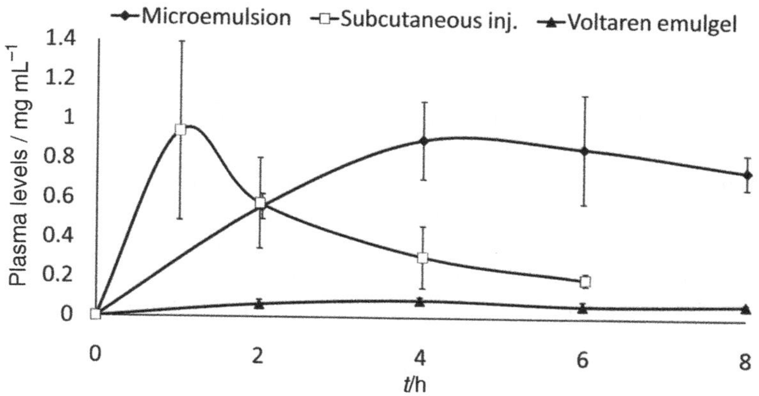

Figure 9.2 Pharmacokinetic profiles of diclofenac in rats after transdermal administration of Voltaren® Emulgel and microemulsion compared to subcutaneous administration. The microemulsion formulation was superior to Voltaren® Emulgel with respect to transdermal delivery and acted for longer time as compared to subcutaneous injection. (Figure redrawn with data from Ref. [59], reprinted with permission of Elsevier.)

[60]. Interestingly, the transdermal administration of the microemulsion to rats resulted in eightfold higher plasma levels of diclofenac than those obtained after application of the Voltaren®. Furthermore, diclofenac plasma levels obtained during 4–8 h following microemulsion application were comparable to the peak plasma levels obtained following subcutaneous administration. After subcutaneous administration, the plasma levels of diclofenac reached a peak of 0.94 μg mL^{-1} at $t = 1$ h and decreased rapidly to 0.19 μg mL^{-1} at $t = 6$ h, while transdermal administration of the diclofenac as microemulsions maintained constant levels of 0.7–0.9 μg mL^{-1} for at least 8 h. No immediate skin damage or irritation was observed for microemulsions in all the animals used (Fig. 9.2).

Bolzinger *et al.* [66] have evaluated bioavailability of niflumic acid on topical administration from bicontinuous microemulsions (with varying drug concentrations) and marketed formulation (Nifluril® ointment (3%)). Niflumic acid microemulsion (1%) was bioequivalent to 3% niflumic acid ointment. Dalmora *et al.* [67] evaluated anti-inflammatory effect of two microemulsions (one containing piroxicam (ME1) and other containing piroxicam-cyclodextrin complex (ME2)) in comparison to piroxicam solution in a cotton granuloma model. Subcutaneous administration of the formulations showed a significant inhibition of inflammation, 68.8 and 70.5% for ME1 and ME2, respectively, whereas piroxicam solution showed only 42.2% inhibition. Furthermore, microemulsions resulted in prolonged anti-inflammatory effect, providing inhibition for 9 days after a single-dose administration. Recently, the *in vivo* advantages of microemulsions have also been demonstrated for celecoxib and rofecoxib [64, 65].

9.3.2.3 *Delivery of steroids*

Steroids such as estrogens, androgens and their derivatives have significant effects on the body and are implicated under various conditions including hormone replacement therapy. Most of the steroidal moieties have poor oral bioavailability due to high first-pass metabolism (estradiol), poor water solubility and permeability (progesterone) or both (finasterisde). Furthermore, most of the steroidal moieties show several untoward effects on oral or systemic delivery. The existing derivatives of steroidal moieties have overcome some of the shortcomings related to the bioavailability but not the untoward effects. The potency of these steroidal components and the problems associated with their current therapy make them ideal candidates for transdermal delivery. Furthermore, steroids/steroidal analogues themselves have potential in the treatment of dermatological conditions such as alopecia and acne as skin expresses steroidal receptors. Microemulsions have tremendous potential in the delivery of these molecules though they are not explored to the fullest. There are few studies dealing with *in vitro* permeation studies and only one study has explored *in vivo* potential. However, since microemulsions have demonstrated potential in targeting sebaceous follicles we believe that they would certainly be explored for the treatment of acne and alopecia.

Estradiol is a principal estrogen that has good oral absorption but high first-pass metabolism. Furthermore, the adverse effects associated with the systemic administration necessitate transdermal therapy. Peltola *et al.* investigated the utility of six microemulsions in improving the transdermal flux of estradiol in comparison to the estradiol solutions in ethanol, isopropanol and phosphate buffer [46] in human cadaver skin. Microemulsion formulations resulted in estradiol flux that was 200- to 700-folds higher than that observed with the control solutions. However, this significant increase in the flux value was not due to the higher permeability but due to the very high concentration gradient across the skin. In fact, the permeability coefficient of the microemulsions was lesser than that of solutions but the increase in the solubility was almost 1500-fold. This clearly demonstrates the different mechanism of improved transdermal delivery of microemulsions as cited earlier. Biruss *et al.* explored in a recent study the potential of eucalyptus oil-based microemulsions in the transdermal delivery of four steroidal moieties, viz. 17-β-estradiol, cyproterone acetate, progesterone and finasteride through porcine skin [68]. A good correlation of transdermal flux values was observed with the self-diffusion coefficients of the steroidal moieties. Furthermore, the investigation also indicated the influence of the physicochemical properties of the drug on the transdermal flux as the composition of microemulsions was the same. Interesting feature of this investigation was the use of essential oil like eucalyptus oil as a dispersed phase of the microemulsions. The eucalyptus oil contains 1,8-cineole as a major component which is also a known permeation enhancer. This appears to be an interesting approach. However, it was observed that the concentration of the APIs in microemulsions declined by 20–40% during the stability studies. Although the reasons were not investigated, this observation raises a concern about the use of essential oils.

Lehmann *et al.* have demonstrated the *in vivo* potential of hydrocortisone-loaded microemulsions in comparison to the marketed product by comparing the blanching effect in humans produced by the various formulations [69]. It was observed that the o/w microemulsion had significantly greater blanching effect than the w/o microemulsion and marketed cream. However, it also resulted in the significant skin irritation in the subjects as compared to the other formulations. This clearly shows the need to evaluate the effect of the microemulsions in terms of tolerability. However, it is noteworthy that the tolerability is mainly governed by the nature of the microemulsion components and by the careful selection of the components, this limitation can be overcome.

9.3.2.4 Delivery of DNA vaccines

DNA vaccines have gained a great interest in recent years and are preferred over traditional vaccines due to several advantages such as better stability, cost-effectiveness for manufacture and storage, safety and the potential to present multiple antigens on one plasmid. Human skin would be an ideal site for the immunisation with DNA vaccines as the skin contains Langerhans cells, immature dendritic cells which are most potent antigen presenting cells. However, stratum corneum barrier has to be traversed in order to achieve topical immunisation. The approaches such as intradermal injections, needle-free jet injections and gene guns have shown potential but their utility is limited [70]. Microemulsions can serve as an interesting approach for topical immunisation with the help of DNA vaccines. Cui *et al.* have successfully demonstrated the potential of microemulsions for topical immunisation. Both the investigations employed fluorocarbon in ethanol microemulsions containing plasmids of interest. In their first investigation, they encapsulated plasmids containing a CMV promoter and β-galactosidase receptor gene or CMV promoter and luciferase gene [70]. *In vivo* studies were performed in the mice to assess plasmid DNA (pDNA) expression in the mice skin and also the immunologic responses. The topical application of pDNA microemulsions demonstrated significant increase in the immune responses and luciferase and β-galactosidase expression as compared to pDNA solutions. The specific serum IgG and IgA titers increased by 45-fold and 1000-fold with microemulsions.

In another investigation [71], authors evaluated the potential of the microemulsions in the topical delivery of anthrax protective antigen (PA) protein encoding DNA vaccine. The pGPA plasmid was encapsulated in the microemulsions and the anti-PA IgG titer values obtained after topical treatment with microemulsion were compared with that obtained with pGPA solution and the standard subcutaneous vaccine therapy. The anti-PA IgG titers obtained with the microemulsions were significantly higher than that of pGPA solutions but were lower than the standard treatment. Extensive investigations are still required in this interesting field.

9.4 Applications in oral drug delivery

Oral route is the most convenient and preferred route of drug delivery and is employed for the delivery of an array of therapeutic agents. However, therapeutic efficacy of 50% of the orally delivered drugs is hampered by their poor water solubility [1]. Furthermore, a majority of the new chemical entities being generated through the drug discovery programmes also exhibit poor water solubility. The problems associated with such drugs include poor oral bioavailability, erratic absorption profile, high intra- and inter-subject variability and lack of dose proportionality [16, 72, 73]. Furthermore, the candidates belonging to BCS Class III such as atenolol, metformin (high water solubility and poor permeability) also exhibit low oral bioavailability and ultimately poor therapeutic efficacy. Similarly, most of the therapeutic peptides such as insulin, calcitonin are difficult to deliver by oral route due to their extreme hydrophilicity, poor permeability and stability in the gastrointestinal environment. The conventional techniques like salt formation, micronisation, use of co-solvents and permeation enhancers, oil solutions and surfactant dispersions which were developed earlier to increase the oral bioavailability have revealed limited utility. Although recently developed strategies like solid dispersion technology and inclusion complexes employing cyclodextrins exhibit good potential, they are specific to the drug [16, 72, 73]. Because of their versatile solubilising potential, microemulsions can be an attractive approach to improve the oral bioavailability of both hydrophobic and hydrophilic drugs.

However, in case of oral delivery, microemulsions suffer from disadvantages like poor palatability due to their lipidic composition, which leads to poor acceptability by patient and hampers their applications in oral delivery. Moreover, due to the high water content of microemulsions they cannot be encapsulated in soft gelatin and hard gelatin capsules for oral delivery [16, 72]. The water content may also lead to precipitation of API on long-term storage, which could affect their utility in oral delivery of certain drugs. Hence, a viable alternative to microemulsions is anhydrous form of microemulsion or microemulsion pre-concentrate or self-microemulsifying drug delivery system (SMEDDS) which is preferred for oral delivery.

9.4.1 Self-microemulsifying drug delivery systems

Self-microemulsifying drug delivery systems, an anhydrous form of microemulsions, are isotropic mixtures of oil, surfactant(s) and co-surfactant(s), which when introduced into aqueous phase under conditions of gentle agitation, spontaneously form fine oil-in-water (o/w) microemulsions. In the body, the agitation required for formation of microemulsions is provided by digestive mobility of GI tract. SMEDDS overcome all the disadvantages of ready-to-use microemulsions such as palatability-related issues as SMEDDS can be filled in soft/hard gelatin capsules unlike ready-to-use microemulsions. The successful commercialisation of products like Sandimmune® Neoral and Norvir® is sufficient to substantiate the potential of SMEDDS in oral delivery. Factors affecting the oral bioavailability and potential of SMEDDS are discussed in the following.

9.4.1.1 Dissolution rate-limited absorption

This is the major problem encountered in the oral delivery of several hydrophobic drugs belonging to BCS class II and IV. The high solubilisation potential and spontaneity of microemulsion formation of the SMEDDS enable rapid dissolution of the therapeutic agent in the gastric environment. In addition to this, the drug is presented in very fine nanodroplets which offer high surface area for absorption [16, 72, 73]. This enables quick absorption of the drug and helps in improving oral bioavailability. In most of the reports, SMEDDS have shown to exhibit superior *in vitro* dissolution profile than the control formulations.

9.4.1.2 Poor membrane permeability

As described earlier, poor permeability is also one of the major factors that limit oral bioavailability of several drugs like atenolol and acyclovir because of which such drugs have to be administered at significantly higher doses than required. Several microemulsion components, for example, oil phases such as oleic acid, monoglycerides of caprylic acid and propylene glycol esters of caprylic acid [17], surfactants such as Labrasol, Vitamin E TPGS, polysorbate 80 [74, 75] and co-surfactants such as PEG 400, Transcutol, alcohol and bile salts [76] have the ability to increase the membrane permeability of the therapeutic agents. Recently, Sha *et al.* evaluated the potential of Labrasol-based SMEDDS to improve permeability of a hydrophilic marker mannitol in Caco-2 cell lines [77]. The SMEDDS formulations resulted in 30- to 60-fold improvement in mannitol permeability by modulating the tight junction proteins such as F-actin and ZO1. It was also observed that positively charged SMEDDS were better than negatively charged SMEDDS [78]. Microemulsions have demonstrated potential to improve the bioavailability of the poorly permeable drug such as acyclovir [79].

9.4.1.3 High degree of pre-systemic and hepatic first-pass metabolism

Oral bioavailability of a vast number of molecules such as anti-hypertensive and cardiovascular agents (β-blockers, calcium channel blockers, ACE inhibitors), anti-hyerlipidemic agents (HMG-CoA reductase inhibitors), anti-diabetic agents (repaglinide), antibiotics (cephalosporins) is limited by pre-systemic and/or hepatic first-pass metabolism. Microemulsions have demonstrated a great potential in improving the bioavailability of such therapeutic agents. Some of the microemulsion components such as surfactants can inhibit the Cytochrome P450 metabolising enzymes [80] whereas some lipidic components such as glyceryl monooleate, long-chain triglycerides have been shown to promote the lymphatic absorption of the therapeutic agents from gastrointestinal tract which prevent the first-pass metabolism of the drugs [81–82]. The SMEDDS have been found to be useful in improving the oral bioavailability of drugs like carvedilol [9], and nitrendipine [83] which undergo high degree of first-pass metabolism.

9.4.1.4 P-glycoprotein (P-gp) efflux

P-glycoprotein is an efflux pump present at several sites in the body including gastrointestinal tract. P-gp prevents the entry of the drugs in the systemic circulation thus hampering the oral bioavailability of the drugs. A considerable number of molecules such as paclitaxel, digoxin and doxorubicin are known P-gp substrates and their bioavailability is hampered due to the P-gp mediated efflux. Many surfactants such as Vitamin E TPGS, Solutol HS 15, Cremophore EL and Polysorbate 80 and oil phases such as Imwitor 742 and Akoline MCM (mono-, and di-glyceride of caprylic acid) have potential to inhibit P-gp effux [84–86]. Hence, SMEDDS can also inhibit the P-gp efflux process.

9.4.1.5 Potential explored in oral delivery

Several investigations are reported in the literature describing the utility of the SMEDDS in oral delivery. The major advantages of SMEDDS include improvement in oral bioavailability, quick onset of action [87] and reduction in intra- and inter-individual variability and food effects [88]. Most of the investigations described so far have focused on the pharmacokinetics of the drug to assess oral bioavailability but not on the pharmacodynamic efficacy and the candidates explored so far belong to discrete therapeutic classes. Table 9.5 enlists examples of various SMEDDS reported so far and their *in vivo* advantage.

Table 9.5 Microemulsions in oral drug delivery

Drug	Theraputic use	Components	*In vivo* results	Reference
Anethole trithione	Chemopreventive	Tween 80, Labrasol, Cremophore MCT	2.5-fold BA increment as compared to tablets	[89]
α-asarone	Andidepressant	Tween 80, Ethyl oleate PEG 400	4.8-fold BA increment as compared to tablets	[53]
Danazol	Steroid derivative	Cremophore, Ethanol, Poloxamers, Maisine 35	1.5- to 7-fold BA increment as compared to suspensions	[90]
Atorvastatin	Antihyperlipidemic	Cremophore RH 40, PG Labrafil 1944 CS	1.5-fold BA increment as compared to tablets	[91]
Simvastatin	Antihyperlipidemic	Cremophore EL, Capryol 90, Transcutol	1.5-fold BA increment as compared to tablets	[8]
Silymarin	Heaptoprotective	Ethyl linoleate, Ethanol, Tween 80	1.88-fold BA increment as compared to PEG solution	[92]
Purearia isoflavone	Antihypertensive	Ethyl oleate, Tween 80, Transcutol	2.5-fold BA increment as compared to tablets	[93]
Paclitaxel	Anticancer	Vitamin E TPGS, Ethanol, Deoxycholate, Vitamin E, Cremophore RH 40, PG	Increased BA as compared to oral Taxol	[94]
Ketoconazole	Antifungal	Oleic acid, Ethanol, Cremophore EL, Transcutol, PEG 6000	7-fold AUC improvement as compared to suspension	[95]
Celecoxib	NSAID	Capmul PG, Tween 20, Akonon MC 8	1.3-fold BA increment as compared to tablets	[87]
Carvedilol	Antihypertensive	Labrafil 1944 CS, Transcutol, Tween 80, Benzoic acid	4.13-fold BA increment as compared to tablets	[9]
Nitrendipine	Antihypertensive	Tween 80, HCO 60, Caprylic acid mono-, di-glycerides	Improvement in BA, Quick onset of action	[83]
N 4472	Antihyperlipidemic	Gelucire 44/14, HCO 60, Ascorbic acid	Higher AUC than solid dispersion	[96]
Halofantrine	Antimalarial	Cremophore EL, Ethanol, Maisine 35, Captex 355	Improvement in BA	[97]
Biphenyl dimethyl dicarboxylate	Heaptoprotective	Tween 80, Triacetin, MCT	4-fold improvement in AUC as compared to suspension	[98]
Acyclovir	Antiviral	Labrasol, Plurol oleique, Labrafac CC	12-fold improvement in BA as compared to tablets	[79]

9.4.2 Oral delivery of peptides

Oral delivery of peptides is a very challenging task due to their extreme hydrophilicity, poor permeability and poor stability in the gastrointestinal environment. Several strategies are being attempted and microemulsions is one of them. Since microemulsions (mainly w/o) can encapsulate hydrophilic moieties in the internal phase, they can offer some degree of protection to the peptides from enzymatic degradation. It has been demonstrated that MCTs and surfactants have the ability to resist hydrolysis of certain drugs in the gastrointestinal tract by enzymes like esterases [7]. Furthermore, as described earlier, via the choice of the oil phase one can improve the permeability of the API which may be of further advantage for oral peptide delivery [17]. The utility of SMEDDS in the oral delivery of lipophilic therapeutic peptide Cyclosporin A is already well established.

Cyclosporin A is a cyclic nonapeptide used as an immunosuppressant in organ transplantation surgery and in the treatment of certain autoimmune diseases. The high molecular weight, hydrophobicity, poor permeability of cyclosporine severely limits its oral bioavailability. Following oral administration, the absorption of the cyclosporine is incomplete and highly variable. Initially, it was formulated as an oily solution (Sandimmune®) that which formed a crude emulsion *in vivo*. However, Sandimmune suffered from disadvantages such as considerable pharmacodynamic inter- and intra-patient variability which were ascribed to the digestion process of lipids used in the formulation. Formulation of cyclosporine as SMEDDS (Sandimmune® Neoral) dramatically improved its *in vivo* performance mainly with respect to bioavailability, inter- and intra-individual variability and also with respect to food effects (Fig. 9.3) [4, 88]. Cyclosporin appears to be the most widely investigated molecule for the design of SMEDDS. In fact, a generic version of SMEDDS (Gengraff®, Abott Labs) is also available on the market. Gao *et al.* formulated a different microemulsion system which was found to be bioequivalent to the Sandimmune® Neoral [99]. In another study, it was observed that positively charged SMEDDS show higher bioavailability compared to that of negatively charged SMEDDS.

The potential of W/O microemulsions was explored first for the delivery of peptides such as vasopressin and insulin. The absorption of peptides was two times higher than that of solutions in small ligated intestinal segments of rats. Furthermore, the bioavailability was higher when straight chain fatty acids were used instead of branched chain esters; thus, indicating the effect of lipid components on the peptide absorption [100]. Constantinides *et al.* formulated several w/o microemulsions of SKF 106760, a water-soluble peptide stable to enzymatic hydrolysis. The bioavailability of this peptide is permeability limited. Interestingly, microemulsion formulations significantly increased the bioavailability (nearly 5- to 58-fold depending upon the microemulsion composition) of SKF 106760 as compared to that of saline solution. Similarly, the bioavailability of calcitonin was also found to increase from 2 to 45% when administered as w/o microemulsions [100, 101].

N-acetylglucosaminyl analogue of muramyl dipeptide (GMDP) can serve as a vaccine adjuvant, immunostimulator, anti-cancer agent and anti-inflammatory agent. The bioavailability is limited due to poor permeability and enzymatic degradation. It was observed that the intraduodenally administered GMDP w/o microemulsion resulted in 10-fold higher bioavailability than that of oral solution [102]. Çelebi *et al.* [103] compared the intragastrically administered epidermal growth factor (EGF) loaded microemulsions with that of intragastrically administered EGF solution and intraperitonially delivered EGF

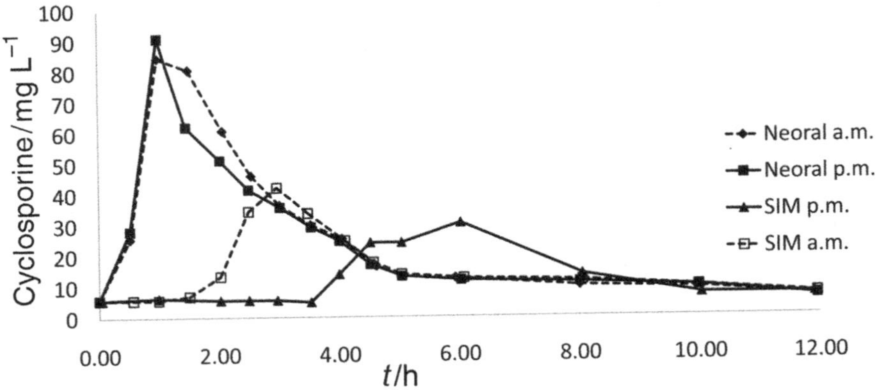

Figure 9.3 Representative cyclosporine blood concentration profiles from a renal transplant patient given the currently marketed formulation Sandimmune (SIM) and the microemulsion formulation Neoral without food (a.m.) or with food (p.m.). (Figure redrawn with data from Ref. [4], reprinted with permission of Elsevier.)

for the reduction of gastric ulcers. It was observed that the microemulsions were superior in performance compared to other formulations. Recently, the ability of microemulsions to improve the delivery of insulin has also been studied by Çilek *et al.* [104]. The intragastric administration of insulin microemulsion with and without aprotinin (an enzyme inhibitor) was compared to that of insulin solution and subcutaneous insulin. The reduction in the blood glucose was significantly more in case of both the microemulsions as compared to insulin solution and was comparable to subcutaneous insulin solution. However, the addition of enzyme inhibitor in microemulsions did not significantly improve the activity. Similar observations were made when bioavailability studies of the insulin were performed.

Most of the reports on the delivery of hydrophilic peptides described to date are preliminary studies and are based on intraduodenal or intragastric administration of microemulsion which is not followed in usual clinical practice. Recently, there are two systematic investigations reported in the literature. Kim *et al.* [105] have evaluated the bioavailability of orally delivered heparin-deoxycholate conjugate loaded microemulsion with that of the solution in mice and monkeys. Interestingly, the quantity of the heparin conjugate available to the body from microemulsion was two times higher than that of solutions in mice whereas it was six times higher in the monkeys. This study nicely demonstrates the inter-species differences in the bioavailability. Furthermore, the histological evaluation of gastrointestinal wall revealed that the microemulsions did not result in any damage to the mucosa.

Leuprolide acetate, a synthetic nonapeptide, is prescribed for the treatment of metastatic prostate cancer and endometriosis. Leuprolide acetate is presented in a number of injectable dosage forms including Lupron® sterile solution for subcutaneous administration (1 and 5 mg mL^{-1}) and Lupron® depot controlled release formulation for intramuscular injection (3.75 and 7.5 mg). Zheng and Fulu [106] have evaluated the *in vivo* effect of leuprolide loaded microemulsions on the genital organs of the male and female rats. In the preliminary pharmacokinetic studies, oral microemulsions of leuprolide administration resulted in 10-fold higher plasma levels of leuprolide as compared to that of saline solution (Fig. 9.4).

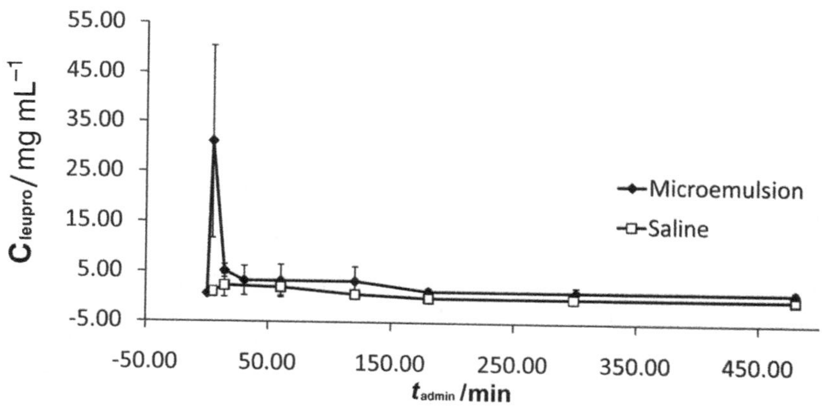

Figure 9.4 Plasma concentrations of leuprolide acetate (C_{leupro}) following single oral administration of leuprolide saline solution and microemulsion at a dose of 3 mg kg^{-1} in rats ($n = 5$; mean ± SD) as a function of the time after administration (t_{admin}). The leuprolide acetate microemulsion was superior to that of saline solution. (Figure redrawn with data from Ref. [106], reprinted with permission of Elsevier.)

Later on, the *in vivo* potential of leuprolide loaded microemulsions was evaluated in a multi-dose level 35-day study and was compared with that of leuprolide subcutaneous depot. It was observed that reduction of the male genital organ weights by once a day treatment (2 mg per rat) was similar to that by twice a day treatment (1 mg per rat) at the same dose level. Inhibition of genital growth was also observed in female rats by treatments with oral leuprolide microemulsion (Table 9.6). The weights of the uterus and ovary following oral administration of leuprolide microemulsion for 35 days were significantly decreased in comparison to untreated rats. The pharmacological activities of oral leuprolide microemulsion in terms of rat organ weights of prostate, seminal vesicle, uterus and ovary were comparable with the commercial product Lupron® depot [106].

9.5 Applications in parenteral drug delivery

Parenteral route is the preferred route of delivery in the cases of emergencies. It is the only resort for non-cooperative and ambulatory patients and for the drugs such as

Table 9.6 Effect of oral leuprolide microemulsion on genital organs of female rats after administration for 35 days ($n = 5$; mean ± SD)

Formula	Dose (per rat)	Uterus weight/g	Ovary weight/g
Placebo control	Untreated	0.56 ± 0.10	0.089 ± 0.032
Lupron® Depot (subcutaneous)	3.75 mg/35 days	0.19 ± 0.02	0.047 ± 0.012
Oral leuprolide microemulsion	2 mg, once a day	0.20 ± 0.03	0.041 ± 0.009

Differences in the Lupron Depot and oral microemulsion are not significant ($P > 0.05$) indicating that oral treatment can be effective. (From Ref. [106], reprinted with permission of Elsevier.)

aminoglycoside antibiotics and peptides which do not show efficacy on oral delivery due to absorption or stability problems. Design of parenteral drug delivery systems is a critical process and has stringent regulatory requirements as the drug is directly injected into the blood stream. The parenteral delivery of several therapeutic agents is very difficult due to their extreme hydrophobicity and also due to the availability of a small number of excipients acceptable by parenteral route. Several techniques have been developed for the parenteral delivery, such as the use of co-solvents like propylene glycol and PEG 400 [18, 19], cyclodextrins [19], oil solutions, micellar solubilisation and the use of emulsions [18, 19]. However, these approaches have their own limitations, such as high concentration requirement for co-solvents which often results in pain on injection, high cost and poor complexation associated with cyclodextrins, poor solubilisation capacity in case of micellar solution and poor physical and microbial stability in case of emulsions [107]. Microemulsions have emerged as an alternative to these conventional delivery approaches.

9.5.1 Advantages of microemulsions in parenteral delivery

The nanostructure of microemulsions ensures that the chances of emboli formation in the blood are very unlikely. Furthermore, the small size of the microemulsions may result in higher blood circulation time which would be useful in certain cases. The excellent thermodynamic stability, high solubilisation capacity, low viscosity and ability to withstand sterilisation techniques make microemulsions an interesting delivery system. Furthermore, microemulsions are likely to be less painful on injection as compared to the co-solvent based formulations as demonstrated by Lee *et al.* [108]. They have also been shown to reduce the toxicity potential of certain drugs like amphotericin B by means of encapsulation [109]. W/O microemulsions can be used for the controlled delivery of hydrophilic therapeutic actives such as antibiotics like aminoglycosides. With the use of suitable excipients such as PEGylated phospholipids, it may be possible to improve either the targeting of the microemulsions to certain organs or the circulation time in the blood which is mainly useful for infections like malaria and in the cancer treatment [110]. The microemulsions can also be stored in form of preconcentrates which can be diluted with the IV fluids such as 0.9% saline solution or 5% dextrose just before the administration to the patients [108]. The preconcentrates would be advantageous in several cases.

9.5.2 Formulation considerations

As described earlier, there are stringent regulations for the parenteral products and only few excipients are acceptable for parenteral delivery. The excipient selected for the parenteral delivery should be biocompatible, sterilisable, non-pyrogenic, non-irritant to nerves and non-haemolytic. Very few excipients fit into all these requirements. For example, the sugar surfactants are biocompatible and have fairly good solubilisation potential but they have been found to be haemolytic [111]. The excipients that are acceptable for parenteral delivery are as follows:

- *Surfactants*: Poloxamer 188, Lecithin, Tween 80, Tween 20, Span 80, Solutol HS 15, Cremophore®EL.
- *Co-surfactants*: Ethanol, Propylene glycol, Benzyl alcohol, Glycofurol, PEG 400.
- *Oils*: Castor oil, soybean oil, MCTs, Ethyl oleate, IPM, Vitamin E.

The use of Cremophore®EL in the microemulsions is avoided nowadays due to several adverse effects such as anaphylactic shocks and histamine release [20]. The other important consideration is the concentration of surfactants and co-surfactants which should be minimal and preferably not exceed 20%. Furthermore, it is necessary to ensure that the microemulsion structure is preserved in the presence of the tonicity adjusting solutions such as 0.9% saline solution and preservatives. The parenteral microemulsions should also be able to withstand tests such as freeze-thaw cycling which ensure their physical stability. It has been shown that the colloidal carriers based on Solutol®HS 15 can withstand freeze-thaw cycling very efficiently whereas lecithins can offer stability to autoclaving [112]. Co-surfactants such as benzyl alcohol cannot be used for intravenous applications but can be used for the small volume parenteral products up to the concentration of 1% w/v. Ethanol at concentrations above 10% usually results in the pain on injection. Co-surfactants such as glycofurol are reported to acceptable for parenteral products but there are no products based on glycofurol available for the human use. The pyrrolidone derivatives are reported to be acceptable for veterinary applications.

9.5.3 Potential explored

Corswant *et al.* [113] reported parenteral bicontinuous microemulsions based on Solutol® HS 15, Soybean lecithin, ethanol, PEG 400 and MCT. The haemodynamic studies in rats indicated that microemulsions had no significant effect on the acid–base balance, blood gases, plasma electrolytes, arterial blood pressure or heart rate. The microemulsions could successfully solubilise drugs like felodipine [113]. Park *et al.* evaluated the potential of flurbiprofen microemulsions based on PEGylated phospholipid, ethanol and ethyl oleate. The pharmacokinetic studies indicated that the half-life, area under the curve (AUC) and mean residence time of flurbiprofen from microemulsions was considerably higher than that of solution. Furthermore, the RES uptake of the flurbiprofen microemulsions was considerably lower than that of the solutions. This indicates that PEGylated phospholipids based microemulsions can give prolonged circulation of the drugs [114]. Thereafter, Lee *et al.* evaluated the potential of polysorbate-based SMEDDS in the delivery of another anti-inflammatory agent, clonixic acid. The pharmacokinetics of clonixic acid SMEDDS was no different than the marketed clonixic acid formulation based on the co-solvents. However, the microemulsions were significantly less painful than the marketed formulation indicating the potential in the delivery [108].

Amphotericin B is an extremely potent anti-fungal agent used in the treatment of systemic fungal infections. However, delivery of amphotericin B is very difficult due to its insolubility in the water as well as commonly used oils. Furthermore, it shows several adverse effects such as nephrotoxicity. Brime *et al.* [109] successfully entrapped amphotericin B in the lecithin-based microemulsions and evaluated their acute toxicity in the mice in comparison to the marketed formulation. Acute toxicity studies indicated that

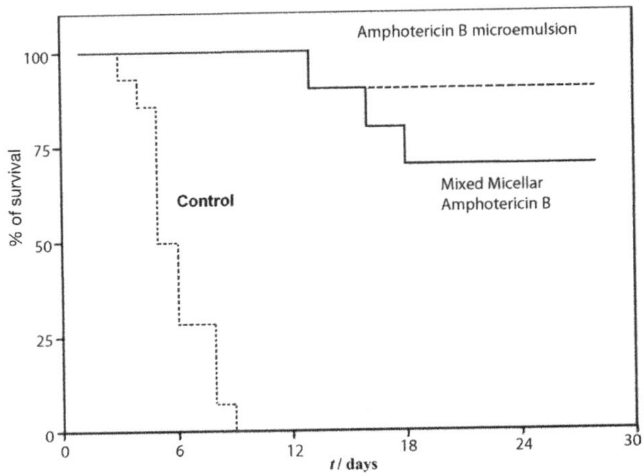

Figure 9.5 Survival of immunocompetent treated 48 h after inoculation with a single intravenous injection of marketed mixed micellar amphotericin B (1 mg kg^{-1}) or amphotericin B microemulsion (2 mg kg^{-1}). Control refers to untreated mice infected with *C. albicans*. Microemulsion of amphotericin B showed survival even at higher dose. (Figure redrawn with data from Ref. [115], reprinted with permission of Elsevier.)

the mortality rate associated with the amphotericin B microemulsion was significantly lesser than that of the mixed micellar formulation. In another investigation, same authors evaluated the *in vivo* potential of the lecithin-based amphotericin microemulsions in the immunocompetent and neutropenic mice infected with systemic candidiasis. Studies indicated that there was threefold increase in the tolerated dose of the amphotericin when it was administered as microemulsion as compared to that of marketed formulation. Furthermore, the microemulsions were superior to marketed formulation with respect to ability to reduce fungal load and to increase the survival period of the mice (Fig. 9.5) [115].

Junping *et al.* formulated microemulsions of vincristine (an anti-cancer agent from natural source) based on the PEGylated phospholipids, vitamin E, cholesterol and oleic acid. The pharmacokinetic parameters of the microemulsion were compared with that of the free drug in the tumour-bearing mice. The microemulsions resulted in significantly higher efficacy, tolerability and survival rate as compared to that of free drug. Furthermore, the vincristine concentrations in the heart, spleen and liver were significantly lower than that of the free drug indicating that microemulsions may exhibit less adverse effects. The tumour concentrations of the vincristine were significantly higher in the case of microemulsions and so was the residence time of the vincristine in the body [110]. This investigation clearly demonstrates the potential of microemulsions in improving the delivery of anti-cancer agents.

Zhang *et al.* have reported formulation of microemulsion of norcanthridin, an anti-cancer agent used for hepatic cancers [116]. The microemulsions were composed of lecithin, ethanol and ethyl oleate. The mean residence time, area under the curve (drug availability) and the liver concentrations of norcanthridin were significantly higher for the microemulsions as compared to that of the marketed formulation.

9.6 Applications in ocular drug delivery

The eye is one of the most vital organs in the body and the intricate structure of the eye necessitates careful design of the formulation strategy. The major disadvantages of the conventional dosage forms like eye drops and ointments are poor ocular drug bioavailability, pulse–drug entry after topical administration, systemic exposure because of nasolacrimal duct drainage and a lack of effective system for drug delivery to the posterior segment of ocular tissue [117–119]. Poor ocular drug bioavailability is the result of ocular anatomical and physiological constraints, which include the relative impermeability of the corneal epithelial membrane, tear dynamics, nasolacrimal drainage and the high efficiency of the blood–ocular barrier and only 1% or less of a topically applied dose gets transported across the cornea and thus reach the anterior segment of the eye. The attempts to improve ocular bioavailability of the drugs have focused on either increasing the ocular contact time of the delivery system and/or drug permeability in the cornea [117–119]. These strategies include the use of bioadhesive hydrogels, temperature or pH-sensitive in situ gels, collagen shields, particulate and vesicular drug delivery systems such as nanoparticles, liposomes and niosomes, or micellar solutions [120, 121]. Because of versatile advantages, microemulsions have emerged as an attractive approach for the ocular delivery of the drugs. Furthermore, conversion of microemulsions into microemulsion-based gels could be of further advantage with respect to improvement in residence time.

9.6.1 *Formulation considerations*

The components selected for designing of ocular microemulsions should be non-irritating and non-invasive to eye and also should not hamper vision. Although an array of excipients are available for the formulation of microemulsions, only few excipients have been reported to be acceptable for ocular delivery. For example, the commonly used co-surfactant Transcutol appeared to be irritating to the eye at a very low concentration of 0.05% w/v [122]. However, at this concentration, it did not cause any visible ocular damage or abnormal clinical signs involving the cornea, iris or conjunctivae at all concentrations. Interestingly, at concentrations between 0.005 and 0.03% w/v, Transcutol considerably increased the corneal permeability of hydrophilic molecules such as levofloxacin, gatifloxacin and enoxacin [122]. Labrasol has been reported to be slight irritating to the eye above 5% concentration as per the manufacturer's instructions. The ocular acceptability of oil phases such as mono-, di-glycerides of caprylic acid is not reported. The excipients having ocular acceptability are as follows:

- *Surfactants*: Poloxamers, Lecithins, Polysorbate 80 and 20, Span 80 and 20, Cremophore® EL, Tyloxapol®
- *Co-surfactants*: Propylene glycol, PEG 200 and 400
- *Oils*: IPM, ethyl oleate, MCTs, castor oil

Another important consideration is the physical stability of microemulsions in the presence of tonicity adjusting agents, preservatives and electrolytes. It is also necessary to evaluate that the microemulsion formulation withstands the sterilisation process such as autoclaving or filtration sterilisation without affecting the physical stability, particle size and pH.

9.6.2 Potential explored

The utility of the microemulsions for the ocular delivery of timolol was described around two decades ago by Gasco and co-workers wherein lecithin-based microemulsions using ion-pairing approach were shown to improve the transport of the timolol [123]. The ocular bioavailability of the timolol from microemulsions was significantly higher than that of the solution. However, microemulsions described contained considerably high concentrations surfactant and co-surfactant and the ocular tolerability was not established. In another investigation, they reported PEG 200-based lecithin microemulsions which were found to be non-irritating to the eye [124]. Keipert and co-workers tried to explore the potential of poloxamers in the formulation of microemulsions [125]. The microemulsions contained triacetin as oil phase and propylene glycol as co-surfactant. However, these microemulsions were irritating to the eye. The same group fabricated different set of microemulsions containing pilocarpine, IPM, Cremophore® EL, PEG 200 and propylene glycol as co-surfactants [126]. The microemulsions were loaded with pilocarpine. The microemulsions were well tolerated by the eye and exhibited superior activity than the solutions.

Lv *et al.* have demonstrated the ability of the microemulsions to improve the hydrolytic stability of a known anti-microbial agent chloramphenicol [5]. The components of microemulsion included IPM, Tween 20 and Span 20, which have good ocular acceptability. The accelerated stability studies on chloramphenicol microemulsions and commercial solution were carried out for 3 months demonstrated significantly lower hydrolysis of chloramphenicol (14.38%) as compared to that of solutions (27.11%). ^{1}H NMR studies on microemulsions confirmed that the chloramphenicol was embedded in the surfactant chains that may provide higher stability.

Fialho and Silva-Cunha [127] recently reported formulation of dexamethasone loaded microemulsion based on Cremophore® EL, propylene glycol and IPM. The tolerability of the microemulsions was established and it did not cause any significant alteration to eyelids, conjuctiva, cornea and iris. The pharmacokinetics of the dexamethasone from microemulsions was compared to the marketed formulation in rabbit eye. The microemulsion formulation acted faster as well as for the longer period of time (Fig. 9.6 and Table 9.7).

Alany and co-workers [128, 129], in two subsequent investigations, have evaluated the tolerability of Span 20, Tween 80, ethyl oleate and various short-chain alcohols and 1,2-diols by egg chorionallantoic membrane (HAT-CAM) test. Interestingly, it was observed that Span 20 and Tween 80 were non-irritant whereas short-chain alcohols and 1,2-diols resulted in considerable irritation. Amongst the alcohols and diols that were screened, ethanol and propylene glycol were least irritating. Hence, authors focused the investigations on co-surfactant-free microemulsions. The potential of co-surfactant-free w/o microemulsions (with varying water content), liquid crystalline phases, o/w emulsions and solutions was evaluated with respect to improvement in the pre-corneal retention of technetium (^{99m}Tc) and the miotic activity of pilocarpine. It was observed that the pre-corneal retention of the

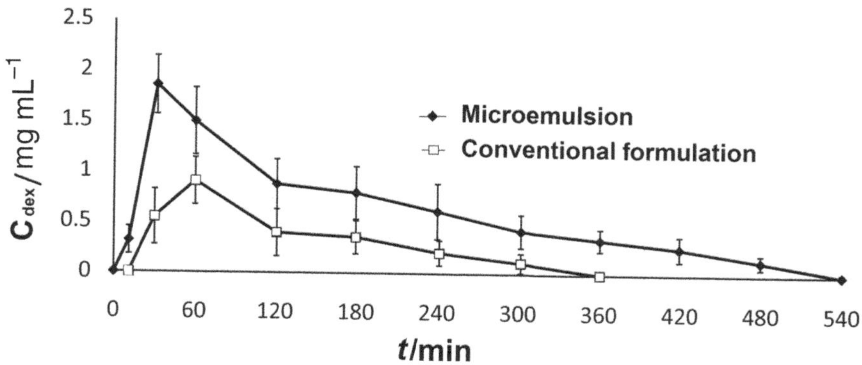

Figure 9.6 Concentration of dexamethasone (C_{dex}) in the aqueous humour after administration of the microemulsion and conventional formulation. The microemulsion was superior to the conventional formulation ($P < 0.05$). (Figure redrawn with data from Ref. [127], reprinted with permission of Blackwell Synergy.)

microemulsions was more than that of the solutions. The miotic activity studies indicated that the microemulsions were significantly superior to the solutions, emulsion and liquid crystalline phase. Furthermore, sustained activity was also observed in the case of the microemulsions.

9.7 Mucosal drug delivery

Mucosal drug delivery includes delivery to mucosal surfaces such as nose, buccal cavity, lungs and vagina for local or for the systemic treatment. The nasal, pulmonary and buccal delivery can be employed for the systemic delivery of the therapeutic agents including peptides which have problems related to the high first-pass metabolism and gastric irritation. The major advantage of mucosal delivery is the avoidance of invasive parenteral delivery which improves the patient compliance. The nasal route can also be useful for the brain targeting in certain cases. Vaginal delivery is usually for the local therapy (e.g. vaginal candidiasis) but can be used for systemic delivery if required. The microemulsion formulations for mucosal delivery should not cause the damage to the mucosal surfaces. The microemulsions are quite under researched for mucosal applications as compared

Table 9.7 Ocular pharmacokinetics of dexamethasone in various formulations

Parameter	Microemulsion	Conventional formulation
T_{max}	30 min	60 min
C_{max}	1.86 ± 0.29 mg mL^{-1}	0.93 ± 0.21 mg mL^{-1}
$AUC_{0\text{–}540\,min}$	325.6 ± 36.51 mg mL^{-1} min^{-1}	121.67 ± 10.16 mg mL^{-1} min^{-1}

The microemulsion formulation was superior to the conventional formulation. (From Ref. [127], reprinted with permission of Blackwell.)

to the other routes of delivery. The ability of microemulsions to increase the membrane permeability and to modulate tight junctions is expected to improve the delivery of the drugs.

9.7.1 Potential explored

Li *et al.* formulated diazepam nasal microemulsions based on Tween 80, propylene glycol, ethanol and ethyl laurate and evaluated their potential as an alternative to the intravenous delivery of diazepam [130]. The relative bioavailablity of the nasally administered diazepam microemulsion was 50% at the same dose level. The nasal microemulsion after the two-fold increment in the diazepam dose was found to be bioequivalent to the parenteral therapy. Zhang *et al.* [131] evaluated the potential of microemulsions to target nimodipine (a drug indicated in senile dementia and memory loss) to brain after nasal delivery. Comparative evaluation of nimodipine microemulsion and nimodipine parenteral formulation indicated that the bioavailability of the nimodipine from microemulsion was 32% relative to the parenteral administration. However, the concentrations of nimodipine in the brain tissues and cerebrospinal fluid were significantly higher than that of parenteral solution. Vyas *et al.* have evaluated the potential of microemulsions and mucoadhesive microemulsions in improving the brain concentrations of sumatriptan and zolmitriptan [132, 133]. The brain targeting of both the anti-migrane drugs was evaluated by using gamma scintigraphic studies after the administration of the radiolabelled microemulsions, radiolabelled mucoadhesive microemulsions and parenteral solutions. Microemulsions and mucoadhesive microemulsions were more efficient in targeting the drugs to the brain as compared to the parenteral formulations. Furthermore, mucoadhesive microemulsions proved to be more advantageous than microemulsions in both the cases. Scherlund *et al.* [134] reported the formulation of the poloxamer-based microemulsions of lidocaine and prilocaine eutectic mixture for periodontal anaesthesia. However, the formulations were not evaluated for the *in vitro* permeation or *in vivo* activity. Ceschel *et al.* [135] evaluated the permeation of *S. desolenea* essential oil components from the porcine buccal mucosa from microemulsions, hydrogels and microemulsion-based gels. The permeation of the essential oil components was more from the microemulsion-based gels as compared to that of hydrogels. Furthermore, the addition of Transcutol to the formulations increased the permeation.

D'Cruz *et al.* [136] have recently reported the formulation of microemulsion-based gel labelled as GM-144 by using Cremophore®EL, Phospholipids, Caprylic acid diglycerides and Poloxamers and xanthan gum. GM-144 completely immobilised sperm in human or rabbit semen in less than 30 s. The *in vivo* contraceptive potency of GM-144 was compared with the standard detergent spermicide, nonoxynol-9 (N-9)-containing formulation (Gynol II) in the rabbits. GM-144 showed remarkable contraceptive activity in the rigorous rabbit model. When compared with control, intravaginal administration of GM-144 and Gynol II resulted in 75 and 70.8% inhibition of fertility, respectively. Thus, GM-144 proved to be a vaginal contraceptive, as effective as the commercially available formulation Gynol II. Hickey and co-workers have reported lecithin-based inverse microemulsions (water-in-propellant) for the pulmonary delivery of hydrophilic solutes [137, 138]. The microemulsions were not investigated for the delivery of the therapeutic agents but the proof of concept was established. The water content of the microemulsions exhibited effect on the pulmonary delivery.

9.8 Microemulsions as templates for the synthesis of pharmaceutical nanocarriers

The advantages of microemulsions such as the ability to entrap hydrophilic as well as hydrophobic moieties, very high interfacial area, the ease of synthesis and scale-up and monodisperse nanostructure make them an ideal template for the synthesis of nanoparticulate systems. The use of microemulsions for various organic reactions (see Chapter 5) and for the formation of inorganic nanostructures (see Chapter 6) is being explored since the last two decades. The potential of microemulsions for synthesising pharmaceutical nanocarriers will be briefly discussed.

9.8.1 *Synthesis of solid lipid nanoparticles*

Solid lipid nanoparticles (SLNs) are of great interest in the drug delivery research and are used in the oral, parenteral and dermal drug delivery. These are solid, submicronic particulate carriers composed of drug dispersed in or adsorbed on to the physiologically acceptable solid lipids. The advantages of SLN in drug delivery have been thoroughly reviewed in the literature and are out of the scope of this chapter [11, 139]. Gasco first explored the use of microemulsions for the synthesis of SLN [31]. The surfactant and co-surfactants can be chosen on the basis of desired route of administration, their biocompatibility and dispersed phase solubilising capacity. The microemulsions are essentially formulated at a temperature above the melting point of solid lipid. The drug is solubilised in the microemulsion and the warm microemulsion is then dispersed quickly into cold water to form SLN loaded with drug. The temperature-dependant behaviour of microemulsions and melted lipids, the thermodynamic stability of microemulsion and the readily formed nanodroplets of the molten lipid in the microemulsion enable nanoengineering of SLN with uniform size distribution. The thermodynamic stability of the microemulsion prevents the nanoparticle aggregation. In SLN obtained from microemulsion, drug is present in the amorphous form. Mumper *et al.* [140] achieved SLN of emulsifying wax and Brij 78 loaded with Gadolinium acetoacetate from a microemulsion template without any dilution. The SLNs were obtained by just cooling the ME (55°C) to room temperature or storing hot ME at 4°C. This clearly demonstrates that the process of SLN synthesis depends on the type of solid lipid, its temperature-dependant behaviour and the type of surfactants and co-surfactants. There are several investigations in the literature employing microemulsions for the synthesis of SLN. Joshi and Patravale have recently established the use of microemulsions for the synthesis of nanostructured lipid carriers (NLCs) which are a variant of SLNs [141].

9.8.2 *Synthesis of nanosuspensions*

Nanosuspensions can be described as colloidal dispersions of nanosized drug particles that are produced by a suitable method and are stabilised by suitable means. They already have been commercialised due to the less stringent regulatory restrictions [12]. Although the use of microemulsions for engineering inorganic nanomaterials has been very well established [142], their potential in generating nanosuspension of API has been realised only recently. The technique is still in its infancy and requires extensive research. Since the biocompatibility is the prerequisite for the use of microemulsions in pharmaceuticals, using them as template to produce nanoparticulate drugs is very limited compared to inorganic nanomaterials (see Chapter 6). This is mainly due to the stringent restrictions on the usage of the organic solvents and surfactants in the drug delivery systems [143].

For the synthesis of nanosuspensions, the internal phase of the microemulsion has to be a pharmaceutically acceptable partially water-miscible solvent like butyl lactate, benzyl alcohol, propylene carbonate or triacetin. These solvents have good solubilising power for the hydrophobic drugs and can be used as a dispersed phase above their miscibility point. The choice of surfactant and co-surfactant is mainly governed by the desired route of administration, their biocompatibility and dispersed phase solubilising capacity. The drug can either be dissolved or solubilised in the internal phase or preformed microemulsions could be saturated with the drug by intimate mixing [143]. Drug nanoparticles are acquired by diluting the microemulsion. The surfactant and co-surfactant present in the microemulsion help in preventing the aggregation of the nanoparticulate drug, thus acting as stabiliser. The nanoparticulate suspension should be made free of the internal phase and surfactants (if necessary) by means of diultrafiltration so as to make it suitable for the administration. However, if all the ingredients used in nanoparticulate suspension production are present in a concentration acceptable for desired route of administration, then simple centrifugation or ultracentrifugation may suffice [143]. Production of drug nanoparticulate suspensions from microemulsion templates has been successfully applied to poorly water-soluble and poorly bioavailable anti-fungal drug griseofulvin wherein significant improvement in dissolution rate of the drug (threefold increase) as compared to the commercial product was observed [144].

9.8.3 Engineering of nano-complexes

The utilisation of microemulsions for carrying out organic and bioorganic reactions has been well described in the literature [145] and is discussed in more detail in Chapter 5. Recently, the use of microemulsions for engineering nanocomplexes of pharmaceutical significance using inverse microemulsions as template has been described. Andersson and Jan-Erik Löfroth [146] have conveniently synthesised a chitosan–heparin nanocomplex by mixing two inverse microemulsions containing chitosan and sodium salt of heparin in the internal phase, respectively. It is hypothesised that after mixing the microemulsions, due to the Brownian motion, the droplets collide with each other and exchange their contents leading to nanocomplex formation in ME (this fusion–fission process is illustrated in Fig. 6.2). The nanosize of the complex is attributed to the microemulsion droplet structure. Preparation of such complexes in inverse ME can enable oral delivery of macromolecules. Exhaustive investigation is still required in this direction to have interesting examples of such nanocomplexes. Recently, the synthesis of nanocrystalline silver–sulfadiazine complex (an antibacterial compound) from inverse microemulsions has been described [147].

9.8.4 Microemulsion polymerisation

Microemulsion polymerisation has shown a great advantage over conventional polymerisation strategies such as emulsion polymerisation with respect to the end particle size, polydispersity and reproducibility of the product characteristics. Although we still face severe problems regarding the polymerisation of microemulsions (see Section 11.2 in Chapter 11), it has been employed for the synthesis of polymeric nanoparticles of pharmaceutical interest. Microemulsion polymerisation involves free-radical polymerisation in a large number of monomer-swollen microemulsion droplets and represents a thermodynamically stable, transparent one-phase reaction system. Generally, the microemulsion droplet is considered as initiation locus for the polymerisation. The type of microemulsion used for the polymerisation depends on the monomer properties [148].

The major concern regarding microemulsion polymerisation from a pharmaceutical perspective is the separation of the polymeric nanoparticles from the other ingredients such as microemulsion components and initiator. However, this problem will not be a limiting factor for the synthesis of polyalkylcyanoacrylate nanoparticles where the polymerisation is initialised by changes in the pH. Furthermore, if the biocompatible components are used for the synthesis of the microemulsion, then the acceptability of these systems would further be enhanced. Rades *et al.* [149–152] have reported the synthesis of polyalkylcyanoacrylate-based insulin nanoparticles in biocompatible microemulsions. The *in vivo* studies indicated that these particles were significantly more efficient with respect to oral insulin delivery as compared to the microemulsions and insulin solution [150]. They have also demonstrated that the nanoparticles can be formulated in anhydrous microemulsions which could be further useful for preserving the peptide efficacy [152].

The synthesis of polyacrylic acid nanoparticles from inverse microemulsions has also been described in the literature. Polyacrylic acid is a well-known bioadhesive agent. Hence, polyacrylic acid nanoparticles loaded with the drug would have enhanced bioavailability and controlled release characteristics. The synthesis of timolol maleate and brimonidine nanoparticles for improved ocular and oral delivery has been reported [153, 154].

It is very important to design a microemulsion polymerisation process that enables successful removal of nanoparticles formed and does not affect the drug efficacy and stability during the polymerisation process. Both these criteria are difficult to meet in the pharmaceutical arena and hence extensive investigations are required to establish the potential of microemulsion polymerisation [153, 154].

9.9 Application in pharmaceutical analysis

The pharmaceutical analysis deals with an array of therapeutic agents either alone or in combinations that are incorporated in dosage forms of extreme characteristics. The main aim of pharmaceutical analysis is to analyse the compounds accurately and as quickly as possible. Amongst several analytical techniques that are used, HPLC is one of the most commonly used techniques for the analysis and separation of an array of therapeutic agents from their degradation products and impurities. In conventional HPLC, often various difficulties are experienced by the pharmaceutical analysts while handling complex mixtures containing therapeutic agents/impurities/degradants of different hydrophilicity and hydrophobicity. In certain cases, use of gradient systems can solve this problem but the time for analyses increases significantly in many cases. Microemulsions have gained a great interest as an eluent in the HPLC analysis. Microemulsions possess a unique property that they can solubilise both polar and non-polar substances. Because of the high aqueous content of o/w microemulsions, they are very compatible with reversed-phase HPLC columns while the hydrophobic oil core gives them the ability to dissolve non-polar solutes and sample matrices [155–158].

In conventional HPLC, separation is governed by solute's partitioning between the mobile and stationary phase: a mixture of solutes will partition to different degree and thus can be separated. Using a microemulsion as a mobile phase alters these partitioning characteristics because a layer of surfactant molecules adsorbs onto the surface of the stationary phase. The retention and separation of solutes in o/w microemulsion liquid chromatography will also be significantly influenced by the implementation of secondary partitioning equilibrium, e.g. solutes partition from the aqueous phase or stationary phase into the droplets [155–158]. The hydrophobic microemulsion core is able to solubilise hydrophobic compounds while hydrophilic compounds are compatible with the aqueous continuous phase. The complexity of the composition of the microemulsion allows a great many manipulations to be made during method development in order to achieve acceptable resolution of complex mixtures. In this way microemulsion eluents enable separation, in isocratic mode, of complex mixtures of hydrophilic and hydrophobic compounds. The advantages of using microemulsions include better resolution and reduced time of analyses.

However, while using microemulsion as an eluent, it should be remembered that several factors such as concentration of surfactants, co-surfactants and organic phase, pH of the aqueous phase, column temperature can influence the resolution and retention of the compounds. The microemulsions described so far are based on sodium dodecyl sulphate, short-chain alcohols such as butanol and octanol and aqueous phase of different pH values [78, 159–161]. The utility of w/o microemulsions in the normal phase HPLC analysis has also been described [162]. The microemulsion liquid chromatography (MELC) has been used for resolution of several API such as paracetamol, loratidine, simvastatin, niacinamide, fosinoprilat from their impurities or degradants or metabolic products [159–162].

Recently, the application of the microemulsion as an eluent in the high-performance thin layer chromatography (HPTLC) has also been described [163]. Microemulsions were used as an eluent for the fingerprinting of licorice extract. As proposed earlier, the microemulsion resulted in much better resolution of the components present in the herbal extract as compared to the conventional eluent. This approach would be very useful in the standardisation of the herbal extracts which often consist of a vast number of components with diverse chemical nature.

9.10 Future perspectives

Although numerous applications of microemulsions have been described, there are still many avenues remaining to be explored. There is still a need for better excipients having acceptability for parenteral and ocular route. The excipients like caprylic acid mono-, di-glycerides and propylene glycol fatty acid esters can be evaluated for their suitability for parenteral and ocular delivery. Although several applications for the transdermal delivery have been established, the potential of microemulsions in the epidermal or dermal targeting has just been realised. The systematic investigations in this direction may provide good therapeutic avenues for skin diseases such as acne, psoriasis and skin cancers. The templating of pharmaceutical nanomaterials with the help of microemulsions is an emerging area and extensive investigations are required in this direction. Pharmaceutical chromatographic analysis is expected to take a great leap with the advent of microemulsions as an eluent.

References

1. Lipinski, C. (2002) Poor aqueous solubility – an industry wide problem in drug discovery. *Am. Pharm. Rev.*, **5**, 82–85.
2. Amidon, G.L., Lennernas, H., Shah, V.P. and Crison, J. (1995) A theoretical basis for a biopharmaceutic drug classification: The correlation of *in vitro* drug product dissolution and *in vivo* bioavailability. *Pharm. Res.*, **12**, 413–420.
3. Waterbeemd, H. (1998) The fundamental variables of the biopharmaceutics classification system (BCS): A commentary. *Eur. J. Pharm. Sci.*, **7**, 1–3.
4. Lawrence M. J. and Rees, G.D. (2000) Microemulsion-based media as novel drug delivery Systems. *Adv. Drug Deliv. Rev.*, **45**, 89–121.
5. Lv, F.F., Li, N. Zheng, L.Q. and Tung, C.H. (2006) Studies on the stability of the chloramphenicol in the microemulsion free of alcohols. *Eur. J. Pharm. Biopharm.*, **62**, 288–294.
6. Gallarate, M., Carlotti, M.E., Trotta, M., Grande, A.E. and Talarico, C. (2004) Photostability of naturally occurring whitening agents in cosmetic microemulsions. *J. Cosmet. Sci.*, **55**, 139–148.
7. Sarciaux, J.M., Acar, L. and Sado, P.A. (1995) Using microemulsion formulations for oral drug delivery of therapeutic peptides. *Int. J. Pharm.*, **120**, 127–136.
8. Kang, B.K., Lee, J.S., Chon S.K., Jeong, S.Y., Yuk, S.H., Khang, G., Lee, H.B. and Cho, S.H. (2004) Development of self-microemulsifying drug delivery systems (SMEDDS) for oral bioavailability enhancement of simvastatin in beagle dogs. *Int. J. Pharm.*, **274**, 65–73.
9. Wei, L., Sun, P., Nie, S. and Pan, W. (2005) Preparation and evaluation of SEDDS and SMEDDS containing carvedilol. *Drug Dev. Ind. Pharm.*, **31**, 785–794.
10. Kreilgaard, M. (2002) Influence of microemulsions on cutaneous drug delivery. *Adv. Drug. Deliv. Rev.*, **54**(Suppl. 1), S77–S98.
11. Müller, R.H., Mäder, K. and Gohla, S. (2000) Solid lipid nanoparticles (SLN) for controlled drug delivery – a review of the state of the art. *Eur. J. Pharm. Biopharm.*, **50**, 161–178.
12. Patravale, V.B., Date, A.A. and Kulkarni, R. (2004) Nanosuspensions: A promising drug delivery strategy. *J. Pharm. Pharmacol.*, **56**, 827–840.
13. Malcolmson, C., Satra, C., Kantaria, S., Sidhu, A. and Lawrence, M.J. (1998) Effect of the nature of oil on the incorporation of testosterone propionate into non-ionic oil-in-water microemulsions. *J. Pharm. Sci.*, **87**, 109–116.
14. Warisnoicharoen, W., Lansley, A.B. and Lawrence, M.J. (2000) Non-ionic oil-in-water microemulsions: The effect of oil type on phase behavior. *Int. J. Pharm.*, **198**, 7–27.
15. Vandamme, T.F. (2002) Microemulsions as ocular drug delivery systems: Recent developments and future challenges. *Prog. Ret. Eye Res.*, **21**, 15–34.
16. Gursoy, N. and Benita, S. (2004) Self-emulsifying drug delivery systems (SEDDS) for improved oral delivery of lipophilic drugs. *Biomed. Pharmacother.*, **58**, 173–182.

17. Lundin, P., Bojrup, M., Ljusberg-Wahren, H., Westrom, B.R. and Lundin, S. (1997) Enhancing effects of monohexanoin and two other medium-chain glyceride vehicles on intestinal absorption of desmopressin (Ddavp). *J. Pharm. Exp. Ther.*, **282**, 585–590.
18. Akers, M.J. (2002) Excipient-drug interactions in parenteral formulations. *J. Pharm. Sci.*, **91**, 2283–2300.
19. Strickley, R.G. (2004) Solubilizing excipients in oral and injectable formulations. *Pharm. Res.*, **21**, 201–230.
20. Tije, A.J., Verweij, J., Loos, W.J. and Spareboom, A. (2003) Pharmacological effects of formulation vehicles implications for cancer chemotherapy. *Clin. Pharmacokinet.*, **42**, 665–685.
21. Stilbs, P., Lindman, B. and Rapacki, K. (1983) Effect of alcohol cosurfactant length on microemulsion structure. *J. Colloid Interface Sci.*, **95**, 583–585.
22. Yaghmur, A., Aserin, A. and Garti, N. (2002) Phase behaviour of microemulsions based on food-grade non-ionic surfactants: Effect of polyols and short-chain alcohols. *Colloid Surf. A*, **209**, 71–81.
23. Tenjarla, S. (1999) Microemulsions: An overview and pharmaceutical applications. *Crit. Rev. Ther. Drug Carrier Syst.*, **16**, 461–521.
24. Alany, R.G., Rades, T., Agatonovic-Kustrin, S., Davies, N.M. and Tucker, I.G. (2000) Effects of alcohols and diols on the phase behaviour of quaternary systems. *Int. J. Pharm.*, **196**, 141–145.
25. Flanagan, J. and Singh, H. (2006) Microemulsions: A potential delivery system for bioactives in food. *Crit. Rev. Food Sci. Nutr.*, **46**, 221–237.
26. Attwood, D. (1994) Microemulsions. In J. Kreuter (ed), *Colloidal Drug Delivery Systems.* Marcel Dekker, New York, pp. 31–71.
27. Eccleston, J. (1994) Microemulsions. In J. Swarbick and J.C. Boylan (eds), *Encyclopedia of Pharmaceuticaltechnology.* Marcel Dekker, New York, pp. 375–421.
28. Kawakami, K., Yoshikawa, T., Moroto, Y., Kanaoka, E., Takahashi, K., Nishihara, Y. and Masuda, K. (2002) Microemulsion formulation for enhanced absorption of poorly soluble drugs I. Prescription design. *J. Control. Release*, **81**, 65–74.
29. Zhang, X., Dong, J., Zhang, G., Hong, X. and Li, X. (2005) The effects of additives on the water solubilization capacity and conductivity in *n*-pentanol microemulsions. *J. Colloid Interface Sci.*, **285**, 336–341.
30. Roy, B.K. and Moulik, S.P. (2002) Functions of hydrotropes (sodium salicylate, proline, pyrogallol, resorcinol and urea) in solution with special reference to amphiphile behaviors. *Colloid Surf. A*, **203**, 155–166.
31. Gasco, M.R. (1997) Solid lipid nanospheres form warm microemulsions. *Pharm. Technol. Eur.*, **9**, 32–42.
32. Djordjevic, L., Primorac, M., Stupar, M. and Krajisnik, D. (2004) Characterization of caprylocaproyl macrogolglycerides based microemulsion drug delivery vehicles for an Amphiphilic drug. *Int. J. Pharm.*, **271**, 11–19.
33. Joubran, R.F., Cornell, D.G. and Parris, N. (1993) Microemulsions of triglyceride and non-ionic surfactant-effect of temperature and aqueous phase composition. *Colloid Surf. A*, **80**, 153–157.
34. Flanagan, J., Kortegaard, K., Neil-Pinder, D., Rades, T. and Singh, H. (2006) Solubilisation of soybean oil in microemulsions using various surfactants. *Food Hydrocolloids*, **20**, 253–260.
35. Pes, M.A., Aramaki, K., Nakamura, N. and Kunieda, H. (1996) Temperature-insensitive microemulsions in a sucrose monoalkanoate system. *J. Colloid Interface Sci.*, **178**, 666–672.
36. Bagwe, R.P., Kanicky, J.R., Palla, B.J., Patanjali, P.K. and Shah, D.O. (2001) Improved drug delivery using microemulsions: Rationale, recent progress, and new horizons. *Crit. Rev. Ther. Drug Carrier Syst.*, **18**, 77–140.
37. Foldvarim, M. (2000) Non-invasive administration of drugs through the skin: Challenges in delivery system design. *Pharm. Sci. Tech. Today*, **3**, 417–425.

38. Benson, H.A.E. (2005) Transdermal drug delivery: Penetration enhancement techniques. *Curr. Drug Deliv.*, **2**, 23–33.
39. Brown, M.B., Martin, G.P., Jones, S.A. and Akomeah, F.K. (2006) Dermal and transdermal drug delivery systems: Current and future prospects. *Drug Deliv.*, **13**, 175–187.
40. Ting, W., Vest, C. and Sontheimer, R. (2004) Review of traditional and novel modalities that enhance the permeability of local therapeutics across the stratum corneum. *Int. J. Dermatol.*, **43**, 538–547.
41. Stott, P.W., Williams, A.C. and Barry, B.W. (2001) Mechanistic study into the enhanced transdermal permeation of a model β-blocker, propranolol, by fatty acids: A melting point depression effect. *Int. J. Pharm.*, **219**, 161–176.
42. Kogan, A. and Garti, N. (2006) Microemulsions as transdermal drug delivery vehicles. *Adv. Colloid Interface Sci.*, **123–126**, 369–385.
43. Gloor, M., Haus, G. and Keipert, S. (2003) Keratolytic activity of microemulsions. *Skin Pharmacol. Physiol.*, **16**, 151–155.
44. Changez M., Varshney, M., Chander, J. and Dinda, A.K. (2006) Effect of the composition of lecithin/*n*-propanol/isopropyl myristate/water microemulsions on barrier properties of mice skin for transdermal permeation of tetracaine hydrochloride: *In vitro*. *Colloid Surf. B*, **50**, 18–25.
45. Kemken, J., Ziegler, A. and Mueller, B.W. (1992) Influence of supersaturation on the pharmacodynamic effect of bupranolol after dermal administration using microemulsions as vehicle. *Pharm. Res.*, **9**, 554–558.
46. Peltola, S., Saarinen-Savolainen, P., Kiesvaara, J., Suhonen, T.M. and Urtti, A. (2003) Microemulsions for topical delivery of estradiol. *Int. J. Pharm.*, **254**, 99–107.
47. Biju, S.S., Ahuja, A. and Khar, R.K. (2005) Tea tree oil concentration in follicular casts after topical delivery: Determination by high-performance thin layer chromatography using a perfused bovine udder model. *J. Pharm. Sci.*, **94**, 240–245.
48. Escribano, E., Obach, M., Arévalo, M.I., Calpena, A.C., Domenech, J. and Queralt, J. (2005) Rapid human skin permeation and topical anaesthetic activity of a new amethocaine microemulsion. *Skin Pharmacol. Physiol.*, **18**, 294–300.
49. Shukla, A., Krause, A. and Neubert, R.H.H. (2003) Microemulsions as colloidal vehicle systems for dermal drug delivery. Part IV: Investigation of microemulsion systems based on a eutectic mixture of lidocaine and prilocaine as the colloidal phase by dynamic light scattering. *J. Pharm. Pharmacol.*, **55**, 741–748.
50. Kreilgaard, M., Pedersen, E.J. and Jaroszewski, J.W. (2000) NMR characterisation and transdermal drug delivery potential of microemulsion systems. *J. Control. Release*, **69**, 421–433.
51. Sintov, A.C. and Shapiro, L. (2004) New microemulsion vehicle facilitates percutaneous penetration *in vitro* and cutaneous drug bioavailability *in vivo*. *J. Control. Release*, **95**, 73–183.
52. Lee, P.J., Langer, R. and Shastri, V.P. (2003) Novel microemulsion enhancer formulation for simultaneous transdermal delivery of hydrophilic and hydrophobic drugs. *Pharm. Res.*, **20**, 264–269.
53. Wang, D.K., Shi, Z.H., Liu, L., Wang, X.Y., Zhang, C.X. and Zhao, P. (2006) Development of self-microemulsifying drug delivery systems for oral bioavailability enhancement of alpha-asarone in beagle dogs. *PDA J. Pharm. Sci. Technol.*, **60**, 343–349.
54. Kreilgaard, M. (2001) Dermal pharmacokinetics of microemulsion formulations determined by *in vivo* microdialysis. *Pharm. Res.*, **18**, 367–373.
55. Kreilgaard, M., Kemme, M.J.B., Burggraaf, J., Schoemaker, R.C. and Cohen, A.F. (2001) Influence of a microemulsion vehicle on cutaneous bioequivalence of a lipophilic model drug assessed by microdialysis and pharmacodynamics. *Pharm. Res.*, **18**, 593–599.
56. Changez, M. and Varshney, M. (2000) Aerosol-OT microemulsions as transdermal carriers of tetracaine hydrochloride. *Drug Dev. Ind. Pharm.*, **26**, 507–512.

57. Changez, M., Chander, J. and Dinda, A.K. (2006) Transdermal permeation of tetracaine hydrochloride by lecithin microemulsion: *In vivo*. *Colloid Surf. B*, **48**, 58–66.
58. Kriwet, K. and Muller-Goymann, C.C. (1995) Diclofenac release from phospholipid drug systems and permeation through excised human stratum corneum. *Int. J. Pharm.*, **125**, 231–242.
59. Kweon, J.H., Chi, S.C. and Park, E.S. (2004) Transdermal delivery of diclofenac using microemulsions. *Arch. Pharm. Res.*, **27**, 351–356.
60. Sintov, A.C. and Botner, S. (2006) Transdermal drug delivery using microemulsion and aqueous systems: Influence of skin storage conditions on the *in vitro* permeability of diclofenac from aqueous vehicle systems. *Int. J. Pharm.*, **311**, 55–62.
61. Park, E.S, Cui, Y., Yun, B.J., Ko, I.J. and Chi, S.C. (2005) Transdermal delivery of piroxicam using microemulsions. *Arch. Pharm. Res.*, **28**, 243–248.
62. Yuan, Y., Li, S., Mo, F.K. and Zhong, D.F. (2006) Investigation of microemulsion system for transdermal delivery of meloxicam. *Int. J. Pharm.*, **321**, 117–123.
63. Chen H., Chang, X., Du, D., Li, J., Xu, H. and Yang, X. (2006) Microemulsion-based hydrogel formulation of ibuprofen for topical delivery. *Int. J. Pharm.*, **315**, 52–58.
64. Subramanian, N., Ghosal, S.K. and Moulik, S.P. (2005) Enhanced *in vitro* percutaneous absorption and *in vivo* anti-inflammatory effect of a selective cyclooxygenase inhibitor using microemulsion. *Drug Dev. Ind. Pharm.*, **31**, 405–416.
65. Desai, K.G.H. (2004) Enhanced skin permeation of rofecoxib using topical microemulsion gel. *Drug Dev. Res.*, **63**, 33–40.
66. Bolzinger, M.A., Carduner, T.C. and Poelman, M.C. (1998) Bicontinuous sucrose ester microemulsion: A new vehicle for topical delivery of niflumic acid. *Int. J. Pharm.*, **176**, 39–45.
67. Dalmora, M.E., Dalmora, S.L. and Oliveira, A.G. (2001) Inclusion complex of piroxicam with β-cyclodextrin and incorporation in cationic microemulsion. *In vitro* drug release and *in vivo* topical anti-inflammatory effect. *Int. J. Pharm.*, **222**, 45–55.
68. Biruss, B., Kahlig, H. and Valenta, C. (2007) Evaluation of an eucalyptus oil containing topical drug delivery system for selected steroid hormones. *Int. J. Pharm.*, **328**, 142–151.
69. Lehmann, L., Keipert, S. and Gloor, M. (2001) Effects of microemulsions on the stratum corneum and hydrocortisone penetration. *Eur. J. Pharm. Biopharm.*, **52**, 129–136.
70. Cui, Z., Fountain, W., Clark, M., Jay, M. and Mumper, R. (2003) Novel ethanol-in-fluorocarbon microemulsions for topical genetic immunization. *Pharm. Res.*, **20**, 16–23.
71. Cui, Z. and Sloat, B.R. (2006) Topical immunization onto mouse skin using a microemulsion incorporated with an anthrax protective antigen protein-encoding plasmid. *Int. J. Pharm.*, **317**, 187–191.
72. Gershanik, T. and Benita, S. (2000) Self-dispersing lipid formulations for improving oral absorption of lipophilic drugs. *Eur. J. Pharm. Biopharm.*, **50**, 179–188.
73. Pouton, C.W. (2000) Lipid formulations for oral administration of drugs: Non-emulsifying, self-emulsifying and 'self-microemulsifying' drug delivery systems. *Eur. J. Pharm. Sci.*, **11**, S93–S98.
74. Koga, K., Kawashima, S. and Murakami, M. (2002) *In vitro* and in situ evidence for the contribution of labrasol and gelucire 44/14 on transport of cephalexin and cefoperazone by rat intestine. *Eur. J. Pharm. Biopharm.*, **54**, 311–318.
75. Rege, B.D., Kao, J. and Pollia, J. (2002) Effects of non-ionic surfactants on membrane transporters in Caco-2 cell monolayers. *Eur. J. Pharm. Sci.*, **16**, 237–246.
76. Hugger, E.D., Audus, K.L. and Borchardt, R.T. (2002) Effects of poly(ethylene glycol) on efflux transporter activity in Caco-2 cell monolayers. *J. Pharm. Sci.*, **91**, 1980–1990.
77. Sha, X., Yan, G., Wu, Y., Li, J. and Fang, X. (2005) Effect of self-microemulsifying drug delivery systems containing labrasol on tight junctions in Caco-2 cells. *Eur. J. Pharm. Sci.*, **24**, 477–486.

78. El-Sherbiny, D.T., El-Enany, N., Belal, F. and Hansen, S. (2007) Simultaneous determination of loratadine and desloratadine in pharmaceutical preparations using liquid chromatography with a microemulsion as eluent. *J. Pharm. Biomed. Anal.*, **43**, 1236–1242.
79. Ghosh, P.K., Majithiya, R.J., Umrethia, M.L. and Murthy, R.S.R. (2006) Design and development of microemulsion drug delivery system of acyclovir for improvement of oral bioavailability. *AAPS Pharmscitech*, **7**, 77.
80. Mountfield, R.J., Senepin, S., Schleimer, M., Walter, I. and Bittner, B. (2000) Potential inhibitory effects of formulation ingredients on intestinal cytochrome P450. *Int. J. Pharm.*, **211**, 89–92.
81. Hauss, D.J., Fogal, S.E., Ficorilli, J.V., Price, C.A., Roy, T., Jayaraj, A.A. and Keirns, J.J. (1998) Lipid-based delivery systems for improving the bioavailability and lymphatic transport of a poorly water-soluble LTB4 inhibitor. *J. Pharm. Sci.*, **87**, 164–169.
82. O'Driscoll, C.M. (2002) Lipid-based formulations for intestinal lymphatic delivery. *Eur. J. Pharm. Sci.*, **15**, 405–415.
83. Kawakami, K., Yoshikawa, T., Moroto, Y., Kanaoka, E., Takahashi, K., Nishihara, Y. and Masuda, K. (2002) Microemulsion formulation for enhanced absorption of poorly soluble drugs II. *In vivo* study. *J. Control. Release*, **81**, 75–82.
84. Cornaire, G., Woodley, J., Hermann, P., Cloarec, A., Arellano, C. and Houin, G. (2004) Impact of excipients on the absorption of P-glycoprotein substrates *in vitro* and *in vivo*. *Int. J. Pharm.*, **278**, 119–131.
85. Hugger, E.D., Novak, B.L., Burton, P.S., Audus, K.L. and Borchardt, R.T. (2002) Comparison of commonly used polyethoxylated pharmaceutical excipients on their ability to inhibit P-glycoprotein activity *in vitro*. *J. Pharm. Sci.*, **91**, 1991–2002.
86. Bernkop-Schnurch, A. and Grabovac, V. (2006) Polymeric efflux pump inhibitors in oral drug delivery. *Am. J. Drug Deliv.*, **4**, 263–272.
87. Subramanian, N., Ray, S., Ghosal, S.K., Bhadra, R. and Moulik, S.P. (2004) Formulation design of self-microemulsifying drug delivery systems for improved oral bioavailability of celecoxib. *Biol. Pharm. Bull.*, **27**, 1993–1999.
88. Holt, D.W., Mueller, E.A., Kovarik, J.M., van Bree, J.B. and Kutz, K. (1994) The pharmacokinetics of sandimmune neoral: A new oral formulation of cyclosporine. *Transplant. Proc.*, **26**, 2935–2939.
89. Jing, O., Shen, Y., Ren, F., Chen, J., Jiang, Z., Peng, B., Leng, Y. and Dong, J. (2006) HPLC determination of anethole trithione and its application to pharmacokinetics in rabbits. *J. Pharm. Biomed. Anal.*, **42**, 613–617.
90. Porter, C., Kaukonen, A.M., Boyd, B.J., Edwards, G. and Charman, W.N. (2004) Susceptibility to lipase-mediated digestion reduces the oral bioavailability of danazol after administration as a medium-chain lipid-based microemulsion formulation. *Pharm. Res.*, **21**, 1405–1412.
91. Shen, H.R. and Zhong, M.K. (2006) Preparation and evaluation of self-microemulsifying drug delivery systems (SMEDDS) containing atorvastatin. *J. Pharm. Pharmacol.*, **58**, 1183–1191.
92. Wu, W., Wang, Y. and Que, L. (2006) Enhanced bioavailability of silymarin by self-microemulsifying drug delivery system. *Eur. J. Pharm. Biopharm.*, **63**, 288–294.
93. Cui, S., Zhao, C., Chen, D. and He, Z. (2005) Self-microemulsifying drug delivery systems (SMEDDS) for improving *in vitro* dissolution and oral absorption of pueraria lobata isoflavone. *Drug Dev. Ind. Pharm.*, **31**, 349–356.
94. Yang, S., Gursoy, N., Lambert, G. and Benita, S. (2004) Enhanced oral absorption of paclitaxel in a novel self-microemulsifying drug delivery system with or without concomitant use of P-glycoprotein inhibitors. *Pharm. Res.*, **21**, 261–270.
95. Heo, M.Y., Piao, Z.Z., Kim, T.W., Cao, O.R., Kim, A. and Lee, B.J. (2005) Effect of solubilizing and microemulsifying excipients in polyethylene glycol 6000 solid dispersion on enhanced dissolution and bioavailability of ketoconazole. *Arch. Pharm. Res.*, **28**, 604–611.

96. Itoh, K., Tozuk, Y., Oguchi, T. and Yamamoto, K. (2002) Improvement of physicochemical properties of N-4472 Part II: Characterization of N-4472 microemulsion and the enhanced oral absorption. *Int. J. Pharm.*, **246**, 75–83.
97. Khoo, S.M., Humberstone, A.J., Porter, C.J.H. and Edwards, G.A. (1998) Formulation design and bioavailability assessment of lipidic self-emulsifying formulations of Halofantrin. *Int. J. Pharm.*, **167**, 155–164.
98. Kim, C.K., Cho, Y.J. and Gao, Z.G. (2001) Preparation and evaluation of biphenyl dimethyl dicarboxylate microemulsions for oral delivery. *J. Control Release*, **70**, 149–155.
99. Gao, Z.G., Choi, H.G., Shin, H.J., Park, K.M. and Lim, S.J. (1998) Physicochemical characterization and evaluation of a microemulsion system for oral delivery of cyclosporin A. *Int. J. Pharm.*, **161**, 75–86.
100. Constantinides, P.P. (1995) Lipid microemulsions for improving drug dissolution and oral absorption: Physical and biopharmaceutical aspects. *Pharm. Res.*, **11**, 1561–1572.
101. Constantinides, P.P., Lancaster, C.M., Marcello J., Chiossone, D.C., Orner, D., Hidalgo, I., Smith, P., Sarkahian, A.B., Yiv, S.H. and Owen, A.J. (1995) Enhanced intestinal absorption of an RGD peptide from water-in oil microemulsions of different composition and particle size. *J. Control Release*, **34**, 109–116.
102. Lyons, K.C., Charman, W.N., Miller, R. and Porter, C.J.H. (2000) Factors limiting the oral bioavailability of *N*-acetylglucosaminyl-*N*-acetylmuramyl dipeptide (GMDP) and enhancement of absorption in rats by delivery in a water-in-oil microemulsion. *Int. J. Pharm.*, **199**, 17–28.
103. Çelebi, N., Türkyilmaz, A., Gönül, B. and Özogul, C. (2002) Effects of epidermal growth factor microemulsion formulation on the healing of stress-induced gastric ulcers in rats. *J. Control Release*, **83**, 197–210.
104. Çilek, A., Çelebi, N., Tırnaksız, F. and Tay, A. (2005) A lecithin-based microemulsion of rh-Insulin with aprotinin for oral administration: Investigation of hypoglycemic effects on non-diabetic and STZ-induced diabetic rats. *Int. J. Pharm.*, **298**, 176–185.
105. Kim, S.K., Lee, E.H., Bagalkot, V., Lee, S., Lee, Y.K., Kim, C.Y., Moon, H.T. and Byun, Y. (2005) Tricaprylin microemulsion for oral delivery of low molecular weight heparin conjugates. *J. Control Release*, **105**, 32–42.
106. Zheng, J.Y. and Fulu, M.Y. (2006) Decrease of genital organ weights and plasma testosterone levels in rats following oral administration of leuprolide microemulsion. *Int. J. Pharm.*, **307**, 209–215.
107. Yalkowsky, S.H., Krzyzaniak, J.F. and Ward, G.H. (1998) Formulation-related problems associated with intravenous drug delivery. *J. Pharm. Sci.*, **87**, 787–796.
108. Lee, J.M., Park, K.M., Lee, S.J., Lee, M.K. and Kim, C.K. (2002) Microemulsion formulation of clonixic acid: Solubility enhancement and pain reduction. *J. Pharm. Pharmacol.*, **54**, 43–49.
109. Brime, B., Moreno, M., Frutos, G., Ballestros, M.P., Frutos, P. (2002) Amphotericin B in oil–water lecithin-based microemulsions: Formulation and toxicity evaluation. *J. Pharm. Sci.*, **91**, 1178–1185.
110. Junping, W., Takayama, K., Nagai, T. and Maitani, Y. (2003) Pharmacokinetics and antitumor effects of vincristine carried by microemulsions composed of PEG-lipid, oleic acid, vitamin E and cholesterol. *Int. J. Pharm.*, **251**, 13–21.
111. Söderlind, E., Wollbratt, M. and Corswant, C.V. (2003) The usefulness of sugar surfactants as solubilizing agents in parenteral formulations. *Int. J. Pharm.*, **252**, 61–71.
112. Jumaa, M. and Mueller, B.W. (2002) Parenteral emulsions stabilized with a mixture of phospholipids and PEG-660–12-hydroxy-stearate: Evaluation of accelerated and long-term stability. *Eur. J. Pharm. Biopharm.*, **54**, 207–212.

113. Corswant, C.V., Thorean, P. and Engstrom, S. (1998) Triglyceride-based microemulsion for intravenous administration of sparingly soluble substances. *J. Pharm. Sci.*, **87**, 200–208.
114. Park, K.M., Lee, M.K., Hwang, K.J. and Kim, C.K. (1999) Phospholipid-based microemulsions of flurbiprofen by the spontaneous emulsification process. *Int. J. Pharm.*, **183**, 145–154.
115. Brime, B., Molero, G., Frutos, P. and Frutos, G. (2004) Comparative therapeutic efficacy of a novel lyophilized amphotericin B lecithin-based oil–water microemulsion and deoxycholate-amphotericin B in immunocompetent and neutropenic mice infected with Candida albicans. *Eur. J. Pharm. Sci.*, **22**, 451–458.
116. Zhang, L., Sun, X. and Zhang, Z.R. (2005) An investigation on liver-targeting microemulsions of norcantharidin. *Drug Deliv.*, **12**, 289–295.
117. Jarvinen, K., Jarvinen, T. and Urtti, A. (1995) Ocular absorption following topical delivery. *Adv. Drug. Deliv. Rev.*, **16**, 3–19.
118. Ding, S. (1998) Recent developments in ophthalmic drug delivery. *Pharm. Sci. Tech. Today*, **1**, 328–335.
119. Urtti, A. (2006) Challenges and obstacles of ocular pharmacokinetics and drug delivery. *Adv. Drug. Deliv. Rev.*, **58**, 1131–1135.
120. Kaur, I.P. and Knawar, M. (2002) Ocular preparations: The formulation approach. *Drug Dev. Ind. Pharm.*, **28**, 473–493.
121. Sultana, Y., Jain, R., Aqil, M. and Ali, A. (2006) Review of ocular drug delivery. *Curr. Drug Deliv.*, **3**, 207–217.
122. Liu, Z., Li, J., Nie, S. and Pan, W. (2006) Effects of transcutol P on the corneal permeability of drugs and evaluation of its ocular irritation of rabbit eyes. *J. Pharm. Pharmacol.*, **58**, 45–50.
123. Gasco, M.R., Gallarate, M., Trotta, M., Bauchiero, L., Gremmo, E. and Chiappero, O. (1989) Microemulsions as topical delivery vehicles: Ocular administration of timolol. *J. Pharm. Biomed. Anal.*, **7**, 433–439.
124. Gallarate, M., Gasco, M.R., Trotta, M., Chetoni, P. and Saettone, M.F. (1993) Preparation and evaluation *in vitro* of solutions and O/W microemulsions containing levobunolol as ion-pair. *Int. J. Pharm.*, **100**, 219–225.
125. Siebenbrodt, I. and Keipert, S. (1993) Poloxamer-systems as potential ophthalmic microemulsions. *Eur. J. Pharm. Biopharm.*, **39**, 25–30.
126. Haβe, A. and Keipert, S. (1997) Development and characterization of microemulsions for ocular application. *Eur. J. Pharm. Biopharm.*, **43**, 179–183.
127. Fialho, S.L. and Silva-Cunha, A. (2004) New vehicle based on a microemulsion for topical ocular administration of dexamethasone. *Clin. Exp. Ophthalmol.*, **32**, 626–632.
128. Alany, R.G., Rades, T., Nicoll, J., Tucker, I.G. and Davies, N.M. (2006) W/O microemulsions for ocular delivery: Evaluation of ocular irritation and precorneal retention. *J. Control. Release*, **111**, 145–152.
129. Chan, J., Maghraby, G., Craig, J.P. and Alany, R.G. (2007) Phase transition water-in-oil microemulsions as ocular drug delivery systems: *In vitro* and *in vivo* evaluation. *Int. J. Pharm.*, **328**, 65–71.
130. Li, L., Nandi, I. and Kim, K.H. (2002) Development of an ethyl laurate-based microemulsion for rapid-onset intranasal delivery of diazepam. *Int. J. Pharm.*, **237**, 77–85.
131. Zhang, O., Jiang, X., Jiang, W., Lu, W., Su, L. and Shi, Z. (2004) Preparation of nimodipine-loaded microemulsion for intranasal delivery and evaluation on the targeting efficiency to the brain. *Int. J. Pharm.*, **275**, 85–96.
132. Vyas, T.K., Babbar, A.K., Sharma, R.K. and Misra, A. (2005) Intranasal mucoadhesive microemulsions of Zolmitriptan: Preliminary studies on brain-targeting. *J. Drug Target.*, **13**, 317–324.
133. Vyas, T.K., Babbar, A.K., Sharma, R.K. and Misra, A. (2006) Preliminary brain-targeting studies on intranasal mucoadhesive microemulsions of sumatriptan. *AAPS Pharmscitech*, **7**, Article 8.

134. Scherlund, M., Malmsten, M., Holmqvist, P. and Brodin, A. (2000) Thermosetting microemulsions and mixed micellar solutions as drug delivery systems for periodontal anesthesia. *Int. J. Pharm.*, **194**, 103–116.
135. Ceschel, G.C., Maffei P., Moretti M.D.L., Demontis, S. and Peana, A.T. (2000) *In vitro* permeation through porcine buccal mucosa of salvia desoleana Atzei and Picci essential oil from topical formulations. *Int. J. Pharm.*, **195**, 171–177.
136. D'Cruz, J., Yiv, S.H. and Uckun, F.M. (2001) GM-144, A novel lipophilic vaginal contraceptive gel-microemulsion. *AAPS Pharmscitech*, **2**, Article 5.
137. Sommerville, M.L., Johnson, C.S., Cain, J.B., Rypacek, F. and Hickey, A.J. (2002) Lecithin microemulsions in dimethyl ether and propane for the generation of pharmaceutical aerosols containing polar solutes. *Pharm. Dev. Tech.*, **7**, 273–288.
138. Sommerville, M.L. and Hickey, A.J. (2003) Aerosol generation by metered-dose inhalers containing dimethyl ether/propane inverse microemulsions. *AAPS Pharmscitech*, **4**, Article 58.
139. Wissing, S., Kayser, O. and Müller, R.H. (2004) Solid lipid nanoparticles for parenteral drug delivery. *Adv. Drug. Deliv. Rev.*, **56**, 1257–1272.
140. Mumper, R.J. and Oyewumi, M.O. (2002) Gadolinium loaded nanoparticles engineered from microemulsion templates. *Drug Dev. Ind. Pharm.*, **28**, 317–328.
141. Joshi, M.D. and Patravale, V.B. (2006) Formulation and evaluation of nanostructured lipid carrier (NLC) based gel of valdecoxib. *Drug Dev. Ind. Pharm.*, **32**, 911–918.
142. Eastoe, J., Hollamby, M.J. and Hudson, L. (2006) Recent advances in nanoparticle synthesis with reversed micelles. *Adv. Col. Int. Sci.*, **128–130**, 5–15.
143. Date, A.A. and Patravale, V.B. (2004) Current strategies for engineering drug nanoparticles. *Curr. Opin. Coll. Int. Sci.*, **9**, 222–235.
144. Trotta, M., Gallarate, M., Carlotti, M.E. and Morel, S. (2003) Preparation of griseofulvin nanoparticles from water-dilutable microemulsions. *Int. J. Pharm.*, **254**, 235–242.
145. Holmberg, K. (2004) Surfactant-templated nanomaterials synthesis. *J. Coll. Int. Sci.*, **274**, 355–364.
146. Andersson, M. and Jan-Erik Löfroth, J. (2003) Small particles of a heparin/chitosan complex prepared from a pharmaceutically acceptable microemulsion. *Int. J. Pharm.*, **257**, 305–309.
147. Nesamony, J. and Kolling, W. (2005) IPM/DOSS/Water microemulsions as reactors for silver sulfadiazine nanocrystal synthesis. *J. Pharm. Sci.*, **94**, 1310–1320.
148. Xu, X.J. and Gan, L.M. (2005) Recent advances in the synthesis of nanoparticles of polymer latexes with high polymer-to-surfactant ratios by microemulsion polymerization. *Curr. Opin. Col. Int. Sci.*, **10**, 239–244.
149. Watnasirichaikul, S., Tucker, I.G., Davies, N.M. and Rades, T. (2000) Preparation of biodegradable insulin nanocapsules from biocompatible microemulsions. *Pharm. Res.*, **17**, 684–689.
150. Watnasirichaikul, S., Tucker, I.G., Davies, N.M. and Davies N.M. (2002) *In-vitro* release and oral bioactivity of insulin in diabetic rats using nanocapsules dispersed in biocompatible microemulsion. *J. Pharm. Pharmacol.*, **54**, 473–480.
151. Watnasirichaikul, S., Rades, T., Tucker, I.G. and Davies N.M. (2002) Effects of formulation variables on characteristics of poly (Ethylcyanoacrylate) nanocapsules prepared from W/O microemulsions. *Int. J. Pharm.*, **235**, 237–246.
152. Krauel, K., Graf, A., Hook, S., Davies, N.M. and Rades, T. (2006) Preparation of poly (Alkylcyanoacrylate) nanoparticles by polymerization of water-free microemulsions. *J. Microencapsul.*, **23**, 499–512.
153. De, T.K. and Hoffman, A.S. (2001) A reverse microemulsion polymerization method for preparation of bioadhesive polyacrylic acid nanoparticles for mucosal drug deliv.: Loading and release of Timolol Maleate. *Art. Cells, Blood Subst. Immob. Biotech.*, **29**, 31–46.
154. De, T.K, Rodman, D.J., Holm, B.A., Prasad, P.N. and Bergey, E.J. (2003) Brimonidine formulation in polyacrylic acid nanoparticles for ophthalmic delivery. *J. Microencapsul.*, **20**, 361–374.

155. Marsh, A., Clark, B. and Altria, K. (2004) Oil-in-water microemulsion high performance liquid chromatographic analysis of pharmaceuticals. *Chromatographia*, **59**, 531–542.
156. Altria, K., Marsh, A. and Clark, B. (2006) High performance liquid chromatographic analysis of pharmaceuticals using oil-in-water microemulsion eluent and monolithic column. *Chromatographia*, **63**, 309–314.
157. Marsh, A., Clark, B. and Altria, K. (2005) Oil-in-water microemulsion LC determination of pharmaceuticals using gradient elution. *Chromatographia*, **61**, 539–547.
158. Altria, K., Broderick, M., Donegan, S. and Power, J. (2005) Preliminary study on the use of water-in-oil microemulsion eluents in HPLC. *Chromatographia*, **62**, 341–348.
159. Jancic, B., Ivanovic, D., Medenica, M., Malenovic, A. and Dimkovic, N. (2005) Development of liquid chromatographic method for fosinoprilat determination in human plasma using microemulsion as eluent. *J. Chromatogr. A*, **1088**, 187–192.
160. Malenovic, A., Medenica, M., Ivanovic, D. and Jancic, B. (2006) Monitoring of simvastatin impurities by HPLC with microemulsion eluents. *Chromatographia*, **63**, S95–S100.
161. Jancic, B., Ivanovic, D., Medenica, M., Jancic, B. and Markovic, S. (2006) Influence of structural and interfacial properties of microemulsion eluent on chromatographic separation of simvastatin and its impurities. *J. Chromatogr. A*, **1131**, 67–73.
162. Mcevoy, E., Donegan, S., Power, J. and Altria, K. (2007) Optimisation and validation of a rapid and efficient microemulsion liquid chromatographic (MELC) method for the determination of paracetamol (Acetaminophen) content in a suppository formulation. *J. Pharm. Biomed. Anal.*, **44**, 137–143.
163. Cui, S., Fu, B., Lee, S.C. and Wang, X. (2005) Application of microemulsion thin layer chromatography for the fingerprinting of licorice (Glycyrrhiza spp.). *J. Chromatogr. B*, **828**, 33–40.

Chapter 10
Microemulsions in Large-Scale Applications

Franz-Hubert Haegel, Juan Carlos Lopez, Jean-Louis Salager and Sandra Engelskirchen

10.1 Introduction

10.1.1 General considerations

The unique properties of microemulsions make them interesting for commercial products and technical processes [1]. In particular, the high solubilising power of microemulsions and microemulsion systems for organic and inorganic compounds favours their use as solvents for a large number of applications. However, cost arguments and the complex behaviour of multicomponent systems forming microemulsion phases have been obstacles to large-scale applications in the past. But the increasing number of patents beginning in the 1990s now shows that many problems could be solved with systematic studies on the structure [2] and the phase behaviour of microemulsions [3, 4]. Increasing knowledge of multiphase systems comprising microemulsion phases led to tailor-made microemulsions, e.g. with enhanced temperature stability [5], low surfactant content [6] and diverse polar oils from natural [7–9] or petrochemical sources [10, 11]. Concentrates have been developed which can be diluted without phase separation thus forming kinetically stable emulsions or even remaining microemulsions over the whole concentration range [12, 13]. Effective surfactants and surfactant mixtures have been found enabling the formation of microemulsions with various oils [14, 15].

Meanwhile, there are a variety of large-scale applications of microemulsion systems. Many products used in daily life contain microemulsions or formulations which are able to form microemulsions (some prominent examples are discussed in Chapters 8 and 9 of this book). Concentrates, surfactants or surfactant mixtures which can be used for microemulsification are frequently applied. All these materials are produced and handled in large quantities. In particular, oil-in-water (o/w) droplet and water-in-oil (w/o) droplet microemulsions are found in many products or technical processes today. Whereas their usage is not very different from ordinary solvents in most cases, the use of bicontinuous microemulsions poses specific problems which will be discussed later on.

10.1.2 Products and processes

O/w-droplet microemulsions are widespread in cleaners, cosmetics and personal care products (see Chapter 8). They are also used as solvents for pharmaceuticals (see Chapter 9) and agrochemicals [16]. In particular, pharmaceutically active ingredients [17] and pesticides [18] which are poorly soluble or insoluble in water can be readily dissolved in o/w-droplet microemulsions. O/w-droplet microemulsions are also used as carriers for biocides in wood preservation [19]. W/o-droplet microemulsions can be found in fuels (see Chapter 11), but also in more sensitive products like cosmetics (see Chapter 8) and food [20, 21]. They can also be used as media for enzymatic reactions.

For a long time, bicontinuous microemulsions were considered to be less suitable for use in products of daily life owing to the limited thermal stability of the single phase. For application in technical processes, however, this problem is often less important. Optimised surfactant systems can in many cases increase the thermal stability to a sufficiently large temperature range for this purpose. Cost arguments connected with the relatively high content of surfactants can often be overcome now by very efficient surfactant systems with polymeric boosters [22]. In technical processes, the application of bicontinuous microemulsions or multiphase microemulsion systems containing a bicontinuous phase already plays an important role. Thus, for example, the production of emulsions can be improved by using bicontinuous microemulsions as intermediates [23]. The temperature dependence of two-phase microemulsion systems can be used for fast and effective separation processes. Extraction of metals [24, 25], dyes [26] and pollutants [27] are examples for improving production processes and reducing environmental problems. However, products containing bicontinuous microemulsions are also commercially available. They are used in cleaning processes, in particular. If conventional detergents are not efficient enough and organic solvents are used instead, bicontinuous microemulsions can often be preferably applied for cleaning. One of the most prominent examples is the removal of ink [28, 29]. Bicontinuous microemulsions can also be used for decontamination of chemical and biological warfare agents [30]. They are further applied for various technical processes, e.g. in textile finishing [31].

All types of microemulsions are also suitable solvents for synthetic processes (see Chapter 5). Fluoropolymer dispersions can favourably be made in microemulsions [32]. In synthetic processes which yield a solid, the structure of the microemulsion can strongly influence the structure and properties of the product [33, 34]. Conductive polymers can be obtained by using o/w-droplet microemulsions of the monomers [35], but it is also possible to increase the content of monomers into the bicontinuous concentration range [36]. The large-scale application of microemulsions in further emerging technologies can be expected. The synthesis of nanoparticles with w/o-droplet microemulsions as hydrolysing agents for metal alkoxides or other strongly reactive precursors is an important example (see Chapter 6 and [37]). It yields products of excellent properties suitable for further improving high-tech commercial products in electroceramics [38]. Patents on microemulsions can also be found for many other applications, e.g. for lubrication in metal processing, corrosion inhibition and protection of buildings from humidity.

In this chapter, three applications at different stages of development will be presented in detail. Whereas the application in soil decontamination is still in the experimental phase, leather degreasing is already an applied technology and enhanced oil recovery (EOR) is one of the oldest large-scale applications of microemulsion systems.

10.1.3 Requirements for large-scale applications

For large-scale applications, microemulsions have often to fulfil further requirements which are not directly connected to the desired phase behaviour or the structure. Harmlessness, biocompatibility, biodegradability or long-term stability of all components may be needed depending on the application. Inertness and tolerance to the contacted target materials is necessary. Last but not least, cost-effectiveness of the components also plays a very important role.

If large quantities are used for technical processes, e.g. for cleaning, the recovery and reuse of the microemulsion or at least of a considerable amount of the most expensive components is desired. Therefore, strategies are needed to separate contaminants from the organic microemulsion components. Separation is usually more complicated than from ordinary solvents and often requires several steps [39, 40]. In particular, the separation of waste materials from the surfactants is usually very difficult or often even impossible. The temperature-dependent phase behaviour of bicontinuous microemulsions, however, can sometimes be beneficially used for separation [41]. Easy separation, at least from the unpolar solvent, can be achieved from microemulsions with supercritical liquids [42].

Other problems connected with large-scale applications are due to the fact that technical or natural materials have to be used. The composition of natural and technical products, which are often made from natural materials, is not exactly the same for different loads. Thus, the phase boundaries can be more or less changed. Technical non-ionic surfactants generally show skewed fish diagrams (see Fig. 1.8 in Chapter 1) due to the presence of surface-active species of different hydrophilicity. Non-ionic surfactants with the same formal degree of ethoxylation may exhibit different behaviour depending on the actual distribution of molecular weight. In particular, the amount of unreacted alcohol has a significant influence on the phase boundaries. As a consequence, quality control of the components used for microemulsion formation is very important.

Whereas the preparation of microemulsions is usually a very simple process on the laboratory scale, some important aspects have to be kept in mind when making large quantities of microemulsions. Weighing the components is very exact and easy for masses up to 100 kg, but less practicable for larger scales. Dosing by volume, however, is considerably less exact and requires good equipment and the strict exclusion of errors. These requirements can usually be fulfilled only in technical plants. If the composition of a microemulsion must be very exact, on-line control of known parameters, like electrical conductivity, turbidity, viscosity or ion activity, with strict temperature control may be helpful. Another important aspect which can be crucial for the preparation and handling of large quantities of microemulsions is the intermediate formation of liquid crystals. This must be avoided by all means because liquid crystals need a long time to dissolve, even when the temperature is raised considerably above the phase boundary of the liquid crystal. Usually, the formation of liquid crystals can be avoided during preparation if the components are added in a certain order. In most cases water should be added as the last component because it is the component with the highest structuring effect.

Depending on the application, the microemulsions or preconcentrates must be sufficiently thermally and chemically stable. Easy handling is often required over a wide range of temperatures. Whereas for short-term applications within a technical plant, these needs can usually be fulfilled, application elsewhere including transport and long-term storage may require further efforts. Phase separation and liquid crystal formation, in particular, as well as chemical or photochemical reactions of components during transport and storage may cause serious problems when working with large quantities of microemulsions.

10.2 Soil decontamination

10.2.1 Requirements

The task seems to be very similar to EOR, namely extracting an organic liquid from a porous medium and separating it from the aqueous phase, but actually there are decisive differences. First of all, the temperature is different. Whereas elevated temperatures are found in oil reservoirs, the temperature of water in an aquifer is usually between 8 and 16°C. Therefore, other surfactants are required for in situ soil remediation. The problem of liquid crystal formation increases dramatically with decreasing temperature. At any rate, the viscosity of the microemulsion is increased. Large pressure gradients, which can be used in rock formations for transporting the liquid, cannot be applied in soil. Since parts of the microemulsions or surfactants will remain in the soil, the components must exhibit sufficient biodegradability. They must not be hazardous for the groundwater. The properties of some organic liquid contaminants are very different from crude oil. The polarity of aromatic hydrocarbons or chlorinated hydrocarbons is much larger and usual surfactants are not sufficiently effective for these contaminants. The extraction of 80% of the organic liquid would be a remarkable result for oil recovery, but hardly sufficient for a remediation process. The recovery of surfactants and other components and their reuse also require higher efficiency of the separation processes.

Even if some large-scale experiments on soil decontamination have already been performed, the application of microemulsion technology for this purpose is still largely a matter of research. Surfactants have often been applied in field-scale experiments or even commercial cases for the mobilisation and solubilisation of organic pollutants. However, the state-of-the-art of surfactant-enhanced subsurface and aquifer remediation is very different depending on the type of contaminant, the actual situation of the contamination, and the processes used for decontamination. Mobilisation of liquid phases can be, but need not be, connected with the formation of microemulsions. This process is favoured by the very low interfacial tension between the flushing solution and the liquid contaminant, as achieved for Winsor III systems, i.e. those which form a middle-phase microemulsion (bicontinuous microemulsion). If solubilisation is involved, swollen micelles or microemulsions are formed during extraction of liquid contaminants from soil. Since there is often no exact boundary between micellar or interfacial processes and microemulsification, many field-scale experiments might fall within the scope of this contribution. However, only a selection of activities involving the development of very effective surfactant systems for the microemulsification of liquid contaminants and the use of preformed microemulsions will be considered here.

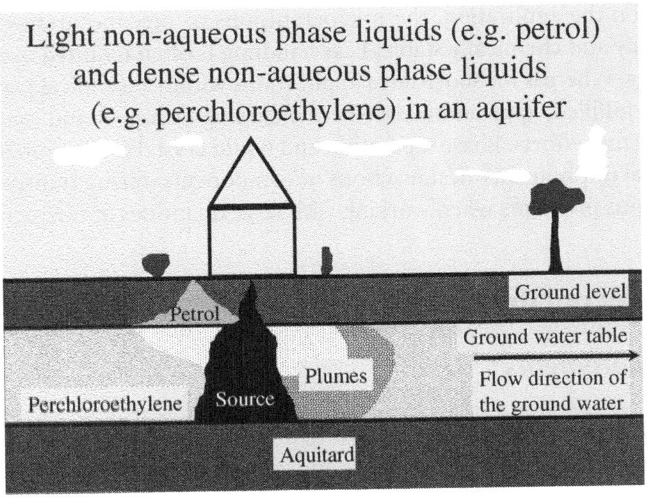

Figure 10.1 Sources and plumes of light non-aqueous phase liquids (LNAPL) and dense non-aqueous phase liquids (DNAPL) in an aquifer. DNAPL can penetrate the ground water table.

10.2.2 Non-aqueous phase liquids

Surfactants and microemulsion systems can be used for ex situ treatment of contaminated soil or in situ soil decontamination. In situ remediation is usually preferred if excavation of the contaminated soil is not possible or expensive, e.g. beneath buildings or for contaminations at great depth. Often bioremediation or natural attenuation is used for decontamination. In most cases, these techniques only permit the effective degradation of contaminants in the plume formed by dissolved pollutants which may be very large. However, for the remediation of a contaminated site, it is also necessary to remove the source where the pollutants may be adsorbed in large quantities or may be present as solid or liquid phases. The latter are called NAPL (non-aqueous phase liquids) and a differentiation is made between LNAPL (light non-aqueous phase liquids) with a lower density than water and DNAPL (dense non-aqueous phase liquids) with a higher density than water (see Fig. 10.1).

The choice of suitable surfactants and additional chemicals for the decontamination of source zones largely depends on the type of pollutant and the structure of the soil (mainly on adsorption behaviour and hydraulic conductivity). Adsorbed and solid pollutants or very viscous liquid phases cannot be mobilised. Preformed microemulsions, co-solvents or co-surfactants can be favourably used for such contaminations in order to enhance the solubilisation capacity of surfactants. NAPL with low viscosity can easily be mobilised and also effectively solubilised by microemulsion-forming surfactant systems. Mobilisation is usually much more efficient. It is achieved by reducing the interfacial tension between NAPL and water. Droplets of organic liquids, which are trapped in the pore bodies, can more easily be transported through the pore necks at lower interfacial tension (see Fig. 10.2). The onset of mobilisation is determined by the trapping number, which is dependent on

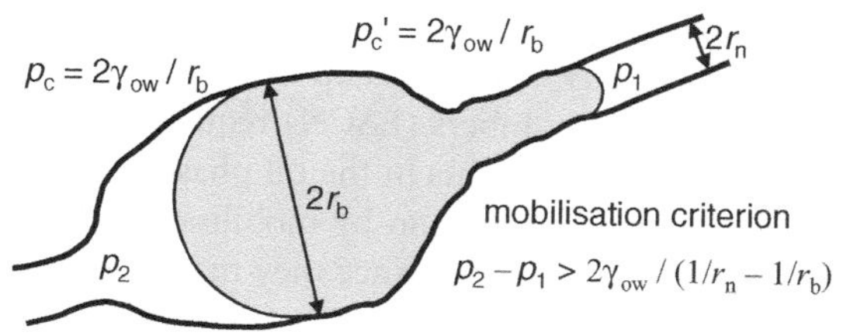

Figure 10.2 Mobilisation criterion for an NAPL droplet trapped in a pore. Pressure difference p_2–p_1 must exceed the difference of capillary pressure $p_c' - p_c$. γ_{ow} = interfacial tension NAPL/water, r_b = radius of pore body, r_n = radius of the pore neck.

soil structure (absolute and relative size of pore bodies and pore necks), flow conditions, density of the contaminant and interfacial tension [43].

Efficient in situ techniques are urgently needed, in particular for DNAPL site remediation [44]. This type of contamination is the most serious and widespread. Many of the sites contaminated with chlorinated hydrocarbons are found in built-up areas, e.g. beneath (disused) dry-cleaning facilities or metal processing plants. DNAPL contamination is very frequent and is often found at great depth beneath the groundwater table in the saturated zone of an aquifer. The pollutant is often concentrated in pools and not, or at least not sufficiently biodegradable, in this position. Microemulsion techniques seem to be most efficient for in situ remediation of DNAPL sources. They may also be interesting for LNAPL sites where the pollutant is less volatile or found in part in the saturated zone owing to fluctuations of the groundwater table.

10.2.3 Microemulsion-forming systems

10.2.3.1 Early research

The findings of early research on surfactant-enhanced aquifer remediation and microemulsification of contaminants, which was strongly influenced by studies on EOR, has been summarised by Harwell *et al.* [45]. In their paper, they also discuss some of the problems related to surfactant injection into the subsurface. First of all, the need for research on biodegradation under aquifer conditions is emphasised. The low temperature and the formation of liquid crystals or gels have also been considered. The measures to obviate liquid crystals or gels are critically discussed. Addition of lower alcohols as co-solvents or co-surfactants implies the use of volatile flammable liquids, which have a considerable potential for contaminating the groundwater and which make the separation and recovery of the active components difficult. Application of branched surfactants, which usually effectively suppress the formation of liquid crystals [46–48], may cause problems owing to their reduced biodegradability [49].

At that time, some large-scale field tests on surfactant and microemulsion technologies had already been performed [50–55]. In most cases, the applied surfactants or microemulsion components were selected in laboratory experiments by determining phase behaviour, interfacial tension, solubilisation capacity, viscosity and extracting power in soil columns.

The expertise with EOR was used for finding suitable microemulsion-forming systems for LNAPL. However, the high polarity of chlorinated hydrocarbons with very low or even negative equivalent alkane carbon numbers (EACN) required novel types of surfactants [56]. The enhanced solubility of surfactants in the oil phase makes most surfactants less effective for solubilisation. DNAPL extraction by mobilisation, however, is problematic owing to the high density of the pollutants, since they may be displaced into deeper soil compartments [57]. This probably happened in at least one field test [58].

Mobilisation of NAPL generally leads to the formation of an oil bank (see Chapter 10.3) in front of the surfactant solution. If the solubilisation capacity of the surfactant solution is too low, large amounts of emulsions will be formed, which can clog the pore space. As the flow in columns is forced, these experiments may not correctly reflect the behaviour of the multiphase system under free flowing conditions in a three-dimensional pore space.

10.2.3.2 Microemulsion-forming surfactants for chlorinated hydrocarbons

All types of microemulsions were obtained in salinity scans with mixtures of Aerosol MA (sodium dihexyl sulphosuccinate) and twin-tailed (Guerbet and Exxon type) alcohol ethoxy and propoxy sulphates for perchloroethylene (PCE), carbon tetrachloride, 1,2-dichlorobenzene and trichloroethylene [59] at 25°C. At lower temperatures, however, stable macroemulsions are formed. Chloroform, 1,2-dichloroethane and other chlorinated hydrocarbons were found to be too polar for those anionic surfactants. Extremely hydrophilic and temperature-insensitive surfactants are necessary for effective solubilisation of chlorinated hydrocarbons yielding Winsor III systems. N-methyl-*N*-D-glucalkaneamide surfactants showed good performance for DNAPL solubilisation even at 16°C [56].

All the systems described thus far require relatively high salinity, which is considered to be critical in soil remediation processes [60]. Another approach for effective microemulsification of organic liquids is the use of co-surfactants. Sodium mono- and dimethyl naphthalene sulphonate were found to be effective co-surfactants in formulations with Aerosol OT (sodium bis(2-ethylhexyl) sulphosuccinate) for diverse chlorinated hydrocarbons and their mixtures between 15 and 25°C [60, 61]. All types of microemulsions could be obtained with this approach.

10.2.3.3 DNAPL remediation

As there are other less sophisticated and less expensive techniques available, surfactant-enhanced aquifer remediation will only be useful for decontamination of LNAPL sites in special cases. However, applicable techniques are still needed for DNAPL sites and microemulsion techniques are really promising. Therefore, most research has concentrated on this type of contaminant in recent years. Integrated concepts have been developed including aspects of soil properties [47, 48, 62, 63], density control [47, 48, 62–64], recovery and reuse of microemulsion components [47], biological degradation of residues of contaminants and injected compounds [48, 65] and costs [47, 48, 64, 65]. Two main approaches have been followed for developing effective surfactant systems which form microemulsions with DNAPL, but do not mobilise the liquid contaminant into deeper aquifer compartments. One approach is based on optimising the solubilisation capacity without reducing the interfacial tension below a critical level [47], and the second is a concept based on neutral buoyancy [66, 67] by the addition of co-solvents which might be injected in the form of emulsions or aqueous solutions followed by surfactant flushing [68]. A third approach to obviate downward migration of DNAPL is the use of concentrated salt solutions injected beneath the DNAPL [69].

10.2.3.4 Supersolubilisation

Whereas Winsor III systems exhibit ultra-low interfacial tensions between the three phases and also very high solubilisation capacity, Winsor I systems have higher interfacial tensions and much lower solubilising power. At the transition between the two types of microemulsion systems, an intermediate behaviour can be found which is called supersolubilisation [47, 70]. The uptake of oils into surfactant aggregates is usually enhanced by one to two orders of magnitude compared to effective micellar systems, but interfacial tension reduction is still moderate. The transition point can be adjusted by varying the salinity or organic components.

Supersolubilisation of NAPL can be achieved by adding lipophilic linkers to the system. Best results for chlorinated hydrocarbons were obtained with both hydrophilic and lipophilic linkers [71]. Lipophilic linkers increase the interaction between surfactant and oil [72], and hydrophilic linkers the interaction between surfactant and water. Systems with Aerosol MA as surfactant, sodium mono- and dimethyl naphthalate as hydrophilic linker, and dodecanol as lipophilic linker display the best performance regarding efficiency, economy and environmental aspects [65].

In order to use supersolubilisation for DNAPL extraction, the reduction of interfacial tension must be well controlled. The critical level of interfacial tension is dependent on size and heterogeneity of the pore space. For example, a value of 4 mN m^{-1} was found for soil from a contaminated site [47]. Since supersolubilising systems exhibit lower interfacial tension, they cannot be directly applied for contaminant extraction. Therefore, a salinity gradient was used for column experiments in preparation for a field test [47, 63]. When the salinity was increased in two steps from 0 to 0.6 wt.% and 1 wt.% $CaCl_2$, a mixture of a sulphated alkyl propoxylate (Isalchem 145-4PO-SO4) and a twin-head aromatic sulphonate (Dowfax 8390) exhibited the usual micellar solubilisation, supersolubilisation and formation of bicontinuous microemulsions with perchloroethylene. Applying this three-step gradient to soil columns contaminated with PCE yielded high extraction values and no mobilisation of DNAPL [47].

10.2.3.5 Concept of neutral buoyancy

DNAPL extraction without vertical displacement of the contaminant can also be achieved by adjusting the density of the microemulsion which is formed in the subsurface to the density of water by the addition of organic components with densities of considerably less than 1 g/cm^3. Co-solvents, co-surfactants and some non-ionic surfactants can be considered for this purpose. Whereas the addition of isopropyl alcohol as co-solvent to the surfactant was successful [48, 62, 73], an attempt with pentanol as co-surfactant was not successful, because partitioning of the co-surfactant into perchloroethylene was too great [73]. A mixture of 5% Aerosol MA-80 (4% active branched sodium dihexyl sulphosuccinate) and up to 8% isopropyl alcohol in 0.6% aqueous NaCl solution was used for the experiments. Furthermore, xanthan gum was added to the surfactant system in order to enhance the viscosity, increase the forces exerted by flow and prevent fingering [64]. Optimum conditions for surfactant-enhanced aquifer remediation at neutral buoyancy were calculated for a medium-scale laboratory experiment from scaling groups regarding the pore structure and the properties of contaminant and surfactant mixture [74].

10.2.3.6 Large-scale experiments

References to large-scale surfactant-enhanced aquifer remediation can be found in the literature [75, 76]. Childs *et al.* also report results from the test site at Dover Air Force Base, Delaware [76]. Large data sets are also available from field tests at Hill Air Force Base, Utah [47, 77, 78]. Other field tests were performed at the Canadian Forces Base, Borden [51, 53], Ontario, the Bachman Road site at Oscoda [79, 80], Michigan, as well as Camp Lejeune, North Carolina [81], Traverse City Coast Guard Base, Michigan [54], Spartan Chemical Company Superfund Site, Michigan [82] and the former Naval Air Station Alameda, California [82, 83]. In the latter case a 97% extraction of DNAPL is reported [83].

10.2.4 Use of preformed microemulsions

Preformed microemulsions can also be used for soil decontamination. The application of bioremediation with microemulsions containing nutrients for oil spills is already a well-known technology [84, 85] and is also proposed for in situ treatment of DNAPL sites [86]. Studies on contaminant extraction, however, are less frequent. In most cases, these systems have been discussed and investigated for adsorbed or highly viscous contaminants which can only be solubilised. Enhancement of solubilisation in microemulsions compared with surfactant solutions was found for pyrene [87] and patented for ex situ treatment of contaminated soil [88]. An interesting cost-effective variation uses partially sulphated castor oil [89].

W/o-droplet microemulsions with non-ionic surfactants containing rapeseed oil methyl ester have been successfully used for in situ extraction of polycyclic aromatic hydrocarbons [40]. However, enhancement of oil content and solubilisation capacity failed with these systems. The use of co-surfactants and co-solvents for suppression of liquid crystal formation was considered to be critical for in situ application.

A complex system containing a branched anionic surfactant, non-ionic surfactants, rapeseed oil methyl ester and an aqueous calcium chloride solution was found to form bicontinuous microemulsions even at low temperatures [46, 90]. This type of microemulsion has been studied for DNAPL extraction on a large scale in an artificial aquifer and later in a joint project with different partners financed by the German Federal Ministry of Education and Research (BMBF) [91]. The project network applied an integrated concept regarding aspects of hydraulics, reuse and biodegradation [92]. Three large-scale experiments each with some hundreds of litres of preformed microemulsion were performed. Whereas extraction of perchloroethylene in the field-scale experiment was not successful owing to an insufficient quantity of microemulsion and the lack of a focussing extraction well, a later test in a large container removed nearly all the perchloroethylene [91]. Biodegradation of the microemulsion components was achieved under field conditions. Even the biodegradation of sodium bis(2-ethylhexyl) sulphosuccinate, which was known only for liquid cultures from the literature [93], was proven under aquifer conditions [91]. Moreover, dehalogenation of perchlorethylene was found after microemulsion application, such as after surfactant flushing at other field sites [80].

Preformed microemulsions containing co-solvents and co-surfactants have been used for laboratory experiments [94] and a field test [55] in Canada. The systems were developed for the extraction of a viscous oil containing up to 16% of chlorinated solvents from a site at Ville Mercier. The contaminant is a DNAPL with a density of 1.05 g/cm^3 and thus exhibits only a small density difference compared to chlorinated solvents [94]. It could not be extracted effectively by the usual Winsor I systems containing *n*-butanol as a co-surfactant. The addition of solvents was necessary for effective solubilisation of the contaminant [94]. A preformed microemulsion containing D-limonene, toluene, *n*-butanol, Hostapur SAS (secondary alkane sulphonate sodium salt) and water was injected into a field test site at Thouin Sand Pit near Montreal. In previous column experiments, a composition of 13.16% D-limonene, 13.16% toluene, 9.21% *n*-butanol, 9.21% Hostapur SAS and 0.3% of sodium ortho-silicate in water was used as a preformed microemulsion for the extraction of DNAPL from Ville Mercier.

10.2.5 Challenges

Applying the microemulsion technology on a large scale at contaminated sites may sometimes cause surprising problems. Thus, the formation of emulsions, gels or liquid crystals, or the mobilisation of fine soil particles may clog the aquifer. Since soil and contaminant distribution are usually very heterogeneous, these problems may not have been found in laboratory experiments. Additional chemical compounds in the soil can induce such problems [75]. Other problems are concerned with the processing of injected systems and effluents. Safety measures will be needed if highly volatile liquids like low alcohols are used. The BMBF project network already encountered some problems when preparing the microemulsion. Natural products like rapeseed oil methyl ester exhibit different solubilisation behaviour depending on the region, the growing conditions and the process of refinement. Slight differences were found for samples of different origins. At this point, the formation of liquid crystals was found for one sample at room temperature. When all samples were checked, a massive presence of liquid crystals at lower temperature was found for a formulation which some time before had worked very well. Intensive research identified the anionic surfactant as the problematic compound. The producer had changed the formulation slightly. Thus, the phase behaviour had to be studied once again. But not only oil and surfactants can cause problems, even the salt ($CaCl_2 \cdot 2H_2O$) used at the field site did not have the exact composition as was found by analysing the microemulsion. The use of technical products may sometimes cause problems for microemulsion formation, which may be very sensitive to slight changes.

Surfactant-enhanced aquifer remediation is relatively expensive. Thus, waste reduction or the reuse of at least a considerable part of the microemulsion components is interesting.

Depending on the technology applied and the contaminant, different technologies have been studied for separating the pollutant from the components of the flushing solutions. Simple methods aim at waste reduction, i.e. separation of the contaminant and the surfactant from waste water. Addition of NaCl at elevated temperatures to a field-generated microemulsion of Winsor type I from a field test at Hill Air Force Base, Utah, and centrifugation were used to reduce the volume of the highly contaminated waste and cut costs for further treatment by 95% [95]. The dependence of the phase behaviour of microemulsions on temperature [9, 40] and salt concentration [41] can also be used in order to recover surfactants and other microemulsion components for reuse.

Other approaches for waste reduction and surfactant recovery are based on liquid–liquid extraction. Whereas solutions of anionic surfactants can be readily extracted by oils in extracting columns [39], solutions of non-ionic surfactants must be extracted by hollow fibre membranes [96] or separated from the oil with a centrifuge [97]. A variation of liquid–liquid extraction processes uses supercritical fluids [98]. Volatile contaminants have been separated from the effluent by air-stripping [99] or other evaporation techniques [91]. These processes are, however, often accompanied by excessive foaming. Membrane technologies are therefore preferred. Pervaporation seems to be the most promising method for separating volatile contaminants from surfactant systems [100]. A review of surfactant recovery and reuse is found in Cheng and Sabatini [101].

Improvement of separation processes is certainly one of the future goals in order to reduce the costs of microemulsion technologies for soil remediation. But in the first place, it will be important to convince the authorities of the benefit of these remediation methods in order to obtain permission and funding for further field tests. Experience with real contaminated sites must be extended [102]. In particular, the influence of heterogeneities of soil and contaminant distribution must be further investigated. Modelling can help to understand the relevant processes [103–105]. Further improvements could be made by using geophysical methods for site characterisation and process monitoring [106, 107].

10.3 Microemulsions in enhanced oil recovery

10.3.1 Why enhanced oil recovery and not alternative fuels?

Despite growing environmental concerns such as global warming, oil (petroleum) is still the most important non-renewable commodity in today's global economy. It is not only one of principal energy sources, but it is also a raw material for many products like plastics, asphalts and lubricants, to name a few. Nobody living in a modern city can imagine life without petroleum derivatives. Nevertheless, some of the main oil reservoirs are becoming quickly depleted leading to an inexorable slow-down of the oil world production. Besides, it is becoming harder and harder to find new giant oil reservoirs to satisfy the ever escalating demand, not to mention the increasing difficulty surrounding oil politics worldwide and the fact that most of the important alternative energy sources are still in the early stages of industrial or massive implementation. All these leave the world on the verge of a crisis: the world's oil dependence and demand is increasing, while reserves are quickly declining.

At this time and in the near future, there is no available substitute for petroleum-based fuels, if only cost issues are taken into account. For instance, biofuels like ethanol or vegetable oil esters cannot compete on purely economic grounds, without considering other undesirable issues, particularly because the fertiliser production and crop handling actually consume a high percentage (about 70%) of the energy content of the biofuel. Worst of all, their large-scale development will end up competing for land space with food production.

Another alternative to attain synthetic liquid fuels is the coal-to-liquid route achieved through the Fischer–Tropsch process [108], a proven technology for which raw materials are plentiful and available almost anywhere. Therefore, it is very likely that this technology will be used to provide some of the liquid fuel in the future. Nevertheless, it first has to be overcome some of its current drawbacks, e.g. relatively high costs and serious ecological concerns. For this reason, this technology has only been used at an industrial level up to now in countries with restricted economies such as World War II Germany or South Africa under apartheid.

Hydrogen and alcohol fuel cell technologies may help in compensating part of the demand for liquid fuel, but its wide-scale application is still at least two decades away, mainly because of technical and economical issues that need to be surmounted.

On the other hand, both oil-field exploring and well drilling are expensive and laborious tasks. More than 100 years after oil production started, it is very likely that the number of fields yet to be discovered is much less than the number of fields with declining production. Exploration and drilling are becoming less appealing. Based upon this fact, it is crucial and urgent to develop, tune and optimise technologies to enhance the ultimate production yield of the current and abandoned oil reservoirs in order to supply the liquid fuel volume that conventional methods will fail to provide in the near future.

However, it is well known that after an oil well is abandoned, a large proportion of the original oil in place (OOIP) still remains in the ground [109–114]. Generally speaking, primary and secondary oil production technologies are able to recover altogether no more than 30% of OOIP [109, 113]. Initially, an oil well usually produces oil and a small amount of water, and then, particularly while applying water flooding, the most common secondary recovery technique, a mixture of oil and water is produced with a rising water cut as the process proceeds through time [113–115]. Finally, the cost of injection becomes too high to continue the secondary recovery process and the well has to be abandoned with 70% or even more of the OOIP still left in the reservoir.

Therefore, in virtue of all the above, tapping some of the already known 70% OOIP leftover via EOR technologies is probably the best economic solution for overcoming the mega world crises that oil production decline might cause quite soon. Working on exhausted oil wells is appealing not only because about 70% of OOIP is known to be there, but also because a considerable amount of information about them is at hand [112–114], at essentially no cost, when compared to exploring and tapping new fields. Furthermore, drilling and completion costs are to a large extent considerably reduced. Needless to say, EOR is not exempt from economical issues that need to be overcome. Although some of the EOR technologies are already well developed, others still need to be applied at pilot and field scale in order to improve their feasibility and reliability [113, 114, 116].

10.3.2 Why microemulsions?

In this chapter, of all the varieties of technologies known as tertiary oil recovery, only the one known as chemical EOR via microemulsions will be dealt with. Although it is the most complex one, it is the only tertiary technology capable of producing a large proportion of the oil still remaining in many abandoned wells, at a reasonable cost [109, 113, 114, 117]. Thus, Chemical EOR via microemulsion (a variety of technologies known as micellar flooding, microemulsion flooding, surfactant flooding, surfactant–polymer flooding etc.) may become one of the few real, technological and cost-effective solutions for the predicament of the human race for liquid fuels.

Some of the main characteristics of the EOR via chemical flooding process were discovered during the 1970s and 1980s [109–112, 115, 117–138]. Technical and cost studies in that time also revealed that this technology could potentially increase oil recovery with up to 40–90% of the OOIP [110–114] at an extra current cost of about 10–15 US$/bbl (i.e. at a maximum 20–30 US$/bbl cost), and probably even less once the technology is fully developed. Taking into account that an average light crude currently sells at over 60 US$/bbl (2007) and that the situation seems likely to last for a while due to structural reasons, EOR would leave a profit of at least 30 US$/bbl. More importantly even, EOR would boost oil world reserves. All these could lead to maintaining the liquid fuel price at a reasonable level, avoiding, thus, a strong impact on world economy, and providing a fair return for producing countries.

After a waterflood, the residual oil remaining in the porous reservoir is trapped by capillary forces. The corresponding interfacial tension between the aqueous and the oil phase is of a few mN m^{-1} [109, 112–115]. Under these conditions, oil production is marginal and the water cut, on the other hand, becomes high [110, 113, 114]. Addition of surface-active substances (i.e. surfactants), however, can lower the interfacial tension by 3–4 orders of magnitude [109, 112–114, 117, 122, 124], which induces the production of more oil and lowers the water cut [113, 139]. Microemulsion flooding tests both at the laboratory and pilot scales have shown that the oil recovery could be more than 60% of the OOIP [110–114], which is about twice the current one. Nevertheless, most of the tests carried out in the field yielded an additional oil recovery of only 10–20% of the OOIP [114, 140], which indicates that the process still has to be improved.

Basically, two types of protocols have been proposed for EOR [110–114, 138]. One of them uses a 'high' volume of diluted surfactant solution, while the other uses a 'small' volume of highly concentrated surfactant solution. Early results obtained by EOR researchers both in the laboratory and pilot showed that by injecting a properly formulated microemulsion and letting it dilute in the reservoir fluids, a very low tension is maintained even at a surfactant concentration lower than 1 wt.% [113, 114]. This indicates that the low surfactant concentration protocol is technologically feasible to recover a large proportion of extra oil [112–114, 138, 140].

Furthermore, if enough surfactant is used, a zero interfacial tension (i.e. total oil–water miscibility) [112, 117, 141] could be reached with the consequent vanishing of the trapping forces and the recovery of almost all the oil and water present in the swept zone [109–111, 114] through the formation of a single phase (i.e. a bicontinuous microemulsion) containing the trapped oil and surfactant solution. Current formulation know-how indicates that a rather large concentration of surfactant (and co-surfactant), say at about 10%, is required to produce such a microemulsion [137, 141]. Moreover, about 10% of the pore volume of the surfactant solution is needed to sweep the whole reservoir and displace the oil effectively. It is worth mentioning that keeping the crude oil and the surfactant solution as a single phase is not an easy task; actually, such a single phase generally ends separating into three phases [115, 138], one of them a microemulsion, but it does not mean that the recovery would not be effective. Therefore, although technologically feasible, the complete solubilisation of all the crude oil in a single-phase microemulsion produced by conventional surfactants might not be an economically attractive operation because it requires a large amount of expensive surfactant.

10.3.3 Basic scientific and technical problems

There are three fundamental and one economic issue that need to be addressed before a microemulsion flooding is applied as a routine of oil recovery technique. The first issue is to design a so-called 'optimum' formulation, to attain either total oil miscibility in a single-phase microemulsion (i.e. zero interfacial tension) or an extremely low interfacial tension, typically 10^{-3} mN m^{-1} or lower, required to mobilise the oil phase through the porous reservoir [109, 112–115]. The second issue is to secure the optimum formulation and to maintain it throughout the whole process while fluids move from the injector to the producer wells. The third fundamental issue is to wrap up a process that involves a sequential injection of a number of fluid volumes, so-called slugs, to displace the oil effectively. The typical slugs, used in chemical EOR, are the preflush, surfactant–co-surfactant, polymer and chase-water slugs. Another slug formed in place is the one known as the oil bank.

The first fundamental issue is easy to reach since it is obtained in the laboratory where all the conditions are reasonably under control, while the second and the third ones are not because they have to be achieved in the field at a much larger scale under a more complex set of conditions and during a much longer time. Before implementing an EOR production routine at industrial scale the three fundamental issues have first to be resolved both at the laboratory and pilot scale; thereafter, a significant economic analysis could be carried out.

10.3.3.1 Attaining optimum formulation

The optimum formulation is a surfactant system with which maximum oil recovery can be achieved. For that purpose, the interfacial tension has to be as low as possible and the oil solubilisation in the microemulsion as large as possible [15, 115, 121, 123, 127, 140, 142–144]. In general, formulation is a concept that tunes the properties of a water–oil–surfactant system such that it can be used for the certain application (see Chapter 3). Extensive studies on the optimum formulation for EOR and various other applications have shown that many variables have to be considered to achieve an ultra-low interfacial tension at relatively low surfactant concentration, or the occurrence of a single-phase bicontinuous microemulsion at high surfactant concentration [15, 143, 144].

It was a century ago that researchers started to study the factors affecting the behaviour of water–oil–surfactant systems but it is only with the introduction of Winsor's R theory (1954) that the formulation effects could be interpreted. Winsor's R theory was the first 'qualitative' description of the formulation, paving the way to an understanding of how intermolecular interactions among the different chemical species present in a system are related to its behaviour. Throughout the following decades, several empirical experimental correlations such as the phase inversion temperature (PIT), semiempirical ones such as the cohesive energy ratio (CER), and models based on thermodynamics such as the surfactant affinity difference (SAD) or the hydrophilic–lipophilic deviation (HLD) [15, 143, 144] led to a quantitative description of the formulation of water–oil–surfactant systems. Thanks to these studies and numerical data [145], it is possible to roughly determine an optimum formulation with a simple pocket calculator (see Chapter 3 for more details).

In spite of many sources of inaccuracy with real products and phases, research on how to reduce the oil–water interfacial tension or to increase the microemulsion solubilisation capability, which is essentially the same, has exhibited a steady improvement from the 70s until to the beginning of the 90s. At the same time, another main target related to EOR, which was to increase the formulation robustness, was not reached because robustness was found to be inherently inverse to performance; in other words, the larger the solubilisation or the lower the interfacial tension, the narrower the formulation variable range over which it takes place [109]. Some apparent exceptions to this general rule [109–111] are still far from being well understood. During the same period of time, several rules for increasing the surfactant performance were proposed, among them (i) to increase both the numerator and denominator of Winsor's R ratio, (ii) to add lipophilic and hydrophilic linkers and low molecular weight polymer additives and (iii) to use surfactant mixture containing surfactants with intramolecular mixtures [143, 144]. Later new amphiphilic structures (so-called extended surfactants) were designed and synthesised. These new surfactants were found to form a single-phase microemulsion with hydrocarbon oils at a surfactant concentration as low as 2 wt.% [14] and to be particularly suited for improving the performance with polar oils for other applications [146–148].

10.3.3.2 Formulation compatibility with reservoir and fluids

When an optimum formulation is sought to be effective to displace the oil, the complex reservoir conditions cause a lot of complications likely to dramatically affect the expected ultra-low interfacial tension value and the phase behaviour (i.e. the surfactant capability to solubilise oil) [109, 110, 119]. For example, the chemical composition of the different slugs, but specially that of the surfactant one, will change in time and space (with the concomitant change in system phase behaviour) once the aqueous solution containing surfactants, co-surfactants, polymer and electrolytes, is brought into contact with the connate aqueous and oil phases and the surface of the porous reservoir [115]. This situation is even more complex because most of the properties and phenomena involved are also sensitive to changes in temperature and pressure [115, 117, 119, 132, 149], which frequently take place in the reservoir during production. All these require that the formulation of each slug be designed such that the variations they might undergo cancel out or compensate each other so that changes do not significantly affect the surfactant performance in displacing the oil through the porous reservoir. In that way, the whole process, not a particular slug or formulation, would be robust. The following are the main factors found to alter the different slug action and performance.

Pressure

Generally, the reservoir pressure is not high enough to produce a significant change in the surfactant or polymer slug. However, the oil phase is normally much more compressible than the aqueous phase [132, 149]. Additionally, it may contain some dissolved gases [141]. Therefore, oil density and oil chemical composition may be quite different from what is observed in the laboratory. Furthermore, the apparent oil equivalent alkane carbon number (EACN) may be affected to some extent by these two phenomena [132, 149, 141]. Since optimal formulation is very sensitive to a small change in EACN [120–124], reservoir pressure may indirectly affect the interfacial tension value or the solubilisation greatly. Thus, when designing the surfactant formulation, it is important to take into account that the samples of 'dead' oil used in laboratory experiments might have an EACN quite different from those of the 'live' oil actually found in the reservoir.

Temperature

Temperature can affect to some extent essentially all the properties of any of the slugs as well as the chemical species they are constituted with. Temperature influences phase behaviour, interfacial tension, electrolyte dissolution, surfactant and alcohol adsorption, and precipitation. It also affects pressure, viscosity, density, and gas concentration in the oil phase, EACN etc. Furthermore, high reservoir temperatures can induce surfactant and polymer chemical degradation. Therefore, temperature is a main factor to take into account when designing the surfactant and the polymer formulas, particularly those that include polyethoxylated non-ionics. Nevertheless, most of the problems caused by the reservoir temperature are relatively easy to solve by formulating the slugs with adequate components [110, 111, 150]. If the reservoir temperature is too high (the chemicals will typically stay in the reservoir for more than 1 year) surfactants and polymers may degrade; therefore, selecting them appropriately is crucial.

Chemical composition

As well as temperature, chemical composition affects all the system's properties. Thus, one needs to consider any parameter that could alter the composition to such an extent that the formulation is no longer 'optimal' and thus might not perform effectively enough.

- *Salinity*: Salinity plays at least two important roles, namely it maintains the integrity of the reservoir and it balances the physicochemical environment so that surfactant formulation stays close to optimal. Thus, ultra-low interfacial tension and oil solubilisation are very sensitive to salinity. Mixing of the surfactant slug with connate water may alter the surfactant formulation mainly due to dilution and to the incorporation of new electrolytes to the formula. Adsorption and desorption of electrolytes, particularly divalent cations, onto or from solid materials such as clay, will also change the salinity of the aqueous phases to some extent and may cause surfactant precipitation, which is however not always an adverse effect [151]. In order to attenuate the undesirable salinity effects on formulation, surfactants able to tolerate salinity changes [109], high salinity [150] and the presence of divalent ions [112] may be used.
- *Surfactant composition*: Surfactants are the critical substances in lowering the interfacial tension. Therefore, all processes likely to change the surfactant composition have to be taken into account. For example, surfactant adsorption and absorption, the dilution of the surfactant containing solution, chemical reactions, degradation and precipitation may occur. All these processes may cause a considerable surfactant concentration drop and thus a chromatographic separation effect, which could result in a surfactant formulation that is no longer 'optimal'. Dispersion, on the other hand, affects the concentration profile. Such effects could be attenuated or compensated by one or more of the following actions: (1) injecting a sacrificial agent slug (see next section) [113, 114];

(2) injecting the surfactant slug with a salinity gradient [109, 114, 117, 131]; (3) injecting two consecutive polymer slugs with quite different salinity, according to the so-called salinity shock [109]; and (4) adding special polymers and other additives [110, 111, 113, 136]. It is also worth mentioning that surfactant precipitation does not always cause a negative effect; sometimes, it helps in enhancing the displacement of the oil through the porous media by indirectly improving the sweep efficiency of the surfactant slug [151]. An important percentage of the surfactant will be lost in the irreducible residual oil that ends up trapped in the porous medium after applying a microemulsion flooding protocol, particularly in globules caught in unconnected pores [109, 115, 117, 141]. This can be remediated by injecting the right surfactant slug volume at a concentration high enough so that the loss would not significantly affect the surfactant slug performance [117].

- *Polymer composition*: Polymers are essential for controlling the surfactant slug mobility and dispersivity; the efficacy of the polymer depends on the polymer composition. Nevertheless, polymers are very likely to undergo different processes that largely change their concentration and structure. Polymers suffer at least one of the following degradations: mechanical, chemical, thermal and bacterial. They also undergo deposition, excessive adsorption, entrapment, inaccessible pore volume. Besides, they frequently present incompatibility with the surfactants [109, 115, 138]. However, polymers have shown to work relatively well in most chemical flooding conditions [113, 114, 116, 139, 140]. For instance, partially hydrolysed polyacrylamides have been found to work well even in a high temperature reservoir [139].
- *Alcohol composition*: Just as electrolytes do, alcohols help to balance the physicochemical environment in order to keep the surfactant formulation close to optimal, according to the so-called $f(A)$ or $\phi(A)$ effects discussed in Chapter 3. Besides, even so some surfactant formulations do not contain alcohols, they are often added into microemulsion systems as co-solvents (particularly in those containing anionic surfactants) to improve the solubility of the main surfactants and prevent the formation of highly viscous meso-phases [114] such as liquid crystals, which are additionally known to stabilise the emulsions that may be formed. Alcohols can also change the surfactant partition coefficients which has a great effect on the oil recovery efficacy [110, 111].

10.3.3.3 Continuous injection process

Typical EOR surfactant–polymer flooding operations include a sequence of slugs (see Fig. 10.3) that are injected into an oil reservoir [109, 113, 114, 119]. The main idea is to wrap up a process to recover most of the oil still in place (at least much more than what waterflooding does) by preserving at the correct level the main characteristics and functions of the different slugs, particularly the surfactant one, as they pass through the reservoir from the injector to the producer wells. Some of the typical slugs and their main features are described below.

The preflush

The preflush (or preliminary injection) is a slug applied to 'prepare' the reservoir to help in protecting the surfactants and polymers against salinity effects and adsorption on the

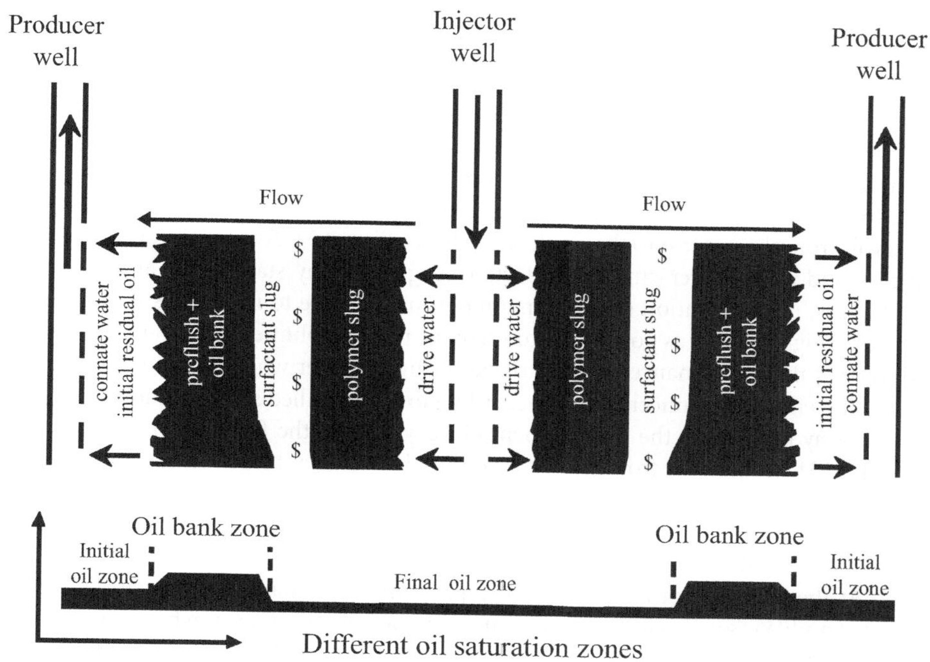

Figure 10.3 Schematic diagram of the oil saturation zones as a function of the different slugs of a typical surfactant–polymer flooding operation.

rock surfaces [114]. In the following some of the commonly designed preflush slugs, which could typically account for 5–10% of the pore volume of the reservoir, are described.

- An alkaline solution is injected as a slug of a surfactant–polymer flooding operation to change the rock surface zeta potential to a negative value. Thus, commonly used anionic surfactants (e.g. petroleum sulphonates) would adsorb less [136]. Alkaline preflush can also alter the rock wettability, although it is not really known whether a water-wet or oil-wet medium is best to prevent surfactant adsorption and chromatographic fractionation. However, alkaline fluids would also induce undesirable effects such as oil emulsification and chemical species precipitation which eventually might cause some plugging. Note that some plugging may improve the sweep efficiency of the surfactant slug, which, in turn, increases oil recovery [151].
- A low salinity water drive could be injected to induce divalent-ion desorption from the porous medium through ion exchange and their subsequent washing away to avoid the main surfactant precipitation as a calcium salt.
- A so-called sacrificial flush is a solution containing cheap surfactant substitutes (e.g. lignosulphonates) likely to adsorb on the rock surface. Such a slug could be injected to prevent or reduce the adsorption from the surfactant slug, thus reducing losses (and cost) and formulation alterations. It also prepares the reservoir fluids to reach the optimal formulation in an easier and faster way.

The surfactant/co-surfactant slug

After the oil bank this is the most important slug because surfactants are the most expensive ingredients, as well as the critical ones from the oil displacement point of view. Once the surfactant slug contacts the oil globules these get mobilised and start moving through narrow necks of the porous reservoir. Afterwards the oil globules coalesce (because optimum formulation is associated with low emulsion stability) to form the oil bank. The oil bank thus formed is then pushed, displaced and propagated by the surfactant slug through the reservoir from the injector to the producer well (see Fig. 10.3). To assure a high oil recovery efficacy and a low water cut, the surfactant slug must stay stable both at the rear and front edges, its formulation should remain optimal or close to it and its sweep efficiency should be high, as close as possible to 100% of the porous volume. For instance, it has been shown that the performance of the surfactant slug is higher when the salinity varies as a decreasing electrolyte concentration, according to the so-called salinity gradient, which is a clever way to increase the displacement efficacy both at the front and rear edges [109, 114, 117, 131, 152]. At the front (leading) edge of the surfactant slug the transition between Winsor II and Winsor III takes place leading to a low interfacial tension between oil and the microemulsion, which improves the displacement of the surfactant slug and oil bank. Because of the ultra-low interfacial tension it also induces the early formation of the oil bank. At the rear edge of the surfactant slug, on the other hand, the transition between Winsor I and Winsor III takes place leading to a low interfacial tension between the aqueous polymer solution and the microemulsion (see supersolubilisation in Section 10.2.3.4), which improves the displacement of both the surfactant and polymer slugs. These features prevent microemulsion trapping, oil redispersion and emulsion formation by attenuating dispersivity and capillary forces [109, 114].

Similarly, alcohols can also induce a dramatic boost in oil recovery. The following two examples illustrate the important role alcohols can play in EOR: (1) Pithapurwala *et al.* [110, 111] found that by changing the alcohol (i.e. the alcohol structure) a dramatically change in oil recovery could be attained (e.g. 11, 14, 27, 92% of the OOIP). (2) Additionally, Shah [109] found that alcohol plays an important role on the oil flattening time, which is the time that an oil drop takes, under a specific environment, to become flat. He found that systems with lower flattening time showed the higher oil recovery efficiency. The volume of the surfactant slug typically ranges from 5 to 50% of the porous volume; it contains surfactant (0.1–10 wt.%), co-surfactant (0–5 wt.%), brine (0–reservoir salinity) and often polymer (0.05–0.2 wt.%) [113, 114, 116].

The pusher slug

The polymer slug that comes after the oil bank and surfactant slugs is the third one of importance. Among other functions, it controls the mobility of the surfactant slug enhancing its efficiency to sweep a large proportion of the whole porous medium and thus increasing the oil recovery. In order to prevent both the redispersion of the mobilised oil and dispersion of the surfactants, both the displacement of the oil bank by the surfactant slug and the surfactant slug by the polymer slug must be stable (i.e. without fingering nor stable emulsion); consequently, each slug must be displaced by a more viscous fluid [114, 115]. The polymer slug is often designed to exhibit a high viscosity at the front contact with the surfactant slug so that an almost piston displacement is warranted. It has been shown [109] that when the injected polymer slug is actually made of two polymer slugs with quite different salinities (in a process called salinity shock), the oil recovery is appreciably increased, while the surfactant loss is greatly reduced. Generally, the volume of the polymer slug ranges from 15 to 100% of the pore volume; it contains polymer (0.05–0.2 wt.%) and brine (0–reservoir salinity) [109, 114, 139, 140]. Additionally, a polymer slug helps in lowering the chemical costs by decreasing the volume of the surfactant slug to be injected.

Driver or chase water slug

The driver or chase water slug is a solution injected after the polymer slug whose viscosity progressively decreases, ending up as brine [109, 115, 119, 133, 141]. This slug is used, instead of a large polymer slug, to cut down the operation and the chemical costs.

Oil bank

Finally, the oil bank is not an injected slug but it is formed when the oil globules displaced by the surfactant slug coalesce, thus forming a two-phase flow pattern. Thereafter, it grows and propagates as it contacts at its leading edge more disconnected oil globules as the surfactant slug displaces it towards the producer wells [142]. The oil bank early formation and its stable propagation through the reservoir are of paramount importance to warranty the recovery of a high percentage of the OOIP [109].

The thickness of each slug decreases as they move radially from the (central) injector to the (peripheral) producer wells, due to both the geometry of the flow and the fact that the components of the slug are retained or incorporated into the oil phase. For example, for a well pattern in which the separation distance between injecting and producing wells is 100 m, a rough calculation, without considering surfactant loss, yields, for a surfactant slug volume of 10% of the pore volume, 31.6 m for the initial thickness and about 5.1 m for the final one. After considering surfactant loss, the surfactant slug would reach the producer wells much slimmer. This could help in designing a criterion to know if the different slug volumes injected into the reservoir are larger than what is really necessary. As a rule of thumb, it may be said that the volume of any of the injected slugs should be large enough to sweep the whole reservoir pore volume without losing its integrity.

10.3.4 Current state-of-the-art in enhanced oil recovery

Laboratory-scale studies, computer simulations and field trials have been performed with the same common goals mentioned above: (1) to enhance the performance of the surfactant formulation in attaining ultra-low interfacial tension and high solubilisation capability; (2) to maintain the optimal formulation during the process; (3) to improve the surfactant–slug sweep efficiency in the porous medium. The underlying idea is that a high oil recovery efficacy should be reached by achieving each of these goals.

10.3.4.1 Laboratory research

A lot of studies on formulating microemulsions for EOR were carried out in the decade 1973–1983 [109–112, 115, 117–131] where crude oil prices were high. About three billon dollars were spent on R&D to attain what is essentially the current know-how. Thereafter, it was essentially only in China that government-supported R&D groups have maintained a high research activity in EOR [113]. In the rest of the world and due to unfavourable economics, only a few laboratories financially supported by national oil companies dedicated some R&D funding to EOR [110, 112–114]. Researches studying microemulsions focused on improving the fundamental understanding [15, 143, 144] and finding many other new applications [15, 27, 143, 144, 153], e.g. soil remediation, cleaning, detergency (also best performed at optimum formulation), water solubilisation of oil (soak-only detergency, oil extraction) and even drug delivery systems. Additional different formulations were developed to lower surfactant loss and mixture chromatographic separation [136].

These research activities, though targeted at other applications, have helped both in widening and deepening the knowledge on surfactant system, which have increased the value of microemulsion know-how, thus making it readily available for the future of EOR, when the oil prices would rise again, as it is currently the case. Some examples of these developments are the anionic–cationic surfactant mixtures that have shown a synergistic effect in lowering interfacial tension and increasing oil solubilisation [143, 144]. Their use in EOR would probably have a positive impact on the percentage of oil recovered. Similarly, some formulations containing anionic surfactants and long-chain alcohols exhibited an increase in solubilisation power with bicontinuous microemulsion at a surfactant concentration as low as 2 wt.% [15, 143, 144]. Further developments from such works are the so-called extended surfactants which are able to lower oil–water interfacial tension at very low surfactant concentration, to form bicontinuous microemulsion with hydrocarbons at a surfactant concentration about 2 wt.% [14], and to exhibit a high solubilisation of polar oils for pharmaceutical and cosmetic applications [146–148]. All these studies led to a better understanding and allowed us to develop very simple models for describing the formulation of surfactant systems [15, 112, 113, 143, 144]. Most of the studies in microemulsion flooding have focused almost exclusively on the performance of the surfactant formulation and on finding cheaper ways of implementing the process. Research in polymer performance, on the other hand, has evolved from other areas and applications different from microemulsion flooding. Since the end of the 70s, both polymer science and technology have undergone a dramatic expansion which can help in improving all the polymer functions and attenuating all the polymer drawbacks. Although polyacrylamide is still the main candidate for chemical EOR [113, 117, 139, 140], new polymers with fancy rheological behaviour could greatly improve the sweep efficiency of the surfactant slug by attenuating phenomena such as fingering and hydrodynamic dispersivity. They can also lower the incompatibility with the different chemical species present in the reservoir and in the different slugs (especially with the surfactant slug) and surfactant loss. Similarly, they can be more tolerant to divalent ions, high salinity and temperature. Such polymers may nowadays look prohibitive from an economic point of view for an application-like EOR in which a large amount of polymer is entirely lost after being injected into a well to perform some particular functions. Nevertheless, in order to improve the current stage of the surfactant performance in recovering higher percentage of the OOIP, it is a must to try to incorporate new polymers in the chemical EOR research, regardless their costs, aiming just to enhance each of their functions and finding the limits.

10.3.4.2 Computer simulations

Computational simulations are always welcome since they can be used to explore features that are impossible or difficult to test or monitor experimentally. Simulations could help to design slug formulas and size, which can greatly impact the economy of the operation [134, 154] so that each slug performs its role throughout the whole reservoir. However, real microemulsion flooding involves a great number of system properties which are interconnected and influenced by the chemical composition, the temperature, the pressure and the porous reservoir [109, 115, 119, 131, 132, 135, 149]. In addition, some of the phenomena, like the fluid flow, are sensitive to gravity [119]. To date the only properties relatively easy to predict and to monitor are the changes induced by temperature and pressure. Since these two parameters are likely to affect the performances of the different slugs, computer simulation of their effect on the surfactant and polymer slug properties as well as on processes in the reservoir are of great help. Unfortunately, due to the complexity of the interconnected processes, interfacial tensions or the phase behaviour are almost impossible to simulate under real reservoir conditions. On the other hand, in spite of intents to provide computational and modelling tools to simulate the phase behaviour or the compositional mass balance of surfactant (UTCHEM) [154], it may be said that computational simulation is still lagging too far behind to be of practical use in an EOR real case scenario.

10.3.4.3 Field research

Several pilot field tests carried out at the end of the 70s, essentially showed three things that were also observed at the laboratory scale. (1) First, EOR by surfactant flooding is a feasible process [109, 113, 114, 117, 138, 140], particularly for low viscosity crudes, low salinity and low temperature reservoirs [116]. (2) Second, a lot of precautions have to be taken into account because of the difficulties that arise from unpredicted formulation changes due to surfactant adsorption, electrolyte desorption, reservoir heterogeneity, insufficient information on the porous medium morphology and fluid pattern [113, 114, 119, 138]. (3) Third, many formulations containing cheap surfactants such as petroleum sulphonates work well [113, 114, 128, 150] though they could probably be improved.

Similarly, many researchers have studied alkali, alkali–polymer and alkali–surfactant–polymer (namely ASP) recovery processes [110, 140, 155]. All these chemical EOR methods are based on the fact that injection of aqueous alkaline solution induces the in situ formation of oil natural surfactants through acid–base chemical reaction [116] in addition to the reservoir rock conditionning effect. In spite of the great effort and the encouraging laboratory results, pilot test outcomes for the first two have been poor [113, 117]. On the other hand, in spite of the relative low chemical cost compared to surfactant–polymer floods [113, 116, 155], hopeful laboratory tests [113, 114, 155, 156] (some of which claim an oil recovery greater than 80% of the OOIP [113, 114]), and some excellent pilot test results [113, 140], the ASP method has not found widespread applications [113, 116]. This suggests that ASP is not a mature technology, probably because the involved phenomena (e.g. change of wettability, spontaneous emulsification, emulsion flow through a porous medium, sensitivity to pH changes) are more complex than what had been first recognised [116, 155].

10.3.5 Future 'GUESSTIMATES'

Even though there is a great potential in the chemical EOR methods to considerably expand the World oil reserves, they still remain marginal [116, 117] more than 30 years after an unprecedented research effort to understand how petroleum is trapped in an oil reservoir and to develop technologies was carried out. The drop of crude oil price in the early 80s has probably been the main reason for this but a second main limiting issue still to be settled is the incomplete understanding of the interconnection between all occurring processes.

Chemical EOR is surely one of the petroleum industry's most complex, costly and risky operations, though so attractive that it is one of the current challenges of the oil industry. Its implementation in an actual field routine demands a great deal of creative thoughts and a great amount of chemicals and energy. Taking this into account, it is believed that EOR technology via microemulsion flooding faces two uncertainties which will determine its future.

1. How could scientific and technological challenges be overcome? The main problems to be dealt with are to maintain the optimum formulation of the surfactant slug during the process and to improve the sweep efficiency of the main surfactant slug.
2. How could operation and chemical costs be lowered? One important cost factor is the surfactant itself and new surfactant formulations are needed, which are capable of solubilising a wide variety of crude oils at low surfactant concentrations, i.e. 2 wt.% or even lower. Another factor that could reduce costs considerably is implementing an EOR operation in the right moment, i.e. primary instead of secondary or tertiary. Generally speaking, from the early studies undergone in the 70s, EOR via microemulsion flooding technology has been designed to be applied after the secondary recovery [112, 115, 119], i.e. as tertiary recovery. It is important mentioning that during a secondary recovery, e.g. by waterflooding, the oil saturation in the porous medium decreases down to a value around 0.3 [112, 113, 115]. Thereafter the oil remains mainly as disconnected globules [109, 113, 115, 129] trapped by capillary forces [109, 115, 117]. Is it necessary to wait till such a condition is reached to apply a chemical EOR technique? How about implementing, for new oil wells, a microemulsion flooding routine just after primary recovery? Unpublished trials [157] suggest that starting a microemulsion flooding routine just after primary recovery could result in an increase in oil recovery efficacy of more than 30% when compared to recovery attained by the traditional microemulsion tertiary flooding protocol. The high increase in oil recovery observed when an oil bank is previously injected [109] supports this inference. Therefore, for new oil wells a microemulsion flooding routine would have to be tried as a 'secondary' recovery routine or enhanced waterflooding. In addition, an implementation after primary recovery could help in increasing the oil ultimate recovery and preserving the reservoir from damages due to the wrong use of different chemical slugs.

Many of the EOR protocol sequences that have been used to date for low surfactant concentrations use a surfactant slug formulated with some polymer to improve its sweep efficiency [109, 138]. Taking this into account and considering the new polymer and surfactant developments, a future surfactant slug would have to be formulated with the new polymers and surfactants. Such a slug would provide the sites (some polymer segments) to the surfactant to stay adsorbed and to promote micelle formation, which would reduce surfactant loss [109 , 136] and increase the salinity tolerance [109]. Besides, in presence of high solubilisation surfactants, such a slug would be able to dissolve the oil in a single-phase microemulsion at a surfactant concentration at most of 2 wt.%. Furthermore, the relatively high viscosity of the slug would probably enhance the cleaning power of the slug.

Because of the diversity, interconnection and extreme complexity of the involved phenomena that take place during an EOR operation, only a few people with the down-the-bench expertise are currently available worldwide, and most of them are close to retirement, hence, great deal of their know-how is likely to be lost soon. It is thus a matter of extreme urgency to train a new generation of scientists and engineers with the proper understanding of both the physical chemistry and actual field issues, to tackle the present and future challenges of chemical EOR. This matter will probably require a concerted commitment between academic and industrial partners.

10.4 Degreasing of leather

10.4.1 Washing processes

When dealing with washing processes such as degreasing two questions have to be asked. First, how much amphiphile, i.e. surfactant, do I need to solubilise the natural fat and under which conditions does the system wash, i.e. degrease most effectively. The first question is a matter of efficiency. To determine the efficiency of a microemulsion system the pivot point for microemulsion phase behaviour, the so-called X-point, has to be determined. Depending on the temperature T and the surfactant mass fraction γ the phase boundaries of a ternary microemulsion system resemble the shape of a fish [158, 159]. The point where fish body (three-phase region) and fish tail (single-phase region) meet is the X-point or fish-tail point, which marks the onset of the single-phase microemulsion region and is defined by the least surfactant mass fraction needed to totally solubilise water and oil (see Chapter 1 for more details). The second question was already addressed by Benson *et al.* [160] together with Kahlweit and Strey [161], who found that the oil removal in a washing process reaches its maximum in the three-phase region of the respective microemulsion system (see Chapter 8, Fig. 8.12). The three-phase region of a microemulsion system is closely connected to the interfacial tension between water and oil σ_{ab}, which becomes minimal at the mean temperature of the three-phase region. For efficient microemulsion systems σ_{ab} drops to ultra-low values of 10^{-3}–10^{-4} mN m^{-1}. Thus, for the efficient degreasing of animal skins one has to know the location of the three-phase region and hence of the X-point.

10.4.2 Leather degreasing via microemulsions

The large-scale application of microemulsions in leather degreasing was motivated by questioning current industrial processes for degreasing due to potentially environmental concerns. Of relevance for leather degreasing were mainly two processes, a solvent-based one and a water-based one. The surfactants showing the best performance in the latter

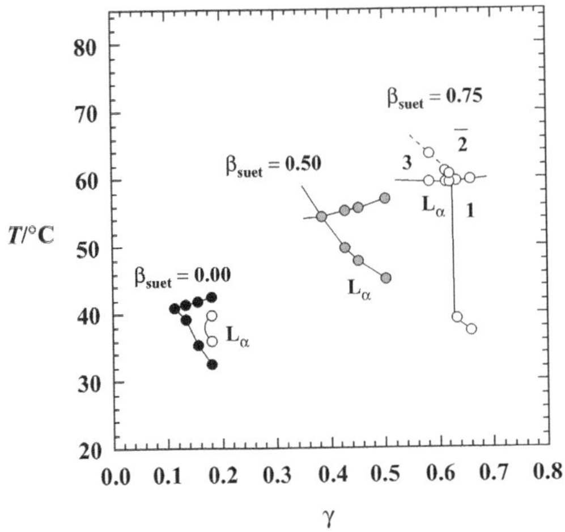

Figure 10.4 Phase diagrams of the systems H_2O/NaCl–rape oleic acid methyl ester/suet–Lutensol® AP9 at $\phi = 0.50$ and $\varepsilon = 0.10$. The parameter β corresponds to the mass fraction of suet in the mixture of suet and rape oleic acid methyl ester. With increasing β, the phase behaviour is shifted to higher temperatures and surfactant mass fractions. Because of the dominance of the L_α-phase a phase diagram at $\beta = 1.00$ could not be obtained.

process are alkyl phenol ethoxylates. As these surfactants decompose during effluent treatment and form chemicals, which have an endocrine-disrupting effect on life-forms, alternatives had to be found, which show a comparable or even better degreasing performance. In the following we present a way how to eco-friendly degrease animal skins via microemulsions and clarify the so far unidentified mechanism of degreasing.

10.4.2.1 Phase behaviour of suet microemulsions

Benchmark

Standard surfactants for water-based degreasing of animal skins are non-ionic alkyl phenol ethoxylates. The phase behaviour of the benchmark Lutensol® AP9 was characterised in the system H_2O/NaCl–rape oleic acid methyl ester/suet–Lutensol® AP9. As natural fat, suet consists mainly of triacylglycerols, which can only be solubilised at high surfactant mass fractions [162]. To enhance the efficiency of the system, rape oleic acid methyl ester was used as co-oil. The mass fraction of suet in the mixture of rape oleic acid methyl ester and suet is given by the parameter β. The salt mass fraction in the water phase ε was kept constant at $\varepsilon = 0.10$ and the phase behaviour was determined at equal volume fractions of water and oil, $\phi = 0.50$. The T–γ cuts presented in Fig. 10.4 show that rape oleic acid methyl ester is efficiently solubilised by Lutensol® AP9. The X-point is located at $\tilde{T} = 40.65$°C and $\tilde{\gamma} = 0.119$ and a liquid crystalline lamellar phase (L_α) extends in the single-phase region. The 'fish' is distorted towards lower γ, which is typical for technical surfactants. Upon increasing the mass fraction of suet to $\beta = 0.50$ the phase behaviour shifts to higher temperatures and surfactant mass fractions. The L_α-phase now extends below the lower phase boundary. This trend continues to $\beta = 0.75$. The X-point is located at $\tilde{T} = 59.40°C$ and $\tilde{\gamma} = 0.628$. The L_α-phase dominates the phase behaviour. As a result the lower phase boundary becomes steep and the X-point is difficult to detect. Because of the dominance of the L_α-phase a system with $\beta = 1.00$ could not be obtained. Thus, the solubilisation of suet by Lutensol®AP9 is rather limited.

Eco-friendly Surfactants

In contrast to alkyl phenol ethoxylates the class of fatty alcohol ethoxylates is biodegradable and eco-friendly. Efficient technical surfactants found in this class are Lutensol®XL 700, Lutensol®TO8, Lutensol®AO7, Lutensol®AO8 and Eusapon®OD. The phase behaviour was characterised in systems of type H_2O/NaCl–suet–technical non-ionic surfactant. The salt mass fraction was kept constant at $\varepsilon = 0.10$ and the oil volume fraction at $\phi = 0.50$. In view of the degreasing process, which is conducted at 30°C, the X-point of the optimal system should be located around 30°C and the formation of the highly viscous L_α-phase should be suppressed. Figure 10.5 presents the T–γ cuts for the respective systems.

In the following the T–γ cuts will be discussed in the order of their efficiency. Because of its comparatively short C-chain length ($i = 10$) Lutensol®XL 700 is the most inefficient one of the surfactants. The X-point is located at $\tilde{T} = 50.67°C$ and $\tilde{\gamma} = 0.678$. A liquid crystalline lamellar phase extends below the lower phase boundary. With increasing C-chain length of the surfactant, i.e. substituting Lutensol®XL 700 by Lutensol®TO8 ($i = 13$), the phase behaviour shifts to lower surfactant mass fractions as expected [163]. Because of the increased number of ethoxylate groups the phase behaviour also shifts to higher temperatures. The L_α-phase again extends below the lower phase boundary. The efficiency can be further increased by substituting Lutensol®TO8 by its longer-chained homologous Lutensol®AO7 and Lutensol®AO8, respectively. The X-points of the systems are located at $\tilde{T} = 57.09°C$ and $\tilde{\gamma} = 0.379$ (AO7) and $\tilde{T} = 61.95°C$ and $\tilde{\gamma} = 0.404$ (AO8).

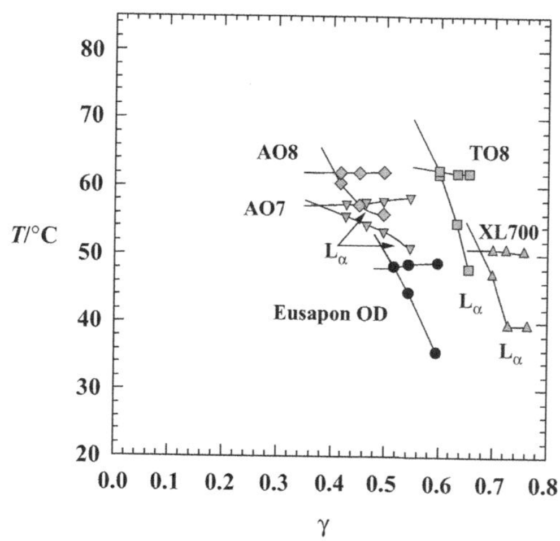

Figure 10.5 Phase diagrams of the systems H_2O/NaCl–suet–technical non-ionic surfactant at $\phi = 0.50$ and $\varepsilon = 0.10$. Although less efficient than Lutensol®AO7 and Lutensol®AO8, Eusapon®OD is the most suitable alternative for Lutensol®AP9 as the X-point is located near the degreasing temperature and no L_α-phase forms.

However, the L_α-phase in both systems extends over a wide γ-range below the lower phase boundary. The most suitable of the characterised surfactants regarding the degreasing process is Eusapon®OD. Although less efficient than Lutensol®AO7 and Lutensol®AO8 no L_α-phase can be observed and the X-point is located near the degreasing temperature, $\tilde{T} = 48.13°C$.

10.4.2.2 Eco-friendly degreasing

The degreasing experiments were performed in the beam house of BASF AG according to recipes for very greasy sheep skins. In seven drums, two skins were degreased at a time. In order to protect the collagen network of the skin the degreasing temperature was kept constant at $T = 30°C$. Prior to the degreasing experiments the New Zealand sheep skins were depickled by adding a short float (30 wt.% of water) containing high amounts of sodium chloride (8 wt.%). For the first degreasing step, surfactant was added (1 wt.%) and the drums were run for approximately 1 h. The short float was then diluted with pure water (70 wt.%). In the following this float will be denoted as diluted float. After drainage of the diluted float the skins were again washed with surfactant (2 wt.%) and water (200 wt.%). In the following this float will be denoted as long float. Finally, the degreased skins were examined by the tanner to evaluate the degreasing performance.

Sheep skins were degreased using Eusapon®OD following the above presented procedure in the beam house. Samples were taken from every float and the phase behaviour was directly characterised by visual inspection. The results are presented in the following beginning with the short float.

Short float

Because of the small amount of water in the short float, samples had to be taken by wringing out the sheep skins. A three-phase state at $T = 37°C$ could be observed, where a microemulsion phase coexisted with a fat-excess phase and a water-excess phase as presented in Fig. 10.6. At this point, the occurrence of the three-phase state in the short float was a first evidence for the good performance of Eusapon®OD. The three-phase state is closely connected to the interfacial tension between water and oil, in this case water and grease (σ_{ab}) which becomes ultra-low for efficient microemulsion systems. At this point the low interfacial tension ensures a good degreasing performance of the system. More

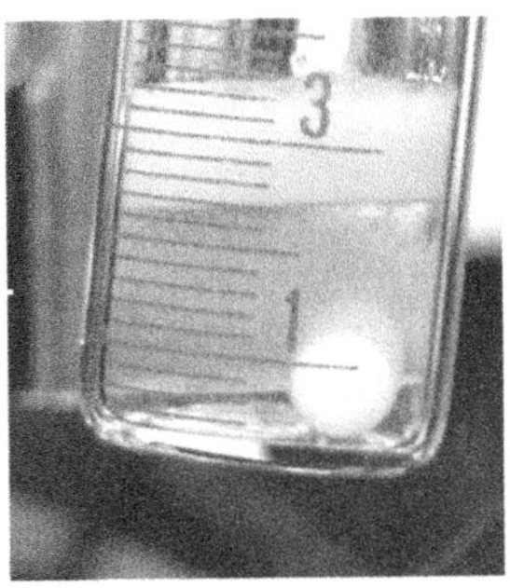

Figure 10.6 Picture of the three-phase state found in the short float at $T = 37°C$. A microemulsion phase coexists with a fat- and a water-excess phase.

Figure 10.7 Picture of the three-phase state found in the diluted and in the long float at $T = 60°C$. A microemulsion phase coexists with a fat- and a water-excess phase.

importantly, it could be verified for the first time that microemulsions play an important role in the degreasing process.

Diluted float/long float

Samples were also taken from the diluted and the long float. The characterisation of the phase behaviour resulted for both floats in a three-phase state, which, in contrast to the short float, is not located at $T = 37°C$, but at $T = 60°C$ as shown in Fig. 10.7. At the degreasing temperature $T = 30°C$ now a fat-in-water microemulsion coexists with a fat-excess phase, which is turned into a stable fat-in-water emulsion via shearing.

10.4.2.3 Correlation of phase behaviour and eco-friendly degreasing

The results obtained from the characterisation of the phase behaviour and in the beam house imply that Eusapon®OD is a suitable alternative allowing for an eco-friendly degreasing of animal skins. However, the understanding of the so far unidentified degreasing mechanism is the key goal for a continuous development of the degreasing process itself. In order to clarify the role of microemulsions in degreasing additional phase behaviour and interfacial tension measurements were conducted.

T–γ cut and T–γ_b cut

Analogue to other natural fat suet shows charge-dependent fluctuations in its composition. Triolein is the main component of nearly all natural fats and was therefore used as a 'model-fat' for further investigations. Figure 10.8 shows the phase diagram of the system H_2O/NaCl–triolein–Eusapon®OD at $\varepsilon = 0.10$ and $\phi = 0.50$. As degreasing takes place at very low values of γ the three-phase region of the system was characterised, as well.

The phase diagram shows that the lower phase boundary of the three-phase region is located above 44°C. This result is contrary to the findings in the degreasing experiments, i.e. the short float, where a three-phase state at 37°C could be observed. In order to understand this discrepancy schematic Gibbs triangles for the characterisation of the phase behaviour and the degreasing experiments are shown in Fig. 10.9. Figure 10.9(a) represents the T–γ cut performed to characterise the phase behaviour of the benchmark Lutensol®AP9 as well as the eco-friendly fatty alcohol ethoxylates at equal volume fractions of water and oil $\phi = 0.50$ (see also Figs. 10.4 and 10.5). However, during the degreasing experiments the

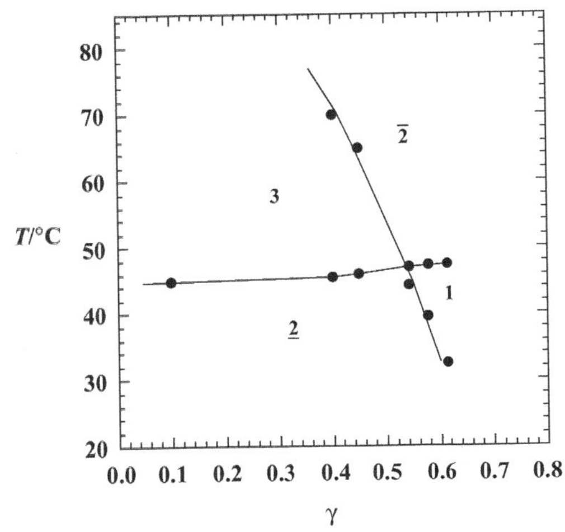

Figure 10.8 Phase diagram of the system H_2O/NaCl–triolein–Eusapon®OD at $\varepsilon = 0.10$ and $\phi = 0.50$. Single-phase region and parts of the three-phase region are shown.

volume fractions of water and oil in the float vary. Starting with small amounts of water in the short float the amount of water is largely increased in the diluted and in the long float. Thus, the cut performed in the degreasing experiments is a so-called T–γ_b cut, where γ_b represents the mass fraction of surfactant in the mixture of oil and surfactant, i.e.

$$\gamma_b = \frac{m_{surfactant}}{m_{surfactant} + m_{fat}}. \quad (10.1)$$

This 'dilution' cut is shown schematically in Fig. 10.9(b). To mimic the conditions in the beam house a T–γ_b cut was conducted for the system H_2O/NaCl–triolein–Eusapon®OD at $\gamma_b = 0.375$ and $\varepsilon = 0.10$. The value of γ_b was derived from the fat content of the float after degreasing. The phase diagrams at constant ϕ (black symbols) and at constant γ_b

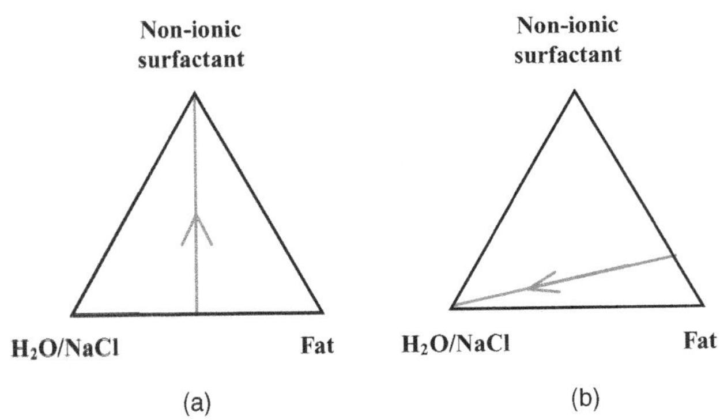

Figure 10.9 Schematic Gibbs triangles for the system H_2O/NaCl–natural fat–non-ionic surfactant. (a) T–γ cut at $\phi = 0.50$; (b) T–γ_b cut with varying water to oil plus water volume fraction.

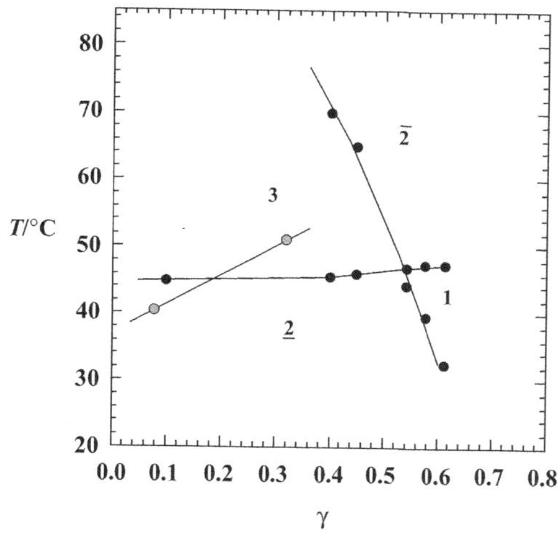

Figure 10.10 Phase diagram of the system H_2O/NaCl–triolein–Eusapon®OD at $\varepsilon = 0.10$ and $\phi = 0.50$ (black symbols) and lower phase boundary of the system H_2O/NaCl–triolein–Eusapon®OD at $\gamma_b = 0.375$ (gray symbols).

(gray symbols) are shown in Fig. 10.10. For the clarification of the degreasing process only the lower phase boundary is important as it is located near the degreasing temperature $T = 30°C$. For this reason the upper phase boundary was not determined. Because of high emulsion stability only two data points could be obtained at constant γ_b. Compared to the T–γ cut the lower phase boundary in the T–γ_b cut shifts to lower temperatures upon dilution with sodium chloride solution, which is caused by the increasing extraction of the hydrophilic fractions of the technical surfactant from the interfacial film into the water phase. As a result the surfactant remaining in the interfacial film becomes more hydrophobic. Because of the temperature shift the location of the lower phase boundary at low γ in the T–γ_b cut now corresponds to the temperature of the three-phase state observed in the short float.

Ultra-low interfacial tension

The reason for the optimal degreasing performance of the microemulsion in the three-phase region is the ultra-low interfacial tension between water and oil σ_{ab}. Figure 10.11 shows the variation of σ_{ab} as a function of T for the system H_2O/NaCl–triolein–Eusapon®OD. In order to mimic the composition of the float a fat to water plus fat mass fraction $\alpha = 0.15$ was chosen, which is the least fat mass fraction required for spinning drop measurements.

The variation of the interfacial tension as a function of T for the Eusapon®OD system shows the typical V-shape. The full curve corresponds to a theoretical description in terms of bending energy [164, 165]. The minimum of the interfacial tension correlates well with the mean temperature of the system and is located at $\sigma_{ab} = 0.13$ mN m^{-1} at $T_m = 51.5°C$. For the benchmark Lutensol®AP9 the interfacial tension between water and oil near the degreasing temperature corresponds to $\sigma_{ab} = 0.43$ mN m^{-1}. Although the interfacial tension between water and triolein is high compared to efficient microemulsion systems, it is still two orders of magnitude lower than the pure water oil interfacial tension (50 mN m^{-1}).

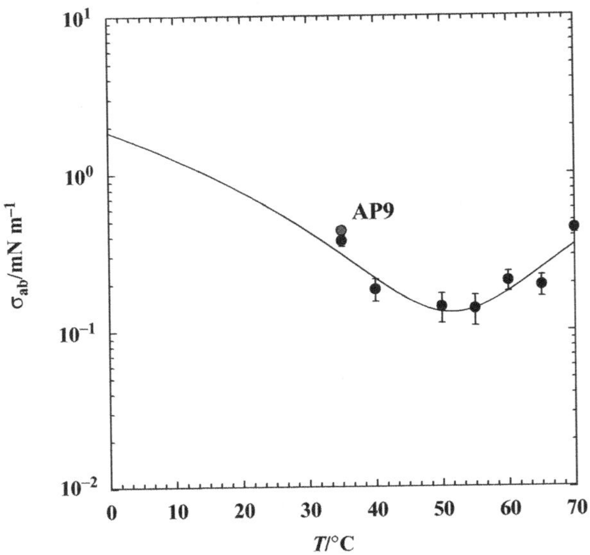

Figure 10.11 Variation of σ_{ab} with temperature for the system H_2O/NaCl–triolein–Eusapon®OD at $\alpha = 0.15$ and $\varepsilon = 0.10$. The value for the Lutensol®AP9 system at $T = 35°C$ is shown as well.

10.4.2.4 The 'Salt-Jump'

Animal skins were degreased in a three-step process according to the procedure described in the section beam house. In a first step, surfactant was added to depickled sheep skins and the drum was run for approximately 1 h. Note, that depickling means the addition of a short float (little water) containing high amounts of sodium chloride. The mass fraction of salt in the water phase of the short float was estimated to be $\varepsilon = 0.21$. After 1 h the short float was diluted with pure water. As a result the sodium chloride mass fraction in the water phase ε was instantaneously reduced to $\varepsilon = 0.07$. During this process the surfactant to fat plus surfactant ratio γ_b was always constant at approximately $\gamma_b = 0.375$.

The dramatic reduction of the salt mass fraction strongly influences the phase behaviour of the float. The resulting variation of the phase behaviour is the key for the optimal degreasing performance and can be understood as follows. In Fig. 10.12, the variation of the phase behaviour in the course of the degreasing process is shown schematically. The presentation in form of T–γ cuts is for clarity reasons. The fat to water plus fat volume fractions ϕ vary during degreasing, thus the presentation in form of Gibbs triangles corresponds to the real experimental conditions.

In Fig. 10.12(a), the phase behaviour in the short float is shown. The three-phase state of the respective systems is located near the degreasing temperature $T = 30°C$. Efficient degreasing is a result of the ultra-low interfacial tension between water and fat. Upon diluting the short float with pure water the salt mass fraction in the water phase is effectively reduced from $\varepsilon = 0.21$ to $\varepsilon = 0.07$. Sodium chloride belongs to the group of lyotropic salts. When the salt mass fraction is reduced the hydration of the surfactant head

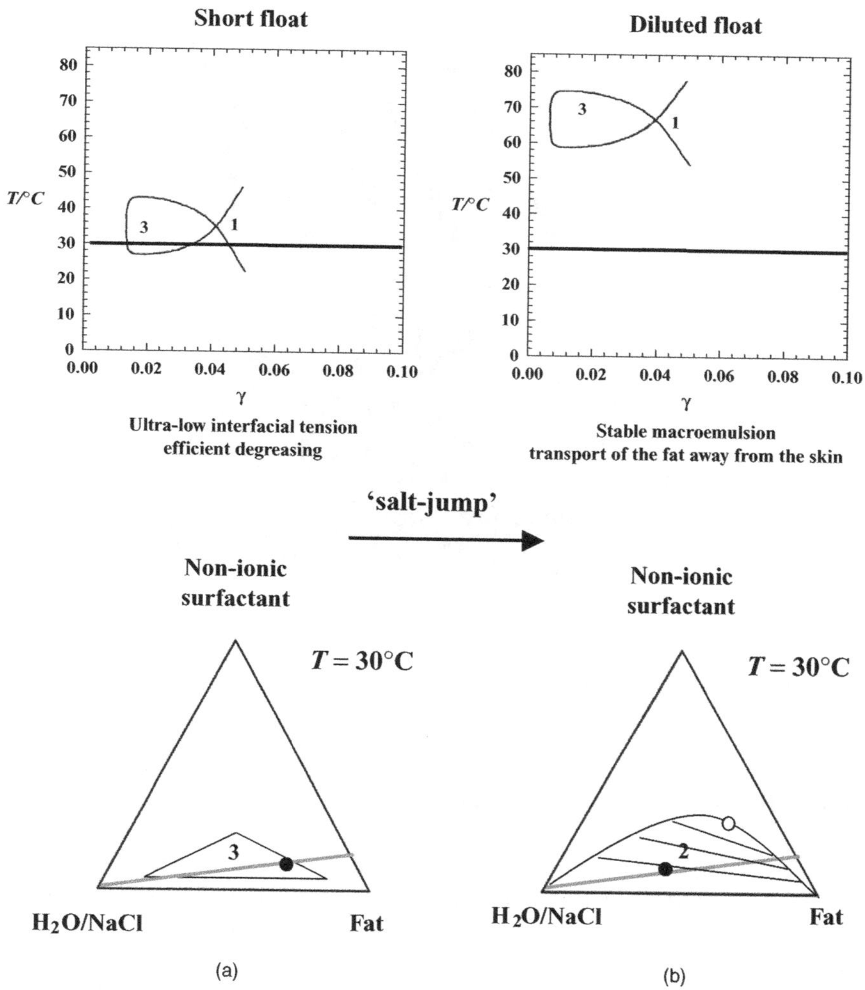

Figure 10.12 Schematic of the variation of the phase behaviour during the degreasing process. In the short float the ultra-low interfacial tension between water and oil ensures efficient degreasing. Upon reducing the salt mass fraction the phase behaviour shifts to higher temperatures. At the degreasing temperature now an oil-in-water microemulsion coexists with an oil-excess phase. Shearing induces the formation of a stable macroemulsion that prevents the depositing of the fat on the skin and ensures the transport of the fat away from the skin. Note that only the Gibbs triangles correspond to the real experimental conditions. The T–γ cuts are shown for clarity.

groups increases and the surfactant becomes effectively more hydrophilic [3]. As a result the phase behaviour shifts to higher temperatures as shown in Fig. 10.12(b) [166–169]. At the degreasing temperature $T = 30°C$, the system is now in the phase state $\underline{2}$, where an o/w microemulsion coexists with an oil excess phase. Shearing induces the formation of a stable macroemulsion, which prevents the fat from depositing on the skin and enables the transport of the fat away from the skin.

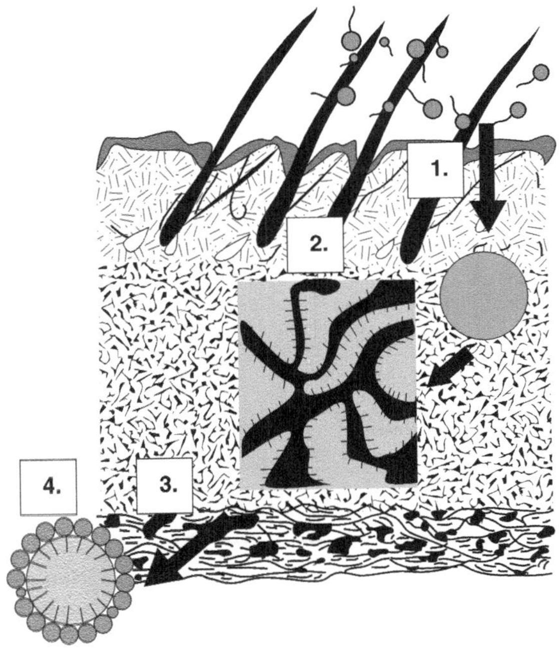

Figure 10.13 Schematic of the four-step process of degreasing according to the interplay of micro- and macroemulsions. (From Ref. [171], reprinted with permission of Umschau Verlag.)

10.4.3 The degreasing mechanism

Efficient degreasing was found to be closely connected to the three-phase state and hence to the ultra-low interfacial tension between water and oil [170]. The so far unidentified mechanism of degreasing of animal skins could be understood and explained. Correlation of results obtained from phase behaviour measurements and degreasing experiments revealed that Eusapon®OD shows the best degreasing performance and lead to the clarification of the four-step process of degreasing as shown in Fig. 10.13 [171]. The first step is the penetration of the surfactant into the skin. In a second step the natural fat is solubilised. A microemulsion phase coexists with a fat- and a water-excess phase and the interfacial tension between water and oil is ultra-low. On the surface of the skin dilution of the microemulsion with pure water, i.e. reduction of the salt concentration in the float, leads to the formation of a stable emulsion via shearing. The stable emulsion prevents the deposition of the fat on the skin and enables the transport of the natural fat away from the skin.

Acknowledgement

SE would like to thank for the performance of the degreasing experiments in the BASF beam house.

References

1. Solans, C. and Kunieda, H. (1997) *Industrial Applications of Microemulsions, Surfactant Science Series 66*. Marcel Dekker, New York.
2. Kahlweit, M., Strey, R., Haase, D., Kunieda, H., Schmeling, T., Faulhaber, B., Borkovec, M., Eicke, H.-F., Busse, G., Eggers, F., Funck, T., Richmann, H., Magid, L., Söderman, O., Stilbs, P., Winkler, J., Dittrich, A. and Jahn, W. (1987) How to study microemulsions. *J. Colloid Interface Sci.*, **118**(2), 436–453.
3. Kahlweit, M., Strey, R., Firman, P., Haase, D., Jen, J. and Schomäcker, R. (1988) General patterns of the phase behavior of mixtures of H_2O, nonpolar solvents, amphiphiles, and electrolytes. 1. *Langmuir*, **4**(3), 499–511.
4. Kahlweit, M., Strey, R., Schomäcker, R. and Haase, D. (1989) General patterns of the phase behavior of mixtures of H_2O, nonpolar solvents, amphiphiles, and electrolytes. 2. *Langmuir*, **5**(2), 305–315.
5. Kunieda, H. and Solans, C. (1997) How to prepare microemulsions: Temperature-insensitive microemulsions. In C. Solans and H. Kunieda (eds), *Industrial Applications of Microemulsions, Surfactant Science Series 66*. Marcel Dekker, New York, pp. 21–45.
6. Jakobs, B., Sottmann, T., Strey, R., Allgaier, J., Willner, L. and Richter, D. (1999) Amphiphilic block copolymers as efficiency boosters for microemulsions. *Langmuir*, **15**(20), 6707–6711.
7. Alander, J. and Warnheim, T. (1989) Model microemulsions containing vegetable oils. Part 1: Nonionic surfactant systems. *J. Am. Oil Chem. Soc.*, **66**(11), 1656–1660.
8. Alander, J. and Warnheim, T. (1989) Model microemulsions containing vegetable oils. 2. Ionic surfactant systems. *J. Am. Oil Chem. Soc.*, **66**(11), 1661–1665.
9. Mönig, K., Haegel, F.-H. and Schwuger, M.J. (1996) Microemulsions with plant oils – systematic investigations on preparation and temperature-induced splitting. *Tenside Surfactants Detergents*, **33**(3), 228–232.
10. Li, X., Washenberger, R.M., Scriven, L.E., Davis, H.T. and Hill, R.M. (1999) Phase behavior and microstructure of water/trisiloxane E_6 and E_{10} polyoxyethylene surfactant/silicone oil systems. *Langmuir*, **15**(7), 2278–2289.
11. Ilic, M., Haegel, F.-H., Zlatanovic, S., Potkonjak, N. and Simonovic, B. (2005) Microemulsion with aromatic oil as cleaner media for electropolymerisation. *Environ. Chem. Lett.*, **3**(2), 82–85.
12. Yu, B. and Mattox, J.R. (1995) (to Rohm and Haas Company) United States Patent 5,444,078.
13. Garti, N., Spernath, A., Aserin, A. and Lutz, R. (2005) Nano-sized self-assemblies of nonionic surfactants as solubilization reservoirs and microreactors for food systems. *Soft Matter*, **1**(3), 206–218.
14. Miñana-Perez, M., Graciaa, A., Lachaise, J. and Salager, J.-L. (1995) Solubilization of polar oils with extended surfactants. *Colloids Surf. A*, **100**, 217–224.
15. Salager, J.-L., Antón, R.E., Sabatini, D.A., Harwell, J.H., Acosta, E.J. and Tolosa, L.I. (2005) Enhancing solubilization in microemulsions – state of the art and current trends. *J. Surfactants Detergents*, **8**(1), 3–21.
16. Tadros, T.F. (1997) Microemulsions in agrochemicals. In C. Solans and H. Kunieda (eds), *Industrial Applications of Microemulsions, Surfactant Science Series 66*. Marcel Dekker, New York, pp. 199–207.
17. Hamied, Y.K., Malhotra, G. and Nayak, V.G. (1999) (to Byron Fine Chemicals Inc.) United States Patent 5,929,030.
18. Foerster, T., Claas, M. and Wollenweber, H.-W. (2001) (to Cognis Deutschland GmbH) United States Patent 6,255,253.

19. Radtke, D. and Wozniak, E. (1991) (to Xylochimie) United States Patent 5,013,748.
20. Dungan, S.R. (1997) Microemulsions in foods: Properties and applications. In C. Solans and H. Kunieda (eds), *Industrial Applications of Microemulsions, Surfactant Science Series 66*. Marcel Dekker, New York, pp. 147–174.
21. El-Nokaly, M., Hiler, G.D. and McGrady, J. (1991) (to The Procter & Gamble Company) United States Patent 5,045,337.
22. Allgaier, J., Willner, L., Richter. D., Jakobs, B. and Sottmann, T. (2004) (to Forschungszentrum Julich GmbH) United States Patent 6,677,293.
23. Schambil, F., Zeidler, U., Shamsai, S., Foerster, T. and Tesmann, H. (2000) (to Henkel Kommanditgesellschaft auf Aktien) United States Patent 6,086,787.
24. Bauer, D., Komornicki, J. and Tellier, J. (1986) (to Societe Nationale Elf Aquitaine) United States Patent 4,587,106.
25. Dantas Neto, A.A., de Castro Dantas, T.N. and Alencar Moura, M.C.P. (2004) Evaluation and optimization of chromium removal from tannery effluent by microemulsion in the Morris extractor. *J. Hazard. Mater.*, **B114**(1–3), 115–122.
26. de Castro Dantas, T.N., Cordeiro Beltrame, L.T., Dantas Neto, A.A. and de Alencar Moura, M.C.P. (2004) Use of microemulsions for removal of color and dyes from textile wastewater. *J. Chem. Technol. Biotechnol.*, **79**(6), 645–650.
27. López-Montilla, J.C., Pandey, S., Shah, D.O. and Crisalle, O.D. (2005) Removal of non-ionic organic pollutants from water via liquid–liquid extraction. *Water Res.*, **39**(9), 1907–1913.
28. Stockigt, D., Oetter, G., Wolff, E., Frank, E. and Schneider, P. (2003) (to BASF Aktiengesellschaft) United States Patent 6,544,348.
29. Bartlett, I.W., Dalton, A.J.P., McGuinness, A. and Palmer, H. (1999) Substitution of organic solvent cleaning agents in the lithographic printing industry. *Ann. Occup. Hyg.*, **43**(2) 83–90.
30. Brown, J.S., Hodge, R.C., McCabe, M.A., McGrady, K.A. and Schilling, A.S. (2006) (to The United States of America as represented by the Secretary of the Navy) United States Patent 7,064,241.
31. Hill, R.M., Kaler, E.W., Ryan, L.D. and Silas, J.A. (2000) (to Dow Corning Corporation) United States Patent 6,013,683.
32. Marchese, E., Visca, M. and Lenti, D. (2003) (to Ausimont S.p.A.) United States Patent 6,660,798.
33. Brinker, C.J., Lu, Y. and Fan, H. (2001) (to Sandia Corporation) United States Patent 6,270,846.
34. Lu, Y.-Y. and Young, C.I. (1997) (to Minnesota Mining and Manufacturing Company) United States Patent 5,624,973.
35. Wolf, G.-D., Jonas, F. and Schomacker, R. (1996) (to Bayer AG) United States Patent 5,575,898.
36. Haegel, F.-H., Schlupen, J., Schultze, J.W. and Winkels, S. (2004) United States Patent 6,695,976.
37. Yamashita, Y. (2005) (to Murata Manufacturing Co., Ltd.) United States Patent Application 20050194573.
38. Pithan, C., Schneller, T., Shiratori, Y., Majumder, S.B., Haegel, F.-H., Dornseiffer, J. and Waser, R. (2006) Microemulsion mediated synthesis of nanocrystalline $BaTiO_3$: Possibilities, potential and perspectives. *Int. J. Mater. Res.*, **97**(5), 499–507.
39. Clarke, A.N., Oma, K.H., Megehee, M.M. and Wilson, D.J. (1993) Soil clean-up by surfactant washing. II. Design and evaluation of the components of the pilot-scale surfactant recycle system. *Separation Sci. Technol.*, **28**(13–14), 2103–2135.
40. Haegel, F.-H., Dierkes, F., Kowalski, S., Mönig, K., Schwuger, M.J., Subklew, G. and Thiele, P. (2000) Soil remediation with microemulsions: Basic properties. In J.F. Scamehorn and J.H. Harwell (eds), *Surfactant-Based Separations: Science and Technology, ACS Symposium Series 740*. American Chemical Society, Washington, DC, pp. 35–56.

41. Cheng, H. and Sabatini, D.A. (2002) Phase-behavior-based surfactant-contaminant separation of middle phase microemulsions. *Separation Sci. Technol.*, **37**(1), 127–146.
42. Wai, C.M. and Enokida, Y. (2006) (to Idaho Research Foundation, Inc.) United States Patent 7,128,840.
43. Pennell, K.D., Pope, G.A. and Abriola, L.M. (1996) Influence of viscous and buoyancy forces on the mobilization of residual tetrachloroethylene during surfactant flushing. *Environ. Sci. Technol.*, **30**(4), 1328–1335.
44. Stroo, H.F., Unger, M., Ward, C.H., Kavanaugh, M.C., Vogel, C., Leeson, A., Marqusee, J.A. and Smith, B.P. (2003) Remediating chlorinated solvent source zones. *Environ. Sci. Technol. A*, **37**(11), 224A–230A.
45. Harwell, J.H., Sabatini, D.A. and Knox, R.C. (1999) Surfactants for ground water remediation. *Colloids Surf. A*, **151**(1–2), 255–268.
46. Hoppe, K., Dierkes, F., Haegel, F.-H., Monig, K. and Subklew, G. (2001) (to Dr. Fauz-Hubert Haegel) United States Patent 6,280,533.
47. Sabatini, D.A., Knox, R.C., Harwell, J.H. and Wu, B. (2000) Integrated design of surfactant enhanced DNAPL remediation: Efficient supersolubilization and gradient systems. *J. Contam. Hydrol.*, **45**(1–2), 99–121.
48. Jayanti, S., Britton, L.N., Dwarakanath, V. and Pope, G.A. (2002) Laboratory evaluation of custom-designed surfactants to remediate NAPL source zones. *Environ. Sci. Technol.* **36**(24), 5491–5497.
49. Scott, M.J. and Jones, M.N. (2000) The biodegradation of surfactants in the environment. *Biochim. Biophys. Acta Biomembr.*, **1508**(1–2), 235–251.
50. Jafvert, C.T. (1996) Surfactants/Cosolvents. GWRTAC E Series Technical Evaluation Report TE-96-02, Ground-Water Remediation Technologies Analysis Center, Pittsburgh, PA. http://www.epa.gov/tio/download/remed/surf_co.pdf.
51. Fountain, J.C. (1992) Field tests of surfactant flooding – mobility control of dense nonaqueous-phase liquids. In D.A. Sabatini and R.C. Knox (eds), *Transport and Remediation of Subsurface Contaminants, ACS Symposium Series 491*. American Chemical Society, Washington, DC, pp. 182–191.
52. Abdul, A.S. and Ang, C.C. (1994) In-situ surfactant washing of polychlorinated-biphenyls and oils from a contaminated field site – Phase II pilot-study. *Ground Water*, **32**(5), 727–734.
53. Fountain, J.C., Starr, R.C., Middleton, T., Beikirch, M., Taylor, C. and Hodge, D. (1996) A controlled field test of surfactant-enhanced aquifer remediation. *Ground Water*, **34**(5), 910–916.
54. Knox, R.C., Sabatini, D.A., Harwell, J.H., Brown, R.E., West, C.C., Blaha, F. and Griffin, C. (1997) Surfactant remediation field demonstration using a vertical circulation well. *Ground Water*, **35**(6), 948–953.
55. Martel, R., Gélinas, P.J. and Saumure, L. (1998) Aquifer washing by micellar solutions: 3. Field test at the Thouin Sand Pit (L'Assomption, Québec, Canada). *J. Contam. Hydrol.*, **30**(1–2), 33–48.
56. Baran, J.R., Pope, G.A., Wade, W.H. and Weerasooriya, V. (1996) Water/chlorocarbon winsor I ↔ III ↔ II microemulsion phase behavior with alkyl glucamide surfactants. *Environ. Sci. Technol.*, **30**(7), 2143–2147.
57. Fountain, J.C., Klimek, A., Beikirch, M.G. and Middleton, T.M. (1991) The use of surfactants for in situ extraction of organic pollutants from a contaminated aquifer. *J. Hazard. Mater.*, **28**(3), 295–311.
58. Oostrom, M., Hofstee, C., Walker, R.C. and Dane, J.H. (1999) Movement and remediation of trichloroethylene in a saturated, heterogeneous porous medium 2. Pump-and-treat and surfactant flushing. *J. Contam. Hydrol.*, **37**(1–2), 179–197.

59. Baran, J.R., Pope, G.A., Wade, W.H., Weerasooriya, V. and Yapa, A. (1994) Microemulsion formation with chlorinated hydrocarbons of differing polarity. *Environ. Sci. Technol.*, **28**(7), 1361–1366.
60. Shiau, B.-J., Sabatini, D.A., Harwell, J.H. and Vu, D.Q. (1996) Microemulsion of mixed chlorinated solvents using food grade (Edible) surfactants. *Environ. Sci. Technol.*, **30**(1), 97–103.
61. Shiau, B.-J., Sabatini, D.A. and Harwell, J.H. (1994) Solubilization and Microemulsification of chlorinated solvents using direct food additive (Edible) surfactants. *Ground Water*, **32**(4), 561–569.
62. Field, J.A., Sawyer, T.E., Schroth, M.H., Humphrey, M.D. and Istok, J.D. (2000) Effect of cation exchange on surfactant-enhanced solubilization of trichloroethene. *J. Contam. Hydrol.*, **46**(1–2), 131–149.
63. Childs, J.D., Acosta, E., Knox, R., Harwell, J.H. and Sabatini, D.A. (2004) Improving the extraction of tetrachloroethylene from soil columns using surfactant gradient systems. *J. Contam. Hydrol.*, **71**(1–4), 27–45.
64. Dwarakanath, V., Kostarelos, K., Pope, G.A., Shotts, D. and Wade, W.H. (1999) Anionic surfactant remediation of soil columns contaminated by nonaqueous phase liquids. *J. Contam. Hydrol.*, **38**(4), 465–488.
65. Acosta, E., Tran, S., Uchiyama, H., Sabatini, D.A. and Harwell, J.H. (2002) Formulating chlorinated hydrocarbon microemulsions using linker molecules. *Environ. Sci. Technol.*, **36**(21), 4618–4624.
66. Shook, G.M. and Pope, G.A. (1999) (to Lockheed Martin Idaho Technologies Company) United States Patent 5,993,660.
67. Kostarelos, K., Pope, G.A., Rouse, B.A. and Shook, G.M. (1998) A new concept: The use of neutrally-buoyant microemulsions for DNAPL remediation. *J. Contam. Hydrol.*, **34**(4), 383–397.
68. Ramsburg, C.A., Pennell, K.D., Kibbey, T.C.G. and Hayes, K.F. (2004) Refinement of the density-modified displacement method for efficient treatment of tetrachloroethene source zones. *J. Contam. Hydrol.*, **74**(1–4), 105–131.
69. Johnson, D.N., Pedit, J.A. and Miller, C.T. (2004) Efficient, near-complete removal of DNAPL from three-dimensional, heterogeneous porous media using a novel combination of treatment technologies. *Environ. Sci. Technol.*, **38**(19), 5149–5156.
70. Wu, B., Harwell, J.H., Sabatini, D.A. and Bailey, J.D. (2000) Alcohol-free diphenyl oxide disulfonate middle-phase microemulsion systems. *J. Surfactants Detergents*, **3**(4), 465–474.
71. Sabatini, D.A., Acosta, E. and Harwell, J.H. (2003) Linker molecules in surfactant mixtures. *Curr. Opin. Colloid Interface Sci.*, **8**(4–5), 316–326.
72. Graciaa, A., Lachaise, J., Cucuphat, C., Bourrel, M. and Salager, J.L. (1993) Improving solubilization in microemulsions with additives. 1. The lipophilic linker role. *Langmuir*, **9**(3), 669–672.
73. Zhou, M. and Rhue, R.D. (2000) Screening commercial surfactants suitable for remediating DNAPL source zones by solubilization. *Environ. Sci. Technol.*, **34**(10), 1985–1990.
74. Shook, G.M., Pope. G.A. and Kostarelos, K. (1998) Prediction and minimization of vertical migration of DNAPLS using surfactant enhanced aquifer remediation at neutral buoyancy. *J. Contam. Hydrol.*, **34**(4), 363–382.
75. Martel, R., Foy, S., Saumur, L., Roy, A., Lefebvre, R., Therrien, R., Gabriel, U. and Gélinas, P.J. (2005) Polychlorinated biphenyl (PCB) recovery under a building with an in situ technology using micellar solutions. *Can. Geotech. J.*, **42**(3), 932–948.
76. Childs, J., Acosta, E., Annable, M.D., Brooks, M.C., Enfield, C.G., Harwell, J.H., Hasegawa, M., Knox, R.C., Rao, P.S.C., Sabatini, D.A., Shiau, B., Szekeres, E. and Wood, A.L. (2006)

Field demonstration of surfactant-enhanced solubilization of DNAPL at Dover Air Force Base, Delaware. *J. Contam. Hydrol.*, **82**(1–2), 1–22.

77. Jawitz, J.W., Dai, D., Rao, P.S.C., Annable, M.D. and Rhue, R.D. (2003) Rate-limited solubilization of multicomponent nonaqueous-phase liquids by flushing with cosolvents and surfactants: Modeling data from laboratory and field experiments. *Environ. Sci. Technol.*, **37**(9), 1983–1991.
78. Wu, B., Shiau, B.-J., Sabatini, D.A. and Harwell, J.H. (2000) Formulating microemulsion systems for a weathered Jet fuel waste using surfactant/cosurfactant mixtures. *Separation Sci. Technol.*, **35**(12), 1917–1937.
79. Ramsburg, C.A., Penell, K.D., Abriola, L.M., Daniels, G., Drummond, C.D., Gamache, M., Hsu, H.-L., Petrovskis, E.A., Rathfelder, K.M., Ryder, J.L. and Yavaraski, T.P. (2005) Pilot-scale demonstration of surfactant-enhanced PCE solubilization at the Bachman road site. 2. System operation and evaluation. *Environ. Sci. Technol.*, **39**(6), 1791–1801.
80. Ramsburg, C.A., Abriola, L.M., Pennell, K.D., Löffler, F.E., Gamache, M., Amos, B.K. and Petrovskis, E.A. (2004) Stimulated microbial reductive dechlorination following surfactant treatment at the Bachman road site. *Environ. Sci. Technol.*, **38**(22), 5902–5914.
81. Holzmer, F.J., Pope, G.A. and Yeh, L. (2000) Surfactant-enhanced aquifer remediation of PCE-DNAPL in low permeability sands. In G.B. Wickramanayake, A.R. Gavaksar and N. Gupta (eds), *Treating Dense Nonaqueous Phase Liquids (DNAPLs): Remediation of Chlorinated and Recalcitrant Compounds.* Battelle Press, Columbus, OH, pp. 187–193.
82. Shiau, B.-J., Hasegawa, M.A., Brammer, J.M., Carter, T., Goodspeed, M., Harwell, J.H., Sabatini, D.A., Knox, R.C. and Szekeres, E. (2002) Field demonstration of surfactant-enhanced DNAPL remediation: Two case studies. In S.D. Warner and S.M. Henry (eds), *Chlorinated Solvent and DNAPL Remediation: Innovative Strategies for Subsurface Cleanup, ACS Symposium Series 837.* American Chemical Society, Washington, DC, pp. 51–72.
83. Hasegawa, M.H., Shiau, B.-J., Sabatini, D.A., Knox, R.C., Harwell, J.H., Lago, R. and Yeh, L. (2000) Surfactant-enhanced subsurface remediation of DNAPLs at the former naval air station Alameda, California. In G.B. Wickramanayake, A.R. Gavaksar and N. Gupta (eds), *Treating Dense Nonaqueous Phase Liquids (DNAPLs): Remediation of Chlorinated and Recalcitrant Compounds (C2-2).* Battelle Press, Columbus, OH, pp. 219–226.
84. Tellier, J., Sirvins, A., Gautier, J.-C. and Tramier, B. (1984) (to Societe Nationale Elf Aquitaine) United States Patent 4,460,692.
85. Atlas, R.M. (1995) Bioremediation of petroleum pollutants. *Int. Biodeterior. Biodegradation*, **35**(1–3), 317–327.
86. Borden, R.C. and Lee, M.D. (2002) (to Solutions Industrial & Environmental Services, Inc.) United States Patent 6,398,960.
87. Clemens, W., Haegel, F.H. and Schwuger, M.J. (1994) Concurrence of solubilization and adsorption in multicomponent suspensions of layer silicates. *Langmuir*, **10**(5), 1366–1369.
88. Haegel, F.-H., Clemens, W., Schwuger, M., Soeder, C.-J., Stickdorn, K. and Webb, L. (1998) (to Forschungszentrum Julich GmbH) United States Patent 5,833,756.
89. Zhao, B., Zhu, L. and Gao, Y. (2005) A novel solubilization of phenanthrene using Winsor I microemulsion-based sodium castor oil sulfate. *J. Hazard. Mater.*, **B119**(1–3), 205–211.
90. Dierkes, F., Haegel, F.-H. and Schwuger, M.J. (1998) Low-temperature microemulsions for the in situ extraction of contaminants from soil. *Colloids Surf. A*, **141**(2), 217–225.
91. Subklew, G., Groeneweg, J., Ulrich, J., Haegel, F.-H., Flucht, R., Fürst, L., Heinen, K., Höltkemeier, A., Poschen, L., Joel, T., Tiehm, A., Schell, H., Stieber, M., Kraßnitzer, S., Müller, A., Schäfer, D., Kohlmeier, E., Seitz, F., Schenk, T., Greiner, A., Dercq, M., Strunk, C., Barczewski, B., Stuhrmann, M., Memminger, B., Bisch, G., Fütterer, N., Patzelt, E., Tkachenko, G. and Philippin, R. (2005) Anwendung von Mikroemulsionen zur in-situ-Sanierung organischer

Untergrundkontaminationen. Projektverbund Mikroemulsion: Abschlussbericht der BMBF-Forschungsvorhaben 02WT0188-0191 [in German]. http://edok01.tib.uni-hannover.de/edoks/e01fb06/511006136.pdf.

92. Subklew, G., Barczewski, B., Kohlmeier, E. and Tiehm, A. (2003) Microemulsions – an efficient solution for the in situ-remediation of DNAPL-contaminated sites. General overview on the research and field test activities of the project network. In G.J. Annokkée, F. Arendt and O. Uhlmann (eds), *ConSoil 2003, Proceedings of the 8th International FZK/TNO Conference on Contaminated Soil*, Gent (Belgium), 12–16 May 2003 Forschungszentrum Karlsruhe, Wissenschaftliche Berichte, FZKA 6943, pp. 1817–1823.
93. Hales, S.G. (1993) Biodegradation of the anionic surfactant dialkyl sulfosuccinate. *Environ. Toxicol. Chem.*, **12**(10), 1821–1828.
94. Martel, R., Gélinas, P.J. and Desnoyers, J.E. (1998) Aquifer washing by micellar solutions: 1. Optimization of alcohol–surfactant–solvent solutions. *J. Contam. Hydrol.*, **29**(4), 319–346.
95. Clark, C.J., II (2005) Reduction of a field generated waste microemulsion by electrolytic addition. *J. Environ. Eng. Sci.*, **4**(1), 83–87.
96. Kitiyanan, B., O'Haver, J.H., Harwell, J.H. and Sabatini, D.A. (2000) The use of liquid–liquid-extraction in hollow fiber membrane for the removal of organic contaminants from aqueous surfactant streams. In J.F. Scamehorn and J.H. Harwell (eds), *Surfactant-Based Separations: Science and Technology, ACS Symposium Series 740.* American Chemical Society, Washington, DC, pp. 76–89.
97. Bonkhoff, K., Haegel, F.-H., Kowalski, S. and Subklew, G. (2000) (to Forschungszentrum Julich GmbH) United States Patent 6,063,281.
98. McMurtrey, R.D., Ginosar, D.M., Moor, K.S., Shook, G.M. and Barker, D.L. (2003) (to Bechtel BWXT Idaho, LLC) United States Patent 6,511,601.
99. Lipe, M., Sabatini, D.A., Hasegawa, M. and Harwell, J.H. (1996) Micellar-enhanced ultrafiltration and air stripping for surfactant-contaminant separation and surfactant reuse. *Ground Water Monit. Remed.*, **16**(1), 85–92.
100. Vane, L.M. and Alvarez, F.R. (2002) Full-scale vibrating pervaporation membrane unit: VOC removal from water and surfactant solutions. *J. Membr. Sci.*, **202**(1–2), 177–193.
101. Cheng, H.F. and Sabatini, D.A. (2007) Separation of organic compounds from surfactant solutions: A review. *Separation Sci. Technol.*, **42**(3), 453–475.
102. Fountain, J.C. (1997) The role of field trials in development and feasibility assessment of surfactant-enhanced aquifer remediation. *Water Environ. Res.*, **69**(2), 188–195.
103. Ouyang, Y., Cho, J.S. and Mansell, R.S. (2002) Simulated formation and flow of microemulsions during surfactant flushing of contaminated soil. *Water Res.*, **36**(1), 33–40.
104. Zhang, R., Wood, A.L., Enfield, C.G. and Jeong, S.-W. (2003) Stochastical analysis of surfactant-enhanced remediation of denser-than-water nonaqueous phase liquid (DNAPL)-contaminated soils. *J. Environ. Qual.*, **32**(3), 957–965.
105. LaForce, T. and Johns, R.T. (2005) Analytical solutions for surfactant-enhanced remediation of nonaqueous phase liquids. *Water Resour. Res.*, **41**(10), W10420.
106. Naudet, V., Revil, A., Rizzo, E., Bottero, J.Y. and Begassat, P. (2004) Groundwater redox conditions and conductivity in a contaminant plume from geoelectrical investigations. *Hydrol. Earth Syst. Sci.*, **8**(1), 8–22.
107. Kemna, A., Binley, A. and Slater, L. (2004) Crosshole IP imaging for engineering and environmental applications. *Geophysics*, **69**(1), 97–107.
108. Fan, L., Yokota, K. and Fujimoto, K. (1992) Supercritical phase Fischer–Tropsch synthesis: Catalyst pore-size effect. *AIChE J.*, **38**(10), 1639–1648.
109. Shah, D.O. (1981) Fundamental aspects of surfactant–polymer flooding process. In *The European Symposium on Enhanced Oil Recovery*, Bournemouth, England, 21–23 September 1981. Elsevier Sequoia S. A., Losanne, Switzerland, pp. 1–40.

110. Sharma, R.C., Pithapurwala A.K. and Shah, D.O. (1983) Selection criteria and formulation of surfactant slug for high temperature, moderate salinity reservoir conditions, SPE11772. Paper Presented at the *SPE International Symposium on Oilfield and Geothermal Chemistry*, Denvel, CO, June 1983.
111. Pithapurwala, A.K., Sharma, R.C. and Shah, D.O. (1986) Effect of salinity and alcohol partitioning on phase-behavior and oil displacement efficiency in surfactant–polymer flooding. *J. Am. Oil Chem. Soc.*, **63**(6), 804–813.
112. Lake, L.W. (1991) *Enhanced Oil Recovery*. Prentice-Hall, Englewood Cliffs, NJ.
113. Li, G.Z., Zhai, L.M., Xu, G.Y., Shen, Q., Mao, H.Z. and Pei, M.S. (2000) Current tertiary oil recovery in China. *J. Disp. Sci. Technol.*, **21**(4), 367–408.
114. Baviere, M. and Canselier, J.P. (1997) Microemulsions in the chemical EOR process. In C. Solans and H. Kunieda (eds), *Industrial Applications of Microemulsions, Surfactant Science Series 66*. Marcel Dekker, New York, pp. 331–353.
115. Reed, R.L. and Healy, R.N. (1977) Some physicochemical aspects of microemulsion flooding: A review. In D.O. Shah and R.S. Schechter (eds), *Improved Oil Recovery by Surfactant and Polymer Flooding*. Academic Press, New York, pp. 383–437.
116. Lake, L.W., Schmidt, R.L. and Venuto, P.B. (1992) A niche for enhanced oil recovery in the 1990s. *Oil Field Rev. Electron. Arch.*, **4**(1), 55–61.
117. Kessel, D.G. (1989) Chemical flooding – status report. *J. Pet. Sci. Eng.*, **2**(2–3), 81–101.
118. Stegemeier, G.L. (1977) Mechanisms of entrapment and mobilization of oil in porous media. In D.O. Shah and R.S. Schechter (eds), *Improved Oil Recovery by Surfactant and Polymer Flooding*. Academic Press, New York, pp. 55–91.
119. Wilson, L. A., Jr. (1977) Physico-chemical environment of petroleum reservoirs in relation to oil recovery system. In D.O. Shah and R.S. Schechter (eds), *Improved Oil Recovery by Surfactant and Polymer Flooding*. Academic Press, New York, pp. 1–26.
120. Cash, L., Cayias, J.L., Fournier, G., MacAllister, D., Schares, T., Schechter, R.S. and Wade, W.H. (1977) Application of low interfacial tension scaling rules to binary hydrocarbon mixtures. *J. Colloid Interface Sci.*, **59**(1), 39–44.
121. Cayias, J.L., Schechter, R.S. and Wade W.H. (1977) Utilization of petroleum sulfonates for producing low interfacial tensions between hydrocarbons and water. *J. Colloid Interface Sci.*, **59**(1), 31–38.
122. Doe, P.H., Wade, W.H. and Schechter, R.S. (1977) Alkyl benzene sulfonates for producing low interfacial tensions between hydrocarbons and water. *J. Colloid Interface Sci.*, **59**(3), 525–531.
123. Bourrel, M., Salager, J.L., Lipow, A.M., Wade, W.H. and Schechter, R.S. (1978) Properties of amphiphile/oil/water systems at an optimum formulation for phase behavior. SPE 7450 Presented at the *53rd Annual Fall Technical Conference and Exhibition of the SPE of AIME*, Houston, TX, October 1–3.
124. Wade, W.H., Morgan, J.C., Schechter, R.S., Jacobson, J.K. and Salager, J.L. (1977) Interfacial tension and phase behavior of surfactant systems. SPE 6844 Presented at the *SPE-AIME 52nd Annual Fall Technical Conference and Exhibition*, Denver, 9–12 October.
125. Hayes, M., El-Emary, M., Schechter, R.S. and Wade, W.H. (1979) Relation between the $EACN_{min}$ concept and surfactant HLB. *J. Colloid Interface Sci.*, **68**(3), 591–592.
126. Hayes, M., Bourrel, M., El-Emary, M., Schechter, R.S. and Wade, W.H. (1979) Interfacial tension and behavior of nonionic surfactants. *Soc. Pet. Eng. J.*, **19**(6), 349–356.
127. Salager, J.L., Bourrel, M., Schechter, R.S. and Wade, W.H. (1979) Mixing rules for optimum phase-behavior formulations of surfactant/oil/water systems. *Soc. Pet. Eng. J.*, **19**(5), 271–278.
128. Salager, J.L., Morgan, J.C., Schechter, R.S., Wade, W.H. and Vasquez, E. (1979) Optimum formulation of surfactant/water/oil systems for minimum interfacial tension or phase behavior. *Soc. Pet. Eng. J.*, **19**(2), 107–115.

129. Bourrel, M., Salager, J.L., Schechter, R.S. and Wade, W.H. (1980) A correlation for phase behavior of nonionic surfactants. *J Colloid Interface Sci.*, **75**(2), 451–461.
130. Baviere, M., Schechter, R. and Wade, W. (1981) The influence of alcohols on microemulsion composition. *J. Colloid Interface Sci.*, **81**(1), 266–279.
131. Bourrel, M., Chambu, C., Schechter, R.S. and Wade, W.H. (1982) The topology of phase-boundaries for oil/brine/surfactant systems and its relationship to oil recovery. *Soc. Pet. Eng. J.*, **22**(1), 28–36.
132. Fotland, P. and Skauge, A. (1986) Ultralow interfacial tension as a function of pressure. *J. Disp. Sci. Technol.*, **7**(5), 563–579.
133. Labrid, J. and Bazin, B. (1989) Alkaline preflush in a low-permeability clayey sandstone. *J. Pet. Sci. Eng.*, **3**(1–2), 111–120.
134. Menzie, D.E., Dutta, S., Shadizadeh, R. and Malik, N. (1989) A method of measuring dispersivity and its use in slug size estimation in a miscible flooding process. *J. Pet. Sci. Eng.*, **3**(1–2), 35–45.
135. Shiyi, Y. and Van Quy, N. (1989) Effects of the main parameters of the chemical flooding process. *J. Pet. Sci. Eng.*, **3**(1–2), 85–96.
136. Yang, C.Z., Jao, W.L. and Huang, Y.H. (1989) The mechanism of adsorption and a new method to reduce surfactant loss in chemical flooding. *J. Pet. Sci. Eng.*, **3**(1–2), 97–109.
137. Morgan J.C., Schechter, R.S. and Wade, W.H. (1977) Recent advances in the study of low interfacial tensions. In D.O. Shah and R.S. Schechter (eds), *Improved Oil Recovery by Surfactant and Polymer Flooding*. Academic Press, New York, pp. 101–118.
138. Gogarty, W.B. (1977) Oil recovery with surfactants: History and a current appraisal. In D.O. Shah and R.S. Schechter (eds), *Improved Oil Recovery by Surfactant and Polymer Flooding*. Academic Press, New York, pp. 27–54.
139. Tielon, C., Zhengyu, S., Fan, Y., Changzhong, H., Ling Q., Jinxing, T. and Queste, S. (1997) A pilot test of polymer flooding in an elevated-temperature reservoir. SPE 36708 First Presented at the 1196 SPE Annual Technical Conference and Exhibition, Denver, Colorado, 6–9 October.
140. Demin, W., Zhenhua, Z., Jiecheng, C., Jingchun, Y., Shutang, G. and Lin, L. (1997) A pilot test of alkaline/surfactant/polymer flooding in daging oil field. SPE 36748 First Presented at the 1196 SPE Annual Technical Conference and Exhibition, Denver, Colorado, 6–9 October.
141. Baviere, M. (1991) *Basic Concepts in Enhanced Oil Recovery Processes, Critical reviews on Applied Chemistry 33*. Elsevier Applied Science, London.
142. Kanicky J.R., Lopez-Montilla, J.C., Pandey, S. and Shah, D.O. (2001) Surface chemistry in the petroleum industry. In K. Holmberg (ed), *Handbook of Applied Surface and Colloid Chemistry*. John Wiley & Sons, New York, pp. 251–267.
143. Salager, J.L. and Antón, R.E. (1999) Ionic microemulsions. In P. Kumar and K.L. Mittal (eds), *Handbook of Microemulsion Science and Technology*. Marcel Dekker, New York, pp. 247–280.
144. Salager, J.L. (1999) Microemulsions. In G. Broze (ed), *Handbook of Detergents, Part A: Properties, Surfactant Science Series 82*. Marcel Dekker, New York, pp. 253–302.
145. Salager, J.L., Antón, R., Andérez, J.M. and Aubry, J.-M. (2001) Formulation des Microémulsions par la Méthode du HLD. *Techniques de L'Ingénieur*, **J2**(157), 1–20 [in French].
146. Scorzza, C., Gode, P., Martin, P., Miñana-Pérez, M., Salager, J.L., Villa, P. and Goethals, G. (2002) Synthesis and surfactant properties of a new 'Extended' glucidoamphiphile made from D-glucose. *J. Surfactants Detergents*, **5**(4), 331–335.
147. Fernandez, A., Scorzza, C., Usubillaga, A. and Salager, J.L. (2005) Synthesis of new extended surfactants containing carboxylate or sulfate polar group. *J. Surfactants Detergents*, **8**(2), 187–191.
148. Fernandez, A., Scorzza, C., Usubillaga,, A. and Salager, J.L. (2005) Synthesis of new extended surfactants derived from a Xylitol Polar Group. *J. Surfactants Detergents*, **8**(2), 193–198.

149. Skauge, A. and Fotland, P. (1990) Effect of pressure and temperature on the phase behavior. *SPE Reservoir Eng.*, **5**, 601–608.
150. Miller, D.J., von Halasz, S.-P., Schmidt, M., Holst, A. and Pusch, G. (1991) Dual surfactant systems for enhanced oil recovery at high salinities. *J. Pet. Sci. Eng.*, **6**(1), 63–72.
151. Agharazi-Dormani, N., Hornof, V. and Neale, G.H. (1990) Effects of divalent ions in surfactant flooding. *J. Pet. Sci. Eng.*, **4**(3), 189–196.
152. Barker, J.W. (1992) Alternatives to the salinity gradient for controlling the effects of dispersion in surfactant floods. *J. Pet. Sci. Eng.*, **7**(1–2), 139–154.
153. López-Montilla, J.C., James, M.A., Crisalle, D.O. and Shah, D.O. (2005). Surfactants and protocols to induce spontaneous emulsification and enhance detergency. *J. Surfactants Detergents*, **8**(1), 45–53.
154. Zerpa, L., Queipo, N.V., Pintos, S. and Salager J.L. (2005) An optimization methodology of alkaline-surfactant–polymer flooding process using field scale numerical simulation and multiple surrogates. *J. Pet. Sci. Eng.*, **47**(3–4), 197–208.
155. Rivas, H., Gutiérrez, X., Ziritt, J.L., Antón, R.E. and Salager, J.L. (1997) Microemulsion and optimal formulation occurrence in pH-dependent systems as found in alkaline-enhanced oil recovery. In C. Solans and H. Kunieda (eds), *Industrial Applications of Microemulsions, Surfactant Science Series 66.* Marcel Dekker, New York, pp. 305–329.
156. Carrero, E., Queipo, N., Pintos, S. and Zerpa, L. (2007) Global sensitivity analysis of alkaline-surfactant–polymer enhanced recovery processes. *J. Pet. Sci. Eng.*, **58**(1–2), 30–42.
157. Shah, D.O. (2007) *Unpublished EOR Research.* CSSE lab at University of Florida, Gainesville.
158. Kahlweit, M. and Strey, R. (1985) Phase-behavior of ternary systems of the type H_2O–oil–nonionic amphiphile (microemulsions). Angew. Chem. Int. Ed. Engl., **24**(8), 654–668.
159. Shinoda, K. (1967) Solvent properties of nonionic surfactants in aqueous solutions. In K. Shinoda (ed), *Solvent Properties of Surfactant Solutions, Surfactant Science Series 2.* Marcel Dekker, New York, pp. 27–63.
160. Benson, H.L., Cox, K.R. and Zweig, J.E. (1985) Nonionic-based detergent systems for cold water cleaning. *Soap Cosmet. Chem. Specialties*, **61**(3), 35–47.
161. Kahlweit, M. and Strey, R. (1987) The phase behavior of H_2O–oil–nonionic amphiphile ternary systems In H.L. Rosano and M. Clausse (eds), *Microemulsion Systems, Surfactant Science Series 24.* Marcel Dekker, New York, pp. 1–13.
162. Engelskirchen, S., Elsner, N., Sottmann, T. and Strey, R. (2007) Triacylglycerol microemulsions stabilized by alkyl ethoxylate surfactants – A basic study: Phase behavior, interfacial tension and microstructure. *J. Colloid Interface Sci.*, **312**(1), 114–121.
163. Burauer, S., Sachert, T., Sottmann, T. and Strey, R. (1999) On microemulsion phase behavior and the monomeric solubility of surfactant. *Phys. Chem. Chem. Phys.*, **1**(18), 4299–4306.
164. Strey, R. (1994) Microemulsion microstructure and interfacial curvature. *Colloid Polymer Sci.*, **272**(8), 1005–1019.
165. Leitão, H., Somoza, A.M., da Gama, M.M.T., Sottmann, T. and Strey, R. (1996) Scaling of the interfacial tension of microemulsions: A phenomenological description. *J. Chem. Phys.*, **105**(7), 2875–2883.
166. Kahlweit, M., Lessner, E. and Strey, R. (1984) Phase behavior of quaternary systems of the type H_2O–oil–nonionic surfactant–inorganic electrolyte. 2. *J. Phys. Chem.*, **88**(10), 1937–1944.
167. Kahlweit, M., Strey, R. and Haase, D. (1985) Phase behavior of multicomponent systems. Water–oil–amphiphile–electrolyte. 3. *J. Phys. Chem.*, **89**(1), 163–171.
168. Schubert, K.V., Busse, G., Strey, R. and Kahlweit, M. (1993) Microemulsions with formamide as polar solvent. *J. Phys. Chem.* **97**(1), 248–254.
169. Schubert, K.-V. and Strey, R. (1991) Small-angle neutron scattering from microemulsions near the disorder line in water/formamide–octane–C_iE_j systems. *J. Chem. Phys.*, **95**(11), 8532–8545.

170. Pabst, G., Lamalle, P., Füngerlings, T., Oetter, G., Erhardt, R., Scherr, G., Strey, R. Sottmann, T. and Engelskirchen, S. (2004) European Patent Application WO2004063396.

171. Herfeld, H. (1987) Entfetten, Fetten und Hydrophobierung bei der Lederherstellung. In H. Herfeld (ed), *Bibliothek des Leders*, Vol. 4, Umschau Verlag, Frankfurt/M [in German], pp. 50–58, 116–137 and 198–199.

Chapter 11
Future Challenges

Cosima Stubenrauch and Reinhard Strey

11.1 Introduction

Reading this book, one might gain the impression that most of the 'microemulsion mysteries' have been solved over the years and that applying microemulsions in fields other than those mentioned in the book is just a question of 'creative thinking'. Unfortunately, or indeed fortunately, that is not the case! What we seek to summarise in this chapter are those applications that are products of our 'creative thinking' but have proved difficult to be put into practice. Although the four examples given below are 'high-risk' projects which might never be successful, their tremendous potential makes the risk worth taking. We would like to point out that the topics covered in this chapter do not comprise the whole range of possible challenging projects to be addressed in the future. What we left out in this overview is non-invasive external stimuli such as electromagnetic fields or simply light despite the fact that these are very promising parameters for tuning the properties of microemulsions and might even be superior to changes in composition, temperature or pressure. For example, a photo-induced phase separation [1] as well as a photo-induced solubilisation [2] has been reported, and a detailed review of the use of photosensitive surfactants was published only recently [3]. However, specific applications in which an external stimulus is used to control the system have not been reported yet. That is why we confined ourselves to those examples where concrete applications were the driving force for new experimental procedures.

11.2 Bicontinuous microemulsions as templates

11.2.1 Why use bicontinuous microemulsions as templates?

As was described in Chapter 1, microemulsions are structured on a nanometre scale with domain sizes of 5–50 nm, which corresponds to surface areas of 300–30 $m^2\ g^{-1}$. Moreover, a microemulsion can have various microstructures, ranging from discrete particles of one phase dispersed in the second (oil droplets in water or vice versa) to a bicontinuous structure consisting of two equal subphases (oil and water). Thus, using microemulsions as templates should lead to a material with a high surface area and a structure equal to the structure of the template. This material would be suitable for a range of applications that require rapid and responsive sampling, selective separations, high throughput catalytic processing or enhanced chemical activity. One popular example of a high surface area material is metallic nanoparticles (see Chapter 6) which are used in cancer treatment, catalysis and optoelectronics. The 'organic' counterpart of metallic nanoparticles, i.e. latex nanoparticles, is used for coating the internal surface of microporous materials and as seed particles for emulsion polymerisations [4]. In both cases droplet microemulsions can be used as templates (see [5, 6] for the synthesis of metallic and [4, 7] for the synthesis of latex nanoparticles, respectively). The successful use of droplet microemulsions as templates for metallic and latex nanoparticles automatically suggests the use of the bicontinuous structure as template which is expected to lead to nanoporous materials with a continuous internal surface. It goes without saying that there are countless potential applications of nanoporous, high surface area materials. The enormous need for new materials in general, and for high surface area materials in particular, is the motivation for this research area. Two examples of the application of high surface area polymers may illustrate the enormous potential.

Firstly, semiconducting high surface area polymers could be used in bulk heterojunction photovoltaic (PV) cells [8]. Four different types of bulk heterojunction PV cells are known, one of which is the polymer–titania PV cell which consists of sintered TiO_2 nanocrystals filled with a conjugated, semiconductive polymer. However, the conjugated polymer cannot fully infiltrate into the TiO_2 matrix, which leads to a significant loss in efficiency. In the current PV cells there are too many large (50–100 nm) domains of empty space, which means that there is no sufficient external quantum efficiency (EQE = number of electrons produced by a PV cell for each incident electron). The main goal is to have a continuous TiO_2 and a continuous polymer phase in the PV cell, i.e. a bicontinuous structure, with an interfacial area between the two semiconductors as large as possible. To achieve this aim the use of bicontinuous microemulsions as templates sounds promising. The improved polymer–titania material could be made by first synthesising the semiconducting, high surface area polymer and then filling it with a titania precursor which is of low viscosity and will therefore easily infiltrate the polymer. The hydrolysis of this precursor would lead to TiO_2 and thus to a truly bicontinuous material with a large internal surface area.

A second promising area is the controlled release of drugs in pharmaceutical applications for which today responsive macro- or microporous polymers are used. Having access to responsive nanoporous polymers with a high surface area would allow us to accelerate the response significantly. A prominent example of such a polymer is poly-*N*-isopropyl-acrylamide (p-NIPAm). The great advantage of p-NIPAm is the fact that it is a water-soluble polymer and can thus be used in medical and pharmaceutical applications. Additionally, it is extremely versatile as it can be easily modified with a wide variety of functional groups, thus making it responsive to a large number of external stimuli [9]. In response to environmental changes (e.g. temperature, pH, light, electrical current) a coil-globule phase transition takes place, which results in dramatic changes in both volume and surface area. In a nanoporous p-NIPAM sponge this would result in a swelling–shrinking transition. The responsiveness could be used to alter the spacing between the surface groups, and to act as an 'on-off' switch for surface processes that require intimate contact between neighbouring species. With biologically active surface groups we would automatically be in the arena of biomimetic materials.

11.2.2 What are the challenges?

Trying to replicate microemulsions via polymerisation, one faces a number of major problems. Firstly, the surfactant monolayers are not rigid enough to preserve the original shape during polymerisation. Secondly, structural changes in a microemulsion are much faster (~1 μs) than the polymerisation (~1 ms per step). In other words, the template structure changes continuously during the polymerisation because the system has enough time to respond to the compositional and volume changes caused by the consumption of monomer [7]. In the worst possible case a phase separation occurs during the polymerisation, while in the best case a polymer with structures much larger than those of the original template (i.e. the microemulsion) is obtained. A prominent example of the 'best case' is the polymerisation of oil-in-water droplet microemulsions to obtain latex nanoparticles. Usually, the resulting particles are 5–10 times larger than the templating microemulsions ([7] and references therein). However, what is only 5–10 times for the particles, is ~50 times for the polymerisation of bicontinuous microemulsions which has so far merely resulted in macroporous gels with small surface areas [7, 10]. In other words, polymerising microemulsions with domain sizes of ~10–50 nm leads to polymers with pore sizes of 0.5–2.5 μm.

How can we address this problem? As already mentioned, the problem is mainly due to the different timescales of structural changes in a microemulsion on the one hand, and to the polymerisation on the other. Thus, the challenge is to slow down or even 'arrest' the structural changes during the polymerisation. One way to 'capture' the template's structure is the use of polymerisable surfactants. In this case the timescales of structural changes and polymerisation are expected to be similar as it is the surfactant itself that is polymerised. Mixtures of polymerisable surfactants were used to template micelles, lyotropic mesophases, and water-in-oil microemulsions and it was shown that the structure of the templates was retained after polymerisation [11]. However, this approach cannot be used for the replication of a bicontinuous microemulsion. In this case not only the separating surfactant layer but also one of the subphases (oil or aqueous phase) needs to be polymerised. Another route towards templating the bicontinuous microstructure is the direct use of polymers [7, 12]. It is possible to obtain bicontinuous microemulsions with ternary polymer blends which consist of two homopolymers and the corresponding diblock copolymer [12]. The microstructure can be tuned by the molecular weights of the polymers and the composition. What we can learn from this study is that polymers do indeed form a bicontinuous structure. The approach, however, is useless if one wants to synthesise a nanoporous material, i.e. a material consisting of one type of polymer only. What we are still looking for is a route that allows us to (a) 'arrest' one of two continuous subphases, (b) polymerise the other continuous subphase and (c) remove the 'arrested' subphase.

The only successful one-to-one replication of a bicontinuous microemulsion to date was achieved by using bicontinuous microemulsion glasses as templates [13, 14]. The microemulsion glass was obtained by replacing water with sugar. This sugar-based microemulsion contains the liquid monomer whose photopolymerisation led to an almost one-to-one replica of the template [14]. The great advantage of this route is the fact that the original microemulsion glass can easily be removed after the polymerisation by simple dissolution of the sugar template. Moreover, only easily recyclable components and no organic solvents are involved in the process. However, there are also three important disadvantages. First, the technique is restricted to the synthesis of oil-soluble polymers only. In other words, we still need a route via which a glass-like oil phase can be obtained which, in turn, serves as a template for the polymerisation of the aqueous phase. As mentioned above, it is especially water-soluble polymers with a high surface area that are desirable for medical and pharmaceutical applications. Second, a nearly water-free polar phase is needed which requires a complicated dehydration procedure. Third, due to the glassy and thus highly viscous state of the polar phase studying the phase behaviour of the respective microemulsions is very time-consuming. The authors themselves state that 'the added complexity of the dehydration and photopolymerisation steps is such that this approach may be suitable only for the preparation specialty polymeric membranes for which no suitable solvents are available'.

11.2.3 What route is the most promising?

The final aim is to provide a synthetic route for both oil- and water-soluble polymers with surface areas that are as large as possible. Once this route is found, one can start tuning the properties of the resulting polymers according to specific needs. While the general route is clear, namely formulating a bicontinuous microemulsion in which the templating phase is in an 'arrested' state while the polymerisation takes place in the remaining low viscous phase, the experimental details are not known yet. One experimental approach was given above, namely arresting the aqueous phase by replacing water with sugar and polymerising the oil phase [13, 14]. However, due to the limitations of this procedure we consider a different route to be the most promising. Instead of using a microemulsion glass a microemulsion gel is aimed to be used as template [15]. In other words, the water (oil) phase is arrested via gellification and not via glassification, thereby producing a hydrogel (an organogel). This approach has three advantages over the use of microemulsion glasses. First, it can be used for the synthesis of both oil- and water-soluble polymers. It is only an appropriate gelator for the oil and the water phase, respectively, that needs to be found in the first place. Second, only 1–4wt.% of a gelator would be necessary, thus avoiding the complicated dehydration procedure required in the case of the sugar template mentioned above. Third, as will be shown below, the phase behaviour of the templating microemulsion is much easier to study especially at temperatures above the sol–gel transition.

The general steps of the proposed templating route are as follows: (a) finding a suitable ternary base system, (b) replacing water (oil) by a polymerisable aqueous (oil) phase, (c) arresting the oil (aqueous) phase via the formation of an organogel (a hydrogel), (d) polymerising the aqueous (oil) phase and (e) removing the organogel (hydrogel). While the first three steps require intensive phase studies, the last two are not difficult from an experimental point of view. However, the characterisation of the templating gelled microemulsion and of the resulting polymer is challenging. In the following, first telling results towards the synthesis of water-soluble polymer gels [15] and a 'road map' for future studies are presented.

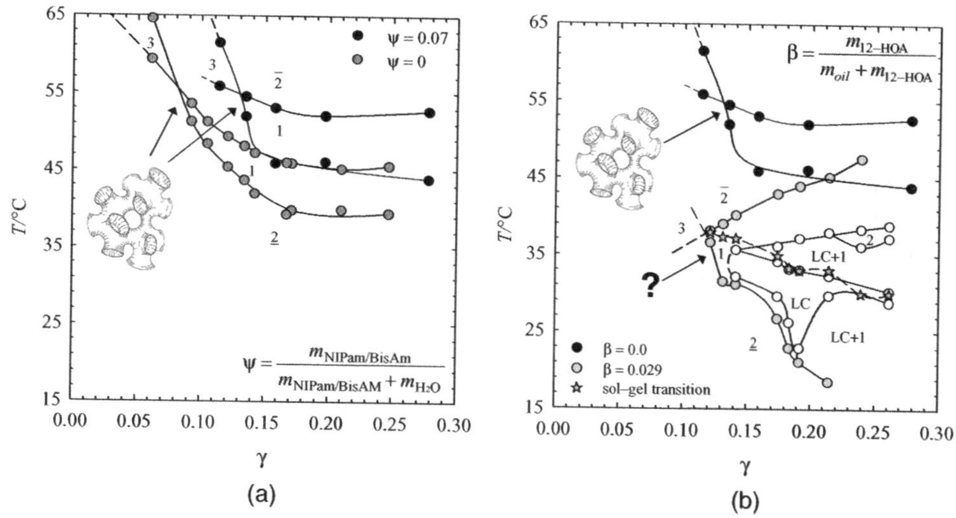

Figure 11.1 'Fish diagrams' of the systems H_2O–*n*-dodecane-$C_{13/15}E_5$ ((a), dark grey circles), H_2O/NIPAm/BisAm–*n*-dodecane–$C_{13/15}E_5$ ((a) and (b), black circles), H_2O/NIPAm/BisAm–*n*-dodecane/12-HOA–$C_{13/15}E_5$ ((b), grey and white circles). All diagrams were measured at $\alpha = 0.5$. The NIPAm + BisAm content in the aqueous phase was 7 wt.% ($\psi = 0.07$) and the gelator content in the oil phase 2.9 wt.% ($\beta = 0.029$). The distorted shape of the phase diagram is due to the use of a technical grade surfactant. Data are taken from [15].

11.2.3.1 *First results*

A suitable ternary base system was found, namely H_2O–*n*-dodecane–$C_{13/15}E_5$ where the surfactant is of technical grade (LutensolAO5®, BASF = non-ionic alkyl pentaethyleneglycol ether with an average chain length of 14C-atoms) [15]. The phase diagrams, one is primarily interested in, are the so-called fish diagrams (Chapter 1), i.e. phase diagrams measured at equal amounts of water and oil ($\alpha = 0.5$) as a function of the total surfactant concentration γ and the temperature T (see Fig 11.1). At low T an o/w-microemulsion coexists with an excess oil phase ($\underline{2}$), while at high T a w/o- microemulsion coexists with an excess water phase ($\overline{2}$). At intermediate T and low γ a bicontinuous microemulsion coexists with water and oil excess phases (3), while at high γ a one-phase bicontinuous microemulsion is formed (1). It is the structure of this microemulsion (schematic drawing in Fig. 11.1) that is the targeted polymer structure.

After measuring the ternary base system (Fig. 11.1(a)), water was replaced by an aqueous solution of the monomer *N*-isopropylacrylamide (NIPAm) and the cross-linker *N,N'*-methylenebisacrylamide (BisAm) and the fish diagram of the new microemulsion system (NIPAm + BisAm = 7 wt.%) was studied. As is seen in Fig. 11.1(a), at $\psi = 0.07$ the phase diagram is shifted towards higher T and higher γ compared to $\psi = 0$. The next step was to arrest the oil phase, i.e. to make an organogel. The challenge was to find a suitable gelator in the first place as the resulting organogel has to be clear to carry out reliable phase studies (most organogels are turbid) and as the addition of the gelator should not destroy the microemulsion (which was the case using block copolymers as gelators).

The gelator 12-hydroxyoctadecanoic acid (12-HOA) fulfils these requirements. Adding 2.9 wt.% of 12-HOA to n^{-}dodecane leads to the phase diagram seen in Fig. 11.1(b). Three features are important. First, the location of the 'fish' and thus of the targeted bicontinuous structure is again significantly affected, which shows that phase studies are indispensable. Second, the sol–gel transition takes place at temperatures between the $\overline{2}$-1 and the 1-$\overline{2}$ phase boundaries. Thus, a one-phase region is accessible that is accessible which indeed is a bicontinuous microemulsion with the same structure as the non-gelled microemulsion [15b]. Third, most of the phase diagram consists of an anisotropic liquid crystalline phase (LC), which substantially limits the γ- and T-range of the isotropic one-phase region – the target!

11.2.3.2 'Road map' for future studies

Phase diagrams

In order to increase the γ- and T-range of the isotropic one-phase region and thus the range suitable for the polymerisation alternative gelators need to be tested. As 12-HOA forms crystalline fibrils and nodes in organogels [16], it is very likely that the presence of 12-HOA induces the formation of LC phases. Thus, a gelator that has no or a lower tendency to form crystalline microdomains needs to be found. Once this is found, phase studies will have to be carried out. If it turns out that this is not feasible, polymerising in the LC region could be a second option as the resulting polymer should also have a high surface area. This, however, is not the primary goal. In any case, extensive phase studies are still required.

Synthesis and characterisation of high surface area polymers

Polymers could be synthesised via photo-polymerisation. After the polymerisation the templating microemulsion gel can be removed simply by raising the temperature above the sol–gel transition: the gel will be destroyed and the gelator-containing phase can be removed by washing the sample, for example, with ethanol. Before and after the polymerisation, studies of the microstructure are indispensable. Has the gelled sample still a bicontinuous structure? In case it does, is the microemulsion gel able to arrest the structure during the polymerisation? The complexity of the template structure arises due to the fact that a microemulsion and a gel are structured on different length scales. While the domain sizes of the former are 10–50 nm, the mesh sizes of gels are in the μm range. To directly image the structure of the templating microemulsion gel transmission electron microscopy (TEM) is the method of choice. However, images have not been provided yet. Complementary techniques to monitor the bicontinuity of the template are NMR self-diffusion measurements, conductivity measurements and small-angle neutron scattering (SANS), which were applied successfully for studying gelled microemulsions only recently [17]. To study the structure of the resulting polymers TEM and NMR self-diffusion measurements [18] can also be used. In addition, the surface area and the pore size distribution need to be determined, e.g. by measuring BET isotherms or by using a solute–exclusion technique using fluorescein isothiocyanate–dextran fractions (which are available in sizes from 2 to 10 nm) as molecular probes [19]. Combining the results obtained by these techniques will allow one to draw a picture of the template and the polymer structure, respectively.

11.3 Nanofoams

11.3.1 Why synthesise nanofoams?

There is a huge variety of macroscopic foams which we use or encounter in our daily life, ranging from bath foam, household sponges, cushions, insulating material in freezers to foamed packaging material. Nanoscopic foams, however, do not exist yet! Examples of nanoporous materials that do exist are aerogels which have interesting mechanical and optical properties. They can be extremely light, optically transparent and thermally extremely well insulating. However, they are currently made by a sol–gel process from SiO_2 with a subsequent energy-intensive drying procedure. The nanostructure of aerogels is reminiscent of open cell foam structures. Therefore, the question arises whether such a foam structure can be obtained via an alternative, less expensive route with various starting materials. In case it can, the resulting nanoporous materials are expected to have properties as interesting as those of aerogels.

11.3.2 What are the challenges?

Is it possible to make a foam that is made of polymeric material and that has a typical bubble size of a few nanometres? Inspection of the respective scientific and patent literature reveals that such foams do not yet exist. Several years of research in cooperation with companies active in foam production have led to the conclusion that all technically viable routes to polymer foams are based on an uncontrolled way of producing bubbles in a polymeric matrix or a polymerising matrix, thus leading to a non-defined foam structure.

11.3.3 What route is the most promising?

An idea worth pursuing is to utilise the well-understood microemulsion structure as a starting material. If we formulate a droplet microemulsion with a diameter of 10 nm and with a number density of 10^{18} droplets per cc and if we then imagine forming (by a process to be discussed) a bubble around each droplet, we will end up with a foam containing around 10^{18} bubbles per cc. Such a foam would be highly reproducible as it is based on the self-assembly of a spontaneously forming microemulsion – a thermodynamic stable phase. At this point the idea came up to make the interior of the microemulsion droplet from a compressed supercritical fluid. We called it the *Principle of Supercritical Microemulsion Expansion* (POSME). The principal experimental procedure [20], which is illustrated in Fig. 11.2, is straightforward. First, a micellar surfactant-in-water solution is prepared. Then a condensable gas at supercritical conditions $T > T_c$ and at ambient pressure $p = 1$ bar is added. After that the container is closed by a piston (see Fig. 11.2(a)). Compressed to $p > p_c$, the gas attains a liquid-like density and – if the surfactant is properly chosen – an o/w microemulsion forms (see Fig. 11.2(b)). Releasing the pressure the gas inside the droplets gradually expands until at pressures near ambient a foam is formed (see Fig. 11.2(c)). Of

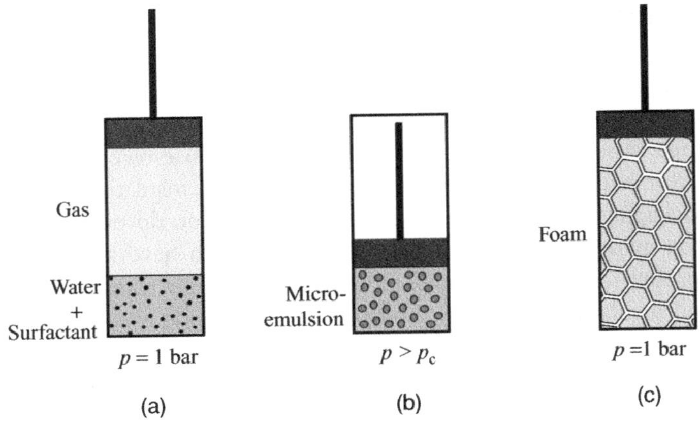

Figure 11.2 Principle of supercritical microemulsion expansion (POSME) [20].

course such a foam would not be stable, because it contains only water. But to simplify the systematic characterisations of such gas-microemulsions one would start with water as hydrophilic part of the system. The last step would be to stepwisely replace the water by a hydrophilic monomer or a cross-linkable polymer. Before or during the last step of POSME the polymerisation or cross-linking of the hydrophilic part of the microemulsion has to take place.

11.3.3.1 First results

Examples of compressible supercritical gases are ethane, propane or CO_2. Using technical grade non-ionic surfactants microemulsions are easily obtained in pressure tight test tubes. In Fig. 11.3, four-phase diagrams of the system water/NaCl–propane/CO_2–Lutensol® XL70 are shown. These phase diagrams are sections through the phase prism at a 1:1 water-to-oil volume ratio (see Chapter 1 for further details). The measurements were carried out at $p = 220$ bar as a function of the temperature T and the total surfactant concentration γ. The aqueous phase contained 5 wt.% of NaCl to minimise spurious ionic effects.

The propane system shown in Fig. 11.3 is clearly subcritical as the critical temperature of propane is about 96°C. An increase of the CO_2 fraction (β) in the mixture of CO_2 and propane shifts the one-phase region (1), i.e. the bicontinuous microemulsion, to lower temperatures. For pure CO_2 the bicontinuous microemulsion (1) exists around 35°C, which is higher than the $T_c = 31$°C of CO_2. In other words, the CO_2 solubilised in the microemulsion is supercritical! Knowing how to tune the phase behaviour of these systems, one can easily shift phase diagrams on the temperature scale by simply choosing an appropriate surfactant. Other tuning parameters are the oil-to-water fraction and the temperature which may be adjusted such that, e.g. a CO_2-in-water droplet microemulsion forms.

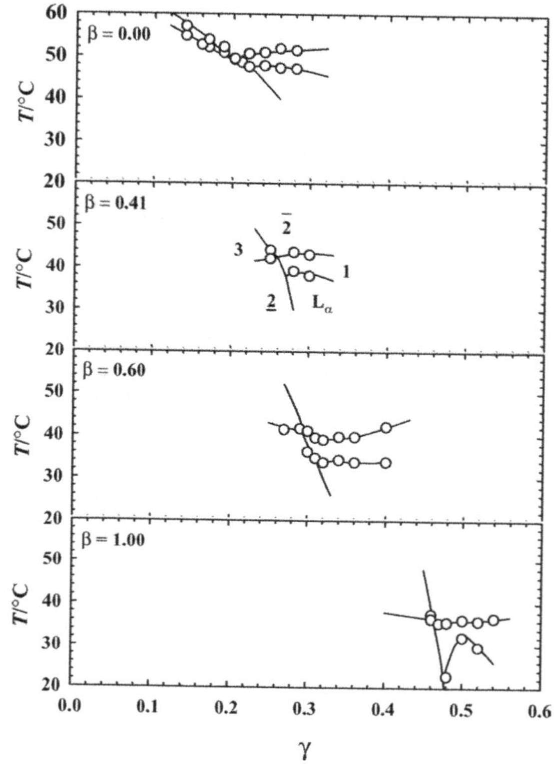

Figure 11.3 Phase diagrams for the system water/NaCl–*n*-propane/CO_2–Lutensol® XL70 measured at a 1:1 water-to-oil volume fraction ($\phi = 0.50$), an electrolyte content of 5 wt.% in the aqueous solution ($\varepsilon = 0.05$), and at $p = 220$ bar as a function of the temperature T and the surfactant mass fraction γ [21]. The fraction of CO_2 in the mixture is given by β.

11.3.3.2 'Road map' for future studies

So far we can generate supercritical carbon dioxide microemulsions with varying microstructures under a pressure of $p = 220$ bar. The next step is to find appropriate surfactants for an efficient solubilisation of CO_2 as today over 48 wt.% surfactant is required to formulate a one-phase region in the system H_2O/NaCl–CO_2–Lutensol XL70 (see Fig. 11.3). Good amphiphilic candidates with CO_2-philic parts are fluorinated surfactants. Once a more efficient surfactant has been identified, the formation of microemulsions under pressure has to be studied.

After an efficient system has been found, the main open question is how to solidify the foams. One route could be to utilise water-soluble polymers or/and polymerise water-soluble monomers. Another route could be to switch to polymeric microemulsions. One would stepwisely replace the water by the preferred polymerisable hydrophilic monomer/polymer. By doing so one can be sure that the nanostructure of microemulsion remains unaffected. The generation of the nanofoam depends on the used monomer. A fixation via irradiation with light would be preferable, when photoactive cross-linkers are in the mixture, because a change in the temperature to activate the system would effect a change of the nanostructure (see Chapter 1).

11.4 Clean combustion of microemulsions

11.4.1 Why use microemulsions for fuel combustion?

Pollutant formation by combustion of fossil and biogenous fuels is still an unsolved problem. Internal combustion engines are significant contributors to air pollution, such as NO_x and particulate matter (i.e. soot), which has a damaging impact on our health and the environment and is suspected to cause global climate changes. Nitrogen oxide was found to enhance photochemical smog and ozone formation in the troposphere as well as acid rain due to its acidic properties. Furthermore, nitrogen oxide contributes to chain reactions which decompose the desired stratospheric ozone. As a consequence increased ultraviolet radiation reaches the earth's surface. The organic carbon-containing fraction adsorbed on soot particles predominantly consists of high-molecular materials, which partially possess a high toxic potential with mutagenic and carcinogenic characteristics.

Under ideal combustion conditions the hydrocarbon compounds C_xH_y are converted into carbon dioxide CO_2 and water vapour H_2O in the oxidation reaction. The nature of the fuel injection, the combustion chamber geometry, the thermal condition of cylinder charge and combustion chamber walls affect the quality of the air/fuel mixture and the charge movement in the combustion chamber. The homogeneity of the air/fuel mixture in the combustion chamber is a measure for the quality of combustion. The local and temporal process sequence as well as the completeness and perfection (pollutant formation) of the combustion in the engine strongly depend on fuel/air mixture generation. The pollutant emissions measurable in the exhaust gas arise from the interrelation of pollutant formation and pollutant degradation in the combustion chamber and in the exhaust system. This particularly applies to the emissions of soot, hydrocarbon and carbon monoxide. Thus, the goal that should be reached is achieving maximum engine output with minimum fuel consumption and, at the same time, minimal emissions. Besides improving the combustion chamber and the injection systems, a change of fuel composition can positively influence the combustion process regarding pollutant emission [22].

The investigations of different groups have shown that the increase of the oxygen content in blend fuels with additives such as ester, ether, acetals, alcohols etc. cause the reduction of soot emissions. However, in general, a simultaneous increase of the nitrogen oxide emissions could be observed (so-called diesel dilemma) [23–25]. The advantages of changing fuel compositions are not only the reduction of pollutant emissions but also the reduction of the CO_2 emission by use of biogenous additives [26]. Biogenous substances are considered as CO_2 neutral as the amount of carbon dioxide emitted during combustion corresponds to the amount accumulated during the growth of the plants. The addition of water leads to a reduction of nitrogen oxide and soot emissions at the same time. In published reports there are several ways which have already been tested, e.g. direct water injection into the combustion chamber [27] or the use of water-containing fuel. Such fuels are already available as water–fuel emulsions [28]. Unfortunately, emulsions do not exhibit long-term stability as they separate into an aqueous and an oily phase after a certain time [29]. The water in an emulsion is distributed as poly-disperse droplets with an average diameter of approximately 50 μm. The combustion of the emulsion droplets is frequently accompanied by the micro-explosion phenomenon, due to the volatility difference between the water and the fuel [30]. Because of the lower combustion temperature the formation of thermal (Zeldovic) NO is reduced [31]. The thermal NO is the largest fraction of the emitted nitrogen oxides, hence the added water can significantly minimise the nitrogen oxide emissions. Not only physical processes such as reduction of the flame temperature and better dispersion of the injected fuel, but also chemical reactions caused by water have a positive effect on the combustion. Water molecules disintegrate in the combustion chamber at higher temperatures and form radicals (·OH, ·O, $\cdot O_2H$), which accelerate the decomposition of hydrocarbon chains in the radical chain branching reaction [31].

In summary, the use of microemulsions offers several advantages for the combustion process. Besides the thermodynamic stability of the aqueous fuel (which automatically means that the energy input for the production is minimal), water, surfactants and freezing point-decreasing components that are homogeneously distributed on a nano-scale can be optimally used for the reduction of soot and NO_x emissions.

11.4.2 What are the challenges?

Microemulsions form spontaneously and exhibit nano-disperse structures. In contrast to emulsions there is no additional energy input necessary for the production of a microemulsion. The formation is thermodynamically favoured due to the ultra-low interfacial tension between the oil and water domains. The microemulsified fuels are in principle thermodynamically stable for an unlimited period of time: only the chemical stability of the single components could be a limiting factor. A further advantage of microemulsions in contrast to emulsions is the fact that the water content can be adjusted over a broad range. Therefore, the combustion process can be customised to specific needs. An important criterion for a microemulsion to be used as fuel is that the one-phase region extends over a wide temperature range (Fig. 11.4). Mixtures of ionic and non-ionic surfactants, which exhibit almost temperature-invariant phase behaviour by optimal composition, are suitable to meet these standards.

Extensive investigations at the Institute for Mechanical Engineering and Vehicle Technology at the University of Applied Science in Trier as well as practical tests showed that soot emissions were lowered by more than 90% upon the application of water–diesel microemulsions (see Fig. 11.5(a)). Interestingly, the NO_x-particulate matter trade-off is avoided, i.e. nitrogen oxide emissions are also lowered significantly (Fig. 11.5(b)). The surfactants used for the formation of the microemulsion are oxidised species, which, as already mentioned above, decrease the soot formation further.

Changing fuel composition by adding surfactants and water decreases its specific energy, which has to be considered in view of effective specific consumption. The estimation of the lower heat value after W. Boie [32] leads to the values listed in Table 11.1 for the examined microemulsions. As can be seen from the numbers the amount of fuel being energy equivalent to diesel is reduced with increasing water content during the combustion of microemulsions.

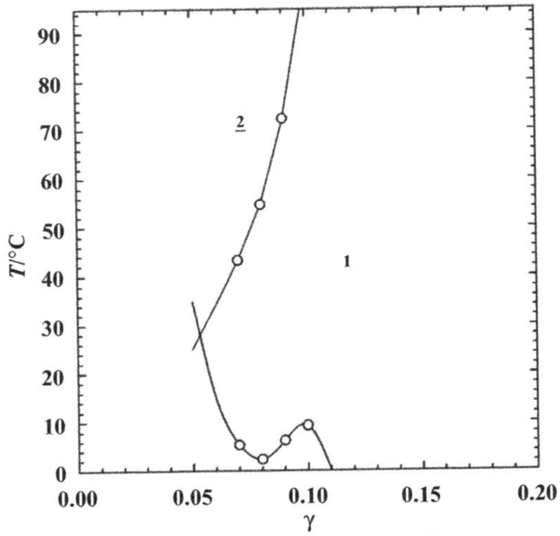

Figure 11.4 Enlarged one-phase area of a phase diagram obtained for the system H_2O/antifreeze–diesel biogenous surfactant/non-ionic surfactant with $\alpha = 0.80$.

Figure 11.6 shows the values for the fuel consumption using microemulsified fuels with varying water content. The full symbols show the total fuel consumption and the empty symbols show the specific fuel consumption taking into account the lower heating value as a function of the engine load condition. The fuel consumption that is equivalent to the amount of diesel energy decreases with rising water content by using microemulsified

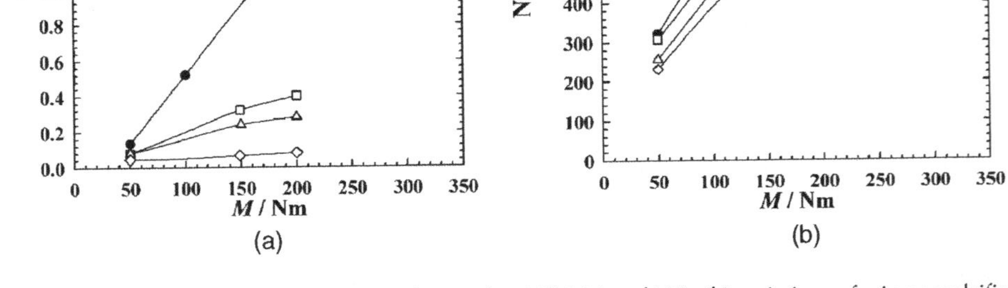

Figure 11.5 Soot in units of the filter smoke number (*FSN*) (a) and NO_x (b) emissions of microemulsified fuels with varying water content for combustion measured as function of torque *M*. The measurements are carried out on a Deutz engine, 82 kW, 4-cylinder, without EGR at constant 1500 rpm as a function of the engine load condition (Institute for Mechanical Engineering and Vehicle Technology, University of Applied Science, Trier).

Table 11.1 Lower heat values of diesel compared to water–diesel microemulsions of various water contents

Fuel	LHV/MJ kg^{-1}
Diesel	43.00
ME 7.5 wt.% H_2O	38.30
ME 15.0 wt.% H_2O	34.52
ME 22.4 wt.% H_2O	30.77

fuels. This shows also an enhancement of the engine efficiency. Note that this small set of data is enough to see a non-linear dependence of the fuel consumption on the quantity of water and on the operating point of the engine. The analysis of the pollutant emissions and fuel consumption shows that simply by the substitution of the conventional fuel with microemulsion fuel the exhaust gas values can be clearly improved over all load ranges.

11.4.3 What route is the most promising?

The measured data shows that by the application of microemulsions as alternative fuel the combustion becomes cleaner and more efficient. Microemulsions can be used successfully in both stationary and dynamically operated engines. In the stationary operation it

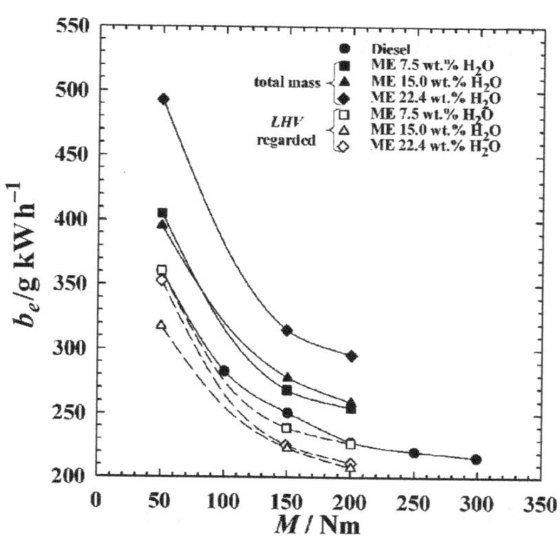

Figure 11.6 Values for specific fuel consumption b_e as function of torque M for the same microemulsions as in Fig. 11.5.

appears reasonable to use the microemulsion with a defined water content. In this case water content can be individually adapted for the respective engine load. In vehicles the engines are operated under a permanent engine load change, therefore the water/fuel ratio must be adapted to the respective operating point. This punctual adjustment ensures minimum of pollutant emissions and favourable fuel consumption. In order to determine the exact composition of the microemulsified aqueous fuel, engine operating maps must be generated with the microemulsions with varying water/fuel ratios. Only an 'on-board-mixing-system' is considered as optimal water dosage in each operating point. This mixing system is supposed to prepare on demand the optimal composition of water, fuel and surfactant for the microemulsion according to the current engine load. This type of mixing system puts up additional requirements for the formulation of microemulsions. The appropriate surfactants can be directly added to the fuel as a concentrate and diesel–concentrate mixture can also be filled into the regular tank. An additional tank for water has to built into the vehicle. The system must function independently of ambient temperatures; therefore, antifreeze has also been present in the aqueous phase, which is no problem though for ethanol, glycerol or ethyleneglycol to name a few. The kinetics of microemulsion formation is of special importance as especially during fast engine load changes the appropriate microemulsion composition has to be immediately available for combustion. For fast changes of the composition of the water–fuel microemulsions the formation kinetics of the order of seconds with a minimum on shearing force. Altogether the formulation of the microemulsion is flexible concerning most potential additives, which reduce on the one hand the pollutant emissions and increase on the other hand the efficiency of the engine. In addition, the aqueous microemulsions can be formulated with both fossil and biogenous fuels, i.e. with very different fuels. The use of the microemulsions in all conceivable kinds of combustion engines makes the selection of optimal burning technologies with minimum pollutant emissions and improved degree of efficiency a route which has not yet received the attention it deserves.

11.5 Solubilisation of triglycerides

Triglycerides are simple lipids consisting of glycerol esterified with three fatty acids, which may differ in chain-length and degree of saturation. As main components of natural fats and oils, complex mixtures of triglycerides can be found in nature being widespread in eukaryotic organisms such as yeasts, moulds and fungi. The efficient solubilisation of triglycerides is interesting and challenging for both basic research as well as large-scale applications. Up to now triglycerides have been solubilised either in the form of kinetically stable emulsions or thermodynamically stable microemulsions [33–40] focusing on single aspects only. In the following, a systematic investigation of phase behaviour, interfacial tension and microstructure of triglyceride microemulsions will be presented to provide the basic information for formulating triglyceride microemulsions.

11.5.1 Road map to the solubilisation of triglycerides

Triolein is one of the main components of many natural oils and fats and was chosen as a starting point for the characterisation of triglyceride phase behaviour. When dealing with

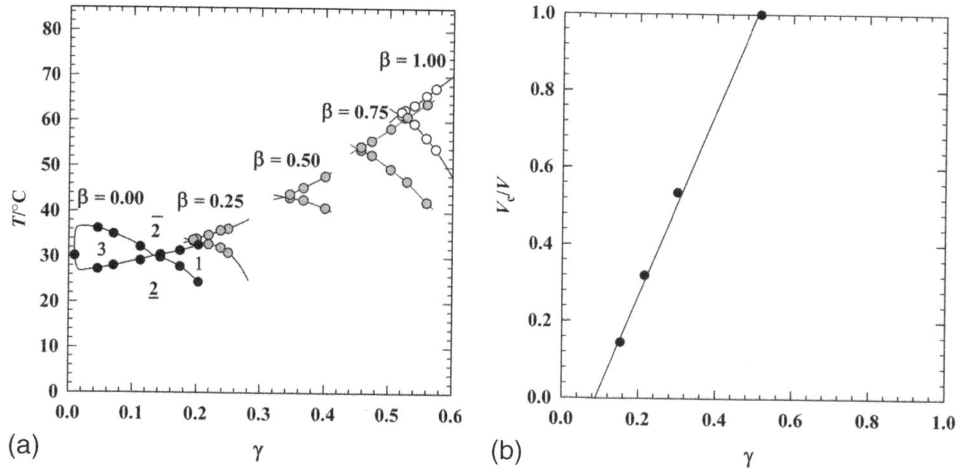

Figure 11.7 (a) Phase diagrams of the systems H_2O–*n*-decane–$C_{10}E_4$ [41] as well as H_2O/NaCl–*n*-decane/triolein–$C_{10}E_4$ for $\beta = 0.25$, 0.50, 0.75 and 1.00 at $\phi = 0.50$ and $\varepsilon = 0.001$ (b). Volume fraction of the middle phase V_c/V as a function of γ for the system H_2O/NaCl–triolein–$C_{10}E_4$ at $\tilde{T} = 61.80°C$. Extrapolation allows the calculation of the monomeric solubility of $C_{10}E_4$ in triolein.

new types of components it has proven useful to systematically substitute one component of a well-defined base system. In this case the base system H_2O–*n*-decane–$C_{10}E_4$ was chosen and *n*-decane was systematically substituted by triolein. The parameter β indicates the mass fraction of triolein in the oil mixture. In Fig. 11.7(a), phase diagrams at various β values are presented. The measurements were carried out at a 1:1 water-to-oil volume fraction as a function of the temperature T and the surfactant mass fraction γ.

The base system shows the typical phase behaviour of non-ionic surfactants. The phase boundaries resemble the shape of a fish with a three-phase region located at lower surfactant mass fraction and a single-phase region located at higher surfactant mass fractions. At lower temperatures one finds an oil-excess phase coexisting with an oil-in-water microemulsion ($\underline{2}$) and at higher temperatures one finds a water-excess phase coexisting with a water-in-oil microemulsion ($\overline{2}$). Upon substituting *n*-decane by triolein the phase behaviour shifts to higher temperatures and surfactant mass fractions. Note that a little amount of sodium chloride was added to the triolein containing systems, which has only marginal influence on the phase behaviour. The upper phase boundary of the different single-phase regions is represented by a single curve, which rises with increasing γ. The lower phase boundary on the other hand shifts parallel to higher T and γ. Considering the fact that the upper phase boundary originates from the binary H_2O/NaCl–$C_{10}E_4$ system, which remains unaffected while replacing *n*-decane by triolein, this behaviour can be easily understood.

In order to determine the monomeric solubility of the non-ionic $C_{10}E_4$ in triolein, the volume fractions of the middle phase V_c/V were measured as function of the overall surfactant mass fraction γ in the three-phase region of the system H_2O/NaCl–triolein–$C_{10}E_4$ at $\tilde{T} = 61.80°C$.

Extrapolation of V_c/V towards $V_c/V = 0$ gives γ_0, which corresponds to the least surfactant mass fraction needed to saturate the oil phase and the water phase as well as

Table 11.2 Melting points of the investigated triglycerides

Triglyceride	C-chain length k	$T_{melt}/°C$
Tricaprin	10	31.5
Trilaurin	12	46.5
Trimyristin	14	58.5
Tripalmitin	16	67.0
Tristearin	18	72.5

the macroscopic interface with surfactant (Fig. 11.7(b)). Taking into account that $\gamma_{mon,a}$ (monomeric solubility in the water phase) $<<$ $\gamma_{mon,b}$ (monomeric solubility in the oil phase) [42], the monomeric solubility of $C_{10}E_4$ in triolein can be derived from

$$\gamma_{mon,b} = \frac{\gamma_0}{\gamma_0 + \alpha(1 - \gamma_0)} \tag{11.1}$$

[43, 44]. The monomeric solubility of $C_{10}E_4$ in triolein at equal volume fractions of water and oil ($\phi = 0.50$) corresponds to $\gamma_{mon,b} = 0.179$, which is around nine times higher than in n-decane ($\gamma_{mon,b} = 0.021$). As a result a high amount of surfactant molecules is dissolved in the oil domains and the oil-excess phase, respectively [45]. Thus, less surfactant is available to form the interfacial film between the nanoscopic H_2O/NaCl and triolein domains. This result is in good agreement with Binks *et al.*, who showed that the critical microemulsion concentration in the oil phase rises upon changing from alkane oils to triglycerides [46].

The characterisation of totally saturated triglycerides with C-chain lengths of the fatty acid residues of $k = 10{:}0$ (tricaprin), 12:0 (trilaurin), 14:0 (trimyristin), 16:0 (tripalmitin) and 18:0 (tristearin) in ternary systems of type H_2O/NaCl–triglyceride–$C_{10}E_4$ revealed that long-chain triglycerides can only be solubilised at high temperatures and surfactant mass fractions (see Fig. 11.8(a)). Note that parts of the phase diagrams of the trimyristin, tripalmitin and tristearin microemulsion systems were characterised in a super-cooled melt (Table 11.2) as with increasing C-chain length the melting point of the respective triglyceride increases as well.

The minimum of the interfacial tension for the system H_2O/NaCl–triolein–$C_{10}E_5$ corresponds to $\overline{\sigma}_{ab} = 2.6\ 10^{-1}$ mN m^{-1} at $\tilde{T} = 43.5°C$. The variation of σ_{ab} as a function of T is similar to the n-alkane systems with the curve showing the typical V-shape (Fig. 11.8(b)). The value of the interfacial tension between water and oil is comparable to the inefficient H_2O–n-octane–C_4E_1 system [48]. However, compared to the pure water–oil interfacial tension (50 mN m^{-1}), it is still two orders of magnitude lower.

The interpretation of small SANS data from systems of type D_2O/NaCl–n-decane/triolein–$C_{10}E_4$ showed that the order of the microstructure systematically decreases with increasing triolein content (Fig. 11.8(c) and Table 11.3). However, the value of the amphiphilicity factor [49, 50] $fa = -0.65$ indicates that the pure triolein microemulsion is still a microemulsion in the narrower sense. The bending constants κ and $\overline{\kappa}$ obtained from phase diagrams and scattering curves furthermore verify that the rigidity of the amphiphilic film decreases with increasing triolein content (Fig. 11.9).

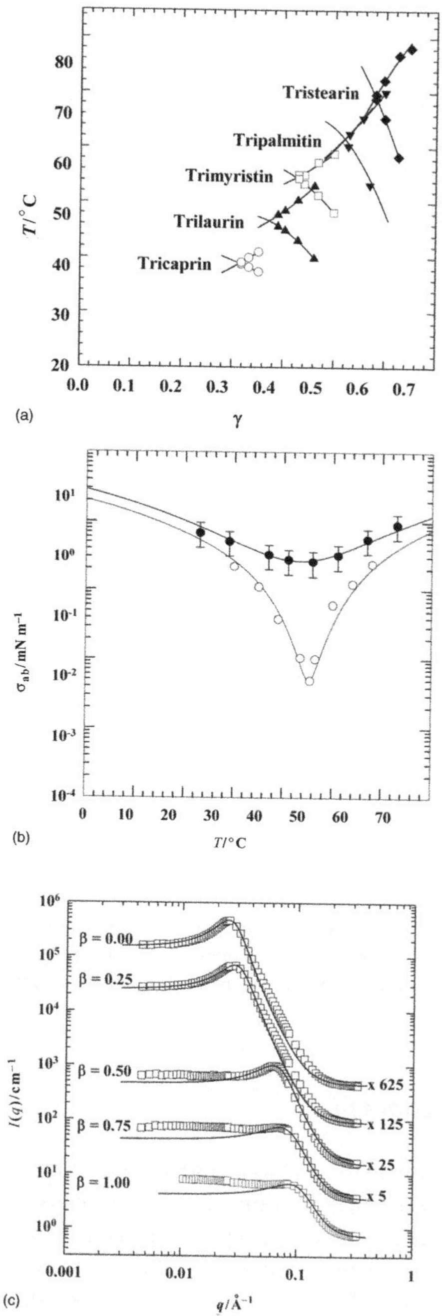

Figure 11.8 (a) Phase diagrams of the systems H_2O/NaCl–triglyceride–$C_{10}E_4$ for $k = 10{:}0$, 12:0, 14:0, 16:0 and 18:0. (b) Variation of the interfacial tension between water and oil σ_{ab} with temperature for the system H_2O/NaCl–triolein–$C_{10}E_5$ (filled symbols). For comparison the variation of σ_{ab} with temperature for the system H_2O– *n*-octane–$C_{10}E_5$ is given as well (empty symbols). (c) Scattering curves for the systems D_2O/NaCl–*n*-decane/triolein–$C_{10}E_4$ for different triolein mass fractions β in the oil phase. The fit corresponds to the Teubner–Strey equation [47].

Table 11.3 Correlation length ξ_{TS}, periodicity d_{TS} and amphiphilicity factor f_a of the systems D_2O/NaCl–*n*-decane/triolein–$C_{10}E_4$ at $\phi = 0.50$ and $\varepsilon = 0.001$

β	ξ_{TS}/Å	d_{TS}/Å	f_a
0.00	124.8	241.1	−0.827
0.25	101.6	211.7	−0.802
0.50	39.2	95.1	−0.741
0.75	28.2	80.1	−0.661
1.00	22.5	65.5	−0.646

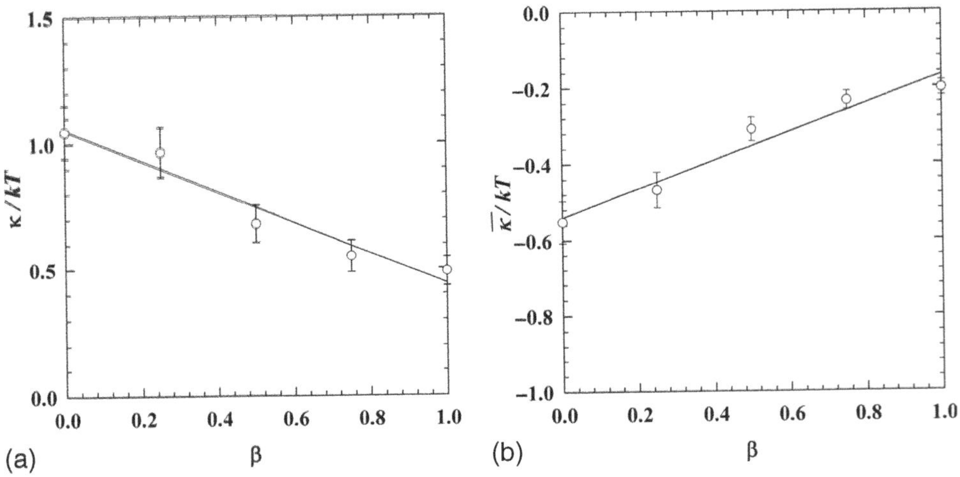

Figure 11.9 Bar bending modulus κ (a) and bar saddle-splay modulus $\bar{\kappa}$ (b) for the systems $H_2O(D_2O)$/NaCl as well as H_2O/NaCl–*n*-decane/triolein–$C_{10}E_4$ as a function of β. With increasing β the bending modulus of the amphiphilic film κ decreases, while the saddle-splay modulus $\bar{\kappa}$ becomes less negative.

11.5.2 The linker concept

Because of high surfactant mass fractions and the dominance of the liquid crystalline lamellar phase the efficient solubilisation of triglycerides is still an unsolved problem. A promising way to enhance the efficiency of triglyceride microemulsions is the linker concept. Lipophilic linkers when added to the oil phase were found to accumulate close to the oil–water interface possibly due to long-range attraction. This so-called interfacial segregation leads to a more continuous transition from bulk oil to bulk water and thus to a better shielding or blurring of the interface. As a result the interfacial tension is lowered and the solubilisation is increased. Typical lipophilic linkers are polar oils such as long-chain ($>C_{10}$) *n*-alcohols, phenols and fatty esters. It was found that the most efficient lipophilic linker is the chain intermediate between oil and surfactant regardless of the polar group.

Hydrophilic linkers act on the water-rich side of the interface by segregating or co-adsorbing with the surfactant while avoiding a strong interaction with the oil phase. Because of a limited number of potential segregation sites the concentration range where linker molecules are able to enhance the solubilisation capacity is also limited. The behaviour of co-added lipophilic and hydrophilic linkers can be interpreted in terms of an assembled surfactant resulting in a very smooth and continuous variation of the polarity from bulk water to bulk oil. Assembled surfactants have already been applied to formulate biocompatible microemulsions [51].

Inspired by the linker concept so-called extended surfactants have been developed. In contrast to commonly used surfactants, extended surfactants possess an additional group of intermediate polarity between the hydrophilic head and the hydrophobic tail ensuring a continuous gradation of polarity throughout the molecule. The stretching of the molecule leads to an increased penetration into the oil and water phases and hence to an increase in solubilisation. Extended surfactants have already been successfully applied in enhancing the solubilisation capacity of triglyceride systems [52–54]. In view of biological applications the second generation of extended surfactant is furthermore characterised by high biocompatibility [55–57]. On the basis of these promising new concepts additional major breakthroughs in the solubilisation of natural oils can be expected for the future.

Acknowledgement

As a lot of the results presented in this chapter are not published yet we would like to thank personally those people who carried out most of the measurements, namely Renate Tessendorf, Dr. Michael Schwan, Lada Bemert and Dr. Sandra Engelskirchen. Additional assistance of Lada Bemert and Dr. Sandra Engelskirchen during writing this chapter is gratefully acknowledged.

References

1. Eastoe, J., Sanchez Dominguez, M., Cumber, H., Wyatt, P. and Heenan, R.K. (2004) Light-sensitive microemulsions. *Langmuir*, **20**(4), 1120–1125.
2. Eastoe, J., Wyatt, P., Sánchez-Dominguez, M., Vesperinas, A., Paul, A., Heenan, R.K. and Grillo, I. (2005) Photo-stabilised microemulsions. *Chem. Commun.*, **22**, 2785–2786.
3. Eastoe, J. and Vesperinas, A. (2005) Self-assembly of light-sensitive surfactants. *Soft Matter*, **1**, 338–347.
4. Co, C.C., de Vries, R. and Kaler, E.W. (2001) Free-radical polymerizations in microemulsions. In J. Texter (ed), *Reactions and Synthesis in Surfactant Systems*. Marcel Dekker, New York, pp. 455–470.
5. Capek, I. (2004) Preparation of metal nanoparticles in water-in-oil (w/o) microemulsions. *Adv. Colloid Interface Sci.*, **110**, 49–74.
6. Uskoković, V. and Drofenik, M. (2005) Synthesis of materials within reverse micelles. *Surf. Rev. Lett.*, **12**, 239–277.
7. Hentze, H.P., Co, C.C., McKelvey, C.A. and Kaler, E.W. (2003) Templating vesicles, microemulsions and lyotropic mesophases by organic polymerization processes. *Top. Curr. Chem.*, **226**, 197–223.

8. Coakley, K.M. and McGehee, M.D. (2004) Conjugated polymer photovoltaic cells. *Chem. Mater.*, **16**, 4533–4542.
9. Lynch, I. and Dawson, K.A. (2003) Effect of a polymeric additive on the pore-size distribution and shrinking process of a hydrogel network. *Macromol. Chem. Phys.*, **204**, 443–450.
10. Palani Raj, W.R., Sasthav, M. and Cheung, H.M. (1995) Polymerization of single phase microemulsions: Dependence of polymer morphology on microemulsion structure. *Polymer*, **36**, 2637–2646.
11. Summers, M., Eastoe, J., Davis, S., Du, Z., Richardson, R.M., Heenan, R.K., Steytler, D. and Grillo, I. (2001) Polymerization of cationic surfactant phases. *Langmuir*, **17**, 5388–5397.
12. Morkved, T.L., Stepanek, P., Krishnan, K., Bates, F.S. and Lodge, T.P. (2001) Static and dynamic scattering from ternary polymer blends: Bicontinuous microemulsions, Lifshitz lines, and amphiphilicity. *J. Chem. Phys.*, **114**, 7247–7259.
13. Gao, F., Ho, C.-C. and Co, C.C. (2004) Sugar-based microemulsion glass templates. *J. Am. Chem. Soc.*, **126**, 12746–12747.
14. Gao, F., Ho, C.-C. and Co, C.C. (2006) Polymerization in bicontinuous microemulsion glasses. *Macromolecules*, **39**, 9467–9472.
15. Stubenrauch, C., Tessendorf, R., Strey, R., Lynch, I. and Dawson, K. (2007) Gelled polymerizable microemulsions. 1. Phase behavior. *Langmuir*, **23**, 7730–7737.
16. Terech, P. and Weiss, R.G. (1997) Low molecular mass gelators of organic liquids and the properties of their gels. *Chem. Rev.*, **97**, 3133–3160.
17. Stubenrauch, C., Tessendorf, R., Salvati, A., Topgaard, D., Sottmann, T., Strey, R., Lyncy, I. (2008) Gelled polymerizable microemulsions. 2. Microstructure. *Langmuir*, ASAP Artick. DOI:10.1021/la800978g.
18. Challa, V., Kuta, K., Lopina, S., Cheung, H.M. and von Meerwall, E. (2003) Microporosity of bicontinuous nanoporous polymeric materials, characterized with restricted diffusion. *Langmuir*, **19**, 4154–4161.
19. Wu, X.S., Hoffmann, A.S. and Yager, P. (1992) Synthesis and characterization of thermally reversible macroporous poly(*N*-isopropylacryl-amide) hydrogels. *J. Polym. Sci. Part A: Polym. Chem.*, **30**, 2121–2129.
20. Strey, R., Sottmann, T. and Schwan, M. (2002) Eur. Patent Appl., Patent No. WO2004058386.
21. Schwan, M. (2005) *Überkritische Mikroemulsionen zur Herstellung nanozellulärer Schäume – Principle of Supercritical Microemulsion Expansion (POSME)*. PhD thesis, University of Cologne.
22. Basshuysen, R.V. and Schäfer, F. (2004) *Internal Combustion Engine Handbook: Basics, Components, Systems, and Perspectives: Basics, Components, Systems and Perspectives: R-345*. Society of Automotive Engineers (SAE), Warrendale.
23. Song, J., Zello, V., Boehman, A.L. and Waller, F.J. (2004) Comparison of the impact of intake oxygen enrichment and fuel oxygenation on diesel combustion and emissions. *Energy Fuels*, **18**, 1282–1290.
24. Song, J., Cheenkachorn, K., Wang, J., Perez, J. and Boehman, A.L. (2002) Effect of oxygenated fuel on combustion and emissions in a light-duty turbo diesel engine. *Energy Fuels*, **16**, 294–301.
25. McCormick, R.L., Ross, J.D. and Graboski, M.S. (1997) Effect of several oxygenates on regulated emissions from heavy-duty diesel engines. *Environ. Sci. Technol.*, **31**, 1144–1150.
26. Nord, K.E. and Haupt, D. (2005) Reducing the emission of particles from a diesel engine by adding an oxygenate to the fuel. *Environ. Sci. Technol.*, **39**, 6260–6265.
27. Bedford, F., Rutland, C., Dittrich, P., Raab, A. and Wirbeleit, F. (2000) Effects of direct water injection on diesel engine combustion. *SAE Technical Papers*, 2000-01-2938.
28. Schmelzle, P., Bernaud, F. and Schulz, P. (2000) Aquazole: An original emulsified water–diesel fuel for heavy-duty applications. *SAE Technical Papers*, 2000-01-1861.
29. Tort, F. and Laluet, L.D.G. (2006) US 2006 007568 A1.

30. Park, J., Huh, K. and Park, K. (2000) Experimental study on the combustion characteristics of emulsified diesel in a rapid compression and expansion machine. *Proc. Inst. Mech. Eng., Part D: J. Automobile Eng.*, **214**(D5), 579–586.
31. Warnatz, J., Maas, U. and Dibble, R.W. (1999) *Combustion: Physical and Chemical Fundamentals, Modelling and Simulation, Experiments, Pollutant Formation.* Springer, Berlin.
32. Boie, W. (1957) *Vom Brennstoff zum Rauchgas.* Teubner, Leipzig.
33. Christov, N.C., Denkov, N.D., Kralchevsky, P.A., Broze, G. and Mehreteab, A. (2002) Kinetics of triglyceride solubilization by micellar solutions of nonionic surfactant and triblock copolymer. 1. Empty and swollen micelles. *Langmuir*, **18**(21), 7880–7886.
34. Boyle-Roden, E. and Khan, M.A. (2001) Quantitative analysis of surface-located triacylglycerol in intact emulsion particles. *J. Agric. Food Chem.*, **49**(4), 2014–2021.
35. Baber, T.M., Vu, D.T. and Lira, C.T. (2002) Liquid–liquid equilibrium of the castor oil plus soybean oil plus hexane ternary system. *J. Chem. Eng. Data*, **47**(6), 1502–1505.
36. Vanapalli, S.A., Palanuwech, J. and Coupland, J.N. (2002) Influence of fat crystallization on the stability of flocculated emulsions. *J. Agric. Food Chem.*, **50**(18), 5224–5228.
37. Parris, N., Joubran, R.F. and Lu, D.P. (1994) Triglyceride microemulsions – effect of nonionic surfactants and the nature of the oil. *J. Agric. Food Chem.*, **42**(6), 1295–1299.
38. vonCorswant, C., Engstrom, S. and Soderman, O. (1997) Microemulsions based oil soybean phosphatidylcholine and triglycerides. Phase behavior and microstructure. *Langmuir*, **13**(19), 5061–5070.
39. Kahlweit, M., Busse, G., Faulhaber, B. and Eibl, H. (1995) Preparing nontoxic microemulsions. Langmuir, **11**(11), 4185–4187.
40. Sottmann, T., Lade, M., Stolz, M. and Schomäcker, R. (2002) Phase behavior of non-ionic microemulsions prepared from technical-grade surfactants. *Tenside Surfactants Detergents*, **39** (1), 20–28.
41. Jacobs, B. (2001) *Amphiphile Blockcopolymere als 'Efficiency Booster' für Tenside: Entdeckung und Aufklärung des Effekts.* PhD thesis, University of Cologne.
42. Kahlweit, M., Strey, R. and Firman, P. (1986) Search for tricritical points in ternary-systems – water oil nonionic amphiphile. *J. Phys. Chem.*, **90**(4), 671–677.
43. Kunieda, H. and Haishima, K. (1990) Overlapping of 3-phase regions in a water nonionic surfactant triglyceride system. *J. Colloid Interface Sci.*, **140**(2), 383–390.
44. Kunieda, H. and Shinoda, K. (1985) Evaluation of the hydrophile-lipophile balance (Hlb) of nonionic surfactants. 1. Multisurfactant systems. *J. Colloid Interface Sci.*, **107**(1), 107–121.
45. Burauer, S., Sachert, T., Sottmann, T. and Strey, R. (1999) On microemulsion phase behavior and the monomeric solubility of surfactant. *Phys. Chem. Chem. Phys.*, **1**(18), 4299–4306.
46. Binks, B.P., Fletcher, P.D.I. and Horsup, D.I. (1991) Effect of microemulsified surfactant in destabilising water-in-oil emulsions containing C12e4. *Colloids Surf.*, **61**, 291–315.
47. Teubner, M. and Strey, R. (1987) Origin of the scattering peak in microemulsions. *J. Chem. Phys.*, **87**(5), 3195–3200.
48. Sottmann, T. and Strey, R. (1997) Ultralow interfacial tensions in water-*n*-alkane–surfactant systems. *J. Chem. Phys.*, **106**(20), 8606–8615.
49. Schubert, K.V., Strey, R., Kline, S.R. and Kaler, E.W. (1994) Small-angle neutron-scattering near Lifshitz Lines – Transition from weakly structured mixtures to microemulsions. *J. Chem. Phys.*, **101**(6), 5343–5355.
50. Leitao, H., da Gama, M.M.T. and Strey, R. (1998) Scaling of the interfacial tension of microemulsions: A Landau theory approach. *J. Chem. Phys.*, **108**(10), 4189–4198.
51. Acosta, E.J., Nguyen, T., Witthayapanyanon, A., Harwell, J.H. and Sabatini, D.A. (2005) Linker-based bio-compatible microemulsions. *Environ. Sci. Technol.*, **39**, 1275–1282.
52. Huang, L., Lips, A. and Co, C.C. (2004) Microemulsification of triglyceride sebum and the role of interfacial structure on bicontinuous phase behavior. *Langmuir*, **20**(9), 3559–3563.

53. Miñana-Pérez, M., Graciaa, A., Lachaise, J. and Salager, J.L. (1995) Solubilization of polar oils with extended surfactants. *Colloids Surf. Physicochem. Eng. Asp.*, **100**,217–224.
54. Witthayapanyanon, A., Acosta, E.J., Harwell, J.H. and Sabatini, D.A. (2006) Formulation of ultralow interfacial tension systems using extended surfactants. *J. Surf. Deterg.*, **9**(4), 331–339.
55. Goethals, G., Fernández, A., Martin, P., Miñana-Pérez, M., Scorzza, C., Villa, P. and Godé, P. (2001) Spacer arm influence on glucido-amphiphilic compound properties. *Carbohydr. Polym.*, **45**, 147–154.
56. Scorzza, C., Godé, P., Goethals, G., Martin, P., Miñana-Pérez, M., Salager, J.L., Usubillaga, A. and Villa, P. (2002) Another new family of 'extended' glucidoamphiphiles. Synthesis and surfactant properties for different sugar head groups and spacer arm lengths. *J. Surf. Deterg.*, **5**(4), 337–343.
57. Scorzza, C., Godé, P., Martin, P., Miñana-Pérez, M., Salager, J.L., Villa, P. and Goethals, G. (2002) Synthesis and surfactant properties of a new 'extended' glucidoamphiphile made from D-glucose. *J. Surf. Deterg.*, **5**(4), 331–335.

Index

N

Printed and bound by CPI Group (UK) Ltd, Croydon, CR0 4YY

15/09/2026

14962393-0005